SpringerWienNewYork

CISM COURSES AND LECTURES

The series presents lecture notes, monographs, edited works and proceedings in the field of Mechanics, Engineering, Computer Science and Applied Mathematics.
Purpose of the series is to make known in the international scientific and technical community results obtained in some of the activities organized by CISM, the International Centre for Mechanical Sciences.

INTERNATIONAL CENTRE FOR MECHANICAL SCIENCES

COURSES AND LECTURES - No. 416

ADAPTIVE FINITE ELEMENTS IN LINEAR AND NONLINEAR SOLID AND STRUCTURAL MECHANICS

EDITED BY

ERWIN STEIN
UNIVERSITY OF HANNOVER, GERMANY

SpringerWienNewYork

This volume contains 214 illustrations

SPIN 11509127

In order to make this volume available as economically and as rapidly as possible the authors' typescripts have been reproduced in their original forms. This method unfortunately has its typographical limitations but it is hoped that they in no way distract the reader.

ISBN-10 3-211-26975-4 SpringerWienNewYork
ISBN-13 978-3-211-26975-6 SpringerWienNewYork

PREFACE

This course with 6 lecturers intends to present a systematic survey of recent research results of well-known scientists on error-controlled adaptive finite element methods in solid and structural mechanics with emphasis to problem-dependent concepts for adaptivity, error analysis as well as h- and p-adaptive refinement techniques including meshing and remeshing. Challenging applications are of equal importance, including elastic and elastoplastic deformations of solids, contact problems and thin-walled structures. Some major topics should be pointed out, namely:

(i) The growing importance of goal-oriented and local error estimates for quantities of interest—in comparison with global error estimates—based on dual finite element solutions;

(ii) The importance of the p-version of the finite element method in conjunction with parameter-dependent hierarchical approximations of the mathematical model, for example in boundary layers of elastic plates;

(iii) The choice of problem-oriented error measures in suitable norms, considering residual, averaging and hierarchical error estimates in conjunction with the efficiency of the associated adaptive computations;

(iv) The importance of implicit local postprocessing with enhanced test spaces in order to get constant-free, i.e. absolute–not only relative–discretization-error estimates;

(v) The coupling of error-controlled adaptive discretizations and the mathematical modeling in related subdomains, such as boundary layers.

The main goals of adaptivity are reliability and efficiency, combined with insight and access to controls which are independent of the applied discretization methods. By these efforts, new paradigms in Computational Mechanics should be realized, namely verifications and even validations of engineering models.

There is an unfortunate shortcoming concerning the late publication about $4\frac{1}{2}$ *years after the course, although 4 of 6 chapters have already been available since 2000. The delay is caused by the fact that co-workers who left institutes worked on improved results for some problems. New updates of the lecture notes were sent by the authors in March 2004. A further recent review of all articles clearly approved the high quality and especially the relevance to the present state of the art.*

I would like to thank my esteemed colleagues and friends Peter Hansbo, Pierre Ladevèze, Rolf Rannacher, Christoph Schwab and Peter Wriggers for following the invitation to lecture in our well-accepted course and contributing to the lecture notes very nicely. We believe that they will provide systematic information and many tools for students, researchers and practitioners.

I would also like to express my gratitude to my former co-worker Stephan Ohnimus who assisted me in organizing the course.

The authors would like to express their appreciation to CISM, especially to Professor S. Kaliszky, for the privilege of lecturing the course and to publish the lecture notes.

Erwin Stein

CONTENTS

Duality Techniques for Error Estimation and Mesh Adaptation in Finite Element Methods

Rolf Rannacher *

Institut für Angewandte Mathematik
Universität Heidelberg
INF 293, D-69120 Heidelberg, Germany
E-Mail: rannacher@iwr.uni-heidelberg.de
URL: http://gaia.iwr.uni-heidelberg.de

Summary. We present a general method for error control and mesh adaptivity in Galerkin finite element discretization of variational problems governed by differential equations. Our approach is based on the variational framework of projection methods and uses concepts from optimal control and sensitivity analysis. By employing global duality arguments and Galerkin orthogonality, we derive a posteriori error estimates in approximating quantities of physical interest such as, for example, point- or mean-values of boundary stresses. In these estimates the cellwise residuals of the computed solution are multiplied by weights which are obtained from the approximate solution of a dual problem. In this way, wo obtain the basis of a feed-back process by which the mesh is successively adapted, i.e. locally refined or coarsened, according to the particular goal of the computation. This method is systematically developed and analized at first for linear elliptic, parabolic as well as hyperbolic model problems. Then, it is extended to nonlinear problems with an application to the Hencky model of static elasto-plasticity. In the last two sections, we present applications in acoustics governed by the linear wave equation and for a nonlinear boundary control problem occuring in super-conductivity.

Contents

* The author acknowledges the support by the German Research Association (DFG) through the SFB 359 "Reactive Flow, Diffusion and Transport" at the University of Heidelberg.

1 Introduction

Consider a *continuous* model governed by a differential operator L and a force term f,

$$Lu = f, \tag{1.1}$$

and a corresponding finite dimensional *discrete* model depending on a parameter $h > 0$,

$$L_h u_h = f_h. \tag{1.2}$$

In controlling this *discretization*, we have to detect the interplay of the various error propagation effects in order to achieve (i) *a posteriori error control*, i.e., control of the error in quantities of physical interest like stress values, mean values of concentrations, drag and lift, etc., and (ii) *solution-adapted meshing*, i.e., design of economical meshes for computing these quantities with best efficiency. Our error analysis is based on the computable "residual"

$$\rho(u_h) = \text{“}f - Lu_h\text{”},$$

which is well defined in the context of a Galerkin finite element method.

Traditionally, *a posteriori* error estimation in Galerkin finite element methods is done with respect to the natural *energy norm* induced by the underlying differential operator,

$$\|u - u_h\|_E \leq c_S \|\rho(u_h)\|_{E^*}, \tag{1.3}$$

for references see the survey articles by Ainsworth and Oden (1997) and Verfürth (1996). This approach seems rather generic as it is directly based on the variational formulation of the problem and allows to exploit its coercivity properties. However, in most applications the error in the energy norm does not provide a useful bound on the error in the quantities of real physical interest. A more versatile method for a posteriori error estimation with respect to relevant error measures like L^2 norm, point values, line averages, etc., is obtained by using duality arguments as common from the *a priori* error analysis of finite element methods. This approach has first been systematically developed by Johnson and his co-workers (Johnson (1993), Eriksson et al. (1995)) and was then extended by the author and his group to a practical feedback method for mesh optimization termed "Dual-Weighted Residual Method" (DWR method) (Becker and Rannacher (1996), Becker and Rannacher (1998). This approach yields *weighted* a posteriori error bounds with respect to arbitrary functionals

$$|J(u) - J(u_h)| \leq \langle \rho(u_h), \omega(z) \rangle, \tag{1.4}$$

where the weight $\omega(z)$ is obtained by approximately solving a *dual problem* $L^* z = J$. The *dual solution* z may be viewed as a generalized Green function with respect to the output functional $J(\cdot)$, and accordingly the weight $\omega(z)$ describes the effect of local variations of the residual $\rho(u_h)$ on the error quantity $J(u) - J(u_h)$, e.g., as the consequence of mesh adaptation.

This article is organized as follows: In Section 2, we describe the main principles of a priori and a posteriori error analysis and introduce the DWR method. Its key feature is the use of *duality arguments* together with *Galerkin orthogonality* for deriving residual-based a posteriori error estimates for arbitrary functionals of the solution. Then, in Section 3, we discuss strategies for mesh adaptation based on these a posteriori error estimates and illustrate the underlying mechanisms by looking at simple model cases. Further, we show how the same approach can also be used for deriving residual-controlled stopping criteria for multigrid iterations. These strategies are further illustrated in Section 4 by applying them to several linear model equations of elliptic type (Lamé-Navier equations), hyperbolic type (scalar transport equation), and parabolic type (heat conduction equation). In Section 5, we show how the DWR method can be applied also for nonlinear problems. This is done at first within an abstract setting and is then be made concrete for a model problem in elasto-plasticity (Hencky model of perfect plasticity). Section 6 contains a more elaborate application of the DWR method to acoustic wave propagation where capturing the transport of local errors is of particular importance. Finally, Section 7 shows that the DWR method has a natural application to optimal control problems which provides a systematic approach to mesh adaptation according to the goal of the optimization.

More involved applications of the DWR method have been considered by Becker (1995) and (2000) (viscous incompressible flows), Braack et al. (1999a), (1999b) and Braack and Rannacher (1999) (chemically reactive flows), Kanschat (1997) (radiative transfer problems), Rannacher and Suttmeier (1999) (elasto-plasticity), Bangerth and Rannacher (1999) (acoustic waves), and Kapp et al. (2000) and (2001) (optimal control). A general survey of the DWR method can be found in Becker and Rannacher (2001) and Bangerth and Rannacher (2003).

2 A Priori and A Posteriori Error Analysis

2.1 Accuracy Control for Quantities of Physical Interest

Solving complex systems of partial differential equations by discretization methods may be considered in the context of *model reduction*: a conceptually *infinite* dimensional model is approximated by a *finite* dimensional one. Here, the quality of the approximation depends on the proper choice of the discretization parameters, e.g., the mesh width, the polynomial degree of the trial functions, and the size of certain stabilization parameters. As the result of the computation, we obtain an approximation to the desired output quantity of the simulation and besides that certain accuracy indicators like cell-truncation errors or cell-residuals. Controlling the error in such an approximation of a continuous model of a physical system requires to determine the influence factors for the *local* error indicators on the target functional. Such a sensitivity analysis with respect to local perturbations of the model is common in optimal control theory and introduces the concept of a *dual* (or *adjoint*) problem.

Let $J(u)$ be a quantity of physical interest derived from the solution u by applying a functional $J(\cdot)$. The goal is to control the error of the discretization with respect to this functional output, i.e., $E(u_h) = J(u) - J(u_h)$ in terms of the computable cell residuals $\rho_K(u_h)$. An example is control of the total error $e_K = u - u_{h|K}$ in some mesh cell K. By superposition, e_K splits into two components, the locally produced *truncation error* and the globally transported *pollution error* (assuming for simplicity that the underlying problem is linear):

$$e_K^{\text{tot}} = e_K^{\text{loc}} + e_K^{\text{trans}}. \tag{2.5}$$

This asks for control of

- error propagation in space (global pollution effect),

– interaction of physical error sources (local sensitivity analysis).

The effect of the cell residual ρ_K on the local error $e_{K'}$, at another cell K', is essentially governed by the Green's function of the continuous problem. Capturing this dependence by *numerical* evaluation is the general philosophy underlying our approach to error control.

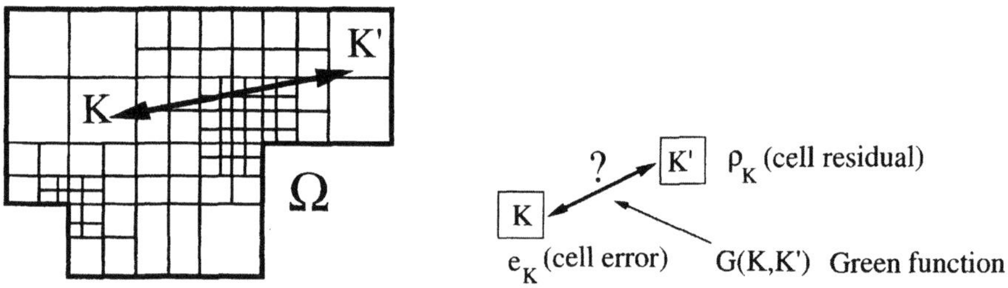

Figure 1. Scheme of error propagation.

In practice it is mostly impossible to determine the complex error interaction by analytical means, it rather has to be detected by computation. This automatically leads to a feed-back process in which error estimation and mesh adaptation goes hand-in-hand leading to economical discretization for computing the quantities of interest. Such an approach seems indispensable for the numerical simulation of large-scale problems. It is particularly designed for achieving high solution accuracy at minimum computational cost.

We illustrate the basic principles underlying error estimation by considering perturbations of a linear algebraic system. For given matrices $L, L_h \in \mathbb{R}^{n\times n}$ and right-hand sides $b, b_h \in \mathbb{R}^n$, consider the system $Lu = b$ together with its perturbation $L_h u_h = b_h$. Let $\langle \cdot,\cdot \rangle$ denote the euclidean inner product and $|\cdot|$ the corresponding norm on $\mathbb{R}^n$.

For estimating the error $e = u - u_h$, there are two main approaches.
a) The *a priori* method uses the *truncation error* $\tau = L_h x - b_h = L_h e$,

$$e = L_h^{-1}\tau \quad \Rightarrow \quad |e| \le c_{s,h}|\tau|, \tag{2.6}$$

with the *discrete* stability constant $c_{s.h} = |L_h^{-1}|$.
b) The *a posteriori* method uses the *residual* $\rho = b - Lu_h = Le$,

$$e = L^{-1}\rho \quad \Rightarrow \quad |e| \le c_s|\rho|, \tag{2.7}$$

with the *continuous* stability constant $c_s = |L^{-1}|$. Alternatively, we may use the solution z of the *dual problem* $L^*z = |e|^{-1}e$, to obtain

$$|e| = \langle e, L^*z\rangle = \langle b - Lu_h, z\rangle = \langle \rho, z\rangle \le |\rho|\,|z| \le c_s^*|\rho|, \tag{2.8}$$

with the *dual* stability constant $c_s^* = |L^{*-1}|$. Of course, this approach does not yield a new result in estimating the error in the l_2-norm. But it also gives us the possibility to bound other error quantities, e.g., general *error moments* $\langle\sigma,e\rangle$. With the solution of the dual problem $L^*z=\sigma$, there holds the *weighted* a posteriori error estimate

$$|\langle\sigma,e\rangle| \le \sum_{i=1}^{n} |\rho_i z_i|. \tag{2.9}$$

An analogous argument can also be applied in the case of a nonlinear equation. Let $L,\, L_h: \mathbb{R}^n \to \mathbb{R}^n$ be (differentiable) vector functions and solve $L(u)=0$ and its perturbation $L_h(u_h)=0$. Then, for the residual $\rho=-L(u_h)$, there holds by Taylor expansion

$$\rho = L(u)-L(u_h) = L'(u)e + R(u,u_h;e), \tag{2.10}$$

with the remainder term

$$R(u,u_h;e) = \int_0^1 \langle L''(u_h+se)e,e\rangle(1-s)\,ds.$$

From this, we obtain the error estimate

$$|e| \le c_s|\rho| + \tfrac{1}{2}c_s \max_{y\in[u,u_h]} |L''(y)|\,|e|^2, \tag{2.11}$$

with the stability constant $c_s=|L'(u)^{-1}|$. Below, we will use this duality technique for generating a posteriori error estimates in finite element Galerkin methods for differential equations.

2.2 A Model Problem

For illustration, we consider the model problem

$$-\Delta u = f \quad \text{in } \Omega, \qquad u=0 \ \text{ on } \partial\Omega, \tag{2.12}$$

on a polygonal domain $\Omega\subset\mathbb{R}^2$. Below, we will need some notation from the theory of functions spaces which is provided in the following subsection. Readers who are familiar with the usual Lebesgue- and Sobolev-space notation may want to skip this and continue with the next subsection on finite element approximation.

Function spaces notation. For an open set $Q\subset\mathbb{R}^d$, we denote by $L^2(Q)$ the Lebesgue space of square-integrable functions defined on Q which is a Hilbert space provided with the scalar product and norm

$$(v,w)_Q = \int_Q vw\,dx, \qquad \|v\|_Q = \Big(\int_Q |v|^2\,dx\Big)^{1/2}.$$

Analogously, $L^2(\partial Q)$ denotes the space of square-integrable functions defined on the boundary ∂Q equipped with the inner product and norm

$$(v,w)_{\partial Q} = \int_{\partial Q} vw\,ds, \qquad \|v\|_{\partial Q} = \Big(\int_{\partial Q} |v|^2\,dx\Big)^{1/2}.$$

The Sobolev spaces $H^1(Q)$ and $H^2(Q)$ consist of those functions $v \in L^2(Q)$ which possess first- and second-order (distributional) derivatives $\nabla v \in L^2(Q)^d$ and $\nabla^2 v \in L^2(Q)^{d\times d}$, respectively. For functions in these spaces, we use the semi-norms

$$\|\nabla v\|_Q = \Big(\int_Q |\nabla v|^2\,dx\Big)^{1/2}, \qquad \|\nabla^2 v\|_Q = \Big(\int_Q |\nabla^2 v|^2\,dx\Big)^{1/2}.$$

The space $H^1(Q)$ is continuously embedded in the space $L^2(\partial Q)$, such that for each $v \in H^1(Q)$ there exists a *trace* $v_{|\partial Q} \in L^2(\partial Q)$. In this notation, we do not distinguish between the function v on Q and its trace on ∂Q. Further, the functions in the subspace $H_0^1(Q) \subset H^1(Q)$ are characterized by the property $v_{|\partial Q} = 0$. By the Poincaré inequality,

$$\|v\|_Q \le c\,\|\nabla v\|_Q, \quad v \in H^1(Q), \tag{2.13}$$

the H^1-semi-norm $\|\nabla v\|_Q$ is a norm on the subspace $H_0^1(Q)$. If the set Q is the set Ω on which the differential equation is posed, we usually omit the subscript Ω in the notation of norms and scalar products, e.g., $\|v\| = \|v\|_\Omega$.

Variational formulation and finite element approximation. The natural solution space for the boundary value problem (2.12) is the Sobolev space $V = H_0^1(\Omega)$ defined above. The variational formulation of (2.12) seeks $u \in V$, such that

$$(\nabla u, \nabla\phi) = (f, \phi) \quad \forall \phi \in V. \tag{2.14}$$

The finite element approximation of (2.14) uses finite dimensional subspaces

$$V_h = \{v \in V : v_{|K} \in P(K),\, K \in \mathbb{T}_h\},$$

defined on decompositions $\mathbb{T}_h$ of Ω into triangles or quadrilaterals K (*cells*) of width $h_K = diam(K)$; we write $h = \max_{K\in\mathbb{T}_h} h_K$ for the *global* mesh width. Here, $P(K)$ denotes a suitable space of polynomial-like functions defined on the cell $K \in \mathbb{T}_h$. We will mainly consider low-order finite elements on quadrilateral meshes where $P(K) = \tilde{Q}_1(K)$ consists of shape functions which are obtained as usual via a bilinear transformation from the space of bilinear functions $Q_1(\hat{K}) = \mathrm{span}\{1, x_1, x_2, x_1x_2\}$ on the reference cell $\hat{K} = [0,1]^2$ (*isoparametric bilinears*). Local mesh refinement or coarsening is realized by using *hanging nodes* as indicated in Figure 2. The variable corresponding to such a hanging node is eliminated from the system by linear interpolation of neighboring variables in order to preserve the conformity of the global ansatz, i.e., $V_h \subset V$ (for more details we refer to Carey and Oden (1984)).

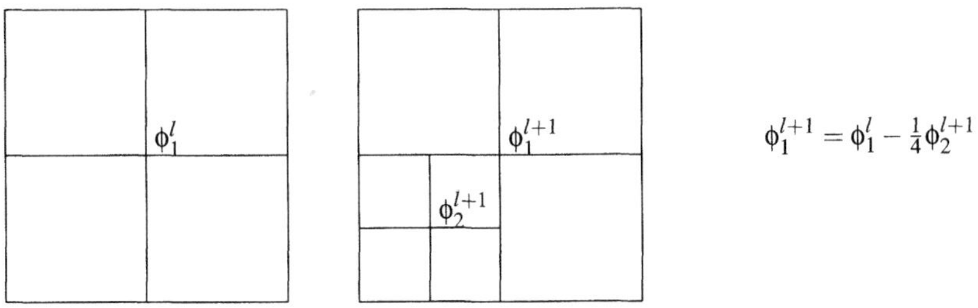

Figure 2. Nodal value interpolation on a 2-d cell patch with hanging nodes.

Now, the discrete problem determines $u_h \in V_h$ by

$$(\nabla u_h, \nabla \phi_h) = (f, \phi_h) \quad \forall \phi_h \in V_h. \tag{2.15}$$

The essential feature of this approximation scheme is the *Galerkin orthogonality* of the error $e = u - u_h$,

$$(\nabla e, \nabla \phi_h) = 0, \quad \phi_h \in V_h. \tag{2.16}$$

A priori error analysis. We begin with a brief discussion of the a priori error analysis for the scheme (2.15). By $i_h u \in V_h$, we denote the natural nodal interpolant of $u \in C(\bar{\Omega})$ satisfying $i_h u(P) = u(P)$ at all nodal points P. There holds (see, e.g., Ciarlet (1984) and Brenner and Scott (1994)):

$$\|u - i_h u\|_K + h_K^{1/2}\|u - i_h u\|_{\partial K} + h_K\|\nabla(u - i_h u)\|_K \le c_i h_K^2 \|\nabla^2 u\|_K, \tag{2.17}$$

with some *interpolation constant* $c_i > 0$.
(i) By the projection property of the Galerkin finite element scheme the interpolation estimate (2.17) directly implies the *energy-error* estimate

$$\|\nabla e\| = \inf_{\phi_h \in V_h} \|\nabla(u - \phi_h)\| \le c_i h^2 \|\nabla^2 u\|. \tag{2.18}$$

(ii) Further, employing a duality argument (so-called *Aubin-Nitsche trick*),

$$-\Delta z = \|e\|^{-1} e \quad \text{in } \Omega, \quad z = 0 \ \text{on } \partial\Omega, \tag{2.19}$$

we obtain

$$\|e\| = (e, -\Delta z) = (\nabla e, \nabla z) = (\nabla e, \nabla(z - i_h z) \le c_i c_s h \|\nabla e\|, \tag{2.20}$$

where the *stability constant* c_s is defined by the a priori bound $\|\nabla^2 z\| \le c_s$. Together with the energy-error estimate (2.18), this implies the improved L^2-error estimate

$$\|e\| \le c_i c_s h^2 \|\nabla^2 u\|. \tag{2.21}$$

(iii) On the bases of the global error estimates (2.18) and (2.21), we can also obtain error bounds for various output functionals, e.g., in $\mathbb{R}^2$:

$$|e(a)| \approx (c+\log h^{-1})\,\|\nabla e\|, \quad |\nabla e(a)| \approx h^{-1}\|\nabla e\|, \quad |(e,\psi)_{\partial\Omega}| \approx h^{-1/2}\|e\|.$$

All these estimates are only sub-optimal with respect to the power of h. This tells us that estimating the error in functional output usually requires special effort and cannot simply be reduced to the standard energy- or L^2-error estimates.

A posteriori error analysis. Next, we seek to derive a posteriori error estimates. Let $J(\cdot)$ be an arbitrary (linear) *error functional* defined on V, and $z \in V$ the solution of the corresponding dual problem

$$(\nabla\phi, \nabla z) = J(\phi) \quad \forall \phi \in V. \tag{2.22}$$

Taking $\phi = e$ in (2.22) and using the Galerkin orthogonality, we obtain after cell-wise integration by parts the error representation

$$\begin{aligned} J(e) &= (\nabla e, \nabla z) \\ &= (\nabla e, \nabla(z-\phi_h)) \\ &= \sum_{K\in\mathbb{T}_h} \Big\{(-\Delta u + \Delta u_h, z-\phi_h)_K - (n\cdot\nabla u_h, z-\phi_h)_{\partial K}\Big\} \end{aligned}$$

This can be rewritten as

$$J(e) = \sum_{K\in\mathbb{T}_h} \Big\{(R(u_h), z-\phi_h)_K - (r(u_h), z-\phi_h)_{\partial K}\Big\}, \tag{2.23}$$

with an arbitrary $\phi_h \in V_h$, and cell-residuals $R(u_h)$ and edge-residuals $r(u_h)$ defined by

$$R(u_h)_{|K} = f + \Delta u_h, \qquad r(u_h)_{|\Gamma} := \begin{cases} \frac{1}{2} n\cdot[\nabla u_h], & \text{if } \Gamma \subset \partial K \setminus \partial\Omega, \\ n\cdot\nabla u_h, & \text{if } \Gamma \subset \partial\Omega, \end{cases}$$

where $[\nabla u_h]$ denotes the jump of ∇u_h across the interelement boundaries Γ. From the error identity (2.23), we can infer an a posteriori error estimate of the form

$$|J(e)| \le \eta(u_h) = \sum_{K\in\mathbb{T}_h} \eta_K, \tag{2.24}$$

with the cell-wise error indicators $\eta_K = |(R(u_h), z-\phi_h)_K - (r(u_h), z-\phi_h)_{\partial K}|$. This estimate can be converted into a standard form as stated in the following proposition.

Proposition 1. *For the finite element approximation of the Poisson equation (2.12), there holds the a posteriori error estimate with respect to a functional* $J(\cdot)$*:*

$$|J(e)| \le \sum_{K\in\mathbb{T}_h} \{\rho_K\,\omega_K + \rho_{\partial K}\,\omega_{\partial K}\}, \tag{2.25}$$

with the cellwise residuals and weights defined by

$$\begin{aligned} \rho_K &= \|R(u_h)\|_K, & \omega_K &= \|z-\phi_h\|_K, \\ \rho_{\partial K} &= h_K^{-1/2}\|r(u_h)\|_{\partial K}, & \omega_{\partial K} &= h_K^{1/2}\|z-\phi_h\|_{\partial K}. \end{aligned}$$

The interpretation of the relation (2.24) is that the weights ω_K describe the dependence of $J(e)$ on variations of the cell residuals ρ_K,

$$\frac{\partial J(e)}{\partial \rho_K} \approx \omega_K \approx \|z - i_h z\|_K \approx h_K^2 \|\nabla^2 z\|_K,$$

where we have chosen $\phi_h = i_h z$. We remark that in a finite difference discretization of the model problem (2.12) the corresponding *influence factors* would behave like $\omega_K \approx \|z\|_K$. In general, the transition form the error identity (2.23) to the error estimate (2.25) causes significant over-estimation of the true error. This crucial point will be discussed in more detail in Section 3.2, below.

In practice the weights ω_K have to be determined computationally. This will be discussed in the next section. We note that the presented approach to a posteriori error estimation easily extends to *non-symmetric* and *non-coercive* problems of the form

$$Au = -\nabla \cdot \{a \nabla u\} + b \cdot \nabla u + cu = f.$$

A posteriori error bounds in global norms. By the same type of argument as used above, we can also derive the traditional global error estimates in the energy and the L^2 norm.
i) Energy-error bound: First, we use the functional

$$J(\phi) = \|\nabla e\|^{-1}(\nabla e, \nabla \phi)$$

in the dual problem. For its solution $z \in V$, there holds the a priori bound

$$\|\nabla z\| \leq 1. \tag{2.26}$$

We obtain the estimate

$$\begin{aligned} \|\nabla e\| &\leq \sum_{K \in \mathbb{T}_h} \{\rho_K \omega_K + \rho_{\partial K} \omega_{\partial K}\} \\ &\leq \Big(\sum_{K \in \mathbb{T}_h} \{h_K^2 \rho_K^2 + h_K \rho_{\partial K}^2\} \Big)^{1/2} \Big(\sum_{K \in \mathbb{T}_h} \{h_K^{-2} \omega_K^2 + h_K^{-1} \omega_{\partial K}^2\} \Big)^{1/2}, \end{aligned}$$

with residual terms and weights as defined above. Now, we use an extension of the interpolation estimate (2.17) (see, e.g., Johnson (1987)),

$$\Big(\sum_{K \in \mathbb{T}_h} \Big\{ h_K^{-2} \|z - \tilde{i}_h z\|_K^2 + h_K^{-1} \|z - \tilde{i}_h z\|_{\partial K}^2 \Big\} \Big)^{1/2} \leq \tilde{c}_i \|\nabla z\|, \tag{2.27}$$

where $\tilde{i}_h z \in V_h$ is a modified nodal interpolation which is defined and stable on $H^1(\Omega)$. This gives us

$$\|\nabla e\| \leq \tilde{c}_i \Big(\sum_{K \in \mathbb{T}_h} \{h_K^2 \rho_K^2 + h_K \rho_{\partial K}^2\} \Big)^{1/2} \|\nabla z\|.$$

Finally, observing (2.26), we conclude the *energy-error* a posteriori error estimate

$$\|\nabla e\| \leq \eta_E = \tilde{c}_i \Big(\sum_{K \in \mathbb{T}_h} \{h_K^2 \rho_K^2 + h_K \rho_{\partial K}^2\} \Big)^{1/2}. \tag{2.28}$$

ii) L^2-norm error bound: Next, we use the functional

$$J(\phi) = \|e\|^{-1}(e,\phi)$$

in the dual problem. If the (polygonal) domain Ω is convex, the dual solution $z \in V$ is in $H^2(\Omega)$ and admits the a priori bound

$$\|\nabla^2 z\| \le c_s = 1. \tag{2.29}$$

This yields analogously as in (i) the error estimate

$$\|e\| \le \Big(\sum_{K\in\mathbb{T}_h} \{h_K^4\rho_K^2 + h_K^3\rho_{\partial K}^2\}\Big)^{1/2}\Big(\sum_{K\in\mathbb{T}_h} \{h_K^{-4}\omega_K^2 + h_K^{-3}\omega_{\partial K}^2\}\Big)^{1/2}.$$

Now, we use the stronger version of the interpolation estimate (2.27),

$$\Big(\sum_{K\in\mathbb{T}_h} \Big\{h_K^{-4}\|z-i_h z\|_K^2 + h_K^{-3}\|z-i_h z\|_{\partial K}^2\Big\}\Big)^{1/2} \le c_i\|\nabla^2 z\|, \tag{2.30}$$

to obtain

$$\|e\| \le c_i\Big(\sum_{K\in\mathbb{T}_h} \{h_K^4\rho_K^2 + h_K^3\rho_{\partial K}^2\}\Big)^{1/2}\|\nabla^2 z\|.$$

Finally, observing (2.29), we conclude the L^2-norm a posteriori error estimate

$$\|e\| \le \eta_{L^2} = c_i c_s\Big(\sum_{K\in\mathbb{T}_h} \{h_K^4\rho_K^2 + h_K^3\rho_{\partial K}^2\}\Big)^{1/2}. \tag{2.31}$$

Local a posteriori error bounds based on energy-error estimates. Finally, we briefly discuss an approach proposed in Oden, Wu and Ainsworth (1993) for estimating the error in local functional output which is based on sharp energy-error estimates. Suppose that we have general upper and lower error bounds available of the form

$$\eta_{low}(u_h) \le \|\nabla e\| \le \eta_{up}(u_h). \tag{2.32}$$

Such estimates may be obtained by using local defect correction techniques (see, e.g., Ainsworth and Oden (1997)). Now, let again a (linear) error functional $J(\cdot)$ on V be given and let $z \in V$ be the corresponding dual solution. Then, there holds the error representation

$$J(e) = (\nabla e, \nabla\varepsilon),$$

where $\varepsilon = z - z_h$, and $z_h \in V_h$ is the Ritz approximation of z defined by

$$(\nabla z_h, \nabla\phi_h) = J(\phi_h) \quad \forall\phi_h \in V_h.$$

By the parallelogram identity, we obtain

$$J(e) = (\theta\nabla e, \theta^{-1}\nabla\varepsilon) = \tfrac{1}{4}\|\nabla(\theta e + \theta^{-1}\varepsilon)\|^2 - \tfrac{1}{4}\|\nabla(\theta e - \theta^{-1}\varepsilon)\|^2,$$

with a suitable balancing parameter $\theta > 0$. In view of (2.32), this implies the a posteriori error estimate

$$\hat{\eta}_{low} \le J(e) \le \hat{\eta}_{up}, \tag{2.33}$$

with the lower bound given by

$$\hat{\eta}_{low} = \tfrac{1}{4}\eta_{low}^2(\theta e + \theta^{-1}\varepsilon) - \tfrac{1}{4}\eta_{up}^2(\theta e - \theta^{-1}\varepsilon),$$

and the upper bound given by

$$\hat{\eta}_{up} = \tfrac{1}{4}\eta_{up}^2(\theta e + \theta^{-1}\varepsilon) - \tfrac{1}{4}\eta_{low}^2(\theta e - \theta^{-1}\varepsilon).$$

This estimate provides sharp upper as well as lower bounds for the functional error $J(e)$ as long as reliable bounds for the energy errors $\|\nabla(\theta e + \theta^{-1}\varepsilon)\|$ and $\|\nabla(\theta e - \theta^{-1}\varepsilon)\|$ are available. This can be achieved as long the functional $J(\cdot)$ is *regular*, i.e. defined on V, which excludes, for example, point-error evaluation. In the latter case, sharp upper and lower energy-error bounds are hard to get. Further it is not clear how to organize effective local mesh adaptation on the basis of the estimate (2.33). Finally, it should be emphasized that the derivation of this error bound requires the underlying energy form to be symmetric and positive definite (i.e. to be a scalar product).

3 Strategies for Mesh Adaptation

We want to discuss some strategies for organizing local mesh adaptation on the basis of a posteriori error estimates as derived in the preceding section. Suppose that for some error quantity $J(u) - J(u_h)$, we have an a posteriori error estimate of the form

$$|J(u) - J(u_h)| \le \eta(u_h) = \sum_{K \in \mathbb{T}_h} \eta_K, \tag{3.34}$$

with local *cell-error indicators* $\eta_K = \eta_K(u_h)$. The prescribed error tolerance is TOL and the maximum number of mesh cells N_{max}. The following three strategies are in use:

- *Error balancing strategy:* Cycle through the mesh and seek to equilibrate the local error indicators according to

$$\eta_K \approx \frac{TOL}{N}, \quad N = \#\{K \in \mathbb{T}_h\}. \tag{3.35}$$

  This process requires iteration with respect to the number of mesh cells N and eventually leads to $\eta(u_h) \approx TOL$.

- *Fixed fraction strategy:* Order cells according to the size of η_K and refine a certain percentage (say 20%) of cells with largest η_K (or those which make up 20% of the estimator value $\eta(u_h)$) and coarsen those cells with smallest η_K. By this strategy, we may achieve a prescribed rate of increase of N (or keep it constant in solving non-stationary problems).

- *Mesh optimization strategy:* Use the representation

$$\eta(u_h) = \sum_{K\in\mathbb{T}_h} \eta_K \approx \int_\Omega h(x)^2\Psi(x)\,dx, \tag{3.36}$$

directly for generating a formula for an optimal mesh-size distribution $h_{opt}(x)$. This approach will be discussed in more detail in the following section.

3.1 The Mesh Optimization Strategy

We will outline a heuristic analysis of the *mesh optimization strategy*. As a by-product, we will also obtain the justification of the error balancing strategy. Let the domain be d-dimens-ional, $\Omega \subset \mathbb{R}^d$, and let $N_{\max}$ and TOL be prescribed. We assume that for $h \to 0$, the normalized cell residuals approach certain mesh-independent limits,

$$h_K^{-d/2}\rho_K \approx \Psi_{u|K}, \tag{3.37}$$

which involve the second derivatives of the solution and the data. This property can be guaranteed on uniformly refined meshes by exploiting super-approximation effects, but it still needs theoretical justification on locally refined meshes as constructed by the strategies described above. For the weights, we know from (2.17) that

$$h_K^{-d/2-2}\omega_K(z) \approx c_i \max_K |\nabla^2 z| = \max_K \Psi_z. \tag{3.38}$$

This suggests to use the relation

$$\eta(u_h) \approx E(h) = \int_\Omega h(x)^2\Psi(x)\,dx, \tag{3.39}$$

with the mesh-independent function $\Psi(x) = \Psi_u(x)\Psi_z(x)$. Further, the number N of cells in the mesh $\mathbb{T}_h$ can be expressed by the formula

$$N(h) = \sum_{K\in\mathbb{T}_h} h_K^d h_K^{-d} \approx \int_\Omega h(x)^{-d}\,dx. \tag{3.40}$$

With this notation, we can formulate the following two mesh-optimization problems which are in some sense adjoint to each other:

$$(I)\quad \eta(u_h) \to \min,\quad N \le N_{max}; \qquad (II)\quad N \to \min,\quad \eta(u_h) \le TOL.$$

These problems can be explicitly solved by functions of the form

$$h_{opt}^{(I)}(x) = \left(\frac{W}{N_{\max}}\right)^{1/d}\Psi(x)^{-1/(2+d)},\quad h_{opt}^{(II)}(x) = \left(\frac{TOL}{W}\right)^{1/d}\Psi(x)^{-1/(2+d)}, \tag{3.41}$$

where

$$W = \int_\Omega \Psi(x)^{d/(2+d)}\,dx < \infty.$$

We want to sketch an argument for this result in the case of problem (I). To this end, we use the classical Lagrange approach. Introducing the Lagrangian functional

$$L(h,\lambda) = E(h) + \lambda\{N(h) - N_{max}\},$$

with the Lagrangian multiplier $\lambda \in \mathbb{R}$, the first-order optimality condition is

$$\frac{d}{dt}L(h+t\phi,\lambda+t\mu)_{|t=0} = 0,$$

for all admissible variations ϕ and μ. This implies that

$$2h(x)\Psi(x) - d\lambda h(x)^{-d-1} = 0, \qquad \int_\Omega h(x)^{-d}\,dx - N_{\max} = 0,$$

and, consequently,

$$h(x) = \Big(\frac{2}{d\lambda}\Psi(x)\Big)^{-1/(2+d)}, \qquad \Big(\frac{2}{d\lambda}\Big)^{d/(2+d)}\int_\Omega \Psi(x)^{d/(2+d)}\,dx = N_{\max}.$$

From this, we deduce the desired relations

$$\lambda \equiv \frac{2}{d}h(x)^{2+d}\Psi(x), \qquad h_{opt}(x) = \Big(\frac{W}{N_{\max}}\Big)^{1/d}\Psi(x)^{-1/(2+d)}.$$

The first identity implies that an optimal mesh-size distribution with local cell-widths h_K is indeed characterized by the equilibration of the cell-indicators $\eta_K \sim h_K^{2+d}\Psi_K$ as assumed by the *error balancing strategy*. The mesh-optimization problem (II) can be treated in an analogous way.

We note that even for rather *irregular* functionals $J(\cdot)$ the quantity W is bounded. For example, the evaluation of derivative point values $J(u) = \partial_i u(P)$ for smooth u in two dimensions leads to $\Psi(x) \approx |x_P|^{-3}$ and, consequently,

$$W \approx \int_\Omega |x-a|^{-3/2}\,dx < \infty.$$

This example will be considered in more detail in the next subsection.

The explicit formulas for $h_{opt}(x)$ have to be used with care in designing a mesh. Their derivation implicitly assumes that they actually correspond to *scalar* mesh-size functions of isotropic meshes, a condition however which is not incorporated into the formulation of the mesh-optimization problems. Anisotropic meshes containing stretched cells require a more involved concept of mesh description and optimization. This question is subject of current research.

3.2 Evaluation of Error Estimates

As a model case, we consider again the Poisson diffusion problem

$$-\Delta u = f \text{ in } \Omega, \quad u = 0 \text{ on } \partial\Omega, \tag{3.42}$$

on a polygonal domain $\Omega \subset \mathbb{R}^2$, say $\Omega = (-1,1)^2$. The corresponding discrete problems read

$$(\nabla u_h, \nabla \phi_h) = (f, \phi_h) \quad \forall \phi_h \in V_h, \tag{3.43}$$

where $V_h \subset V = H_0^1(\Omega)$ are the finite element subspaces as defined above. In Section 2.2, we have derived an a posteriori identity for the error with respect to a (linear) functional $J(\cdot)$:

$$J(e) = \sum_{K \in \mathbb{T}_h} \Big\{ (R(u_h), z - \phi_h)_K - (r(u_h), z - \phi_h)_{\partial K} \Big\}, \tag{3.44}$$

with an arbitrary approximation $\phi_h \in V_h$ of the dual solution $z \in V$. The dual solution corresponding to the prescribed *error functional* $J(\cdot)$ is defined by the auxiliary problem

$$(\nabla \phi, \nabla z) = J(\phi) \quad \forall \phi \in V. \tag{3.45}$$

From the identity (3.44), we obtain the first error estimate

$$|J(e)| \leq \eta_{glob}(u_h), \tag{3.46}$$

with the (global) error estimator

$$\eta_{glob}(u_h) = \Big| \sum_{K \in \mathbb{T}_h} \Big\{ (R(u_h), z - \phi_h)_K - (r(u_h), z - \phi_h)_{\partial K} \Big\} \Big|.$$

In using this estimate, we have to investigate how to evaluate the weighting factor $z - \phi_h$. The ultimate goal is to achieve an optimal *effectivity index* for the error estimator $\eta(u_h)$

$$I_{\text{eff}} = \lim_{TOL \to 0} \frac{\eta(u_h)}{|J(e)|} = 1\,.$$

We run some computational experiments in order to see what can be expected. Unfortunately, the result is rather negative, i.e., asymptotic sharpness does not seem to be achievable with acceptable effort. With all the cheaper methods considered the effectivity index I_{eff} never really tends to one, but in most relevant cases stays well below two, what may actually be considered as good enough. There are two separate aspects to be considered: the sharpness of the global error bound $\eta(u_h)$ and the effectivity of the local error indicators η_T which are used in the mesh refinement process.

The very first step is to convert the global *error identity* (3.44) into a localized *error estimate* by taking the absolute sign insight the sum,

$$|J(e)| \leq \eta(u_h) = \sum_{K \in \mathbb{T}_h} \eta_K\,, \tag{3.47}$$

with the local *error indicators* defined by

$$\eta_K = \Big| (R(u_h), z - \phi_h)_K - (r(u_h), z - \phi_h)_{\partial K} \Big|. \tag{3.48}$$

However, already by this localization the asymptotic sharpness of the global error estimate may get lost. This is seen, for instance, in the case $J(u) = u(0)$ when the exact as well as the approximate solution are anti-symmetric with respect to the x-axis meaning that $e(0) = 0$, but $\eta(u_h) \neq 0$.

Next, we will analyze various ways of directly evaluating the error identity (3.44). All strategies are based on solving the dual problem numerically. Let $z_h^{(1)} \in V_h$ denote the approximation to z obtained on the current mesh by the same bilinear finite element ansatz as used for computing u_h.

1. Approximation by higher order methods: The dual problem is solved by using *biquadratic* finite elements on the current mesh yielding an approximation $z_h^{(2)}$ to z. The error estimator obtained by using this approximation in (3.48) is denoted by $\eta^{(1)}(u_h)$. It is composed of the local cell-error indicators

$$\eta_K^{(1)} = \left|(R(u_h), z_h^{(2)} - \phi_h)_K - (r(u_h), z_h^{(2)} - \phi_h)_{\partial K}\right|,$$

where $\phi_h = i_h z_h^{(2)} \in V_h$ is the bilinear nodal interpolation of $z_h^{(2)}$. It is seen by theoretical analysis as well as by numerical experiments that $\eta^{(1)}(u_h)$ behaves like (Becker and Rannacher (1996)):

$$\lim_{TOL\to 0} I_{\text{eff}} = 1. \tag{3.49}$$

However, computing approximations to z by using a higher order finite element scheme does not appear very economical in estimating the error in the low-order scheme.

2. Approximation by higher order interpolation: A simplification is achieved by simple patchwise *biquadratic* interpolation of the bilinear approximation $z_h^{(1)}$ on the current mesh yielding an approximation $i_h^{(2)} z_h^{(1)}$ to z. This construction requires some special care on elements with hanging nodes, in order to preserve the higher order accuracy of the interpolation process. The resulting global error estimator is denoted by $\eta^{(2)}(u_h)$. It consists of local cell-error indicators of the form

$$\eta_K^{(2)} = \left|(R(u_h), i_h^{(2)} z_h^{(1)} - z_h^{(1)})_K - (r(u_h), i_h^{(2)} z_h^{(1)} - z_h^{(1)})_{\partial K}\right|,$$

Numerical experiments show that in this case (Becker and Rannacher (1996)):

$$\liminf_{TOL\to 0} I_{\text{eff}} > 1 \ \ (\sim 1-2). \tag{3.50}$$

3. Approximation by difference quotients: We use the cell-wise interpolation estimate (2.17) to estimate the weights ω_K and $\omega_{\partial K}$ by

$$\omega_T + \omega_{\partial K} = \|z - i_h z\|_K + h_K^{1/2} \|z - i_h z\|_{\partial K} \le c_i h_K^2 \|\nabla^2 z\|_K.$$

The second derivative $\nabla^2 z$ is then replaced by a suitable second-order difference quotient $\nabla_h^2 z_h^{(1)}$ of the approximate dual solution $z_h^{(1)} \in V_h$. The computation of such second-order difference quotients may be costly particularly on cell patches with hanging nodes. A useful substitute is the approximation

$$\|\nabla^2 z\|_K \approx h_K^{-1/2} \|n\cdot[\nabla z_h^{(1)}]\|_{\partial K}, \tag{3.51}$$

which corresponds to the jump-residual $\rho_{\partial K}$ of u_h the evaluation of which has to be implemented anyway. The resulting global error estimator is denoted by $\eta^{(3)}(u_h)$. It consists of local cell-error indicators of the form

$$\eta_K^{(3)} = c_i h_K^{-1/2} \{\rho_K + \rho_{\partial K}\} \|n\cdot[\nabla z_h^{(1)}]\|_{\partial K},$$

where the cell-residuals are defined as above by $\rho_K = \|R(u_h)\|_K$, $\rho_{\partial K} = h_K^{-1/2}\|r(u_h)\|_{\partial K}$. Numerical experiments show that also in this case (Becker and Rannacher (1996)):

$$\liminf_{TOL\to 0} I_{\text{eff}} > 1 \quad (\sim 1-8). \tag{3.52}$$

4. Approximation by local residual problems: Following the idea already used in the energy-error estimator of Bank and Weiser (Ainsworth and Oden (1997)), on each element K the local Neumann problems

$$(\nabla v_K, \nabla \phi_h)_K = (R(u_h), \phi_h)_K - (r(u_h), \phi_h)_{\partial K} \quad \forall \phi_h \in V_K, \tag{3.53}$$

are solved, where $V_K = \{q \in \tilde{Q}_2(K),\, q \perp \tilde{Q}_1(K)\}$. The related error indicator based on solving patch-wise Dirichlet problems is not considered here, since it turns out to be too expensive in practice. Using this construction, the local cell-error indicators (3.48) are replaced by

$$\eta_K = |(\nabla v_K, \nabla(z-\phi_h))_K| \le \|\nabla v_K\|_K \|\nabla(z-\phi_h)\|_K.$$

Employing again the approximation (3.51), we obtain a global error estimator denoted by $\eta^{(4)}(u_h)$ which consists of locale cell-error indicators of the form

$$\eta_K^{(4)} = c_i h_K^{1/2} \|\nabla v_K\|_K \|n\cdot[\nabla z_h]\|_{\partial K}.$$

The effectivity indices observed for these strategies in computing the point-value $u(0)$ are listed in Table 1 (from Becker and Rannacher (1996)). The performance of the estimator $\eta^{(4)}(u_h)$ is similar to that of $\eta^{(3)}(u_h)$ and is therefore not listed in the table. These results indicate that even for the simplest model situations, an asymptotic efficiency index $I_{\text{eff}} = 1$ is achievable only at the expense of unacceptably high cost, e.g., by approximating the dual solution using higher-order elements.

Table 1. Effi ciency of *weighted* error indicators for the point-error $J(e) = |e(0)|$.

L	N	$\eta^{(1)}/J(e)$	$\eta^{(2)}/J(e)$	$\eta^{(3)}/J(e)$
1	4^3	6.667	12.82	2.066
2	4^4	1.253	13.51	45.45
3	4^5	1.052	3.105	9.345
4	4^6	1.007	2.053	7.042
5	4^7	1.003	1.886	6.667

3.3 Examples

We consider again the Poisson diffusion problem

$$-\Delta u = f \text{ in } \Omega, \quad u = 0 \text{ on } \partial\Omega, \tag{3.54}$$

on a domain $\Omega \subset \mathbb{R}^2$ and its approximation by *bilinear* finite elements as described above.

Example 1. First, we consider a highly localized error functional. As concrete example, we choose again the square domain $\Omega = (-1,1)^2$ and the error functional

$$J(u) = \partial_1 u(0).$$

In this case, the dual solution does not exist in the sense of $H_0^1(\Omega)$, such that for practical use, we have to regularize the functional like

$$J_\varepsilon(u) = |B_\varepsilon|^{-1} \int_{B_\varepsilon} \partial_1 u \, dx = \partial_1 u(0) + O(\varepsilon),$$

where $B_\varepsilon = \{x \in \Omega | \, |x| < \varepsilon\}$, and $\varepsilon = \text{TOL}$ is a suitable error tolerance. The corresponding dual solution z behaves like $|\nabla^2 z(x)| \approx d(x)^{-3}$, where $d(x) = |x| + \varepsilon$. From the general a posteriori error estimate (2.25), we obtain for the present case

$$|J_\varepsilon(e)| \approx \eta_\omega(u_h) = c_i \sum_{K \in \mathbb{T}_h} \frac{h_K^3}{d_K^3} \{\rho_T + \rho_{\partial T}\}. \tag{3.55}$$

Table 2. (Example 1) Computing $\partial_1 u(0) \sim J_\varepsilon(u)$ using the weighted estimator $\eta_\omega(u_h)$.

$\eta_\omega(u_h)$					
TOL	N	L	$\lvert\partial_1 e(0)\rvert$	η	I_{eff}
4^{-3}	940	9	4.10e-1	1.42e-2	0.30
4^{-4}	4912	12	4.14e-3	3.50e-3	0.80
4^{-5}	20980	15	2.27e-4	9.25e-4	4.16
4^{-6}	86740	17	5.82e-5	2.38e-4	4.16

Now, we want to use the results of Section 3.1 for determining a priori an optimal mesh-size distribution. To this end, let us again assume that the residuals behave like $h_K^{-1}\{\rho_K^{(1)}(u_h) + \rho_K^{(2)}(u_h)\} \approx \Psi_{u|K}$, with a mesh-independent function Ψ_u, such that

$$\eta_\omega(u_h) \approx \int_\Omega h(x)^2 \Psi(x) \, dx, \quad \Psi(x) = \frac{\Psi_u(x)}{d(x)^3}.$$

Hence, from (3.41), we obtain

$$h_{opt}(x) \approx \left(\frac{W}{N_{max}}\right)^{1/2} \Psi(x)^{-1/4} \approx N_{max} d(x)^{-3/4},$$

with $W = \int_\Omega \Psi(x) \, dx < \infty$. This implies the relation

$$N_{opt} = \sum_{K \in \mathbb{T}_h} h_K^2 h_K^{-2} = \left(\frac{N_{opt}}{TOL}\right)^{1/2} \sum_{K \in \mathbb{T}_h} h_K^2 d_K^{-3/2} \approx \left(\frac{N_{opt}}{TOL}\right)^{1/2},$$

Figure 3. (Example 1) Refined mesh and approximate dual solution for computing $\partial_1 u(0)$ using the a posteriori error estimator $\eta_\varepsilon(u_h)$, with TOL $= 4^{-4}$.

and, consequently, $N_{opt} \approx TOL^{-1}$. This is better than what could be achieved on uniformly refined meshes. In fact, mesh refinement on the basis of global energy-error control results in almost uniform refinement, i.e., $N_{opt} \approx TOL^{-2}$. This predicted asymptotic behavior is well confirmed by the results of our computational test. Figure 3 shows the balanced mesh for TOL $= 4^{-4}$ and the approximation to the dual solution z_ε, with $\varepsilon =$ TOL, computed on this mesh. The corresponding errors for a sequence of tolerances are listed in Table 2. For the problem considered, the weighted a posteriori estimate (3.55) for the point error is obviously asymptotically optimal and the predicted dependence $N_{opt} \approx \mathrm{TOL}^{-1}$ is confirmed; for more details, we refer to Becker and Rannacher (1996).

Example 2. In the second example, we consider again the point evaluation $\partial_1 u(P)$ at some point $P \in \Omega$, but on a slit domain; see Figure 4. We compare the performance of our weighted error estimator $\eta_\omega(u_h)$ with the traditional *energy-error* estimator $\eta_E(u_h)$ derived in (2.28). Obviously, the weighted error estimator concentrates mesh cells at the slit in order to suppress the pollution effect of the corner singularity but also at the evaluation point, while the energy-norm estimator induces a stronger refinement at the slit.

Example 3. The third example is meant as an illustrative exercise. For problem (3.54) on a smoothly bounded domain $\Omega \subset \mathbb{R}^2$, we consider the functional

$$J(u) = \int_{\partial\Omega} \partial_n u \, ds \quad \left(= \int_\Omega f \, dx \right),$$

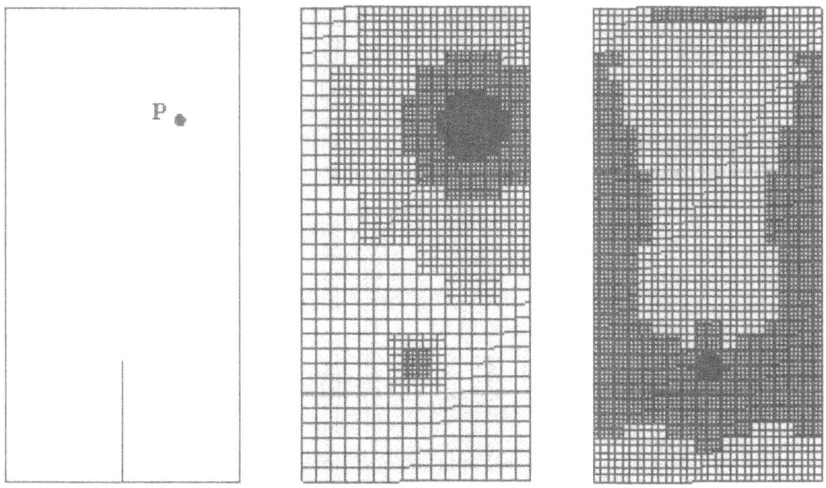

Figure 4. (Example 2) Refined meshes with about 5,000 elements for computing $\partial_1 u(P)$ using the weighted error estimator $\eta_\omega(u_h)$ (middle) and the energy error estimator $\eta_E(u_h)$ (right).

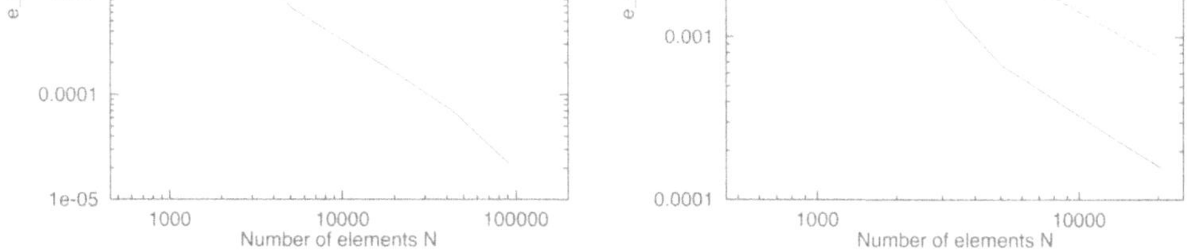

Figure 5. (Example 2) Comparison of efficiency between $\eta_E(u_h)$ and $\eta_\omega(u_h)$ on the slit domain.

and ask the question: *What is an optimal mesh-size distribution for computing* $J(u)$*?* The corresponding dual problem

$$(\nabla\phi,\nabla z) = (1,\partial_n\phi)_{\partial\Omega} \quad \forall\phi\in V\cap W^{2,1}(\Omega),$$

has a measure solution with density of the form $z\equiv 1$ in Ω, $z=0$ on $\partial\Omega$. In order to avoid the use of measures, we consider the regularized functional

$$J_\varepsilon(\phi) = |S_\varepsilon|^{-1}\int_{S_\varepsilon}\partial_n\phi\,dx = \int_{\partial\Omega}\partial_n\phi\,ds + O(\varepsilon),$$

where $\varepsilon = \mathrm{TOL}$, $S_\varepsilon = \{x\in\Omega,\ \mathrm{dist}\{x,\partial\Omega\}<\varepsilon\}$, and $\partial_n\phi$ the generic continuation of $\partial_n\phi$ to S_ε. The regularized dual solution is $z_\varepsilon = 1$ in $\Omega\setminus S_\varepsilon$, $z_\varepsilon(x) = \varepsilon^{-1}\mathrm{dist}\{x,\partial\Omega\}$ on S_ε. This implies that

$$J_\varepsilon(e) \le c_i \sum_{K\in\mathbb{T}_h,K\cap S_\varepsilon\ne\emptyset} h_K^2\{\rho_K+\rho_{\partial K}\}\,\|\nabla^2 z_\varepsilon\|_K,$$

i.e., there is no contribution to the error from cells in the interior of Ω. Hence, independent of the form of the forcing f, the optimal strategy is to refine the elements adjacent to the boundary and to leave the others unchanged, assuming of course, that the force term f is integrated exactly.

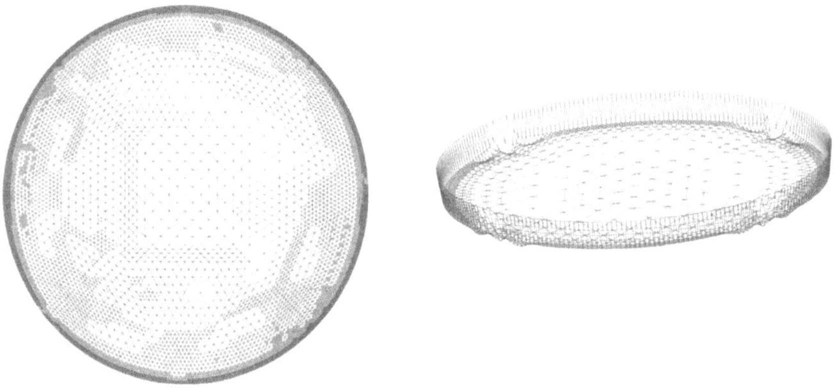

Figure 6. (Example 3) Refi ned mesh and approximate dual solution obtained by the weighted error estimator $\eta_\omega(u_h)$ after 7 refi nement steps.

3.4 Estimation of Iteration Error

The evaluation of the a posteriori error estimate (3.34) assumes that the discrete solution $u_h\in V_h$ is computed exactly. However, in general, we only have an approximation $\tilde u_h\in V_h$ obtained by some iteration process. In the following, we will present an approach to controlling the total error

$$\tilde e = u-\tilde u_h = u-u_h+u_h-\tilde u_h = e+e_h,$$

in a similar way as already used for the pure discretization error e. Here, we concentrate on the L^2-norm error; the energy-norm error or any other functional of the error, $J(\tilde{e})$, can be treated in an analogous manner.

Associated to the subspaces $V_h \subset V$, we introduce the L^2 projection $P_h : V \to V_h$ and the *discrete* Laplacian $\Delta_h : V_h \to V_h$ defined by

$$(P_h v, \phi_h) = (v, \phi_h), \qquad -(\Delta_h v_h, \phi_h) = (\nabla v_h, \nabla \phi_h) \quad \forall \phi_h \in V_h.$$

This notation allows us to define the algebraic residual

$$r_h(\tilde{u}_h) = P_h f + \Delta_h \tilde{u}_h,$$

with the property $r_h(u_h) = 0$. The goal is now to derive an a posteriori error estimate in terms of the error estimator $\eta_{L^2}(\tilde{u}_h)$ derived in Section 2.2 and the algebraic residual $r_h(\tilde{u}_h)$.

Multigrid iteration. We will analyze the iteration error in the context of a multigrid method. Under certain restrictions this solution scheme can be interpreted as a projection method possessing Galerkin-orthogonality properties. This allows to estimate the error in the multigrid iteration along the line of the error analysis of the finite element method.

Let $\mathbb{T}_l$ $(l = 0, \ldots, L)$ be a sequence of nested meshes with mesh-size parameters h_l, and let V_l be the corresponding finite element spaces such that $V_l \subset V_{l-1}$. Further, let $u_l \in V_l$ be the *exact* discrete solutions with the approximations $\tilde{u}_l \in V_l$ and the associated discrete residuals

$$r_l(\tilde{u}_l) = P_l f + \Delta_l \tilde{u}_l \in V_l \quad (l = 0, \ldots, L).$$

The multigrid algorithm for computing the approximate solutions employs *smoothing* operations (Jacobi-, Gauss-Seidel-, ILU-iteration, etc.)

$$S_l : V_l \to V_l \quad (\textit{smoother}),$$

and mesh-transfer operations

$$R_l^{l-1} : V_l \to V_{l-1} \quad (\textit{restriction}), \quad P_{l-1}^l : V_{l-1} \to V_l \quad (\textit{prolongation}).$$

In the context of a finite element discretization the most natural choices are

$$R_l^{l-1} = P_{l-1}, \qquad P_{l-1}^l = id.$$

Using this notation, we can describe the steps of a multigrid cycle for computing the approximation $\tilde{u}_L$ on the *finest* mesh level $\mathbb{T}_L$ as follows:

Multigrid Cycle: For some starting value $\tilde{u}_L^0 \in V_L$ on the finest mesh (usually $\tilde{u}_L^0 = 0$), the iteration step $\tilde{u}_L^k \to \tilde{u}_L^{k+1}$ consists of the following stages:

1. Pre-smoothing: $S_L(\tilde{u}_L^k)$.

2. Residual computation: $r_L = P_L f + \Delta_L S_L(\tilde{u}_L^k)$.

3. Restriction: $R_L^{L-1} r_L$.

4. Approximate solution of defect equation: $-\Delta_{L-1}\tilde{v}_{L-1} = R_L^{L-1} r_L$.
   This is accomplished by recursively calling the multigrid solver on mesh level $\mathbb{T}_{L-1}$. On the *coarsest* mesh level $\mathbb{T}_0$ the defect equation $-\Delta_0\tilde{v}_0 = R_1^0 d_1$ is solved exactly.

5. Prolongation: $P_{L-1}^L \tilde{v}_{L-1}$.

6. Correction: $u_L^{k+1} = S_L(\tilde{u}_L^k) + P_{L-1}^L \tilde{v}_{L-1}$.

This iteration is continued until some stopping criterion is reached. Depending on how many multigrid cycles ($q = 1, 2, \ldots$) are performed for the defect equations $-\Delta_{l-1}\tilde{v}_{l-1} = R_l^{l-1} r_l$ on each mesh level, we speak of a *V-cycl* ($q = 1$) or *W-cycle* ($q = 2$); using more than $q = 2$ cycles would be inefficient. The *F-cycle* is a compromise between V-cycle and W-cycle. In order to preserve the character of a Galerkin (projection) method of the multigrid scheme, we do not allow damping in the correction step and *post-smoothing*.

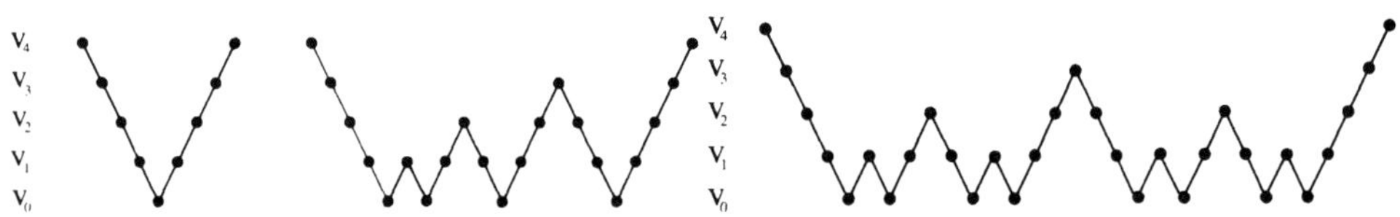

Figure 7. Schemes of the multigrid V-cycle (left), F-cycle (middle) and W-cycle (right).

Under the foregoing assumptions, we have the a posteriori error estimate (Becker, Johnson and Rannacher (1995))

$$\|u - \tilde{u}_L\| \le c_s \Big\{ \eta_{L^2}(\tilde{u}_L) + c_i' \sum_{l=1}^{L} \Big(\sum_{K\in\mathbb{T}_{l-1}} h_{l-1,K}^4 \|r_l(\tilde{v}_l)\|_K^2 \Big)^{1/2} \Big\}, \tag{3.56}$$

with the error estimator of the finite element scheme

$$\eta_{L^2}(\tilde{u}_L) = c_i \Big(\sum_{K\in\mathbb{T}_h} h_L^4 \big\{ \|R(\tilde{u}_L)\|_K^2 + \|r(\tilde{u}_L)\|_{\partial K}^2 \big\} \Big)^{1/2},$$

and the algebraic residuals $r_l(\tilde{v}_l) = P_l f + \Delta_l \tilde{v}_l$ on the intermediate mesh levels. The constants c_i, c_i' are again *interpolation constants* and the *stability constant* can be set to $c_s = 1$. Since the multigrid iteration converges, $\tilde{u}_L^k \to u_L \quad (k \to \infty)$, the second sum in (3.56) tends to zero.

On the basis of the a posteriori error estimate (3.56), we can design a stopping criterion for the multigrid iteration for computing the approximate solution $\tilde{u}_L \in V_L$ on the finest mesh level $\mathbb{T}_L$. Let some desirable tolerance TOL for the L^2 error be given. Then, the multigrid iteration is carried on until the following condition is satisfied:

$$c_i' \sum_{l=1}^{L} \Big(\sum_{K\in\mathbb{T}_{l-1}} h_{l-1,K}^4 \|r_l(\tilde{v}_l)\|_K^2 \Big)^{1/2} \le \eta_{L^2}(\tilde{u}_L) \le \tfrac{1}{2} TOL. \tag{3.57}$$

The iteration error may be decreased by increasing the number k of multigrid cycles on mesh level $\mathbb{T}_L$ or by directly reducing the defect norms $\|r_l(\tilde{v}_l)\|_K$ (for instance, by increasing the number of smoothing steps).

We want to demonstrate the performance of this stopping strategy by a simple numerical test. The Poisson problem (3.54) is solved on the unit square for a smooth solution using bilinear finite elements on uniformly refined meshes. The multigrid scheme uses the V-cycle with one damped Jacobi iteration for smoothing. The resulting multigrid scheme has a rather poor convergence rate, but the performance of the proposed a posteriori stopping criterion is well illustrated.

Table 3. Multigrid V-cycle with $k = 8$ levels.

k	e	e^*	e_L	e_L^*	$(e^* + e_L^*)/e$
0	6.2960	–	6.2949	5.5000	0.8735
1	3.2089	0.0006	3.2078	2.7995	0.8726
2	1.6652	0.0008	1.6641	1.5000	0.8772
3	0.8719	0.0010	0.8708	0.7665	0.8802
4	0.4594	0.0011	0.4583	0.4049	0.8838
5	0.2433	0.0011	0.2422	0.2149	0.8879
6	0.1296	0.0011	0.1285	0.1145	0.8926
7	0.0695	0.0011	0.0684	0.0612	0.8984
8	0.0376	0.0012	0.0365	0.0328	0.9058
9	0.0206	0.0012	0.0195	0.0177	0.9156
10	0.0115	0.0012	0.0104	0.0095	0.9287
11	0.0067	0.0012	0.0056	0.0051	0.9462
12	0.0041	0.0012	0.0030	0.0028	0.9677
13	0.0027	0.0012	0.0016	0.0015	0.9913
14	0.0020	0.0012	0.0008	0.0008	1.0125
15	0.0016	0.0012	0.0004	0.0004	1.0280
16	0.0014	0.0012	0.0002	0.0002	1.0371
17	0.0012	0.0012	0.0001	0.0001	1.0416
18	0.0012	0.0012	0.0000	0.0000	1.0434

In Table 3 the following quantities are shown:

$$\varepsilon = \|u - \tilde{u}_L\|, \qquad \varepsilon_L = \|\tilde{u}_L - u_L\|,$$

$$\varepsilon^* = \eta_{L^2}(\tilde{u}_L), \qquad \varepsilon_k^* = c_i' \sum_{l=1}^{L} \Big(\sum_{K \in \mathbb{T}_{l-1}} h_{l-1,K}^4 \|r_l(\tilde{v}_l)\|_K^2 \Big)^{1/2}.$$

Appropriate values for the constants $c_i \approx 0.45$ and $c_i' \approx 0.2$ have been determined by calibration on coarser meshes. According to the criterion (3.57) the multigrid iteration on mesh level $\mathbb{T}_8$ is stopped when

$$\varepsilon_L^* \leq \varepsilon^*, \tag{3.58}$$

i.e. after 14 iterations. We see that this criterion is rather sharp. Carrying the multigrid iteration beyond this point does not improve much on the accuracy of the discrete solution $\tilde{u}_k$ as an approximation to the true solution u.

4 Application to Differential Equations of Various Types

In this chapter, we demonstrate how our general approach to a posteriori error estimation may be applied to linear problems of elliptic (Lamé-Navier system), hyperbolic (scalar transport equation) and parabolic (heat conduction equation) type.

4.1 An Elliptic Model Problem: Lamé-Navier System

The fundamental problem of linear elasticity theory in classical notation reads

$$\begin{aligned} -\nabla\cdot\sigma = f, \quad \sigma = A\varepsilon(u) \quad \text{in } \Omega, \\ u = 0 \quad \text{on } \Gamma_D, \quad n\cdot\sigma = g \quad \text{on } \Gamma_N. \end{aligned} \tag{4.59}$$

It describes the deformation of an elastic body occupying a bounded domain $\Omega \subset \mathbb{R}^d$ ($d = 2$ or 3) which is fixed along a part Γ_D of its boundary $\partial\Omega$, under the action of a body force with density f and a surface traction g along $\Gamma_N = \partial\Omega\setminus\Gamma_D$. The loading is supposed to be small in order to neglect nonlinear effects. Here, u denotes the displacement, $\varepsilon(u) = \frac{1}{2}(\nabla u + \nabla u^T)$ the strain and σ the stress tensor. We assume a linear–elastic isotropic material law,

$$\sigma = A\varepsilon(u) = 2\mu\varepsilon^D(u) + \kappa\nabla\cdot uI,$$

with material dependent constants $\mu > 0$ and $\kappa > 0$, and ε^D is the deviatoric part of ε. For notational simplicity, the domain Ω is assumed as polyhedral in order to ease the approximation of the boundary. More general situations may be treated by the usual technical modifications.

The *primal* variational formulation of problem (4.59) is obtained by writing the equilibrium law in weak form,

$$(\sigma, \varepsilon(\phi)) = (f,\phi) + (g,\phi)_{\Gamma_N} \qquad \forall \phi \in V,$$

where $V = \{v \in H^1(\Omega)^d,\ v = 0 \text{ on } \Gamma_D\}$. Then, substituting the stress tensor results in

$$a(u,\phi) = (f,\phi) + (g,\phi)_{\Gamma_N} \qquad \forall \phi \in V, \tag{4.60}$$

with the energy form $a(\cdot,\cdot) = (A\varepsilon(\cdot), \varepsilon(\cdot))$. For our analysis, we assume the boundary part Γ_D to be non-trivial such that by Korn's inequality, there holds

$$a(v,v) = \|v\|_E^2 \geq \gamma\|v\|_1^2, \quad v \in V, \tag{4.61}$$

with some constant $\gamma > 0$. The case $\Gamma_D = \emptyset$ requires some modification. Then, the variational problem (4.60) is uniquely solvable (see Ciarlet (1991)). If the boundary $\partial\Omega$ is smooth or Ω convex polyhedral, this variational solution is known to possess H^2-regularity, provided that the force f and boundary traction g are sufficiently regular.

Primal finite element scheme. Starting from the primal variational formulation (4.60) the choice of finite element subspaces $V_h \subset V$ is straightforward. We consider here mainly the lowest-order approximation by linear or (isoparametric) d-linear shape functions as described above in the 2-dimensional case. The meshes are assumed to match the decomposition $\partial\Omega = \Gamma_u \cup \Gamma_\sigma$. The discrete solutions $u_h \in V_h$ are determined by the equation

$$a(u_h, \phi_h) = (f, \phi_h) + (g, \phi_h)_{\Gamma_\sigma} \qquad \forall \phi_h \in V_h. \tag{4.62}$$

This problem is seen to be uniquely solvable by the same arguments as used in establishing the solvability of the continuous problem. As in the scalar model case above, subtracting equation (4.62) from (4.60) implies the *Galerkin orthogonality* relation

$$a(e, \phi_h) = 0, \quad \phi_h \in V_h, \tag{4.63}$$

for the error $e = u - u_h$. From this, we infer again the following a priori energy-norm estimate:

$$\|e\|_E \leq \inf_{\phi \in V_h} \|u - \phi_h\|_E \tag{4.64}$$

Now, the a posteriori error analysis starts with selecting a functional $J(\cdot)$ which is to be used for error control. We assume $J(\cdot)$ to be well defined on V, such that the corresponding *dual problem*

$$a(\phi, z) = J(\phi) \qquad \forall \phi \in V, \tag{4.65}$$

possess a unique solution $z \in V$. In the case of an irregular functional as corresponding, for example, to point evaluation, this requires regularization as described in the examples of Section 3.3. Taking $\phi = e$ in (4.65) and using again Galerkin orthogonality, we obtain the error representation

$$J(e) = a(e, z) = a(e, z - \phi_h), \tag{4.66}$$

with an arbitrary $\phi_h \in V_h$. Now, we proceed analogously as in the scalar model case. Splitting the global integration over Ω into the contributions of the mesh cells $K \in \mathbb{T}_h$ and integrating cell-wise by parts yields

$$J(e) = \sum_{K \in \mathbb{T}_h} \Big\{ (-\nabla \cdot A\varepsilon(e), z - \phi_h)_K + (n \cdot A\varepsilon(e), z - \phi_h)_{\partial K} \Big\}.$$

Then, observing that $-\nabla \cdot A\varepsilon(u) = f$ and the continuity of $n \cdot A\varepsilon(u)$ across interelement edges, we obtain the error identity

$$J(e) = \sum_{K \in \mathbb{T}_h} \Big\{ (R(u_h), z - \phi_h)_K - (r(u_h), z - \phi_h)_{\partial K} \Big\}, \tag{4.67}$$

with the cell residual $R(u_h)_{|K} := f + \nabla \cdot A\varepsilon(u_h)$ and the edge residual $r(u_h)$ defined by

$$r(u_h)_{|\Gamma} := \begin{cases} \frac{1}{2} n \cdot [A\varepsilon(u_h)], & \text{if } \Gamma \subset \partial K \setminus \partial\Omega, \\ n \cdot A\varepsilon(u_h), & \text{if } \Gamma \subset \Gamma_D, \\ n \cdot A\varepsilon(u_h) - g, & \text{if } \Gamma \subset \Gamma_N. \end{cases}$$

This implies the following result.

Proposition 2. *For the approximation of the Lamé-Navier problem (4.59) by the finite element scheme (4.62), there holds the a posteriori error estimate*

$$|J(e)| \leq \eta(u_h) = \sum_{K\in\mathbb{T}_h} \{\rho_K\omega_K + \rho_{\partial K}\omega_{\partial K}\}, \tag{4.68}$$

with the local residuals ρ_K *and weight factors* ω_K *defined by*

$$\rho_K = \|R(u_h)\|_K, \qquad \omega_K := \|z-\phi_h\|_K,$$
$$\rho_{\partial K} := h_K^{-1/2}\|r(u_h)\|_{\partial K}, \qquad \omega_{\partial K} := h_K^{1/2}\|z-\phi_h\|_{\partial K}.$$

The practical evaluation of the weights ω_K and $\omega_{\partial K}$ uses again the strategies described above in Section 3.2. The weighted error estimate 4.68 has to be compared with the traditional *energy* error estimate, analogous to (2.28),

$$\|e\|_E \leq \eta_E(u_h) := c_s c_i \Big(\sum_{K\in\mathbb{T}_h} \{\rho_K^2 + \rho_{\partial K}^2\}\Big)^{1/2}, \tag{4.69}$$

with the local residuals $\rho_K, \rho_{\partial K}$ as defined above.

Numerical test. The described approach is applied for a typical model problem in plain linear elasticity. A square elastic disc with a crack is subjected to a constant boundary traction acting along half of the upper boundary (see Figure 8). Along the right-hand and lower parts of the boundary the disc is clamped and along the remaining part of the boundary (including the crack) it is left free. This problem is interesting as its solution develops a singularity at the tip of the crack where a stress singularity occurs with an asymptotic behavior (expressed in terms of polar coordinates (r,θ)) of the form $\sigma \approx r^{-1/2}$. The material parameters are chosen as commonly used for aluminium, i.e., $2\mu \sim \lambda \sim 0.16N/m^2$. The surface traction is of size $g \equiv 0.1N/m^2$.

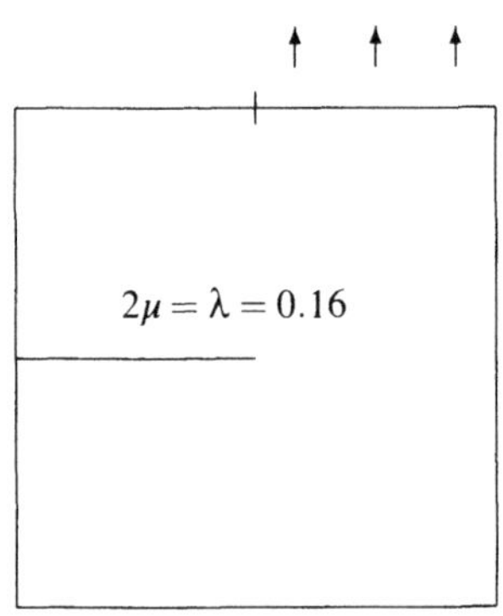

Figure 8. Geometry sketch of the square disc test problem.

We want to compute the mean normal stress over the clamped portion of the boundary,

$$J(u) = \int_{\Gamma_u} n{\cdot}A\varepsilon(u){\cdot}n\,ds. \qquad (4.70)$$

Table 4 and Figure 9 show the results obtained by the *weighted* error estimator $\eta_\omega(u_h)$ and the energy-error estimator $\eta_E(u_h)$ in both cases using the *fixed fraction strategy* for mesh refinement. The *effectivity index* is again defined by $I_{eff} = \eta_\omega(u_h)/|J(e)|$. Obviously, the *weighted* error estimator represents the true error quite well and leads to more economical meshes than the energy-error estimator.

Table 4. Adapted meshes based on the *weighted* error estimator (left) and the *energy-error* estimator (right).

L	N	$J(u_h)$	$J(e)/J(u)$	I_{eff}
1	256	0.017080	0.0283	1.80
2	484	0.019542	0.0180	1.96
3	1060	0.021138	0.0113	1.95
4	2113	0.022157	0.0070	1.96
5	4435	0.022795	0.0044	1.92
6	8830	0.023198	0.0027	1.86
7	15886	0.023428	0.0017	1.79
8	29947	0.023593	0.0010	1.79
9	52288	0.023697	0.0006	1.86

L	N	$J(u_h)$	$J(e)/J(u)$
1	256	0.017080	0.0283
2	544	0.018174	0.0237
3	1180	0.019363	0.0188
4	2659	0.020528	0.0139
5	6193	0.021538	0.0096
6	13423	0.022319	0.0064
7	31336	0.022811	0.0043
8	65332	0.023153	0.0029
9			

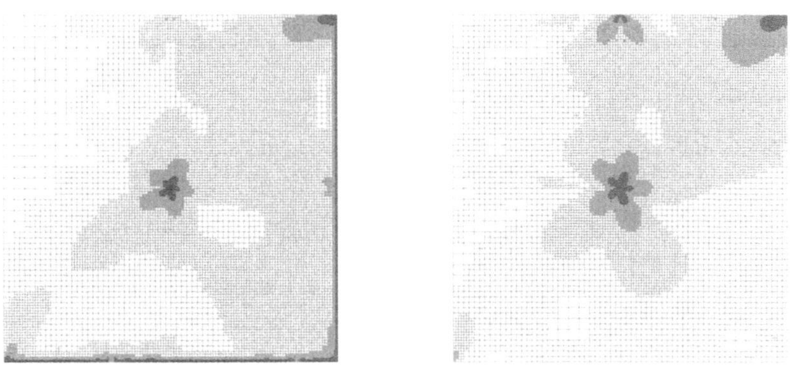

Figure 9. Finest meshes obtained by the weighted estimator (left) and the energy-error estimator (right).

4.2 A Hyperbolic Model Problem: Scalar Transport Equation

As a simple model case, we consider the scalar transport equation

$$\beta\cdot\nabla u = f, \tag{4.71}$$

on a domain $\Omega \subset \mathbb{R}^2$ with inflow boundary condition $u = g$ along the *inflow boundary* $\partial\Omega_- = \{x \in \partial\Omega, n\cdot\beta < 0\}$. Accordingly, $\partial\Omega_+ = \partial\Omega \setminus \partial\Omega_-$ is the *outflow boundary.* The transport vector β is assumed as constant for simplicity. Then, the natural solution space is $V = \{v \in L^2(\Omega),\ \beta\cdot\nabla v \in L^2(\Omega)\}$. This problem is discretized using the Galerkin finite element method with *streamline diffusion* stabilization (see Hughes and Brooks (1982) and Johnson (1987)). On quadrilateral meshes $\mathbb{T}_h$, we define subspaces

$$V_h = \{v \in H^1(\Omega), v_{|T} \in \tilde{Q}_1(K), K \in \mathbb{T}_h\}$$

consisting of *bilinear* finite elements. The discrete solution $u_h \in V_h$ is defined by

$$(\beta\cdot\nabla u_h - f, \phi_h + \delta\beta\cdot\nabla\phi_h) + (n\cdot\beta(g - u_h), \phi_h)_{\partial\Omega_-} = 0 \qquad \forall \phi_h \in V_h, \tag{4.72}$$

where the stabilization parameter is determined locally by $\delta_K = h_K$. In this formulation the inflow boundary condition is imposed in the weak sense. This facilitates the use of a duality argument in generating a posteriori error estimates. Let $J(\cdot)$ be a given functional for controlling the error $e = u - u_h$. Following our general approach, we consider the corresponding dual problem

$$A_h(\phi, z) = (\beta\cdot\nabla\phi, z + \delta\beta\cdot\nabla z) - (n\cdot\beta\phi, z)_{\partial\Omega_-} = J(\phi) \qquad \forall \phi \in V, \tag{4.73}$$

which is a transport problem with transport in the negative β-direction. We note that the stabilized bilinear form $A_h(\cdot,\cdot)$ is used in the duality argument, in order to achieve an optimal treatment of the stabilization terms; for a detailed discussion of this point see Houston, Rannacher and Süli (1999). The error representation reads

$$J(e) = (\beta\cdot\nabla e, z - \phi_h + \delta\beta\cdot\nabla(z - \phi_h)) - (n\cdot\beta e, z - z_h)_{\partial\Omega_-},$$

for arbitrary $\phi_h \in V_h$. This results in the following result.

Proposition 3. *For the approximation of the transport problŽem (4.71) by the finite element scheme (4.72), there holds the a posteriori error estimate*

$$|J(e)| \le \eta(u_h) = c_i \sum_{K \in \mathbb{T}_h} \{\rho_K\omega_K + \rho_{\partial K}\omega_{\partial K}\}, \tag{4.74}$$

with the cell residuals and weights defined by

$$\rho_K = \|f - \beta\cdot\nabla u_h\|_K, \qquad \omega_K = \{\|z - \phi_h\|_K + \delta_K\|\beta\cdot\nabla(z - \phi_h)\|_K\},$$
$$\rho_{\partial K} = h_K^{-1/2}\|n\cdot\beta(u_h - g)\|_{\partial K\cap\partial\Omega_-}, \qquad \omega_{\partial K} = h_K^{1/2}\|z - \phi_h\|_{\partial K\cap\partial\Omega_-}.$$

We note that this a posteriori error bound explicitly contains the mesh size h_K and the stabilization parameter δ_K as well. This gives us the possibility to simultaneously adapt both parameters, which may be particularly advantageous in capturing sharp layers in the solution.

We begin with a simple thought experiment to illustrate the features of the error estimate (4.74). Let $\Omega = (0,1)^2$ and $f \equiv 0$. We take the functional $J(u) = (1, n{\cdot}\beta u)_{\partial\Omega_+}$. The corresponding dual solution is $z \equiv 1$, so that $J(e) = 0$. This implies

$$(1, n{\cdot}\beta u_h)_{\partial\Omega_+} = (1, n{\cdot}\beta u)_{\partial\Omega_+} = -(1, n{\cdot}\beta g)_{\partial\Omega_-},$$

recovering the well-known global conservation property of the scheme.

Next, we take again the unit square $\Omega = (0,1)^2$ and $f \equiv 0$, and consider the case of a constant transport coefficient $\beta = (1, 0.5)^T$ and inflow data $g(x,0) = 0$, $g(0,y) = 1$. The quantity to be computed is part of the outflow as indicated in Figure 10:

$$J(u) = \int_\Gamma \beta{\cdot}nu\,ds.$$

The mesh refinement is organized according to the *fixed fraction strategy* described above. In Table 5, we show results for this test computation. The corresponding meshes and the primal as well as the dual solution are presented in Figure 10. Notice that there is no mesh refinement enforced along the upper line of discontinuity of the dual solution since here the residual of the primal solution is almost zero.

Figure 10. Configuration and grids of the test computation for the model transport problem (4.71) (left), primal solution (middle) and dual solution (right) on an adaptively refined mesh.

Table 5. Convergence results of the test computation for the model transport problem (4.71).

L	N	$J(e)$	η	I_{eff}
0	256	2.01e-2	2.38e-2	1.18
2	634	1.09e-2	1.21e-2	1.11
4	1315	6.25e-3	7.88e-3	1.26
6	2050	4.21e-3	5.37e-3	1.27
8	2566	3.90e-3	5.01e-3	1.28
10	3094	3.41e-3	4.71e-3	1.38

4.3 A Parabolic Model Problem: Heat Conduction Equation

We consider the heat-conduction problem

$$\begin{aligned} \partial_t u - \nabla\cdot(a\nabla u) &= f \quad \text{in } Q_T, \\ u_{|t=0} &= u^0 \quad \text{in } \Omega, \quad u_{|\partial\Omega} = 0 \quad \text{on } I, \end{aligned} \tag{4.75}$$

on a space-time region $Q_T = \Omega \times I$, where $\Omega \subset \mathbb{R}^d, d \geq 1$, and $I = [0,T]$; the coefficient a may vary in space. This model is used to describe diffusive transport of energy or certain species concentrations. The results of this section are taken from Hartmann (1998); see also Bangerth and Rannacher (2003).

The discretization of problem (4.75) is by a Galerkin method in space-time. We split the time interval $[0,T]$ into subintervals $I_n = (t_{n-1}, t_n]$ according to

$$0 = t_0 < \ldots < t_n < \ldots < t_N = T, \qquad k_n := t_n - t_{n-1}.$$

At each time level t_n, let $\mathbb{T}_h^n$ be a regular finite element mesh as defined above with local mesh width $h_K = \operatorname{diam}(K)$, and let $V_h^n \subset H_0^1(\Omega)$ be the corresponding finite element subspace with d-linear shape functions.

Extending the spatial mesh to the corresponding space-time slab $\Omega \times I_n$, we obtain a global space-time mesh consisting of $(d+1)$-dimensional cubes $Q_K^n := K \times I_n$. On this mesh, we define the global finite element space

$$V_h^k = \left\{ v \in W,\ v(\cdot,t)_{|Q_K^n} \in \tilde{Q}_1(K),\ v(x,\cdot)_{|Q_K^n} \in P_r(I_n)\ \forall\, Q_K^n \right\},$$

where $W = L^2((0,T); H_0^1(\Omega))$ and $r \geq 0$. For functions from this space and their time-continuous analogues, we use the notation

$$v^{n+} = \lim_{t \to t_n+0} v(t), \quad v^{n-} = \lim_{t \to t_n-0} v(t), \quad [v]^n = v^{n+} - v^{n-}.$$

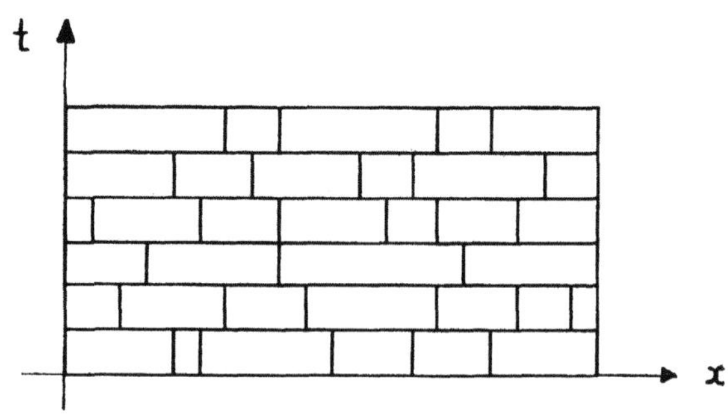

Figure 11. Sketch of a space-time mesh with hanging nodes

The discretization of problem (4.75) is based on a variational formulation which allows the use of piecewise discontinuous functions in time. This method, termed *dG(r) method* (*discontinuous* Galerkin method in time), determines approximations $u_h \in V_h^k$ by requiring

$$A(u_h, \phi_h) = 0 \qquad \forall \phi_h \in V_h^k, \tag{4.76}$$

with the semi-linear form

$$A(u, \phi) = \sum_{n=1}^{N} \int_{I_n} \Big\{ (\partial_t u, \phi) + (a\nabla u, \nabla\phi) - (f, \phi) \Big\} dt + \sum_{n=1}^{N} ([u]_{n-1}, \phi_{n-1}^{+}),$$

where $u_0^- = u_0$. We note that the continuous solution u also satisfies equation (4.76) which again implies Galerkin orthogonality for the error $e = u - u_h$ with respect to the bilinear form $A(\cdot,\cdot)$. Since the test functions $\phi_h \in V_h^k$ may be discontinuous at times t_n, the global system (4.76) decouples and can be written in form of a time-stepping scheme,

$$\int_{I_n} \Big\{ (\partial_t u_h, \phi_h) + (a\nabla u_h, \nabla\phi_h) \Big\} dt + ([u_h]^{n-1}, \phi_h^{(n-1)+}) = \int_{I_n} (f, \phi_h)\, dt,$$

for all $\phi_h \in V_h^n$, $n = 1, \dots, N$. In the following, we consider only the lowest-order case $r = 0$, the *dG(0) method*, which is similar to the backward Euler scheme. We concentrate on the control of the spatial L^2 error $\|e^{N-}\|$ at the end time $T = t_N$. To this end, we use a duality argument in space-time,

$$\begin{aligned} \partial_t z - a\Delta z &= 0 \quad \text{in } \Omega \times I, \\ z_{|t=T} &= \|e^{N-}\|^{-1} e^{N-} \quad \text{in } \Omega, \quad z_{|\partial\Omega} = 0 \ \text{ on } I, \end{aligned} \tag{4.77}$$

which can also be written in variational form as

$$A(\phi, z) = J(\phi) = \|e^{N-}\|^{-1} (\phi^{N-}, e^{N-}) \qquad \forall \phi \in W. \tag{4.78}$$

Then, our general approach leads us to the a posteriori error representation

$$\begin{aligned} \|e^{N-}\| = \sum_{n=1}^{N} \sum_{K \in \mathbb{T}_h^n} \Big\{ & (R(u_h), z - \phi_h)_{K \times I_n} - (r(u_h), z - \phi_h)_{\partial K \times I_n} \\ & - ([u_h]^{n-1}, (z - \phi_h)^{(n-1)+})_K \Big\}, \end{aligned} \tag{4.79}$$

with an arbitrary $\phi_h \in V_h^k$. Here, the local residuals are given by

$$R(u_h) = f + \nabla\cdot(a\nabla u_h) - \partial_t u_h, \qquad r(u_h)_{|\Gamma} := \begin{cases} \frac{1}{2} n\cdot[a\nabla u_h], & \text{if } \Gamma \subset \partial K \setminus \partial\Omega, \\ n\cdot a\nabla u_h, & \text{if } \Gamma \subset \partial\Omega, \end{cases}$$

From this, we infer the following following result.

Proposition 4. *For the approximation of the heat conduction problem (4.75) by the space-time finite element scheme (4.76), there holds the a posteriori error estimate*

$$\|e^{N+}\| \le \sum_{n=1}^{N} \sum_{K \in \mathbb{T}_h^n} \Big\{ \sum_{i=1,2,3} \rho_K^{n,i}(u_h)\, \omega_K^{n,i}(z) \Big\}, \tag{4.80}$$

where the cellwise residuals and weights are defined by

$$\rho_K^{n,1} = \|R(u_h)\|_{K\times I_n}, \qquad \omega_K^{n,1} = \|z-\phi_h\|_{K\times I_n},$$
$$\rho_K^{n,2} = \tfrac{1}{2}h_K^{-1/2}\|n\cdot[a\nabla u_h]\|_{\partial K\times I_n}, \qquad \omega_K^{n,2}(z) = h_K^{1/2}\|z-\phi_h\|_{\partial K\times I_n},$$
$$\rho_K^{n,3} = k_n^{-1/2}\|[u_h]_{n-1}\|_K, \qquad \omega_K^{n,3} = k_n^{1/2}\|(z-\phi_h)^{(n-1)+}\|_K.$$

with an arbitrary approximation $\phi_h \in W_h$ *of* z.

Numerical test. The performance of the error estimator (4.80) is illustrated by a simple test in two space dimensions where the constructed exact solution represents a smooth rotating bump on the unit square; for details see Hartmann (1998). Figure 12 shows a sequence of adapted meshes at successive times obtained by controlling the spatial L^2 error at the end time $t_N = 0.5$. We clearly see the effect of the weights in the error estimator which suppress the influence of the residuals during the initial period.

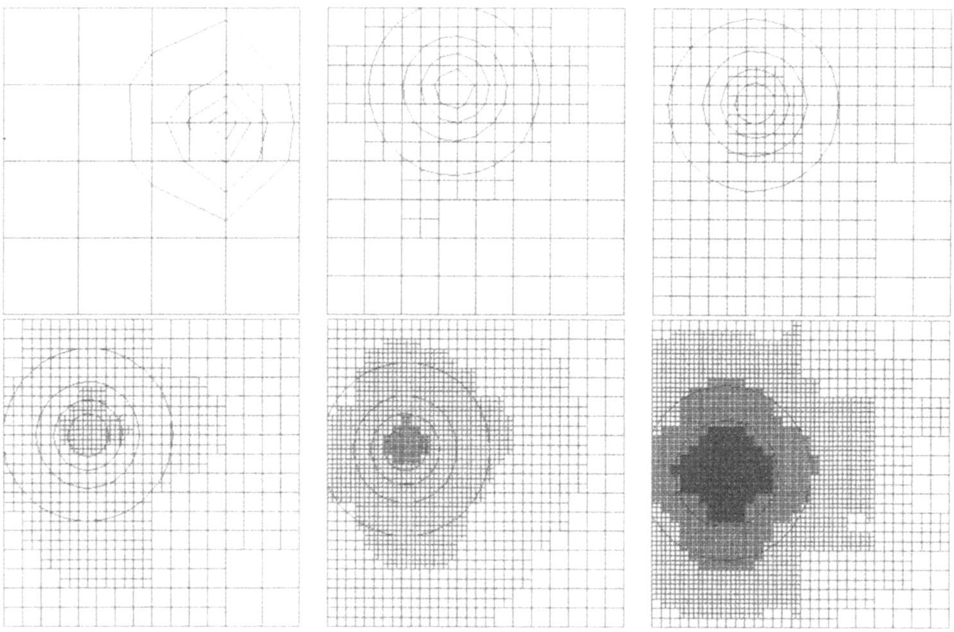

Figure 12. Sequence of refi ned meshes for controlling the end-time error $\|e^{N-}\|$ shown at six consecutive times $t_n = 0.125000, \ldots, 0.5$.

5 Extension to Nonlinear Problems

In this section, we show how the approach to a posteriori error estimation discussed so far only for linear problems can also be applied to nonlinear problems.

5.1 A General Paradigm for Residual-Based A-Posteriori Error Estimation

We outline the general approach in an abstract setting following the paradigm introduced by Eriksson et al. (1995). Let V be a Hilbert space with scalar product $(\cdot,\cdot)$ and corresponding norm $\|\cdot\|$, and $A(\cdot;\cdot)$ a semi-linear form continuously defined on $V \times V$. We seek a solution $u \in V$ to the variational problem

$$A(u;\phi) = 0 \qquad \forall \phi \in V. \tag{5.81}$$

This problem is approximated by a Galerkin method using a sequence of finite dimensional subspaces $V_h \subset V$ parameterized by a discretization parameter h. The discrete problems determine approximations $u_h \in V_h$ by

$$A(u_h;\phi) = 0 \qquad \forall \phi \in V_h. \tag{5.82}$$

With the tangent form $A'(\cdot;\cdot,\cdot)$, we have the following orthogonality relation

$$\int_0^1 A'(u_h + se; e, \phi_h)\,ds = A(u;\phi_h) - A(u_h;\phi_h) = 0, \quad \phi_h \in V_h, \tag{5.83}$$

for the error $e = u - u_h$. This suggests the use of the bilinear form

$$L(u,u_h;\phi,z) = \int_0^1 A'(u_h + se; \phi, z)\,ds, \tag{5.84}$$

depending on u as well as on u_h, in the duality arguments. Suppose that the quantity $J(u)$ has to be computed, where (for simplicity) $J(\cdot)$ is a linear functional defined on V. For representing the error $J(e)$, we use the dual problem

$$L(u,u_h;\phi,z) = J(\phi) \qquad \forall \phi \in V, \tag{5.85}$$

which is assumed to possess a (unique) solution $z \in V$. Then, taking $\phi = e$ in (5.85) and using Galerkin orthogonality (5.83), we obtain the general error representation

$$J(e) = L(u,u_h;e,z) = L(u,u_h;e,z-\phi_h) = -A(u_h;z-\phi_h), \tag{5.86}$$

with an arbitrary $\phi_h \in V_h$.

Since the bilinear form $L(u,u_h;\cdot,\cdot)$ contains the unknown solution u as coefficient, its evaluation requires approximation. The simplest way is to replace u by u_h, yielding a perturbed dual solution $\tilde{z} \in V$ defined by

$$L(u_h,u_h;e,z) = A'(u_h;\phi,\tilde{z}) = J(\phi) \quad \forall \phi \in V. \tag{5.87}$$

Proposition 5. *Let $\tilde{z} \in V$ be the solution of the linearized dual problem (5.87), then there holds the a posteriori error estimate*

$$|J(e)| \le \inf_{\phi_h \in V_h} |A(u_h;\tilde{z}-\phi_h)| + \tfrac{1}{2} \sup_{v \in [u,u_h]} |A''(v;e,e,\tilde{z})|, \tag{5.88}$$

where $A''(\cdot;\cdot,\cdot,\cdot)$ is the second derivative of the form $A(\cdot;\cdot)$.

Proof. Setting $\phi = e$ in the dual problem (5.87) yields $J(e) = A'(u_h;e,\tilde{z})$. By Taylor expansion, we have

$$0 = A(u;z) = A(u_h;\tilde{z}) - A'(u_h;e,\tilde{z}) - \int_0^1 A''(u_h + se;e,e,z)\,s\,ds,$$

and, consequently, using again Galerkin orthogonality (5.83),

$$J(e) = A(u_h;\tilde{z} - \phi_h) + \int_0^1 A''(u_h + se;e,e,\tilde{z})\,s\,ds,$$

with an arbitrary $\phi_h \in V_h$. From this, we infer (5.88). Q.E.D.

Clearly, the remainder term in (5.88) vanishes if $A(\cdot;\cdot)$ is affine-linear. In general it is quadratic in e and is therefore supposed to be small compared to the error $J(e)$ and is neglected. Controlling the effect of this linearization on the accuracy of the resulting error estimator may be a delicate task and depends strongly on the particular problem considered. Our experiences indicate that this problem is not critical if the continuous solution is stable. The crucial problem is the numerical computation of the perturbed dual solution $\tilde{z}$ by solving a discretized dual problem

$$L(u_h,u_h;\phi,\tilde{z}_h) = J(\phi) \qquad \forall \phi \in V_h. \tag{5.89}$$

This results in a practically useful error estimator

$$J(e) \approx \eta(u_h), \tag{5.90}$$

in which the weight $z - \phi_h$ is replaced by some approximation as described in Section 3.2.

5.2 Nested Solution Approach

For solving the nonlinear problem (5.81) by a Galerkin finite element method (5.82), we employ the following iterative scheme. Starting from a coarse initial mesh $\mathbb{T}_0$, a hierarchy of refined meshes

$$\mathbb{T}_0 \subset \mathbb{T}_1 \subset \cdots \subset \mathbb{T}_i \subset \cdots \subset \mathbb{T}_L,$$

and corresponding finite element spaces V_l, $l = 1,\dots,L$, is generated by a nested solution process.

1. *Initialization:* For $j = 0$, start on coarse mesh $\mathbb{T}_0$ with $u_0^{(0)} = u_{-1} \in V_0$.
2. *Defect correction iteration:* For $j \geq 1$, start with $u_j^{(0)} = u_{j-1} \in V_j$.
3. *Iteration step:* Evaluate the defect

$$(d_i^{(j)},\phi) = f(\phi) - A(u_i^{(j)};\phi), \quad \phi \in V_i,$$

and solve the correction equation

$$\tilde{A}'(u_i^{(j)};v_i^{(j)},\phi) = (d_i^{(j)},\phi) \quad \forall \phi \in V_i,$$

by Krylov-space or multigrid iterations using the hierarchy of precedingly constructed meshes $\{\mathbb{T}_i,\dots,\mathbb{T}_0\}$. Update $u_i^{(j+1)} = u_i^{(j)} + v_i^{(j)}$, set $j = j+1$ and go back to (2). This process is repeated until a limit $u_i \in V_i$, is reached with some required accuracy.

4. *Error estimation:* Solve the (linearized) discrete dual problem

$$z_i \in V_i: \quad A'(u_i;\phi,z_i) = J(\phi) \quad \forall \phi \in V_i$$

and evaluate the a posteriori error estimate

$$|J(e_i)| \approx \eta(u_i).$$

If $\tilde{\eta}(u_i) \le TOL$, or $N_i \ge N_{max}$, then stop. Otherwise cell-wise mesh adaptation yields the new mesh $\mathbb{T}_{i+1}$. Then, set $i = i+1$ and go back to (1).

5.3 Application to a Model Problem in Elasto-Plasticity Theory

The fundamental problem in the static deformation theory of linear–elastic perfect–plastic material (so-called *Hencky* model) reads

$$\begin{aligned} &\nabla\cdot\sigma = -f, \quad \varepsilon(u) = A:\sigma + \lambda \quad \text{in } \Omega, \\ &\lambda:(\tau-\sigma) \le 0 \quad \forall\, \tau \text{ with } F(\tau) \le 0, \\ &u = 0 \text{ on } \Gamma_D, \quad \sigma\cdot n = g \text{ on } \Gamma_N, \end{aligned} \tag{5.91}$$

where u, ε, and σ are the displacement vector and the strain and stress tensor, respectively, while λ denotes the plastic growth. This system describes the deformation of an elasto-plastic body occupying a bounded domain $\Omega \subset \mathbb{R}^d$ ($d = 2$ or 3) which is fixed along a part Γ_D of its boundary $\partial\Omega$, under the action of a body force with density f and a surface traction g along $\Gamma_N = \partial\Omega \setminus \Gamma_D$. The displacement u is supposed to be small in order to neglect geometric nonlinear effects, so that the strain tensor can be written as $\varepsilon(u) = \frac{1}{2}(\nabla u + \nabla u^T)$. The material tensor A is again assumed to be symmetric and positive definite. We assume a linear–elastic isotropic material law $\sigma = 2\mu\varepsilon^D(u) + \kappa\nabla\cdot uI$, with material dependent constants $\mu > 0$ and $\kappa > 0$, while the plastic behavior follows the von Mises flow rule

$$F(\sigma) = |\sigma^D| - \sigma_0 \le 0,$$

with some $\sigma_0 > 0$. Here, ε^D and σ^D denote the deviatoric parts of ε and σ, respectively.

The *primal* variational formulation of problem (5.91) seeks a displacement $u \in V$, where $V = \{u \in H^1(\Omega)^d,\ u_{|\Gamma_D} = 0\}$, such that

$$A(u;\phi) = (C(\varepsilon(u)),\varepsilon(\phi)) - (f,\phi) - (g,\phi)_{\Gamma_N} = 0 \qquad \forall \phi \in V. \tag{5.92}$$

Here, the nonlinear tensor-function $C(\varepsilon(u)) = \Pi(2\mu\varepsilon^D(u)) + \kappa\nabla\cdot uI$, uses the projection

$$\Pi(2\mu\varepsilon^D(u)) = \begin{cases} 2\mu\varepsilon^D(u) & , \text{ if } \ |2\mu\varepsilon^D(u)| \le \sigma_0, \\ \dfrac{\sigma_0}{|\varepsilon^D(u)|}\varepsilon^D(u) & , \text{ if } \ |2\mu\varepsilon^D(u)| > \sigma_0. \end{cases}$$

The finite element approximation of problem (5.92) seeks $u_h \in V_h$, such that

$$A(u_h;\phi_h) = 0 \qquad \forall \phi_h \in V_h, \tag{5.93}$$

where V_h is the finite element space of (isoparametric) bilinear shape functions as described above. To a computed displacement u_h, a stress σ_h is associated by

$$\sigma_h = \Pi(2\mu\varepsilon^D(u_h)) + \kappa\nabla\cdot u_h I.$$

Details of the solution process can be found in Suttmeier (1997), and Rannacher and Suttmeier (1999). Given a (linear) error functional $J(\cdot)$, our general analysis yields the a posteriori error representation

$$J(e) = -A(u_h;z-\phi_h) + R(u,u_h;e,e,z), \tag{5.94}$$

with an arbitrary approximation $\phi_h \in V_h$ to the solution $z \in V$ of the linearized dual problem

$$(C'(u_h)\varepsilon(\phi),\varepsilon(z)) = J(\phi) \qquad \forall\phi \in V. \tag{5.95}$$

Here, the Jacobian $C'(\cdot)$ is defined by

$$C'(\tau)\varepsilon := \begin{cases} C\varepsilon, & \text{if } \; |2\mu\tau^D| \le \sigma_0, \\ \dfrac{\sigma_0}{|\tau^D|}\Big\{I - \dfrac{(\tau^D)^T\tau^D}{|\tau^D|^2}\Big\}\varepsilon^D + \kappa\,\mathrm{tr}(\varepsilon)I, & \text{if } \; |2\mu\tau^D| > \sigma_0. \end{cases}$$

Accordingly, the remainder term is $R(u,u_h;e,e,z) = O(e^2)$ in regions where the form $A(\cdot;\cdot)$ is twice differentiable, i.e. outside the elastic-plastic transition zone $\{|2\mu\tau^D| = \sigma_0\}$. The residual term in the error identity (5.94) has the form

$$-A(u_h;z-\phi_h) = \sum_{K\in\mathbb{T}_h} \Big\{(R(u_h),z-\phi_h)_K - (r(u_h),z-\phi_h)\Big\}, \tag{5.96}$$

with the cell and edge residuals (using the notation from above)

$$R(u_h)_{|K} = f - \nabla\cdot C(\varepsilon(u_h)), \quad r(u_h)_{|\Gamma} = \begin{cases} \tfrac{1}{2}n\cdot[C(\varepsilon(u_h))], & \text{if } \Gamma\subset\partial K\setminus\partial\Omega, \\ n\cdot C(\varepsilon(u_h)) - g, & \text{if } \Gamma\subset\Gamma_N, \\ n\cdot C(\varepsilon(u_h)), & \text{if } \Gamma\subset\partial\Omega\setminus\Gamma_N. \end{cases}$$

Using this notation, we have the following result.

Proposition 6. *For any given (linear) error functional $J(\cdot)$, let $z \in V$ be the solution of the linearized dual problem (5.95). Then, there holds the a posteriori error estimate*

$$|J(e)| \approx \eta_\omega(u_h) = \sum_{K\in\mathbb{T}_h} \{\rho_K\omega_K + \rho_{\partial K}\omega_{\partial K}\}, \tag{5.97}$$

up to a higher-order remainder term, where the cellwise residual terms and weights are defined by

$$\begin{aligned} \rho_K &= \|R(u_h)\|_K, & \omega_K &= \|z-\phi_h z\|_K, \\ \rho_{\partial K} &= h_K^{-1/2}\|r(u_h)\|_{\partial K}, & \omega_{\partial K} &= h_K^{1/2}\|z-\phi_h\|_{\partial K}, \end{aligned}$$

with an arbitrary $\phi_h \in V_h$ of z.

We compare the *weighted* error estimate (5.97) with two heuristic ways of estimating the stress error $e_\sigma = \sigma - \sigma_h$.

(1) *ZZ-error indicator:* The heuristic error indicator of Zienkiewicz and Zhu (1987) (see also Ainsworth and Oden (1997)) uses higher–order stress recovery by local averaging, $\sigma \approx M_h\sigma_h$,

$$\|e_\sigma\| \approx \eta_{ZZ}(u_h) = \Big(\sum_{K\in\mathbb{T}_h} \|M_h\sigma_h - \sigma_h\|_K^2\Big)^{1/2}, \tag{5.98}$$

(2) *An energy-error indicator:* The energy-error estimator of Johnson and Hansbo (1992) is based on decomposing the domain Ω into *discrete* plastic and elastic zones, $\Omega = \Omega_h^p \cup \Omega_h^e$. Accordingly the error estimator has the form

$$\|e_\sigma\| \approx \eta_E(u_h) = c_i\Big(\sum_{K\in\mathbb{T}_h} \eta_K^2\Big)^{1/2}, \tag{5.99}$$

with the local error indicators defined by

$$\eta_K^2 := \begin{cases} h_K^2\{\rho_K + \rho_{\partial K}\}^2, & \text{if } K \subset \Omega_h^e, \\ \{\rho_K + \rho_{\partial K}\}\|M_h\sigma_h - \sigma_h\|_K, & \text{if } K \subset \Omega_h^p. \end{cases}$$

Numerical test. A geometrically two-dimensional square disc with a hole is subjected to a constant boundary traction acting upon two opposite sides. We use the two-dimensional plain-strain approximation, i.e., the components of $\varepsilon(u)$ in z-direction are assumed to be zero. In virtue of symmetry the consideration can be restricted to a quarter of the domain shown in Figure 13; for the precise parameters in this model see Rannacher and Suttmeier (1999). Among the quantities to be computed is the component σ_{22} of the stress tensor at point a_2. The result on a very fine mesh with about 200,000 cells is taken as reference solution u_{ref}. We note that in the plastic zone the material is almost incompressible, i.e., $\nabla \cdot u \approx 0$. In this case, a more robust and accurate discretization is obtained by introducing an independent *pressure* variable $p = -\kappa\nabla \cdot u$. For the resulting *mixed formulation* of the problem a result analogous to Proposition 6 holds true (Rannacher and Suttmeier (1999)).

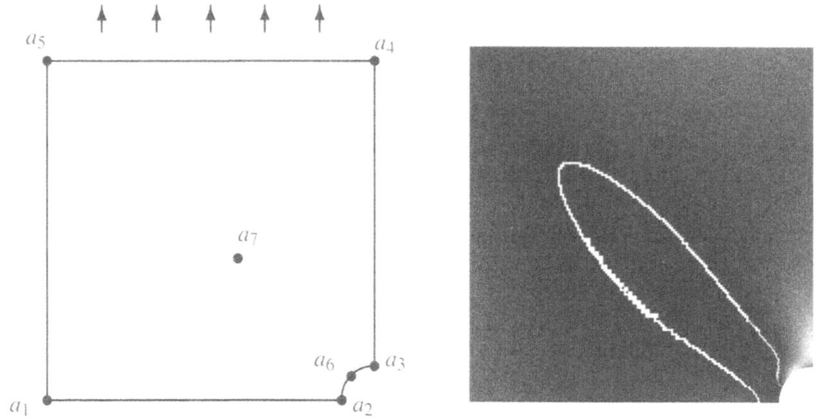

Figure 13. Geometry of the elasto-§lastic benchmark problem and plot of $|\sigma^D|$ (plastic region black, transition zone white) computed on a mesh with $N \approx 10{,}000$ cells.

The result of this test is summarized in Figure 14. We see that the weighted a posteriori error estimate leads to more economical meshes, particularly if high accuracy is required. For more details as well as for further results also for the time-dependent *Prandtl-Reuss* model in perfect plasticity, we refer to Rannacher and Suttmeier (1999).

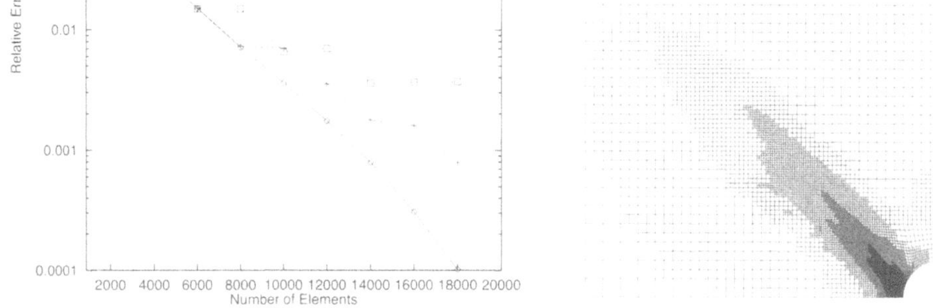

Figure 14. Relative error for computation of σ_{22} at point a_2 using different estimators and optimized grid with $N \approx 10{,}000$ cells.

6 Application to Acoustic Wave Propagation

We consider the acoustic wave equation

$$\begin{aligned}
&\partial_t^2 u - \nabla\cdot a\nabla u = 0 \quad \text{in } Q_T = \Omega \times I, \\
&u_{|t=0} = u^0, \ \partial_t u_{|t=0} = v^0 \ \text{in } \Omega, \\
&a\partial_n u_{|\partial\Omega} = 0 \ \text{on } I,
\end{aligned} \tag{6.100}$$

on a space-time region $Q_T = \Omega \times I$, with $\Omega \subset \mathbb{R}^d$ $(d \geq 1)$, and $I = (0,T)$. Here, we assume constant density, $\rho \equiv 1$, but allow the elastic coefficient a to vary in space. This equation frequently occurs in the simulation of acoustic waves in gaseous or fluid media, seismics, electrodynamics and many other applications. In many cases, initial data and coefficients are rough or discontinuous which limits the global regularity of solutions. The material of this section is taken from Bangerth and Rannacher (1999).

We approximate problem (6.100) by a *velocity-displacement* formulation which is obtained by introducing a new velocity variable $v = \partial_t u$. The corresponding natural solution space is $W = H^1(I, H^1(\Omega)) \times H^1(I, L_2(\Omega))$. Then, the pair $w = \{u, v\} \in W$ satisfies the linear variational equation

$$A(w,\tau) = 0 \qquad \forall \tau = \{\phi, \psi\} \in W^*, \tag{6.101}$$

with the bilinear form

$$A(w,\tau) = (\partial_t u, \phi)_{Q_T} - (v, \phi)_{Q_T} + (\partial_t v, \psi)_{Q_T} + (a\nabla u, \nabla\psi)_{Q_T},$$

and a suitable space W^* of test functions τ for which $A(w,\tau)$ is defined. The precise definition of the space W^* is not important for what follows and is therefore not further discussed. This formulation of the wave equation, but without weak enforcement of initial values, and discretization thereof was introduced by Hulbert and Hughes (1990); see Bangerth and Rannacher (1999) for more references. It takes into account the case of reduced solution regularity, as the coupling $v = \partial_t u$ is enforced only in the L^2-sense.

Since solutions to the wave equation often have very localized features, such as wave fronts, efficient algorithms need to employ some kind of adaptivity in the choice of computational grids. On the other hand, the quantities one is interested in from an engineering viewpoint also are often local rather than global. It is therefore not always necessary to resolve all features of a wave field, if one is only interested in a certain portion. These two aspects of localization are observed in our method of error estimation.

6.1 Space-Time Discretization

The variational problem (6.101) is the basis for a Galerkin discretization in space-time. We use the Rothe approach of first discretizing in time and then in space on each discrete time level. This has the advantage of having the freedom to choose the computational mesh differently at each time level. We decompose the time interval $I = [0,T]$ into subintervals $I_n = (t_{n-1}, t_n]$ with length $k_n = t_n - t_{n-1}$. On each time slab $Q^n = \Omega \times I_n$, we use meshes $\mathbb{T}_h^n$ consisting of $(d+1)$-dimensional cubes $Q_K^n = K \times I_n$ with spatial width h_K; these meshes may vary between the time levels in order to allow for grid refinement moving with the wave field; see Figure 11.

The discrete *trial spaces* $W_h = V_h \times V_h$ in space-time domain consist of functions which are $(d+1)$-linear on each space-time cell Q_K^n and globally *continuous* on Q_T. This prescription requires the use of *hanging nodes* if the spatial mesh changes across a time level t_n. The corresponding discrete *test spaces* $W_h^* \subset W^*$ consist of functions which are constant in time on each cell Q_K^n, while they are d-linear in space and globally continuous on Ω. We further assume that the test space W^* is chosen large enough to contain the elements from $w + W_h$. The Galerkin approximation of problem (6.101) seeks pairs $w_h = \{u_h, v_h\} \in W_h$ satisfying

$$A(w_h, \tau_h) = 0 \quad \forall \tau_h = \{\phi_h, \psi_h\} \in W_h^*. \tag{6.102}$$

This time-discretization scheme is termed *cG(1)-method* (*continuous* Galerkin method) in contrast to the *dG*-method (*discontinuous* Galerkin method) where the trial functions are allowed to be discontinuous in time. We note that from this scheme, we can recover the standard Crank-Nicolson scheme in time combined with a spatial finite element method:

$$\begin{aligned} (u_h^n - u_h^{n-1}, \phi_h) - \tfrac{1}{2}k_n(v_h^n + v_h^{n-1}, \phi_h) &= 0 \qquad \forall \phi_h \in V_h, \\ (v_h^n - v_h^{n-1}, \psi_h) + \tfrac{1}{2}k_n(a\nabla(u_h^n + u_h^{n-1}), \nabla\psi_h) &= 0 \qquad \forall \psi_h \in V_h. \end{aligned} \tag{6.103}$$

The system (6.103) splits into two equations, a discrete Helmholtz equation and a discrete L^2 projection. We chose this time stepping scheme because it is of second order accurate and (in the case of time semi-discretization) energy conserving, i.e.

$$\|v^n\|^2 + \|\sqrt{a}\nabla u^n\|^2 = \|v^0\|^2 + \|\sqrt{a}\nabla u^0\|^2. \tag{6.104}$$

This conservation property carries over to the spatially discretized equations provided that the meshes do not change between time levels.

Notice that in the construction of W_h, we have associated the triangulations $\mathbb{T}_h^n$ with discrete time levels instead of time slabs. In the interior of time slabs, we let w_h be from the union of the finite element spaces defined on the triangulations at the two adjacent time levels. This construction is necessary to allow for trial functions that are continuous also at discrete time levels even if grids change between time steps. Associating triangulations with time slabs instead of time levels would yield a time stepping scheme which includes jump terms due to the discontinuity at discrete time levels.

For later use, we define a subdivision $S(\mathbb{T}_h^{n-1}, \mathbb{T}_h^n)$ of Ω, as follows:

$$S(\mathbb{T}_h^{n-1}, \mathbb{T}_h^n) = \{\omega \subset \Omega \mid \exists K \in \mathbb{T}_h^{n-1}, K' \in \mathbb{T}_h^n : \omega = K \cap K'\}.$$

In two dimensions, this set is the subdivision of Ω by the union of mesh lines of $\mathbb{T}_h^{n-1}$ and $\mathbb{T}_h^n$. Due to the irregular structure of the elements of $S(\mathbb{T}_h^{n-1}, \mathbb{T}_h^n)$, computations are only feasible with reasonable effort if the grids $\mathbb{T}_h^{n-1}$ and $\mathbb{T}_h^n$ are related in some way. We note that for hierarchically refined grids, $S(\mathbb{T}_h^{n-1}, \mathbb{T}_h^n)$ is the set of most refined cells from the two grids. For more details, we refer to Bangerth and Rannacher (1999). Since the continuous solution w also satisfies (6.102), we have again Galerkin orthogonality for the error $e = \{e_u, e_v\}$.

6.2 A-Posteriori Error Estimation

We want to control the error $e = \{e_u, e_v\}$ using a functional of the form

$$J(e) = (j, e_u)_{Q_T},$$

with some density function $j = j(x,t)$. To this end, we use again a duality argument in space-time written in variational form as

$$A(\tau, \xi) = J(\tau) \quad \forall \tau \in W^*, \tag{6.105}$$

where the dual solution is of the form $\xi = \{-\partial_t z, z\}$. This means that z satisfies the backward-in-time wave equation

$$\begin{aligned} \partial_t^2 z - \nabla\cdot\{a\nabla z\} &= j \quad \text{in } Q_T, \\ z_{|t=T} = 0, \ -\partial_t z_{|t=T} &= 0 \quad \text{in } \Omega, \\ a\partial_n z_{|\partial\Omega} &= 0 \quad \text{on } I. \end{aligned} \tag{6.106}$$

Then, from our general theory, we have the abstract result

$$(j, u)_{Q_T} = -A(w_h, \xi - \xi_h), \tag{6.107}$$

with an arbitrary element $\xi_h \in W_h$. Recalling the definition of the bilinear form $A(\cdot,\cdot)$, we obtain

$$\begin{aligned} |(j,u)_{Q_T}| \le \sum_{n=1}^{N} \sum_{K\in\mathbb{T}_h^n} \Big| &(\partial_t u_h - v_h, \partial_t z - \psi_h)_{K\times I_n} \\ &- (\partial_t v_h - \nabla\cdot a\nabla u_h, z - \phi_h)_{K\times I_n} - \tfrac{1}{2}(an\cdot[\nabla u_h], z - \phi_h)_{\partial K\times I_n}\Big|, \end{aligned}$$

with arbitrary elements $\phi_h, \psi_h \in V_h$, while again $[\nabla u_h]_{|\Gamma} = \nabla u_h$ on cell edges $\Gamma \subset \partial\Omega$. From this, we infer the following result.

Proposition 7. *For the approximation of the acoustic wave equation problem (6.100) by the space-time finite element scheme (6.102), there holds the a posteriori error estimate*

$$|J(e)| \le \eta_\omega(u_h, v_h) = \sum_{n=1}^{N} \sum_{K\in\mathbb{T}_h^n} \{\rho_K^{n,1}\omega_K^{n,1} + \rho_K^{n,2}\omega_K^{n,2} + \rho_{\partial K}^n \omega_{\partial K}^n\}, \tag{6.108}$$

where the cellwise residuals and weights are defined by

$$\begin{aligned} \rho_K^{n,1} &= \|\partial_t u_h - v_h\|_{K\times I_n}, & \omega_K^{n,1} &= \|\partial_t z - \psi_h\|_{K\times I_n}, \\ \rho_K^{n,2} &= \|\partial_t v_h - \nabla\cdot a\nabla u_h\|_{K\times I_n}, & \omega_K^{n,2} &= \|z - \phi_h\|_{K\times I_n}, \\ \rho_{\partial K}^n &= \tfrac{1}{2} h_K^{-1/2} \|an\cdot[\nabla u_h]\|_{\partial K\times I_n}, & \omega_{\partial K}^n &= h_K^{1/2}\|z - \phi_h\|_{\partial K\times I_n}. \end{aligned}$$

We will compare the error estimator (6.108) with a simple heuristic *energy-error* indicator which measures the spatial smoothness of the computed solution u_h,

$$\eta_E(u_h) = \Big(\sum_{n=1}^{N} \sum_{K\in\mathbb{T}_h^n} (\rho_{\partial K}^n)^2\Big)^{1/2}. \tag{6.109}$$

6.3 Numerical Tests

We will illustrate the mechanism underlying our a posteriori error estimator by two test computation, one for a simple spatially one-dimensional problem and one for a more demanding two-dimensional problem.

Example 1 We begin with an example in one space dimension. Let in (6.100) be $d = 1$, $a \equiv 1$, and $\Omega = (-1, 1)$. We choose initial and boundary conditions as shown in Figure 15, with half-width $s = 0.1$ and end-time $T = 2.7$.

$$u^0(x) = \exp(-|x_s|^2)(1-|x_s|^2)\Theta(1-|x_s|),$$
$$v^0(x) = 0,$$
$$\partial_x u(1) = 0, \quad \partial_x u(-1) = 0,$$

with $x_s = x/s$ and the jump function

$$\Theta(y) = \begin{cases} 0 & \text{for } y < 0, \\ 1 & \text{for } y \geq 0. \end{cases}$$

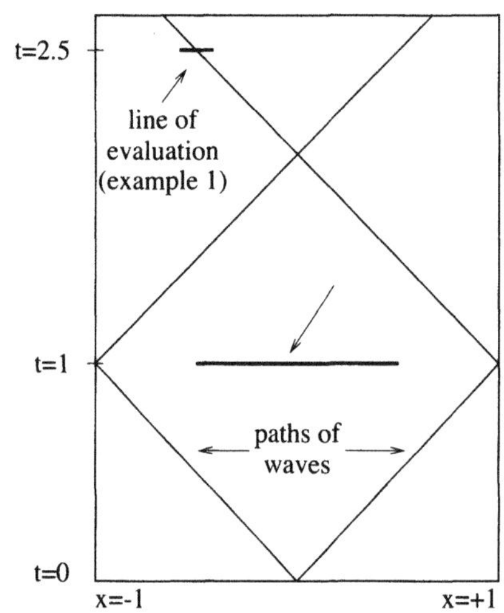

Figure 15. Confi guration of Example 1

We are now interested in the evaluation of only one branch of the solution, for example the one initially traveling to the right. This is a rather frequent case if one wants to compare the outcome of a simulation with measurements. Here, we choose to localize the measurements at $t = 2.5$ and around $x = -0.5$ and use as target functional the quantity

$$J(\tau) = \int_{-0.6}^{-0.4} \phi(x, 2.5)\,dx,$$

with $\tau = \{\phi, \psi\}$. Note that the solution's two peaks are centered around $x = \pm 0.5$ at $t = 2.5$, with diameter 0.2 as in the initial distribution. The solution u^* of the corresponding dual problem is shown in Figure 16 on the left. As can easily be seen, the integral kernel of the functional $J(\cdot)$, i.e. the characteristic function of $[-0.6, 0.4] \times \{2.5\}$, serves as source term for the dual solution. The dual solution therefore is discontinuous due to the discontinuous integral kernel.

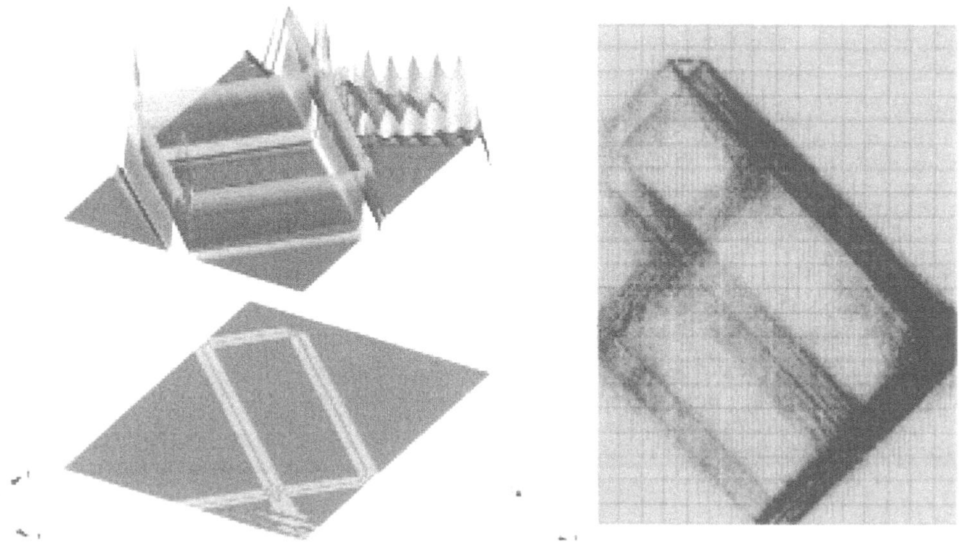

Figure 16. Example 1. Left: primal solution u and dual solution u^*. (Notice that the dual solution is computed backward in time.) Right: resulting space-time grid after three cycles of refi nement.

The resulting space-time grid after three refinement cycles is shown in Figure 16. As can be seen, the error estimator does not only do the obvious thing, which would be to track just one branch of the two waves, but also takes into account errors occurring in the whole space-time domain. It is therefore far better than a priori refining the mesh by hand, because one would not refine the other branch at all. On the other hand, the branch leading directly to the region of evaluation is much more refined, due to the larger weighting in the a posteriori error estimate. Also note the very coarse grid for $t > 2.5$, it is coarsened in each refinement cycle and will end up in only one cell on each time level after some cycles. This is the reason for the bad resolution of the solution at late times. Since the one-dimensional case with constant coefficients is too easily solved to four or five digits of accuracy, we do not show convergence charts.

Example 2. Next, we consider a more realistic example in two space dimensions. Let in (6.100) be $\Omega = (-1,1)^2 \subset \mathbb{R}^2$ and the initial values

$$u^0(x) = \exp(-|x_s|^2)(1-|x_s|^2)\Theta(1-|x_s|), \qquad v^0 = 0,$$

with x_s and the jump function $\Theta(\cdot)$ as defined in the first example, and $s = 0.01$. Again, we choose the elasticity coefficient discontinuous, $a = 1$ for $y < 0.2$, and $a = 9$ for $y \geq 0.2$. A typical wave pattern is shown in Figure 17.

In geophysics and seismics it is an important task to accurately model the signal arrival time at a given point. In our case, we are interested in the situation at the point $x_0 = (0.75, 0)$. As shown in the layout (see Figure 17) the three first waves arriving at this point are the Huygens' wave, the direct wave, and the one reflected from the discontinuity. The first one travels into the medium of higher wave velocity, travels some distance parallel to the discontinuity and then back towards the point of measurement. Among all waves it is the one which has the least action along its path and is therefore called Huygens' wave. From extrapolation of computed data, we estimate its arrival time to be approximately $\tau_H \approx 0.618$, while the arrival times of the other ones are $\tau_d \approx 0.75$ for the direct wave and $\tau_r \approx 0.85$ for the reflected wave. A quantity related to the arrival time is

$$J(\tau) = \int_{t_1}^{t_2} t\phi(x_0,t)\,dt, \quad \tau = \{\phi,\psi\},$$

with a time interval $[t_1,t_2]$ suitably chosen around the signal and such that it does not include other signals. This interval is usually chosen in accordance with experimental data. We take $t_1 = 0.55$ and $t_2 = 0.68$, to catch the first wave only. Accordingly, we choose $T = t_2$, to stop the computation at the first possible time.

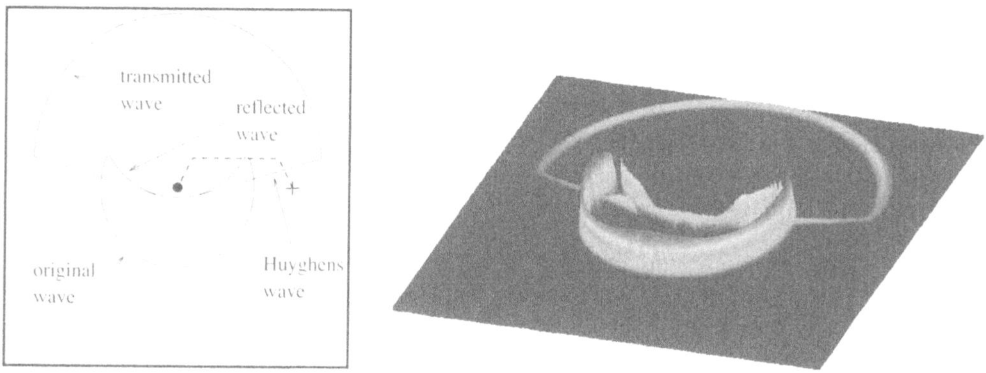

Figure 17. Example 2. Left: layout of the wave pattern after some time, a bullet and a cross indicate the positions of source and receiver, respectively. The dotted line indicates the discontinuity in the coeffi cient, while the thick dashed line denotes the path of least action. Right: plot of the solution at $t = 0.45$.

In Figure 18, we show the computational grids at times $t = 0.15$, $t = 0.45$, and $t = T$, as generated by refinement by the energy error indicator (6.109) and by the dual estimator (6.108). It is readily seen that the latter only tracks that part of the wave field that travels to the right. A closer look at a more complete sequence of grids than shown here reveals that the most refined parts of the grids indeed track the path of least action (the dashed line in Figure 17) which marks the path of the first signal to arrive at the receiver. The first grid shown is at a time where the wave to arrive first is still traveling upward, while in the second it is already traveling downward again. These complicated features of wave propagation are clearly reflected in the grids.

In Figure 19 the convergence of $J(w_h)$ towards the inferred value $J(w) \approx 0.618$ is shown. Since the grids only tracked the interesting part of the wave it is not surprising that it accomplishes the same accuracy with a significantly lower number of space-time cells than the grids refined with the simplified energy error indicator (6.109). Note that the dip in each curve is due to the error, $J(w_h) - 0.618$, changing its sign, which happens to bring $J(w_h)$ close to the exact value. Leaving aside these two data points, the grids as refined by (6.108) show a higher order of convergence than the grids as refined by (6.109). It should be mentioned that refinement by the two methods starts from the same grid, but that in the first step the error estimator (6.108) coarsens more cells than it refines, which leads to an overall decrease of space-time cells.

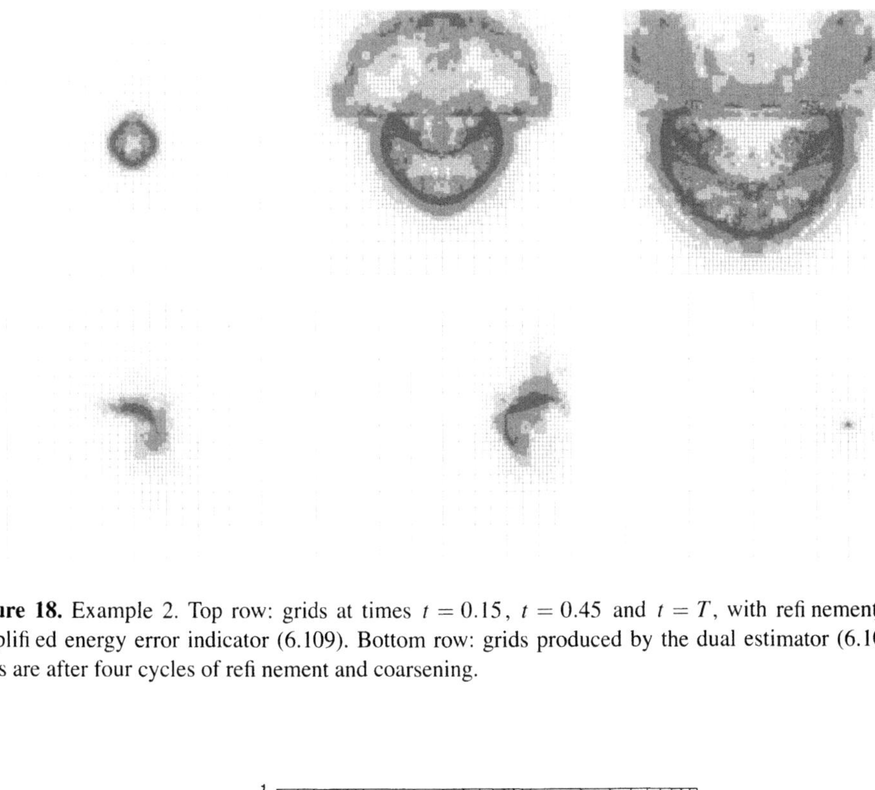

Figure 18. Example 2. Top row: grids at times $t = 0.15$, $t = 0.45$ and $t = T$, with refi nement by the simplifi ed energy error indicator (6.109). Bottom row: grids produced by the dual estimator (6.108). All grids are after four cycles of refi nement and coarsening.

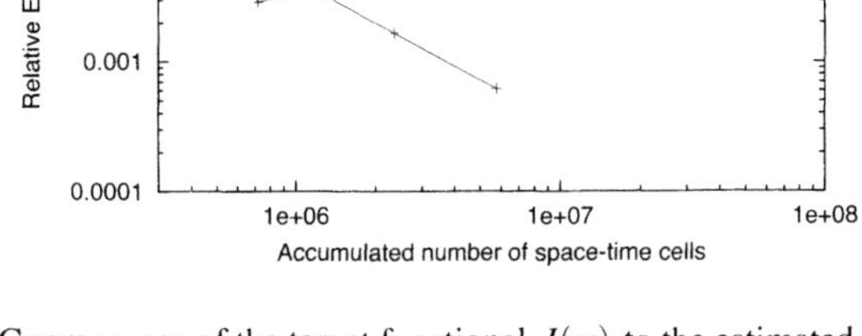

Figure 19. Example 2. Convergence of the target functional $J(w)$ to the estimated exact value 0.618. Note that the dip in each curve is due to a change in the sign of the error.

7 Application to Optimal Control Problems

As the last application, we consider the finite element approximation of optimal control problems. Let a *state variable* $u \in V$ and a *control variable* $q \in Q$ be related by a *state equation* (e.g., a partial differential equation) written in variational form

$$A(u,q;\phi) = 0 \quad \forall \phi \in V, \tag{7.110}$$

where V and Q are suitable function spaces. We seek to minimize a given *cost functional*,

$$J(u,q) \to min!, \tag{7.111}$$

over the set M of all *admissible* pairs $\{u,q\} \in V \times Q$,

$$M = \big\{\{u,q\} \in V \times Q \text{ satisfying (7.110)}\big\}.$$

For solving this problem, one has to discretize the state equation, e.g. by a Galerkin method,

$$A(u_h,q_h;\phi_h) = 0 \qquad \forall \phi_h \in V_h, \tag{7.112}$$

in finite dimensional subspaces $V_h \subset V$ and $Q_h \subset Q$, resulting in a finite dimensional optimization problem posed on an approximate admissible set $M_h \approx M$. This perturbation certainly has an effect on the quality of the optimization. Since discretization of partial differential equations is expensive, the approximation of the state equation (7.110) requires adaptivity. In the following, we will discuss the question of how to organize mesh adaptation in finite element discretization of such problems. The material of this section is taken from Kapp et al. (2000).

The starting point is the reformulation of the optimal control problem as a boundary value problem by employing the Lagrangian approach (so-called *indirect method in constraint optimization*). We introduce the Lagrangian functional

$$L(u,q,\lambda) = J(u,q) + A(u,q;\lambda),$$

with the adjoint variable $\lambda \in V$ (*Lagrangian multiplier*). Then, the first-order optimality condition (*Euler-Lagrange equations*)

$$\nabla L(u,q,\lambda)(\psi,\chi,\pi) = 0 \qquad \forall \{\psi,\chi,\pi\} \in V \times Q \times V, \tag{7.113}$$

determines possible solutions of the optimal control problem as stationary points of $L(\cdot,\cdot,\cdot)$. This variational equation is now approximated by a Galerkin method using finite dimensional subspaces $V_h \times Q_h \times V_h \subset V \times Q \times V$,

$$\nabla L(u_h,q_h,\lambda_h)(\phi_h,r_h,\tau_h) = 0 \qquad \forall \{\psi_h,\chi_h,\pi_h\} \in V_h \times Q_h \times V_h. \tag{7.114}$$

7.1 A Boundary Control Problem

As proto-typical model case, we consider a boundary-control problem (simplified *Ginzburg-Landau model* following Du et al. (1992)). The state equations are

$$-\Delta u + s(u) = 0 \quad \text{on } \Omega, \tag{7.115}$$
$$\partial_n u = q \text{ on } \Gamma_C, \quad \partial_n u = 0 \text{ on } \partial\Omega \setminus \Gamma_C,$$

defined on an open bounded domain $\Omega \subset \mathbb{R}^2$, with boundary $\partial\Omega$. The nonlinearity is $s(u) = u^3 - u$. The control q acts on the boundary component Γ_C, while the observations $u_{|\Gamma_O}$ are taken on a boundary component Γ_O (see Figure 20 for our reference configurations). The cost functional is given by

$$J(u,q) = \tfrac{1}{2}\|u - u_O\|^2_{\Gamma_O} + \tfrac{1}{2}\alpha\|q\|^2_{\Gamma_C}, \qquad (7.116)$$

with a prescribed observable u_O and a regularization parameter $\alpha \geq 0$. Here, the natural function space for the state variable u and the adjoint variable λ is the Sobolev space $V = H^1(\Omega)$, while the control q is determined in the boundary Lebesgue space $Q = L^2(\Gamma_C)$.

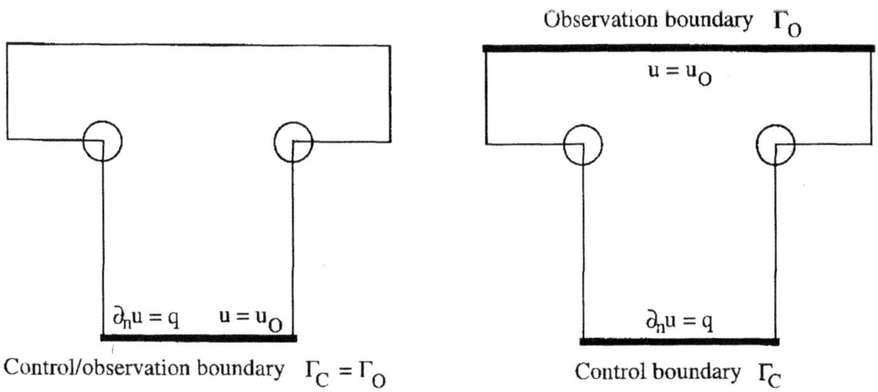

Figure 20. Configuration of the boundary control model problem on a T-domain: Configuration 1 (left), Configuration 2 (right).

In this special case the Lagrangian functional has the form

$$L(u,\lambda,q) = J(u,q) + (\nabla u, \nabla\lambda) - (s(u),\lambda) + (\partial_n u - q, \lambda)_{\Gamma_C}, \qquad (7.117)$$

and the corresponding Euler-Lagrange equation determines stationary points $\{u,q,\lambda\} \in V \times Q \times V$ by the system

$$\begin{aligned} (\nabla u, \nabla\pi) + (s(u),\pi) - (q,\pi)_{\Gamma_C} &= 0 \qquad \forall \pi \in V, \\ (\alpha q - \lambda, \chi)_{\Gamma_C} &= 0 \qquad \forall \chi \in Q, \\ (\nabla\psi, \nabla\lambda) + (\psi, s'(u)\lambda) + (\psi, u - u_O)_{\Gamma_O} &= 0 \qquad \forall \psi \in V. \end{aligned} \qquad (7.118)$$

For approximating system (7.118), we use again a Galerkin finite element discretization as described above. For quadrilateral meshes $\mathbb{T}_h$, let $V_h \subset V$ be the generic finite element spaces consisting of piecewise (isoparametric) bilinear shape functions and, for simplicity, let the space of discrete controls $Q_h \subset Q$ consist just of the traces of functions in V_h. Then, the discrete

Euler-Lagrange equations determine approximations $\{u_h, q_h, \lambda_h\} \in V_h \times Q_h \times V_h$ by

$$\begin{aligned} (\nabla u_h, \nabla \pi_h) + (s(u_h), \pi_h) - (q_h, \pi_h)_{\Gamma_C} &= 0 \qquad \forall \pi_h \in V_h, \\ (\alpha q_h - \lambda_h, \chi_h)_{\Gamma_C} &= 0 \qquad \forall \chi_h \in Q, \\ (\nabla \psi_h, \nabla \lambda_h) + (\psi_h, s'(u_h)\lambda_h) + (\psi_h, u_h - u_O)_{\Gamma_O} &= 0 \qquad \forall \psi_h \in V_h. \end{aligned} \tag{7.119}$$

For solving this saddle–point problem, we use a discretized Newton iteration which is defined on the continuous level in the function space $V \times Q \times V$. The discretization is controlled by an residual-based a posteriori error estimate which will be discussed below. The resulting linearized algebraic problems are solved by a GMRES method with multigrid preconditioning. For algorithmic details, we refer to Kapp et al. (2000).

7.2 A-Posteriori Error Estimation

The conventional approach to a posteriori error estimation in optimal control is based on the concept of controlling the accuracy in satisfying the state equation, i.e. *control of admissibility*. Using the *energy norm* this error control of the state equation alone leads to the heuristic a posteriori error indicator

$$\eta_E(u_h) = c_i \sum_{K \in \mathbb{T}_h} \{\rho_K^{(u)2} + \rho_{\partial K}^{(u)2}\}, \tag{7.120}$$

where the residual terms are given by

$$\rho_K^{(u)} = \|R_h^{(u)}\|_K, \qquad \rho_{\partial K}^{(u)} = h_\Gamma^{-1/2} \|r_h^{(u)}\|_{\partial K},$$

with the cell and edge residuals

$$R_{h|K}^{(u)} = \Delta u_h - s(u_h), \qquad r_{h|\Gamma}^{(u)} = \begin{cases} \frac{1}{2} n \cdot [\nabla u_h], & \text{if } \Gamma \subset \partial K \setminus \partial\Omega, \\ \partial_n u_h, & \text{if } \Gamma \subset \partial\Omega \setminus \Gamma_C, \\ \partial_n u_h - q_h, & \text{if } \Gamma \subset \Gamma_C. \end{cases}$$

This estimator is suggested by the corresponding result derived for the (linear) Poisson problem in Section 2.2. If one wants to include also error control for the adjoint variable, the a posteriori error indicator becomes

$$\eta_E(u_h, \lambda_h) = c_i \sum_{K \in \mathbb{T}_h} \{\rho_K^{(u)2} + \rho_{\partial K}^{(u)2} + \rho_K^{(\lambda)2} + \rho_{\partial K}^{(\lambda)2}\}, \tag{7.121}$$

where the additional residual terms are given by

$$\rho_K^{(\lambda)} = \|R_h^{(\lambda)}\|_K, \qquad \rho_{\partial K}^{(\lambda)} = h_\Gamma^{-1/2} \|r_h^{(\lambda)}\|_{\partial K},$$

with the cell and edge residuals

$$R_{h|K}^{(\lambda)} = \Delta \lambda_h - s(u_h)\lambda_h, \qquad r_{h|\Gamma}^{(\lambda)} = \begin{cases} \frac{1}{2} n \cdot [\nabla \lambda_h], & \text{if } \Gamma \subset \partial K \setminus \partial\Omega, \\ \partial_n \lambda_h, & \text{if } \Gamma \subset \partial\Omega \setminus \Gamma_C, \\ c_0 - u_h + \partial_n \lambda_h, & \text{if } \Gamma \subset \Gamma_C. \end{cases}$$

However, the use of these heuristic error indicators seems questionable since they do not take into account the sensitivity of the cost functional with respect to the local perturbations introduced by discretization. Capturing these dependencies should be accomplished by a *good* error estimator.

Preparing for this, consider the model Poisson equation

$$-\Delta u = f \quad \text{in } \Omega, \quad u = 0 \quad \text{on } \partial\Omega, \tag{7.122}$$

and its approximation by a standard Galerkin finite element method as described above. The solution $u \in V = H_0^1(\Omega)$ of (7.122) is characterized by attaining minimum energy, i.e.,

$$L(u) = \|\nabla u\|^2 - 2(f, u) \quad \to \quad min! \quad \text{on } V. \tag{7.123}$$

The Galerkin finite element approximation $u_h \in V_h$, defined by

$$(\nabla u_h, \nabla \phi_h) = (f, \phi_h) \qquad \forall \phi_h \in V_h,$$

satisfies $(\nabla u_h, \nabla u) = \|\nabla u\|^2$. Using this relation, we conclude that

$$\begin{aligned} L(u_h) - L(u) &= \|\nabla u_h\|^2 - 2(f, u_h) - \|\nabla u\|^2 + 2(f, u) \\ &= \|\nabla u_h\|^2 - 2(\nabla u, \nabla u_h) + \|\nabla u\|^2 = \|\nabla e\|^2. \end{aligned}$$

Hence, energy-error control means control of the error with respect to the *energy functional* $L(\cdot)$. As we have seen above, this can be accomplished by using the linear error functional

$$J(\phi) = (\nabla e, \nabla \phi)$$

in the corresponding dual problem (2.22). Obviously, the dual solution is then simply given by $z = e$. The resulting a posteriori error estimate then takes the form

$$\|\nabla e\|_\Omega^2 \le \sum_{K \in \mathbb{T}_h} \{\rho_K^{(u)} \omega_K^{(u)} + \rho_{\partial K}^{(u)} \omega_{\partial K}^{(u)}\}, \tag{7.124}$$

where the residual terms and weights are given by

$$\begin{aligned} \rho_K^{(u)} &= \|R(u_h)\|_K, & \omega_K^{(u)} &= h_K^{-1}\|u - \phi_h\|_K, \\ \rho_{\partial K}^{(u)} &= h_K^{-1/2}\|r(u_h)\|_{\partial K}, & \omega_{\partial K}^{(u)} &= h_K^{1/2}\|u - \phi_h\|_{\partial K \setminus \partial\Omega}, \end{aligned}$$

with arbitrary $\phi_h \in V_h$. By this example, we have discovered a general principle in a posteriori error estimation for much more general variational problems: If the problem comes from the Euler-Lagrange equation of an *energy functional* and if the same functional is also taken for error control, then the corresponding dual solution coincides with the error itself, i.e., $z = e$.

7.3 A-Posteriori Error Control for Euler-Lagrange Equations

Now, we consider the special situation that the variational problem considered is obtained as Euler-Lagrange equation of an *energy functional* $L(\cdot)$ defined on some Hilbert function space X, i.e. using the notation of Section 5.1, $A(x; y) = L'(x; y)$. We assume that the functional $L(\cdot)$ is three-times differentiable with derivative forms $L'(\cdot;\cdot)$, $L''(\cdot;\cdot,\cdot)$, and $L'''(\cdot;\cdot,\cdot,\cdot)$ continuously defined on V. The Euler-Lagrange equation

$$L'(x;y)=0 \qquad \forall y\in X, \tag{7.125}$$

determines possible extremal points of $L(\cdot)$. On finite dimensional subspaces $X_h\subset X$, the approximate problems read

$$L'(x_h;y_h)=0 \qquad \forall y_h\in X_h. \tag{7.126}$$

We want to develop an a posteriori error analysis for this abstract situation where the error $e=x-x_h$ is controlled by the generating energy functional $L(\cdot)$. By Taylor expansion, we have

$$L(x)-L(x_h) = L'(x_h;e)+\tfrac{1}{2}L''(x_h;e,e)+\tfrac{1}{2}\int_0^1 L'''(x_h+se;e,e,e)\,(s-1)^2\,ds.$$

Further, there holds

$$L'(x;e) = L'(x_h;e)+L''(x_h;e;e)-\int_0^1 L'''(x_h+se;e,e,e)\,(s-1)\,ds,$$

and consequently, observing that $L'(x;e)=0$,

$$\tfrac{1}{2}L''(x_h;e;e) = -\tfrac{1}{2}L'(x_h;e)+\tfrac{1}{2}\int_0^1 L'''(x_h+se;e,e,e)\,(s-1)\,ds.$$

Combining these relations and observing that $(s-1)^2+s-1=s(s-1)$ implies

$$L(x)-L(x_h) = \tfrac{1}{2}L'(x_h;e)+\tfrac{1}{2}\int_0^1 L'''(x_h+se;e,e,e)\,s(s-1)\,ds.$$

Hence, employing Galerkin orthogonality, we obtain

$$L(x)-L(x_h) = \tfrac{1}{2}L'(x_h;e-y_h)+\tfrac{1}{2}\int_0^1 L'''(x_h+se;e,e,e)\,s(s-1)\,ds.$$

for an arbitrary $y_h\in X_h$. We summarize our result as follows.

Proposition 8. *For the approximation of the problem (7.125) by the Galerkin scheme (7.126), there holds the a posteriori error representation*

$$L(x)-L(x_h) = \tfrac{1}{2}\min_{y_h\in X_h} L'(x_h;x-y_h)+R(x,x_h;e,e), \tag{7.127}$$

with the remainder term given by

$$R(x,x_h;e,e)=\tfrac{1}{2}\int_0^1 L'''(x_h+se;e,e,e)\,s(s-1)\,ds.$$

Notice that the remainder $R(x,x_h;e,e)$ vanishes if the functional $L(\cdot)$ is quadratic. Again, in the a posteriori error estimate (7.127) the quantity $x-y_h$ has to be approximated as described above by using the computed solution $x_h\in X_h$. We emphasize that in this particular case the evaluation of the a posteriori error estimate with respect to the *energy* functional $L(\cdot)$ does not require the explicit solution of a dual problem.

7.4 Application to the Optimal Control Problem

Now, we apply the idea laid out above to the particular situation of the boundary control problem. The functional of interest is the Lagrangian functional of the optimal control problem,

$$L(x) = J(u,q) + (\nabla u, \nabla\lambda) + (s(u),\lambda) - (q,\lambda)_{\Gamma_C},$$

defined for triples $x = \{u,q,\lambda\}$ in the Hilbert space $X = V \times Q \times V$. Following the formalism of the previous section, we seek to estimate the error $e = \{e_u, e_q, e_\lambda\}$ with respect to the Lagrangian functional $L(\cdot)$. In the present case, there holds

$$\begin{aligned} L(v) - L(v_h) &= J(u,q) + (\nabla u, \nabla\lambda) + (s(u),\lambda) - (q,\lambda)_{\Gamma_C} \\ &\quad - J(u_h,q_h) - (\nabla u_h, \nabla\lambda_h) - (s(u_h),\lambda_h) + (q_h,\lambda_h)_{\Gamma_C} \\ &= J(u,q) - J(u_h,q_h), \end{aligned}$$

since $\{u,q,\lambda\}$ and $\{u_h,q_h,\lambda_h\}$ satisfy the state equations (7.118) and (7.119), respectively. Hence, error control with respect to the Lagrangian functional $L(\cdot)$ and the cost functional $J(\cdot)$ is equivalent. Now, the general error identity (7.127) implies that

$$|J(u,q) - J(u_h,q_h)| \le \inf_{y_h\in X_h} |L'(x_h; x - y_h)| + R(x,x_h;e,e,e), \tag{7.128}$$

where $x_h = \{u_h,\lambda_h,q_h\}$ and $x = \{u,\lambda,q\}$. From (7.118), we see that

$$\begin{aligned} L'(v_h; v - \phi_h) &= (u_h - u_O, u - \psi_h)_{\Gamma_O} + (\nabla(u - \psi_h), \nabla\lambda_h) + (u - \psi_h, s'(u_h)\lambda_h) \\ &\quad + (\nabla u_h, \nabla(\lambda - \pi_h)) + (s(u_h), \lambda - \pi_h) - (q_h, \lambda - \pi_h)_{\Gamma_C} \\ &\quad + (\lambda_h - \alpha q_h, q - \chi_h)_{\Gamma_C}, \end{aligned}$$

and (for details see Rannacher (2000))

$$R(x,x_h;e,e,e) \le 6\int_\Omega \Big\{ \max\{|u,|u_h|\}|e_u|^3 + |e_u|^2|e_\lambda| \Big\}\, dx. \tag{7.129}$$

Splitting the global integrals into the contributions from each single cell $K \in \mathbb{T}$ and each cell edge $\Gamma \subset \partial\Omega$, respectively, and integrating locally by parts yields

$$\begin{aligned} L'(v_h; v - \phi_h) &= \sum_{\Gamma\subset\Gamma_O} (u_h - u_O + \partial_n\lambda_h, u - \psi_h)_\Gamma + \sum_{\Gamma\subset\partial\Omega\setminus\Gamma_O} (\partial_n\lambda_h, u - \psi_h)_\Gamma \\ &\quad + \sum_{\Gamma\subset\Gamma_C} (\partial_n u_h - q_h, \lambda - \pi_h)_\Gamma + \sum_{\Gamma\subset\partial\Omega\setminus\Gamma_C} (\partial_n u_h, \lambda - \pi_h)_\Gamma \\ &\quad + \sum_{\Gamma\subset\Gamma_C} (\lambda_h - \alpha q_h, q - \chi_h)_\Gamma \\ &\quad + \sum_{K\in\mathbb{T}_h} \Big\{ (-\Delta u_h + s(u_h), \lambda - \pi_h)_K + \tfrac{1}{2}(n\cdot[\nabla u_h], \lambda - \pi_h)_{\partial K\setminus\partial\Omega} \Big\} \\ &\quad + \sum_{K\in\mathbb{T}_h} \Big\{ (u - \psi_h, -\Delta\lambda_h + s'(u_h)\lambda_h)_K + \tfrac{1}{2}(u - \psi_h, n\cdot[\nabla\lambda_h])_{\partial K\setminus\partial\Omega} \Big\}. \end{aligned}$$

The cell residuals $R_h^{(u)}, R_h^{(\lambda)}$ are defined by $R_{h|K}^{(u)} = -\Delta u_h + s(u_h)$, $R_{h|K}^{(\lambda)} = -\Delta\lambda_h + s'(u_h)$, and the edge residuals $r_h^{(u)}, r_h^{(\lambda)}, r_h^{(q)}$ by

$$r_{h|\Gamma}^{(u)} = \begin{cases} \frac{1}{2}[\partial_n\phi_h], & \text{if } \Gamma \subset \partial K \setminus \partial\Omega, \\ \partial_n u_h, & \text{if } \Gamma \subset \partial\Omega \setminus \Gamma_C, \\ \partial_n u_h - q_h, & \text{if } \Gamma \subset \Gamma_C, \end{cases} \qquad r_{h|\Gamma}^{(\lambda)} = \begin{cases} \frac{1}{2}[\partial_n\phi_h], & \text{if } \Gamma \subset \partial K \setminus \partial\Omega, \\ \partial_n\lambda_h, & \text{if } \Gamma \subset \partial\Omega \setminus \Gamma_O, \\ c_0 - u_h + \partial_n\lambda_h, & \text{if } \Gamma \subset \Gamma_O, \end{cases}$$

and $r_{h|\Gamma}^{(q)} = \alpha q_h - \lambda_h$, if $\Gamma \subset \Gamma_C$. From this, we infer the following result.

Proposition 9. *For the approximation of the Euler-Lagrange system (7.118) by the Galerkin finite element scheme (7.119), there holds the a posteriori error estimate*

$$|J(u,q) - J(u_h,q_h)| \le \eta_\omega(u_h,\lambda_h,q_h) = \sum_{K\in\mathbb{T}_h} \eta_K(u_h,\lambda_h,q_h), \tag{7.130}$$

with the local error indicators

$$\eta_K(u_h,\lambda_h,q_h) = \rho_K^{(u)}\omega_K^{(\lambda)} + \rho_{\partial K}^{(u)}\omega_{\partial K}^{(\lambda)} + \rho_K^{(\lambda)}\omega_K^{(u)} + \rho_{\partial K}^{(\lambda)}\omega_{\partial K}^{(u)} + \rho_{\partial K}^{(q)}\omega_{\partial K}^{(q)},$$

and the cellwise residuals and weights

$$\begin{aligned} \rho_K^{(u)} &= \|R_h^{(u)}\|_K, & \omega_K^{(\lambda)} &= \|\lambda - \pi_h\|_K, \\ \rho_{\partial K}^{(u)} &= h_\Gamma^{-1/2}\|r_h^{(u)}\|_{\partial K}, & \omega_{\partial K}^{(\lambda)} &= \|\lambda - \pi_h\|_{\partial K}, \\ \rho_K^{(\lambda)} &= \|R_h^{(\lambda)}\|_K, & \omega_K^{(u)} &= \|u - \psi_h\|_K, \\ \rho_{\partial K}^{(\lambda)} &= h_\Gamma^{-1/2}\|r_h^{(\lambda)}\|_{\partial K}, & \omega_{\partial K}^{(u)} &= \|u - \psi_h\|_{\partial K}, \\ \rho_{\partial K}^{(q)} &= \|r_h^{(q)}\|_{\partial K\cap\Gamma_C}, & \omega_{\partial K}^{(q)} &= \|q - \chi_h\|_{\partial K\cap\Gamma_C}. \end{aligned}$$

In the computational examples presented below, we evaluate the weights by using the local interpolation estimate

$$h_K^{-1}\|\phi - i_h\phi\|_K + h_K^{-1/2}\|\phi - i_h\phi\|_{\partial K} \le c_i h_K^2 \max_K |\nabla^2\phi|. \tag{7.131}$$

The derivatives $\nabla^2 u$, $\partial_s^2 q$, and $\nabla^2\lambda$ are then approximated by corresponding difference quotients of the computed solution $\{u_h,q_h,\lambda_h\}$.

7.5 Numerical Results

We perform computations for the model equation (7.118) on a series of locally refined meshes. In all cases, the weights in the a posteriori error estimates are evaluated by using difference approximation as described in the previous section with interpolation constants set to $c_i = 0.1$. The mesh refinement follows the *fixed fraction* strategy. The *effectivity index* $I_{eff} = \eta(u_h,q_h)/\text{Error}$ with $\text{Error} = |J(u_{ref},q_{ref}) - J(u_h,q_h)|$, is used as a measure for the quality of our a posteriori error estimator. The reference solutions are obtained on meshes with about $2\cdot 10^5$ cells.

Test for Configuration 1. We begin with Configuration 1 shown in Figure 20. The cost functional is chosen with $u_O = \sin(0.19x)$ and $\alpha = 0$. This test case represents an extreme situation since here the observation is evaluated right at the control boundary Γ_C, i.e., the information relevant for the optimization problem does not have to pass through the domain. Hence, the corner singularities do not much affect the optimization process and should therefore not induce mesh refinement. This is accomplished by our weighted error estimator. In this case, the heuristic energy-error indicators (7.120) and (7.121) produce inefficient meshes as seen in Figures 21, 22 and 23. In fact, in (7.130) the boundary indicators are dominant over the interior indicators. Finally, Figure 23 compares the efficiency of the meshes generated by the dual–weighted error estimator and the energy-error indicator. Clearly, the first one yields significantly more economical meshes: $N \sim 3500$ cells compared to $N \sim 100000$ for obtaining the value $J(u_h, q_h) \approx 0.011948$.

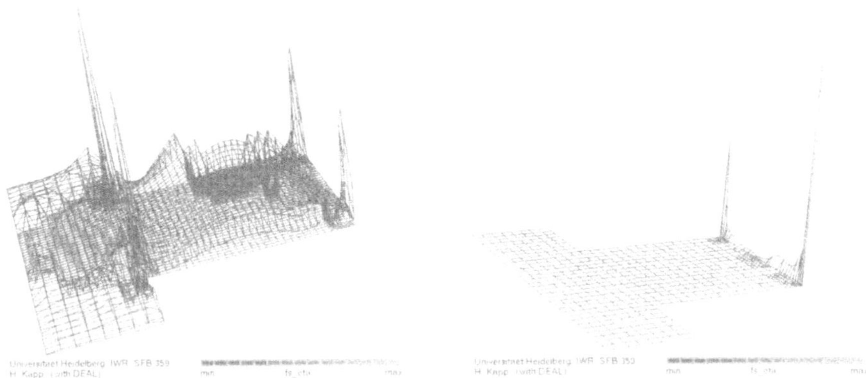

Figure 21. Confi guration 1: Size of cell indicators η_K in the error estimators $\eta_E(u_h)$ (left) and $\eta_\omega(u_h, \lambda_h, q_h)$ (right).

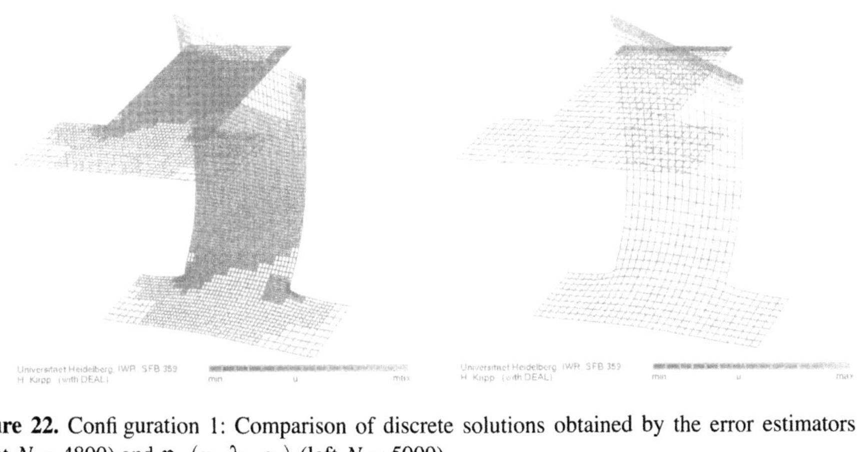

Figure 22. Confi guration 1: Comparison of discrete solutions obtained by the error estimators $\eta_E(u_h)$ (right, $N \sim 4800$) and $\eta_\omega(u_h, \lambda_h, q_h)$ (left, $N \sim 5000$).

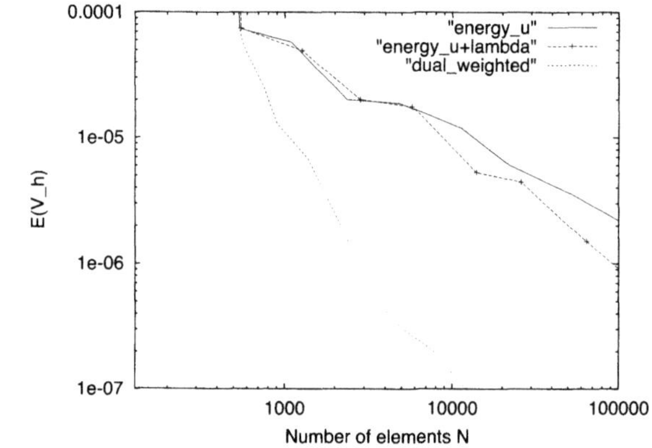

Figure 23. Configuration 1: Comparison of efficiency of meshes generated by the error indicators $\eta_E(u_h)$ (solid line), $\eta_E(u_h, \lambda_h)$ (crosses), and $\eta_\omega(u_h, \lambda_h, q_h)$ (dashed line).

Test for Configuration 2. For the last test, we take the observables as $u_O \equiv 1$ and set the regularization parameter to $\alpha = 0.1$. Now, depending on the nonlinearity $s(u)$, there may be several stationary points of the Euler-Lagrange equations. By varying the starting values for the Newton iteration, we can approximate these solutions. A trivial solution corresponds to $u \equiv u_O$ and is actually the global minimum of the cost functional. The two other computed solutions correspond to a local minimum and a local maximum.

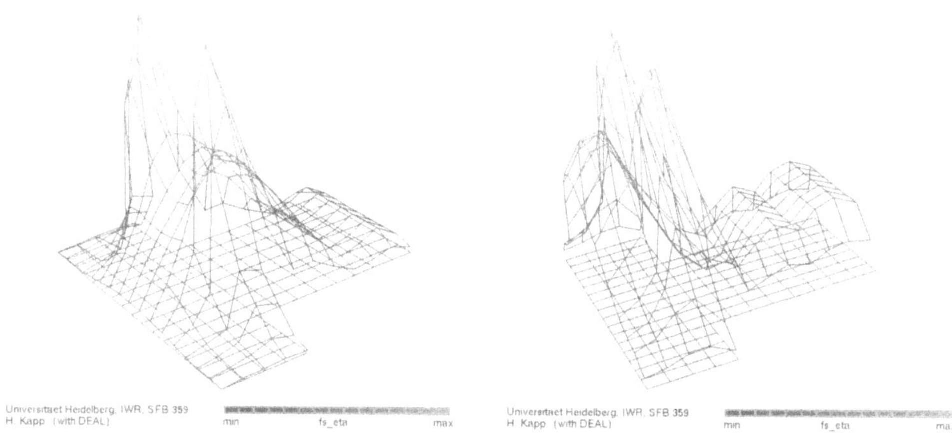

Figure 24. Configuration 2: Distributions of the cell indicators η_K in the error estimators $\eta_E(u_h)$ (left) and $\eta_\omega(u_h, \lambda_h, q_h)$ (right).

Figure 24 shows the distribution of the local cell indicators η_K for the two error estimators; the corresponding meshes are seen in Figure 25. Obviously, the weighted error estimator induces a stronger refinement along the observation and control boundaries which seems more relevant for the optimization process than resolving the corner singularities. However, the total contribution of the interior cell indicators is dominant over that of the boundary indicators. This explains why the gain in efficiency (about 25%) of the dual-weighted error estimator over the energy-norm indicator is less significant here compared to the previous example.

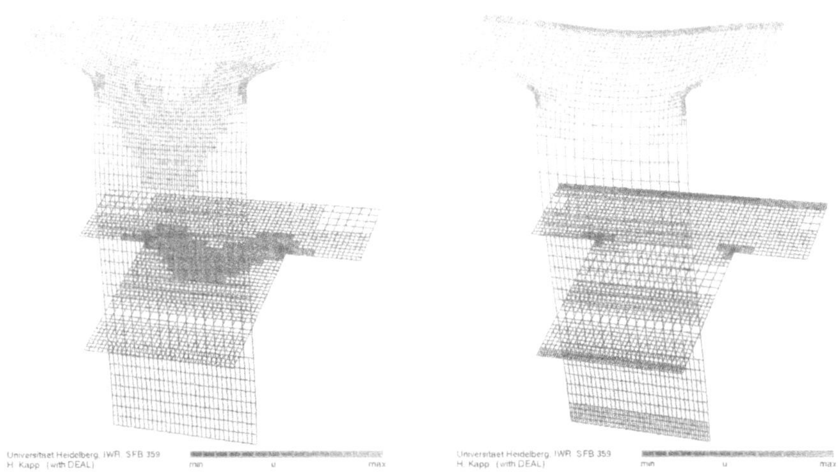

Figure 25. Configuration 2: Comparison of discrete solutions obtained by the error estimators $\eta_E(u_h)$ (right, $N \sim 3300$) and $\eta_\omega(u_h, \lambda_h, q_h)$ (left, $N \sim 3000$).

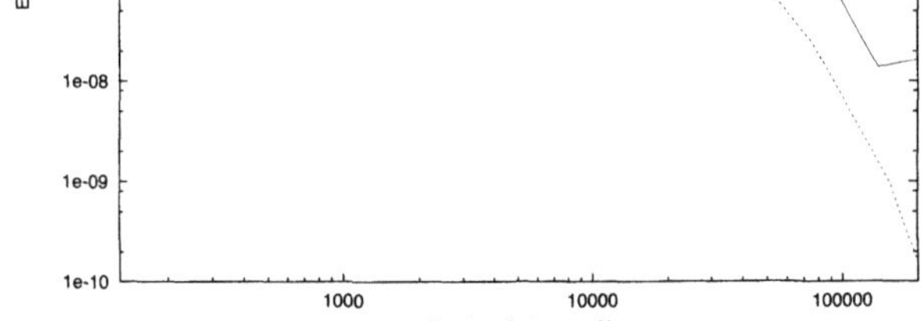

Figure 26. Configuration 2: Comparison of efficiency of meshes generated by the error indicators $\eta_E(u_h)$ (solid line), $\eta_E(u_h, \lambda_h)$ (crosses), and $\eta_\omega(u_h, \lambda_h, q_h)$ (dashed line).

References

Ainsworth, M., and Oden, J.T. (1997). A posteriori error estimation in fi nite element analysis. *Comput. Methods Appl. Mech. Engrg.* 142:1–88.

Babuška, I., and Rheinboldt, W.C. (1978). Error estimates for adaptive fi nite element computations. *SIAM J. Num. Anal.* 15:736–754.

Bangerth, W., and Kanschat, G. (1999). Concepts for Object-Oriented Finite Element Software – the `deal.II` library. Preprint 99-43 (SFB 359), Universität Heidelberg.

Bangerth, W., and Rannacher, R. (1999). Finite element approximation of the acoustic wave equation: error control and mesh adaptation. *East-West J. Numer. Math.* 7:263–282.

Bangerth, W., and Rannacher, R. (2003). *Adaptive Finite Element Methods for Differential Equations.* Lectures in Mathematics, ETH Zürich, Birkhäuser, 2003.

Becker, R. (1995). *An Adaptive Finite Element Method for the Incompressible Navier-Stokes Equations on Time-Dependent Domains.* Doctor Thesis, Preprint 95-44 (SFB 359), Universität Heidelberg.

Becker, R. (2000). An optimal-control approach to a posteriori error estimation for fi nite element discretizations of the Navier-Stokes equations. *East-West J. Numer. Math.*, 9:257–274.

Becker, R. (2001). Mesh adaptation for stationary fbw control. *J. Math. Fluid Mech.*, 3:317–341.

Becker, R., Braack, M., Rannacher, R., and Waguet, C. (1999). Fast and reliable solution of the Navier-Stokes equations including chemistry. *Comput. Visual. Sci* 2:107–122.

Becker, R., Braack, M., and Rannacher, R. (1999). Numerical simulation of laminar fames at low Mach number with adaptive fi nite elements. *Combust. Theory Modelling* 3:503–534.

Becker, R., Johnson, C., and Rannacher R. (1995). Adaptive error control for multigrid fi nite element methods. *Computing* 55:271–288.

Becker, R., Kapp, R., and Rannacher, R. (2000). Adaptive fi nite element methods for optimal control of partial differential equations: basic concepts. *SIAM J. Control Optim.*, 39, 113–132, 2000.

Becker, R., and Rannacher, R. (1998). Weighted a posteriori error control in FE methods. Presented at ENUMATH-95, Paris, August 1995, published in Proc. ENUMATH-97 (Bock, H.G., et al., eds.), pp. 621–637, World Scient. Publ., Singapore.

Becker, R., and Rannacher R. (1996). A feed-back approach to error control in fi nite element methods: basic analysis and examples. *East-West J. Numer. Math.* 4:237–264.

Becker, R., and Rannacher R. (2001). An optimal control approach to a posteriori error estimation in fi nite element methods. In *Acta Numerica 2001* (Iserles, A., ed.,), pp. 1–101, Cambridge University Press.

Braack, M. (1998). *An Adaptive Finite Element Method for Reactive Flow Problems.* Doctor Thesis, Preprint 98-25 (SFB 359), Universität Heidelberg.

Braack, M., and Rannacher R. (1999). Adaptive fi nite element methods for low-Mach-number fbws with chemical reactions. In H. Deconinck, editor, *30th Computational Fluid Dynamics, Lecture Series*, Vol. 1999-03, von Karman Institute for Fluid Dynamics, 1999.

Brenner, S., and Scott, R.L. (1994). *The Mathematical Theory of Finite Element Methods.* Berlin-Heidelberg-New York: Springer.

Carey, G.F., and Oden, J.T. (1984). *Finite Elements, Computational Aspects.* Vol. III, Prentice-Hall.

Ciarlet, P.G. (1984). *Finite Element Methods for Elliptic Problems.* Amsterdam: North-Holland.

Ciarlet, P.G. (1991). *Mathematical Elasticity, Vol. 1: Three-Dimensional Elasticity.* Amsterdam: North-Holland.

Duvaut, G., and Lions, J. L. (1976). *Inequalities in Mechanics and Physics.* Berlin-Heidelberg-New York: Springer.

Du, Q., Gunzburger, M.D., and Peterson, J.S. (1992). Analysis and approximation of the Ginzburg-Landau model of superconductivity. *SIAM Review* 34:54–81.

Du, Q., Gunzburger and Hou, L.S. (1996). Finfte-dimensional approximation offa class of constrained nonlinear optimal control problems. *SIAM J. Control Optimization* 34:1001–1043.

Eriksson, K., and Johnson, J. (1988). An adaptive finite element method for linear elliptic problems. *Math. Comp.* 50:361–383.

Eriksson, K., Estep, D., Hansbo, P., and Johnson, J. (1995). Introduction to adaptive methods for differential equations. In *Acta Numerica 1995* (Iserles, A., ed.,), pp. 105–158, Cambridge University Press.

Führer, C., and Kanschat, G. (1997). Error control in radiative transfer. *Computing* 58:317–334.

Hartmann, R. (1998). *A posteriori Fehlerschätzung und adaptive Schrittweiten- und Ortsgittersteuerung bei Galerkin-Verfahren für die Wärmeleitungsgleichung.* Diploma Thesis, Institut für Angewandte Mathematik, Universität Heidelberg.

Houston, P., Rannacher, R., and Süli, E. (1999). A posteriori error analysis for stabilized finite element approximation of transport problems. Report No. 99/04, Oxford University Computing Laboratory, Oxford, 1999, to appear in *Comput. Meth. Appl. Mech. Engrg.*

Hughes, T.J.R., and Brooks, A.N. (1982). Streamline upwind/Petrov Galerkin formulations for convection dominated flows with particular emphasis on the incompressible Navier-Stokes equation. *Comput. Meth. Appl. Mech. Engrg.* 32:199–259.

Hughes, T.J.R., Franca, L.P., and Balestra, M. (1986). A new finite element formulation for computational fluid dynamics: V. Circumvent the Babuska-Brezzi condition: A stable Petrov-Galerkin formulation for the Stokes problem accommodating equal order interpolation. *Comp. Methods Appl. Mech. Engrg.* 59:89–99.

Hulbert, G.M., and Hughes, T.J.R. (1990). Space-time finite element methods for second-order hyperbolic equations. *Comp. Meth. Appl. Mech. Engrg.* 84:327–348.

Johnson, C. (1987). *Numerical Solution of Partial Differential Equations by the Finite Element Method.* Cambridge-Lund: Cambridge University Press.

Johnson, C. (1993). Discontinuous Galerkin finite element methods for second order hyperbolic problems. *Comp. Meth. Appl. Mech. Engrg.* 107:117–129.

Johnson, C. (1993). A new paradigm for adaptive finite element methods. In Proc. *MAFELAP 93* (Whiteman, J., ed.), John Wiley.

Johnson, C. (1976). Existence theorems for plasticity problems. *J. Math. Pures Appl.* 55:431–444.

Johnson, C., and Hansbo, P. (1991). Adaptive finite element methods for small strain elasto-plasticity. Preprint, Dep. of Math., Chalmers Univ. of Technology.

Johnson, C., and Hansbo, P. (1992). Adaptive finite element methods in computational mechanics. In *Computer Methods in Applied Mechanics and Engineering, Vol. 101*, Amsterdam: North-Holland.

Kanschat, G. (1996). *Parallel and Adaptive Galerkin Methfds for Radiative Transfer Problems.* Doctor Thesis, Preprint 96-29 (SFB 359), Universität Heidelberg.

Kanschat, G. (1997). Efficient and reliable solution of multi-dimensional radiative transfer problem. In *Proc. Workshop Multiscale Phenomena and Their Simulation*, Singapore: World Scient. Publ.

Lions, J.L. (1971). Optimal Control of Systems Governed by Partial Differential Equations. Berlin: Springer.

Machiels, L., Patera, A.T., and Peraire, J. (1998). Output bound approximation for partial differential equations; Application to the incompressible Navier-Stokes equations. In *Proc. Workshop Industrial and Environmental Applications of Direct and Large Eddy Numerical Simulation* (Biringen, S., ed.), Springer.

Machiels, L., Peraire, J., and Patera, A.T. (1999). A posteriori finite element output bounds for the incompressible Navier-Stokes equations; Application to a natural convection problem. MIT FML Report 99-4.

Oden, J.T., Wu, Weihan, and Ainsworth, M. (1993). An a posteriori error estimate for finite element approximations of the Navier-Stokes equations. *Comput. Meth. Appl. Mech. Engrg.* 111:185-202.

Rannacher, R. (2000). A posteriori error analysis for nonlinear finite element approximations. Preprint, Institut für Angewandte Mathematik, Universität Heidelberg.

Rannacher, R., and Suttmeier, F.-T. (1999). A posteriori error estimation and mesh adaptation for finite element models in elasto-plasticity. *Comput. Meth. Appl. Mech. Engrg.* 176:333–361.

Suttmeier, F.-T. (1997). Adaptive Finite Element Approximation of Problems in Elasto-Plasticity Theory. Doctor Thesis, Preprint 97-11 (SFB 359), Universität Heidelberg.

Verfürth, R. (1996). *A Review of A Posteriori Error Estimation and Adaptive Mesh-Refinement Techniques*. New York-Stuttgart: John Wiley/Teubner.

Zienkiewicz, O.C., and Zhu, J.Z. (1993). A simple error estimator and adaptive procedure for practical engineering analysis. *Int. J. Numer. Methods Engrg.* 24:337–357.

Hierarchical Model and Solution Adaptivity of Thin–walled Structures by the Finite–Elements–Method

Erwin Stein, Rainer Niekamp, Stephan Ohnimus and Matthias Schimdt

Institute for Structural and Computational Mechanics, University of Hanover, Appelstraße 9A, 30167 Hannover, Germany

e-mails: stein@ibnm.uni-hannover.de

Keywords: Adaptivity, Discretization and Model Error, Hierarchical Modeling, Error Estimation by Equilibration, Large Scale Systems, Elastoplasticity, Parallel Computing

Abstract. Hierarchical discretization and model adaptivity is achieved by locally computed quantitative (absolute) global error estimators with strict upper bounds for the discretization error of the FE–solution with respect to the boundary value problem of the current mathematical model and also for a model within a nested sequence of models from lower to higher complexity for dimensions, kinematics and material behavior in order to account for boundary layers, stress concentrations, stiffness jumps etc. as well as for changes of material behavior (e.g. from elastic to elastoplastic deformations).

1 Introduction

1.1 Features and Challenges

It is a reasonable and necessary requirement for reliable engineering analysis as the basis for dimensioning and safety conditions that the global or local approximation errors (discretization errors) of the numerical method – here of the Finite Element Method (FEM) – as well as the modeling error of the considered mathematical model should be nearly equal and of a prescribed magnitude in all regular and disturbed areas of the domain. For other tasks only specific results might be of interest, controlled by goal-oriented error estimates.

Due to the fact that simplified mathematical models in solid and structural mechanics – e.g. a consistent 2D- or $2\frac{1}{2}$D-shear-elastic bending theory of thin plates and shells instead of 3D theory of elasticity or 3D-theory of elastoplasticity are usually valid only in regular or undisturbed interior subdomains of a mechanical system, model adaptivity has to be coupled with approximation adaptivity. This means that hierarchically expanded mathematical models have to be implemented in conjunction with optional dimensional extensions – e.g. from 2D- to 3D-theory in space – which is necessary in disturbed subdomains like boundary layers, thickness jumps and regions with concentrated loads. Therefore, a sequence of hierarchically reduced, i.e. simplified mathematical models derived from an appropriate master model is needed for regional model and dimensional adaptivity, controlled by the model error in connection with the discretization error. It has to be mentioned that multilevel theories on macro-, meso- or micro-levels of the material structure – with direct discretization on the lowest level and homogenization to macrolevels – is applied in research areas so far. For the addressed model adaptivity the problem of identification of material parameters is crucial. In total, only verifications including error analysis but not yet validations of simplified models w.r.t. a reference (master) model are practically possible at present.

The finite element test- and trial- spaces should be flexible for adequate directional h- and p- refinements, controlled by proper and effective global or goal-oriented a posteriori error estimators. They should be free of constants, i.e., not only relative, which needs implicit postprocessing on the level of element patches or single elements. For the quantitative approximation of the residual discretization error, improved element boundary traction, derived from equilibrated residua by local Neumann problems, and enhanced test functions are used. A certain class of error estimators like residual, hierarchical and averaging type, can be presented in different norms which either yield good approximation properties, e.g. without bounds, or bad approximation and effectivity behavior, but sharp bounds. Residual error estimators for elliptic problems in general have upper bound properties for a posteriori error estimators whereas hierarchical and averaging type estimators only provide upper bounds for the primal finite element method under special conditions. Both discretization adaptivity and model adaptivity, often combined with dimensional adaptivity, are coupled and need an overall concept for error analysis and practical strategies for adaptivity. These subjects are treated in detail in this section of the book. In Figure 1 a comprehended scheme of integrated approximation and model adaptivity is shown.

A posteriori error estimates are also derived for geometrically linear elastoplastic deformation processes under stability requirements for the system and the material. An operator split is usually used for elastoplastic deformations, applying Ritz-Galerkin method in space and finite difference method in time. A consistent error analysis with coupling effects in space and time (with the implicit time as the load variable) is not possible by this procedure, and especially the accumulation error in time can not be estimated in a proper way. This would be possible by a space-time Ritz-Galerkin method, for viscoelasticity see [15, 22]. In this section a so-called pilgrim step method is used by one or two backstepping in time depending on the spatial error and the indicated local time error.

A further topic of this section is the node-regular refinement of hexahedral elements for 3D-problems. For face-oriented element refinements one gets in the first step 3% of degenerated elements inside a hexahedron in the form of trihedrons or heptahedrons which vanish in the second step of each mesh refinement. Also transition elements from 2D shell- to 3D body- meshes are presented. Another strategy of mesh refinements for tetrahedrons is the use of irregular nodes but regular elements by introducing and eliminating so-called hanging nodes. This technique can be extended from 2D- to 3D-meshing without basic problems and is widely used. It should be mentioned that the use and refinement of tetrahedral elements doesn't provide the above described problems but they are less suitable for thin-walled structures like shells.

The solution of large-dimensional multi-level 3D-problems emphasizes parallel treatment, even the break-even-point with serial processes becomes larger and larger for continuously improving hardware technologies. This needs problem- dependent effective iterative solvers with problem dependent preconditioner usually using multi- level solution techniques, and desirably hp-adaptive discretization. Those complex methods can only be realized by an object oriented language like C++ with strict logical separation of topology, geometry, physics, numerical methods and solvers. This was achieved by the Parallel Finite Element Program ParaFEP.

1.2 Methodology and references

A very important goal is a robust, reliable and efficient global or sectional error estimator for the approximated solution of a given setting of mathematical models. A further crucial problem is the appropriate choice of a hierarchically nested sequence of models. For this purpose, a posteriori computed absolute – i.e. constant-free – residual error estimators are presented using isotropic or directionally anisotropic enhanced hierarchical test functions and equilibrated, i.e. improved C^0-continuous element boundary tractions w.r.t. normal directions of the elements; these tractions can be computed elementwise due to the C^0-continuity and the $L_2(\partial\Omega)$ – more precisely the $H^{-1/2}(\partial\Omega)$ – property and by a regularization of the overdetermined equilibrium conditions on element level. The local part of the global approximation error estimator is gained by the difference of the improved FE-solution on element level and the related FE-solution, see e.g. [25, 26, 8, 1, 2, 16, 39, 41, 42, 43, 14]. Absolute anisotropic error estimators (by directionally higher test functions) can be computed, and such anisotropic h- and p- adaptivity can be applied. This possibly leads to an overall hp-adaptive strategy.

We have to distinguish between global an goal-oriented – e.g. functional or local discretization – errors. The latter type includes solutions of dual problems for the far field influence , e.g. for a functional see [34, 7, 30].

The strategy of residual error estimation by equilibration of element boundary tractions can also be realized for local estimators of displacement- and stress-errors by solving a dual problem in the same fashion, [29, 30]. An adaptive FEM-program was developed and applied to 2D-modeled elastic plate and shell problems with strong 3D-effects as well as for material nonlinearities including dimensional and model adaptivity.

First, the concepts of a posteriori error analysis – especially error estimators with bounds – are compared, namely residual type global and local estimators and advanced averaging type estimators for elastic and elastoplastic deformations. Instead of the relative residual error estimator, local equilibration of the residua using improved boundary tractions yields approximated errors. By this, the upper bound is improved and such yields sharper error bounds. Also directional error estimators can be derived by testing with higher test function in the directions of the nodal kinematical freedoms which are more efficient for thin-walled structures than isotropic ones, namely using different p-orders in normal and tangential directions of a plate or shell.

The model error in a sequence of hierarchical mathematical models and related dimensions can be as important as the discretization error, especially in boundary layers of thin-walled structures, at thickness jumps and at load concentrations. Therefore, integrated error controlled adaptivity of finite- element approximations and of models in relevant subdomains is a challenging task for complex engineering structures and is presented in this section. This yields effective discretization and model errors and related adaptive techniques. The regional change of the current mathematical model is activated if the model error is larger than the discretization error [29, 41, 42, 44]. The discretization error itself has to be sufficiently small according to the a priori and a posteriori error estimators.

Adaptive mesh refinements and solution algorithms are also shown for 3D-problems using hexahedrons with regular-node refinements where degenerated elements, so- called trihedrons, appear inside the refined elements and vanish in the next step. Their implementation is realized both for serial computers with an AMS-FEP Adaptive Model and Solution Finite Element Program (AMS-FEP) and for MIMD- parallel computers by the Parallel Adaptive Finite Element Program (PARAFEP) in C++ [28]. Some illustrative examples show the numerical and physical features of the methodology.

1.3 Survey

This section of the CISM-lectures treats four main topics in an engineering- orientated fashion, namely: (i) Concepts and representation of relative and absolute global and local a posteriori error estimators with applications to adaptive finite element approximations in linear elasticity, using residual type and averaging type error estimators; chapters 2 to 6, (ii) Integrated solution- and model-adaptivity by model expansion within an hierarchical sequence of models, based on equilibrated residua and extended test spaces and applied to linear elastic structural problems; chapters 6 and 7, (iii) Residual and averaging error estimates and adaptivity in linear elastoplasticity, using an operator split in space and time and including some benchmark results; chapters 8 to 10, (iv) Node-regular refinement techniques for hexaled and 2D-3D transition elements, implementing hpd-adaptivity and additionally hierarchically graded meshes; chapters 11 and 12.

2 Reliability aspects for adaptive modeling and finite element discretization

An important task for engineers is the reliable prediction of local and global mechanical behavior of structures under various loading conditions, including damage and failure, according to national or supranational norms like EUROCODES for Europe. This needs adequately simplified physical structures and systems and related mathematical models, based on experimental evidence, proper homogenizations and the primal laws of mechanics. Engineering models result from physical idealizations and related mathematical homogenizations on different scales of consideration (from nano- to macro-scales where the macroscale is the C1-manifold of continuum mechanics), following the rule "as simple as possible and as complicated as necessary". For a certain class of problem we define a mathematical master model, e.g 3D-elastoplasticity, from which a sequence of more simplified or reduced models can be introduced stepwise in an hierarchical or parallel order, determined by further kinematical and/or physical hypotheses (e.g. the 2D-elastic Reissner-Mindlin plate theory), see the scheme in Figure 1, and [29, 41, 42, 44, 45]. The analytical solutions of such a nested sequence of models should provide consistent transitions by asymptotic analysis, e.g. concerning the types of operators and singularities, depending from kinematics, constitutive equations, topology, geometry, boundary conditions and loading. The global or local discretization error of the approximated solution of a specific model under considerations can be estimated by locally computed and summed **error estimators** (defined by ensuring an upper and/or a lower **error bound**), [21], or by **error indicators** (without bounds) for the choice of further h- or p- refinements which can provide sharper estimates i.e. better effectivity of the adaptive process. As pointed out in chapter 1, relative a posteriori estimators can be computed locally without postprocessing, i.e. explicitly whereas absolute (constant-free) estimators need postprocessing on element or patch levels using higher order test functions with respect to the present FE-approximation in order to compute approximated errors, according to the Galerkin-orthogonality of the error w.r.t. to the present variational approximation.

A prescribed global or local accuracy of an approximated solution and its derivatives, usually of the displacements, principal stresses or of stress resultants, should be achieved at minimum cost by quasi-optimal error estimators and related hp- refinements which is still a matter of research. However, we do want convergence not only with respect to a specific simplified model but also to the hierarchically expanded more complex model towards a reasonable master model, e.g for elastoplasticity with hardening, such that **model-adaptivity** (including dimension-adaptivity of the test- and trial- spaces) has to be applied in certain subdomains, **integrated with solution-adaptivity** for each related model. It's obvious that a reduction of the discretization error below the admitted error tolerance of the hierarchically expanded master model is meaningless [47, 46]. As will be shown later in chapter 6.2.1 for an elastic clamped plate in bending, 3D-elasticity as an updated model becomes the more predominant in the total domain the smaller an absolute global approximation error tolerance of a wanted quantity is prescribed, starting from adequate 2D- or $2\frac{1}{2}$D-elastic theories for thin plates and shells which are first consistent approximations of the related 3D-theory of elasticity. The model error is relatively small in undisturbed subdomains but it is significant in boundary layers of elliptic problems. There, St. Venant's principle of decaying local disturbances holds under some further restrictions such that neighbored subdomains with different elliptic operators are admitted without progressing disturbances.

Therefore, the goal of hierarchical model and solution error analysis is the automatic detection of subdomains in which, dependent from the admitted global or local error tolerances of interesting physical quantities, the mathematical model is updated. Model adaptivity is realized here by an **expansion strategy**, progressing from simple to more complicated mathematical models which is more efficient than using a **reduction strategy**, starting from the master model and detecting areas in which certain kinematical and physical hypotheses are valid, Figure 1, [29, 45]. This also corresponds to usual en-

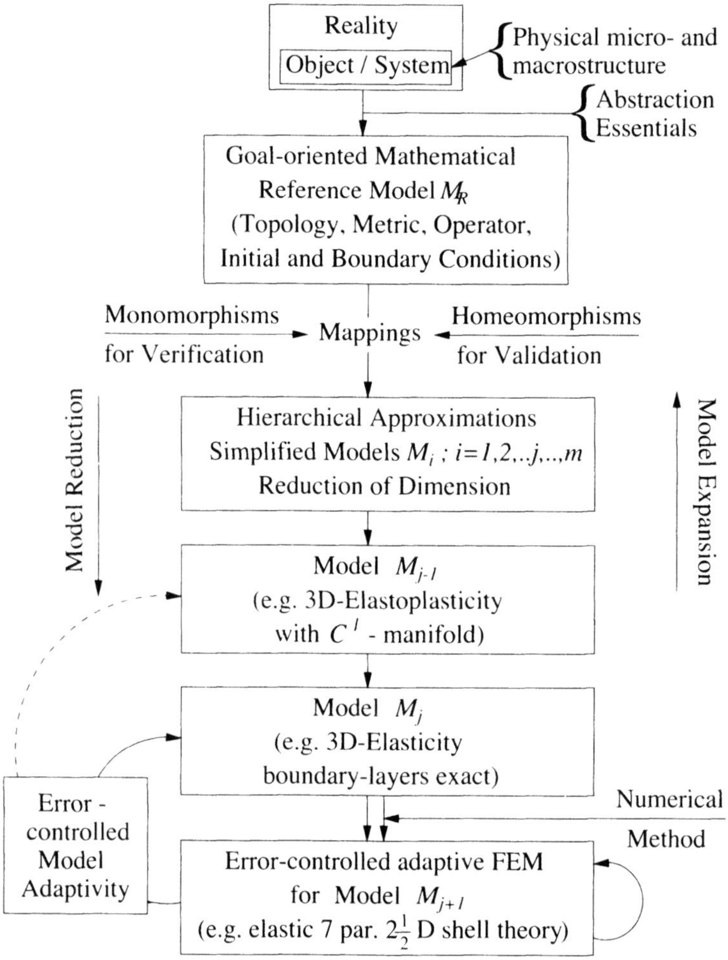

Figure 1: Scheme of the sequence of physical structures, mathematical models and numerical methods

gineering working as remarked at the beginning, namely to proceed from small to higher complexity if necessary.

The expansion strategy from 2D- to 3D-theories needs the implementation of 3D-tensor products for trial- and test- functions of finite elements also for 2D- or 1D-mathematical models (as $2\frac{1}{2}$D or $1\frac{1}{2}$D- versions) in order to admit dimensional adaptivity of the testspaces without locking effects and such without transition elements. As widely known 2D-finite shell elements, e.g. of the **Reissner-Mindlin** type, should be derived from the 3D-continuum theory, especially for composites [21], and implemented at least in a $2\frac{1}{2}$D-version, including constant, linear or higher hierarchical trial functions in thickness direction. Those elements are anisotropic and can be directionally p- adapted in all three directions according to anisotropic error estimators. Moreover, $2\frac{1}{2}$D-test spaces are necessary to avoid locking phenomena for coupling with 3D-elements, derived from 3D-operators. Strong mechanical effects in thickness direction – e.g. in composite shells – can be described by an h-version, using more numerical elements, or in a p-version by increasing the order of the hierarchical polynomials in normal direction of the shell, controlled by corresponding error estimators.

3 Concepts for relative and absolute a posteriori error estimation of FEM

We have to distinguish between **upper bound estimators** for primal variational functional and **lower bound estimators**, derived from interpolation properties or from dual variational principles. The next distinction has to be made between locally computed relative (explicitly computed) **global error estimators** on one hand, especially residual error estimators of the **Babuška- Rheinboldt** type in the energy norm [5, 6] and improved (implicit) residual estimators derived from C^0- continuous of element boundary tractions on element level computed with local **Neumann problems** as well as **goal-oriented error estimators** for specific local quantities or functions or functionals of them (e.g. the Rice-integral for crack propagation) on the other hand, Babuška& Strouboulis [7], Johnson & Hansbo [23], Rannacher & Suttmeier [38], Ramm & Cirak [34], Ohnimus, Stein & Walhorn [30], Stein & Rüter [44].

Goal oriented estimators consist of the usual local part of the global estimator described above and the pollution error which catches the influence of the system onto the local error. This pollution part can be computed numerically by analysing Green's influence function for a finite local support. Local errors are designed to get optimal refinements at minimal cost for the specific local or goal-oriented quantity of interest, without regarding the size of error elsewhere, whereas global errors yield adaptive meshes in a way that the chosen norm of the considered global error reaches a prescribed value in the whole system.

An important distinction concerns **relative and absolute error estimators**. The interpolation constants for global upper and lower residual error bounds of elliptic problems can be computed with some effort – but this is usually not done, and they may differ by a factor of about 100 in case of non-robust problems, e.g. with large ellipticity constants – whereas absolute error estimators provide approximated quantitative errors which can be computed implicitly by local equilibration of residua using local Neumann or Dirichlet problems. In both cases, enhanced test functions of the Ansatz space (at least two) are used, and an energy-equivalent projection of the errors onto the unit cube can be applied in order to reduce the computational effort considerably [29, 42]. The local Neumann problems yield upper error bounds whereas the local Dirichlet problems, performed for each element with a surrounding row of elements fixed at the outer edge, provide lower bounds.

3.1 Residual error estimation with equilibrated boundary tractions

We developed the error analysis by postprocessing on element level with equilibrated residua where the regularization of improved boundary tractions is different from such by Ladevèze et al [25, 26], Bank & Weiser [8] and Ainsworth & Oden [1]. There are also differences amongst these authors in getting improved local solutions by using various regularizations and – in our case – directionally higher test functions, such yielding error norms of different quality. We used the name "Posterior Equilibrium Method" (PEM) for this implicit error analysis and the adaptive algorithm in contrast to the Babuška- Rheinboldt adaptive FEM with explicit relative error estimators which we called "Residual Error Method" (REM). It should be mentioned that also goal-oriented error estimates by adding a dual solution can be gained by PEM, see [30].

An important idea for getting approximated absolute error estimators is the usage of locally expanded test spaces, (virtual deformations modes), in order to estimate the local error $\boldsymbol{e}(\boldsymbol{x}) = \boldsymbol{u}(\boldsymbol{x}) - \boldsymbol{u}_h(\boldsymbol{x})$ of the finite element solution in the energy norm, with $\boldsymbol{u} \in \boldsymbol{V}, \boldsymbol{u}_h \in \boldsymbol{V}_h$. We are looking for an improved solution $\boldsymbol{u}_{h+} \in \boldsymbol{V}_{h+}$ with the error $\boldsymbol{e}_{h+} = \boldsymbol{e}_h + \boldsymbol{e}_+$, $\boldsymbol{u}_+ \in \boldsymbol{V}_+$ and $\boldsymbol{V}_h \cup \boldsymbol{V}_+ = \boldsymbol{V}_{h+} \subset \boldsymbol{V}$ with the extended solution space $\boldsymbol{V}_{h+} = \boldsymbol{V} \ominus \boldsymbol{V}_h$, see Figure 2.

The error $\boldsymbol{e}_h$ in the energy norm, $||\boldsymbol{e}_h||_{E(\Omega)}$, then can be estimated by a local variational problem with extended trial- and test- spaces $\boldsymbol{V}_{h+} \subset \boldsymbol{V}$, in case of orthogonal ansatz-spaces usually with two higher p-orders or corresponding h- refinements (a symmetric and an antisymmetric one). In engineering terms, the residua of the strong equilibrium conditions in element domains and the jumps of tractions at element interfaces are weighted with higher test functions in order to get the approximated error.

4 Variational formulation of elasticity and primal finite element method

4.1 Theory of elasticity and primal finite element method

The primal elastic variational formulation needs, as the input, the kinematic field conditions and the geometric boundary conditions defined for the classical point continuum, a C^1–differentiable manifold. The geometrically linear case is treated here.

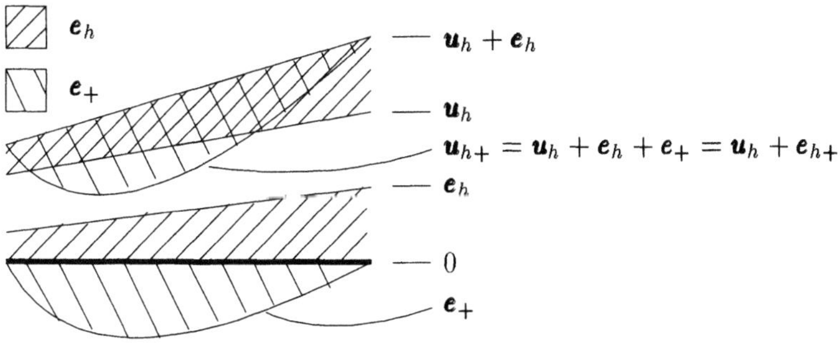

Figure 2: Displacement error $\boldsymbol{e}_h$ of current solution $\boldsymbol{u}_h$ and improved solution $\boldsymbol{u}_{h+}$ with error $\boldsymbol{e}_+$ as part of displacement error $\boldsymbol{e}_{h+}$

With the displacement vector $\boldsymbol{u}(\boldsymbol{x}) = u_i(\boldsymbol{x})\boldsymbol{e}_i$, $\boldsymbol{x} \in \Omega \subset \boldsymbol{R}^3$, the geometrical strains are defined by the symmetric tensor

$$\boldsymbol{\varepsilon}^{\text{geom}}(\boldsymbol{u}) := \boldsymbol{\nabla}_s \boldsymbol{u} = \frac{1}{2}(u_{i,j} + u_{j,i})\boldsymbol{e}_i \otimes \boldsymbol{e}_j, \quad \boldsymbol{e}_i \cdot \boldsymbol{e}_j = \delta_{ij} \ , \tag{1}$$

with 6 independent components and the engineering shear strains $\gamma_{ij} = 2\varepsilon_{ij}$ for $i \neq j$. In matrix notation this reads $\underline{\varepsilon} = (\varepsilon_{11}, \varepsilon_{22}, \varepsilon_{33}, \gamma_{12}, \gamma_{13}, \gamma_{23})^T$. The kinematic field conditions are

$$\underline{\varepsilon}^{\text{geom}} = \underline{\varepsilon}^{\text{phys}}, \tag{2}$$

where $\underline{\varepsilon}^{\text{phys}}$ are the physical strains in the constitutive equations, given by Eq. 5. For mixed BVP's the Dirichlet – and Neumann – parts of the boundary $\partial\Omega = \Gamma$ are given as

$$\boldsymbol{u} = \bar{\boldsymbol{u}} = \boldsymbol{0} \text{ on } \Gamma_u \text{ and } \boldsymbol{\sigma} \cdot \boldsymbol{n} = \bar{\boldsymbol{t}} \text{ on } \Gamma_t \ ; \ \Gamma_u \cup \Gamma_t = \Gamma = \partial\Omega \ ; \ \Gamma_u \cap \Gamma_t = \emptyset, \tag{3}$$

with $\boldsymbol{\sigma}$ the symmetric stress tensor and $\boldsymbol{n}$ the outward normal vector on Γ_t. The strong form of the local equilibrium is given by the equilibrium conditions

$$\text{div}\boldsymbol{\sigma}(\boldsymbol{u}(\boldsymbol{x})) + \boldsymbol{b} = \boldsymbol{0} \text{ and } \boldsymbol{\sigma} = \boldsymbol{\sigma}^T \ \forall \, \boldsymbol{x} \in \Omega \subset \boldsymbol{R}^3 \ , \tag{4}$$

with body force $\boldsymbol{b} \in [L_2(\Omega)]^3$. The matrix notation of $\boldsymbol{\sigma}$ is $\underline{\sigma} = (\sigma_{11}, \sigma_{22}, \sigma_{33}, \sigma_{12}, \sigma_{13}, \sigma_{23})^T$. The constitutive equations for isotropic 3D–elasticity are given by

$$\underline{\sigma} = \underline{\underline{C}}_{3\text{D}} \underline{\varepsilon}^{\text{phys}} \ , \tag{5}$$

with the elasticity tensor $\underline{\underline{C}}_{3\text{D}}$ in matrix notation

$$\underline{\underline{C}}_{3\text{D}} := \frac{E}{(1-2\nu)(1+\nu)} \begin{pmatrix} 1-\nu & \nu & \nu & 0 & 0 & 0 \\ \nu & 1-\nu & \nu & 0 & 0 & 0 \\ \nu & \nu & 1-\nu & 0 & 0 & 0 \\ 0 & 0 & 0 & (1-2\nu)/2 & 0 & 0 \\ 0 & 0 & 0 & 0 & (1-2\nu)/2 & 0 \\ 0 & 0 & 0 & 0 & 0 & (1-2\nu)/2 \end{pmatrix} \tag{6}$$

with the Lamé–parameters $\underline{\underline{C}}_{3\text{D}} = \underline{\underline{C}}_{3\text{D}}^T$ and $\lambda = \frac{E\nu}{(1+\nu)(1-2\nu)}$, $\mu = \frac{E}{2(1+\nu)}$. The strain energy density is $\mathcal{W}_s = \frac{1}{2}\underline{\varepsilon}T\underline{\underline{C}}_{3\text{D}}\underline{\varepsilon}$ with $\underline{\sigma} = \frac{\partial \mathcal{W}_s}{\partial \underline{\varepsilon}}$ such that $d\mathcal{W}_s$ exists and $\boldsymbol{\sigma}$ has potential properties. By the conditions $0 < \mu < \infty$ and $-1 < \nu < 0.5$, $\underline{\underline{C}}_{3\text{D}}$ and such $\mathcal{W}_s$ is positiv definite, such that the Lamé–operator, Eq. 7, is locally stable.

Introducing Eqs. 1 and 2 into Eq. 5 and this into Eq. 4 yields the symmetric strong elliptic Lamé–equations

$$(\lambda + \mu)\text{grad}(\text{div}\, \boldsymbol{u}(\boldsymbol{x})) + \underbrace{\mu(\text{div}(\text{grad}\boldsymbol{u}(\boldsymbol{x}))}_{\mu\Delta\boldsymbol{u}} + \boldsymbol{b} = \boldsymbol{0}\ \forall \boldsymbol{x} \in \Omega, \tag{7}$$

with the Laplace operator Δ. For sufficiently smooth $\boldsymbol{b} \in [L_2(\Omega)]^3$, $\boldsymbol{t} \in [H^{1/2}(\Gamma_t)]^3$ and smooth boundary conditions the self-adjoint symmetric Lamé–operator in Eq. 7 has unique solutions with the property $\boldsymbol{u}(\boldsymbol{x}) \in [H^2(\Omega)]^3; \boldsymbol{u} = \boldsymbol{0}$ on Γ_u, H^2 being the Hilbert space with the first and second derivatives of $\boldsymbol{u}(\boldsymbol{x})$ being square-integrable. For polygonally bounded domains which fulfill Lipschitz conditions and for loads which fulfill the conditions $\boldsymbol{b} \in [H^{-1}(\Omega)]^3$, $\boldsymbol{t} \in [H^{-1/2}(\Gamma_t)]^3$, the solutions are also unique but in a weaker topologies than $[H^2(\Omega)]$.

Applying this 3D-theory to thin-walled plates and shells by introducing the Reissner Mindlin hypotheses for displacements and stresses over the thickness we get a simplified 2D-theory with approximated constant transverse shear angles of the cross-sections. Admitting furthermore a constant strain in thickness direction according to Poisson's ratio we get in essence a $2\frac{1}{2}$D-model with thickness stretches due to Poisson's ratio and the elasticity matrix

$$\underline{\underline{C}}_{2\text{D}} := \frac{E}{1-\nu^2} \begin{pmatrix} 1 & \nu & & & & \\ \nu & 1 & & & & \\ & & 1-\nu^2 & & & \\ & & & (1-\nu)/2 & & \\ & & & & \frac{5}{6}(1-\nu)/2 & \\ & & & & & \frac{5}{6}(1-\nu)/2 \end{pmatrix}. \tag{8}$$

The extension of the 2D-model in thickness direction is important for avoiding thickness-locking of the primal FEM in case of dimensional expansion within model adaptivity to 3D-elasticity. In this case $2\frac{1}{2}$D-3D-transition elements are not necessary.

For transversally isotropic and laterally orthotropic, homogenized elastic composite material with Young's moduls $E_1^\star = E_1/\beta$ in the main fiber direction and $E_2^\star = E_2/\beta$

normal to this, and furthermore $\beta = (1+\nu)(1-\nu-2\nu^2 E_2/E_1)$ with Poisson's ratio ν as well as the inplane shear modulus G_{12} and the averaged transverse shear modulus $G_{13} = G_{23}$, we get the orthotropic elasticity matrix

$$\underline{\underline{C}}_{3D}^{orth} := \begin{pmatrix} E_1^\star(1-\nu^2) & E_2^\star\nu(1+\nu) & E_2^\star\nu(1+\nu) & 0 & 0 & 0 \\ E_2^\star\nu(1+\nu) & E_2^\star(1-\nu^2\frac{E_2}{E_1}) & E_2^\star\nu(1+\nu\frac{E_2}{E_1}) & 0 & 0 & 0 \\ E_2^\star\nu(1+\nu) & E_2^\star\nu(1+\nu\frac{E_2}{E_1}) & E_2^\star(1-\nu^2\frac{E_2}{E_1}) & 0 & 0 & 0 \\ 0 & 0 & 0 & G_{12} & 0 & 0 \\ 0 & 0 & 0 & 0 & G_{12} & 0 \\ 0 & 0 & 0 & 0 & 0 & G_{23} \end{pmatrix} . \tag{9}$$

The local direction of fibre reinforcement usually deviates form the local orientation of a finite element with the angle φ. Therfore we need a rotation matrix $\underline{\underline{T}}(\varphi)$ which transforms the elasticity matrix into that of the finite element orientation by a congruence transformation, $\underline{\underline{C}}_{3D}^{orth} \rightarrow \underline{\underline{T}}^T \underline{\underline{C}}_{3D}^{orth} \underline{\underline{T}}$ with

$$\underline{\underline{T}}(\varphi) := \begin{pmatrix} cos^2\varphi & sin^2\varphi & 0 & cos\,\varphi\, sin\,\varphi & 0 & 0 \\ sin^2\varphi & cos^2\varphi & 0 & -cos\,\varphi\, sin\,\varphi & 0 & 0 \\ 0 & 0 & 1 & 0 & 0 & 0 \\ -2sin\,\varphi\, cos\,\varphi & 2sin\,\varphi\, cos\,\varphi & 0 & -cos^2\varphi - sin^2\varphi & 0 & 0 \\ 0 & 0 & 0 & 0 & cos\,\varphi & sin\,\varphi \\ 0 & 0 & 0 & 0 & -sin\,\varphi & cos\,\varphi \end{pmatrix} . \tag{10}$$

The primal **variational form** of Eqs. 1 to 7 can be gained by multiplying the local equilibrium conditions, Eq. 4, with geometrically admissible test functions $\boldsymbol{v}$ (or virtual displacements $\delta\boldsymbol{u}$ in the test space $\boldsymbol{V}, \boldsymbol{v} \equiv \delta\boldsymbol{u} \in \boldsymbol{V}$, see Eq. 14, and integrating over the domain Ω which is – in view of FEM – already partitioned into n_e finite element domains $\Omega \approx \bigcup_{n_e} \Omega_e$. Integrating this weak form of equilibrium by parts and applying the divergence theorem, we get the principle of virtual work $\delta\mathcal{W} = \delta\mathcal{A}$, the so called material free weak form of equilibrium: the input is geometry in all subdomains Ω_e, at element interfaces Γ_e and Dirichlet–boundaries Γ_{eu}, and the output is equilibrium in each Ω_e, on element boundaries Γ_e and at Neumann boundaries Γ_{et}. Introducing the elasticity matrix equations Eq. 6 into the work principle – i.e. potential properties of the stresses as well as symmetry and positive definiteness of $\underline{\underline{C}}$ we get the primal variational form of linear elasticity

$$a(\boldsymbol{u}, \boldsymbol{v}) = l(\boldsymbol{v}) \quad \forall \boldsymbol{v} \in \boldsymbol{V} , \tag{11}$$

with the symmetric bilinar form (energy norm)

$$a(\boldsymbol{u}, \boldsymbol{v}) = \sum_e \int_{\Omega_e} \boldsymbol{\varepsilon}(\boldsymbol{u}) : \boldsymbol{C} : \boldsymbol{\varepsilon}(\boldsymbol{v}) \, d\boldsymbol{x} \tag{12}$$

and the linear form for mixed boundary conditions

$$l(\boldsymbol{v}) = \left\{ \sum_e \int_{\Omega_e} \boldsymbol{b} \cdot \boldsymbol{v} \, d\boldsymbol{x} + \int_{\Gamma_t} \bar{\boldsymbol{t}} \cdot \boldsymbol{v} \, d\boldsymbol{s} \right\} . \tag{13}$$

The test space of virtual displacements reads

$$\boldsymbol{V} := \{\boldsymbol{v} | \boldsymbol{v} \in [H^1(\Omega)]^3 \ ; \ \underbrace{\boldsymbol{v}^+ - \boldsymbol{v}^-}_{[\boldsymbol{v}]} = \boldsymbol{0} \ ; \ \text{at } \Gamma_e \ ; \ \boldsymbol{v} = \boldsymbol{0} \text{ at } \Gamma_{eu} \subset \Gamma_u\} \ . \tag{14}$$

Here we see a keypoint of the variational form and the related FEM: although $\boldsymbol{v}_h$ and $\boldsymbol{u}_h$ are in $H^1(\Omega)$ the solution $\boldsymbol{u}$ is in $H^2(\Omega)$ in case of wellposedness of the problem and sufficient regularity. Of course, the physically given loads have to be square-integrable and smooth, i.e. $\boldsymbol{b} \in [L_2(\Omega_e)]^3, \bar{\boldsymbol{t}} \in [H^{1/2}(\Gamma_{et})]^3$.

In the **Finite– Element– Method**, the Ritz-Galerkin method for finite subdomains, we introduce trial functions $\boldsymbol{u}_h \in \boldsymbol{V}_h \subset \boldsymbol{V}$, defined in Ω_e and for symmetry of the discretized problem also $\boldsymbol{v}_h \in \boldsymbol{V}_h \subset \boldsymbol{V}$, from a finite dimensional subspace $\boldsymbol{V}_h$ and such restrict – so to say – the variational calculus in its action. Then we get the approximated variational form

$$a(\boldsymbol{u}_h, \boldsymbol{v}_h) = l(\boldsymbol{v}_h) \quad \forall \boldsymbol{v}_h \in \boldsymbol{V}_h \subset \boldsymbol{V} \ , \tag{15}$$

where also Ω is approximated as $\Omega \approx \bigcup_{n_e} \Omega_e$. The approximated solution $\boldsymbol{u}_h$ is unique and computable from a positive definit symmetric linear system of equations if the static system is well-posed. Introducing the solution error

$$\boldsymbol{e} := \boldsymbol{u} - \boldsymbol{u}_h, \tag{16}$$

inserting it into the discrete test version of Eq. 11, namely $a(\boldsymbol{u}, \boldsymbol{v}_h) = l(\boldsymbol{v}_h)$, and replacing $l(\boldsymbol{v}_h)$ by $a(\boldsymbol{u}_h, \boldsymbol{v}_h)$ from Eq. 14, we get the important Ritz-Galerkin orthogonality condition

$$a(\boldsymbol{e}, \boldsymbol{v}_h) = 0 \quad \forall \boldsymbol{v}_h \in \boldsymbol{V}_h \quad \text{or} \quad a(\boldsymbol{e}, \boldsymbol{w}_h) = 0 \ \forall \boldsymbol{w}_h \in \boldsymbol{V}_h, \tag{17}$$

with $\boldsymbol{w}_h$ as a further equivalent test function. From Eq. 15 it follows with Cea's lemma that the Galerkin method is optimal with respect to the energy norm.

Now we can derive a variational form for computing approximatively the error $\boldsymbol{e}$ by subtracting Eq. 15 from Eq. 11 and using $\boldsymbol{w}_h$, which yields the main equation of error analysis by using the orthogonality Eq. 17

$$a(\boldsymbol{e}, \boldsymbol{v}) = a(\boldsymbol{e}, \boldsymbol{v} - \boldsymbol{w}_h) = l(\boldsymbol{v} - \boldsymbol{w}_h) - a(\boldsymbol{u}_h, \boldsymbol{v} - \boldsymbol{w}_h) \ \forall \boldsymbol{v} \in \boldsymbol{V} \ , \ \forall \boldsymbol{w}_h \in \boldsymbol{V}_h. \tag{18}$$

Only by introducing further test functions $\boldsymbol{w}_h \in \boldsymbol{V} \ominus \boldsymbol{V}_h$ it is possible to get the upper bound estimate for the interpolation constant c_u of the important Babuška- Miller upper and lower estimate, Eq. 20. The error in the energy norm, η_E, see also Eq. 24, can be computed from Eq. 18 as

$$\eta_2^2 = a(\boldsymbol{e}, \boldsymbol{e}) = \sum_e \int_{\Omega_e} \boldsymbol{\varepsilon}(\boldsymbol{e}) : \boldsymbol{C} : \boldsymbol{\varepsilon}(\boldsymbol{e}) \ d\boldsymbol{x} \geq 0. \tag{19}$$

4.2 Residual error estimation and equivalent error representations

For strong elliptic problems, the error $||\boldsymbol{e}|| := a(\boldsymbol{e}, \boldsymbol{e})^{1/2}_{E(\Omega)}$ within the whole domain Ω can be estimated from above and below by projection of the total residual error onto an element patch $\mathcal{P}_j$ (with a central node j) with upper and lower bounds, given by two constants c_l and c_u, being mesh-independent [23], presented in matrix notation

$$c_l^2 \sum_j \left\{ h_j^2 \int_{\mathcal{P}_j} \boldsymbol{R}_h^T \boldsymbol{R}_h \, d\boldsymbol{x} + h_j \int_{\partial \mathcal{P}_j} \boldsymbol{J}_h^T \boldsymbol{J}_h \, d\boldsymbol{s} \right\} \leq a(\boldsymbol{e}, \boldsymbol{e}) \tag{20}$$

$$\leq c_u^2 \sum_j \left\{ h_j^2 \int_{\mathcal{P}_j} \boldsymbol{R}_h^T \boldsymbol{R}_h \, d\boldsymbol{x} + h_j \int_{\partial \mathcal{P}_j} \boldsymbol{J}_h^T \boldsymbol{J}_h \, d\boldsymbol{s} \right\},$$

with $c_l, c_u \in \boldsymbol{R}_+$ and the residuum of the strong form of equilibrium

$$\boldsymbol{R}_h = \operatorname{div}\boldsymbol{\sigma}(\boldsymbol{u}_h) + \boldsymbol{b} \qquad \forall \boldsymbol{x} \in \mathcal{P}_j \subset \Omega \,, \tag{21}$$

as well as the jumps $\boldsymbol{J}_h$ of the tractions $\boldsymbol{t}_h = \boldsymbol{\sigma}(\boldsymbol{u}_h) \cdot \boldsymbol{n}$ at element interfaces Γ_e and at Neumann boundaries Γ_{et} of elements

$$\boldsymbol{J}_h := \left\{ \begin{array}{ll} \frac{1}{2}(\boldsymbol{t}_h^+ + \boldsymbol{t}_h^-) = \frac{1}{2}((\boldsymbol{\sigma n})^+ + (\boldsymbol{\sigma n})^-) & \text{at} \quad \Gamma_e \\ \boldsymbol{t}_h^+ = \boldsymbol{\sigma n} & \text{at} \quad \Gamma_{et} \\ \boldsymbol{0} & \text{at} \quad \Gamma_{eu} \end{array} \right. . \tag{22}$$

The upper bound interpolation constant c_u can be computed due to Johnson and Hansbo [23] as

$$c_u^2 = \max_e^{n_e} \sup_{\boldsymbol{v} \in \boldsymbol{V}} \inf_{\boldsymbol{w}_h \in \boldsymbol{V}_h} \max \left(\frac{\|\boldsymbol{v} - \boldsymbol{w}_h\|_{\Omega_e}^2}{h_e^2 a(\boldsymbol{v}, \boldsymbol{v})_{\Omega_e}}, \frac{\|\boldsymbol{v} - \boldsymbol{w}_h\|_{\partial\Omega_e}^2}{h_e a(\boldsymbol{v}, \boldsymbol{v})_{\Omega_e}} \right). \tag{23}$$

The estimation of this constant c_u is given in section 6 using local substitute problems and higher test spaces $\boldsymbol{v}_{h+} \in \boldsymbol{V}_{h+} \subset \boldsymbol{V}$, see also [41, 42]. The upper bound residual error estimator then reads [5]

$$\eta^2 = \sum_e \eta_e^2 \leq c_u^2 \sum_e \left\{ h_e^2 \int_{\Omega_e} \boldsymbol{R}_h^T \boldsymbol{R}_h \, d\boldsymbol{x} + h_e \int_{\partial\Omega_e} \boldsymbol{J}_h^T \boldsymbol{J}_h \, d\boldsymbol{s} \right\}. \tag{24}$$

More complicated is the estimation of the lower interpolation constant c_l, using approximation theory, especially Friedrichs inequality. For well-posed V- elliptic problems, $c_u \approx c_l \approx 1$ holds, whereas for non-robust, i.e. parameter- dependent problems the relation $c_u \approx 100 c_l$ is possible, e.g. for strongly anisotropic elastic properties.

4.3 Equivalent representations of residual and equilibrated error estimators

The estimated error $\boldsymbol{e} = \boldsymbol{u} - \boldsymbol{u}_h$ for the current approximated solution can be formulated in the following three equivalent forms. Find the error $\boldsymbol{e} \in \boldsymbol{V}$ such that it holds, see Eq. 18,

$$a(\boldsymbol{e},\boldsymbol{v}-\boldsymbol{w}_h)_\Omega = \sum_e \{l(\boldsymbol{v}-\boldsymbol{w}_h)_{\Omega_e} - a(\boldsymbol{u}_h,\boldsymbol{v}-\boldsymbol{w}_h)_{\Omega_e}\} \tag{25}$$

$$= \sum_e \left\{ \int_{\partial\Omega_e} \boldsymbol{J}_h(\boldsymbol{v}-\boldsymbol{w}_h)d\boldsymbol{s} + \int_{\Omega_e} \boldsymbol{R}_h(\boldsymbol{v}-\boldsymbol{w}_h)d\boldsymbol{x} \right\} \tag{26}$$

$$= \sum_e \left\{ \int_{\partial\Omega_e} \overset{\star}{\mathbf{t}}{}_h^T (\boldsymbol{v}-\boldsymbol{w}_h)d\boldsymbol{s} - a(\boldsymbol{u}_h,\boldsymbol{v}-\boldsymbol{w}_h)_{\Omega_e} + \int_{\Omega_e} \boldsymbol{b}^T \boldsymbol{v}_h \, d\boldsymbol{x} \right\} \tag{27}$$

$$\forall \boldsymbol{v} \in \boldsymbol{V}; \qquad \forall \boldsymbol{w}_h \in \boldsymbol{V}_h \ ,$$

with the orthogonality condition $a(\boldsymbol{e},\boldsymbol{w}_h) = 0$; $\forall\, \boldsymbol{w}_h \in \boldsymbol{V}_h$. The residual $\boldsymbol{R}_h$ and the stress jumps $\boldsymbol{J}_h$ are given in Eqs. 21 and 22. The new introduced boundary tractions $\overset{\star}{\mathbf{t}}_h$, Eqs. 25 - 27 , have to be C^0-continuous w.r.t. the normal directions of the interfaces and have to fulfill the equilibrium conditions at + and - element interfaces according to the Cauchy theorem, i.e.

$$\overset{\star}{\mathbf{t}}{}_h^+ - \overset{\star}{\mathbf{t}}{}_h^- = \mathbf{0} \qquad \forall\, \Gamma_e \qquad \text{for } \bar{\boldsymbol{t}}_e = 0\ . \tag{28}$$

These improved tractions are approximately determined on element level by averaging the traction of neighborhood elements or by the equivalent variational problem for each element Ω_e

$$\int_{\partial\Omega_e} \overset{\star}{\mathbf{t}}{}_h^T \ \boldsymbol{v}_h \, d\boldsymbol{s} \ = \ a(\boldsymbol{u}_h,\boldsymbol{v}_h)_{\Omega_e} - \int_{\Omega_e} \boldsymbol{b}^T \boldsymbol{v}_h \, d\boldsymbol{x} \qquad \forall\, \boldsymbol{v}_h \ \in \ \boldsymbol{V}_h(\Omega_e), \tag{29}$$

i.e. by equilibrating the residua, see also Eq. 91, section 6 there, the corresponding error estimator with equilibrated boundary tractions $\overset{\star}{\mathbf{t}}_h$ is presented.

4.4 Error estimation by algebraic defect correction

¿From an engineering point of view the calculation starts with a reduced model in a reduced dimension. After FE–discretization of the displacements we get a system of algebraic equations in matrix notation

$$\boldsymbol{K}_h \hat{\boldsymbol{u}}_h = \hat{\boldsymbol{p}}_h \quad , \quad \boldsymbol{u}_h = \boldsymbol{N}\hat{\boldsymbol{u}}_h \quad , \quad \boldsymbol{u}_h \in \boldsymbol{V}_h \subset \boldsymbol{V}, \tag{30}$$

with energy-equivalent nodal forces $\hat{\boldsymbol{p}}_h$ according to the types and directions of the nodal displacements $\hat{\boldsymbol{u}}_h$.

The coupled discretization, dimensional and model error $\boldsymbol{e}$ of the displacements is defined with respect to an expanded approximation space as well as the enlarged dimension and enhanced constitutive equations as

$$\boldsymbol{e} = \boldsymbol{e}_h + \boldsymbol{e}_+ \quad , \quad \boldsymbol{e} \in \boldsymbol{V} \quad , \quad \boldsymbol{e}_h \in \boldsymbol{V}_h \quad , \quad \boldsymbol{e}_+ \in \boldsymbol{V} \ominus \boldsymbol{V}_h . \tag{31}$$

For the **model adaptivity** we introduce a **correction – stiffness matrix** $\Delta \boldsymbol{K}_h$

$$\tilde{\boldsymbol{K}}_h = \boldsymbol{K}_h + \Delta \boldsymbol{K}_h \qquad \text{with} \qquad ||\Delta \boldsymbol{K}_h||_E < ||\tilde{\boldsymbol{K}}_h||_E , \tag{32}$$

$\tilde{\boldsymbol{K}}_h$ is the enhanced stiffness matrix including the model–correction with respect to the expanded model in subdomains. For hierarchically expanded test–spaces the error estimation yields the algebraic system of equations for nodal values

$$\begin{bmatrix} \boldsymbol{K}_h + \Delta \boldsymbol{K}_h & \boldsymbol{L}_+ \\ \boldsymbol{L}_+^T & \boldsymbol{H}_+ \end{bmatrix} \begin{pmatrix} \hat{\boldsymbol{u}}_h + \hat{\boldsymbol{e}}_h \\ \hat{\boldsymbol{e}}_+ \end{pmatrix} = \begin{pmatrix} \hat{\boldsymbol{p}}_h \\ \hat{\boldsymbol{p}}_+ \end{pmatrix} , \tag{33}$$

where $\boldsymbol{L}_+$ is the coupling matrix between the approximation space $\boldsymbol{V}_h$ and the higher order approximation space $\boldsymbol{V}_+$. Then Eq. 33 yields

$$\begin{bmatrix} \boldsymbol{K}_h + \Delta \boldsymbol{K}_h & \boldsymbol{L}_+ \\ \boldsymbol{L}_+^T & \boldsymbol{H}_+ \end{bmatrix} \begin{pmatrix} \hat{\boldsymbol{e}}_h \\ \hat{\boldsymbol{e}}_+ \end{pmatrix} = \begin{pmatrix} -\Delta \boldsymbol{K}_h \cdot \hat{\boldsymbol{u}}_h \\ \hat{\boldsymbol{p}}_+ - \boldsymbol{L}_+^T \cdot \hat{\boldsymbol{u}}_h \end{pmatrix} . \tag{34}$$

The **model–error** is associated with the residual $-\Delta \boldsymbol{K}_h \hat{\boldsymbol{u}}_h$, whereas the **discretization**– and the **dimensional– errors** are associated with the residual $\hat{\boldsymbol{p}}_+ - \boldsymbol{L}_+^T \hat{\boldsymbol{u}}_h$.

Eq. 34 represents a global problem, and the task is to find effective algorithms in order to use the information of the undisturbed reduced system for the computation of the defect correction due to model e.g. adaptivity in the subdomains. Further information and ideas for this problem are given by Hackbusch [19, 20].

5 Absolute residual error estimates by enhanced test functions on element patches with Dirichlet boundary conditions

5.1 Representation of the method using projections onto unit patches

Our goal is the computation of global error estimators, realized on element patches with Dirichlet boundary conditions. The technical idea is a bijective invariant energy transformation from the unit patch to the real mesh, see Ohnimus [29, 42, 43]. The domain Ω_e of the current element is extended onto the patch $\mathcal{P}_e$ with 8 additional elements for 2D–problems and 26 elements for 3D–problems. The residual $\boldsymbol{R}$, Eqs. 21 and 24, is defined in Ω_e, the traction jumps $\boldsymbol{J}_i$ on $\partial\Omega_e$ and $\boldsymbol{J}_t$ on $\partial\Omega_t$. The error $\boldsymbol{e}$ is defined on $\mathcal{P}_e$ with setting $\boldsymbol{e} = \boldsymbol{0}$ on the outer boundary $\Gamma_{\mathcal{P}_e}$, i.e. for Dirichlet boundary conditions, see figure 3. Now this fixed patch $\mathcal{P}_e$ is mapped onto the parameter space and yields the domain $\mathcal{I}$ with the fixed boundary $\Gamma_{\mathcal{I}}$. The shape functions are expanded from the parametric element $i \subset \mathcal{I}$ to the whole patch $\mathcal{I}$ regarding the fixed boundary. The stiffness matrix

$\tilde{\boldsymbol{K}}$ of this **patch** $\mathcal{I}$ can be easily computed from that of the interior element domain i by multiplication of the coefficients with factors 1,2,4 or 8, see table 1. $\tilde{\boldsymbol{K}}$ is symmetric and positive definite. The patch $\mathcal{P}_e$ can also be chosen without the 4 corner elements.

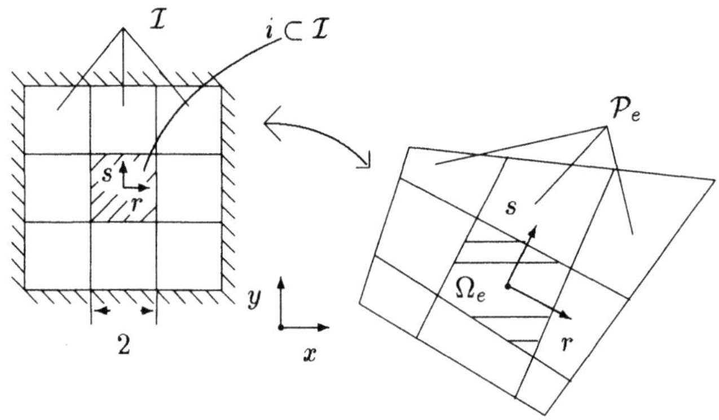

Figure 3: Isoparametric transformation from a patch $\mathcal{P}_e$ surrounding Ω_e into the parametric patch $\mathcal{I}$ with Dirichlet boundary conditions, surrounding the parameter element $i \subset \mathcal{I}$

Another simplified error estimation – hierarchical error estimation – consists in restricting the bilinear form to the higher order approximation space only, i.e. regarding only the 2nd equation of Eq. 33, i.e.. This can be achieved by bubble functions in the domain of the parametric element i and at its boundaries. This modification of the locally transformed stiffness matrix for the element i yields a regular and positive definite stiffness matrix associated with a domain without boundary conditions. The regularity of $\tilde{\boldsymbol{K}}$ results from the fact that the higher order terms of the approximation space don't include rigid body modes. For the bilinear form $a(\boldsymbol{e}, \boldsymbol{v})$ we get with Eq. 17

$$a(\boldsymbol{e}, \boldsymbol{v}) = a(\boldsymbol{e}, \boldsymbol{\pi}_+ \boldsymbol{v}) \approx a(\boldsymbol{\pi}_+ \boldsymbol{e}, \boldsymbol{\pi}_+ \boldsymbol{v}) \qquad \forall \boldsymbol{v} \in \boldsymbol{V}. \tag{35}$$

Table 1: Factors for the coefficients of the local 3D– stiffness matrix for element $i \subset \mathcal{I}$ in the unit patch $\mathcal{I}$

factor	associated coefficient
1	inside domain i
2	on a surface of domain i
4	at an edge of the domain i
8	in a corner of the domain i

Eq. 18, applied to the patch $\mathcal{P}_e$, yields

$$a(\boldsymbol{e}, \boldsymbol{\pi}_+ \boldsymbol{v})_{\mathcal{P}_e} = l(\boldsymbol{\pi}_+ \boldsymbol{v})_{\mathcal{P}_e} - a(\boldsymbol{u}_h, \boldsymbol{\pi}_+ \boldsymbol{v})_{\mathcal{P}_e} \; ; \; \forall \boldsymbol{v} \in \boldsymbol{V}_{\mathcal{P}_e} \tag{36}$$

and

$$\int_{\mathcal{P}_e} \nabla_x(\boldsymbol{e})^T \boldsymbol{C} \nabla_x(\boldsymbol{v}) dV = \int_{\Omega_e} \boldsymbol{R}^T \boldsymbol{\pi}_+ \boldsymbol{v} dV + \int_{\partial\Omega_e} J^T \boldsymbol{\pi}_+ \boldsymbol{v} dO. \tag{37}$$

The bijective transformation of Ω_e onto the corresponding cube element $i \subset \mathcal{I}$ needs the Jacobian matrix $\boldsymbol{\mathcal{J}}$

$$\boldsymbol{\mathcal{J}} := [\mathcal{J}_{ij}] = \left[\frac{\partial x_i}{\partial r_j}\right] = \nabla_r \boldsymbol{x} \tag{38}$$

with the following property for a vector-valued differentiable function $\boldsymbol{f}(\boldsymbol{x})$ with approximately constant Jacobian matrix $\boldsymbol{\mathcal{J}}$ within Ω_e in matrix notation

$$\nabla_x^T \boldsymbol{f}(\boldsymbol{x}) = \nabla_x^T \boldsymbol{\mathcal{J}} \tilde{\boldsymbol{f}}(\boldsymbol{r}) = \nabla_x^T \nabla_r \boldsymbol{x} \tilde{\boldsymbol{f}}(\boldsymbol{r}) = \nabla_r^T \tilde{\boldsymbol{f}}(\boldsymbol{r}) \underbrace{\boldsymbol{I}}_{\nabla_x \boldsymbol{x}} . \tag{39}$$

Therefore, the energy-equivalent transformation from the real to the unit patch only holds with sufficient accuracy if the distortion of the elements is small. The bijective transformations of the error $\boldsymbol{e}$ and the test-function $\boldsymbol{v}$ from $\mathcal{P}_e$ to the energy invariant error $\tilde{\boldsymbol{e}}$ in $\mathcal{I}$ are

$$\boldsymbol{e} = \boldsymbol{\mathcal{J}} \tilde{\boldsymbol{e}} \; ; \; \boldsymbol{v} = \boldsymbol{\mathcal{J}} \tilde{\boldsymbol{v}} \;\; \text{with} \;\; \tilde{\boldsymbol{e}} \, , \, \tilde{\boldsymbol{v}} \in \boldsymbol{V}_{\mathcal{P}_e}. \tag{40}$$

For an upper bound of the error estimation we have to set

$$Max(\det \boldsymbol{\mathcal{J}}_{\Omega_e}) := \max_{\boldsymbol{x} \in \Omega_e} \; (\; \det \boldsymbol{\mathcal{J}}(\boldsymbol{x}) \;). \tag{41}$$

Inserting the mapped fields into Eq. 36 results in

$$\int_{\mathcal{P}_e} \nabla_x(\boldsymbol{\mathcal{J}}\tilde{\boldsymbol{e}})^T \boldsymbol{C} \nabla_x(\boldsymbol{\mathcal{J}}\tilde{\boldsymbol{v}}) dV = \tag{42}$$

$$= \int_{\Omega_e} \boldsymbol{R}^T \boldsymbol{\pi}_+ \boldsymbol{\mathcal{J}} \tilde{\boldsymbol{v}} dV + \int_{\partial\Omega_e} \boldsymbol{J}^T \boldsymbol{\pi}_+ \boldsymbol{\mathcal{J}} \tilde{\boldsymbol{v}} dS$$

and is transformed into

$$Max(\det \boldsymbol{\mathcal{J}}_{\Omega_e}) \cdot \int_{\mathcal{I}} \nabla_r(\tilde{\boldsymbol{e}})^T \boldsymbol{C} \nabla_r(\tilde{\boldsymbol{v}}) dV = \tag{43}$$

$$= \int_{\mathcal{I}} \tilde{\boldsymbol{R}}^T \boldsymbol{\pi}_+ \tilde{\boldsymbol{v}} dV + \int_{\partial\mathcal{I}} \tilde{\boldsymbol{J}}^T \boldsymbol{\pi}_+ \tilde{\boldsymbol{v}} dO$$

with the bijective transformation of the residual and traction jumps

$$\tilde{\boldsymbol{R}} = \boldsymbol{R}^T \boldsymbol{\mathcal{J}} \det(\boldsymbol{\mathcal{J}})_{\Omega_e} \quad , \quad \tilde{\boldsymbol{J}} = \boldsymbol{J} \boldsymbol{\mathcal{J}} \det(\boldsymbol{\mathcal{J}})_{\partial\Omega_e} \tag{44}$$

such that an energy-invariant formulation of the weak form for the unit patch $\mathcal{I}$ is achieved by

$$a(\boldsymbol{e}, \boldsymbol{v})_{\mathcal{P}_e} = a(\boldsymbol{e}, \boldsymbol{\pi}_+ \boldsymbol{v})_{\mathcal{P}_e} \leq Max(\det \boldsymbol{\mathcal{J}}_{\Omega_e}) a(\tilde{\boldsymbol{e}}, \boldsymbol{\pi}_+ \tilde{\boldsymbol{v}})_{\mathcal{I}} \leq \tag{45}$$

$$\leq \int_{\mathcal{I}} \tilde{\boldsymbol{R}}^T \boldsymbol{\pi}_+ \tilde{\boldsymbol{v}} dV + \int_{\partial\mathcal{I}} \tilde{\boldsymbol{J}}^T \boldsymbol{\pi}_+ \tilde{\boldsymbol{v}} dO,$$

$$\forall \, \tilde{\boldsymbol{v}} \in \boldsymbol{V} \quad \text{and} \quad \tilde{\boldsymbol{e}} \in \boldsymbol{V}.$$

It is evident that the real elements should not deviate too much from the square element, i.e. $\det(\boldsymbol{\mathcal{J}})_{\Omega_e}$ has to be restricted.

5.2 Residual error estimation

¿From the weak form of the unit patch, Eq. 45, the bilinar form is written in matrix notation as

$$a(\tilde{\boldsymbol{e}}, \boldsymbol{\pi}_+ \tilde{\boldsymbol{v}})_{\mathcal{I}} = a(\tilde{\boldsymbol{e}}, \tilde{\boldsymbol{v}})_{\mathcal{I}} = \hat{\tilde{\boldsymbol{e}}}^T \tilde{\boldsymbol{K}} \hat{\tilde{\boldsymbol{v}}}, \tag{46}$$

with the interpolations and the stiffness matrix

$$\tilde{\boldsymbol{e}} = \boldsymbol{N}\hat{\tilde{\boldsymbol{e}}} \quad , \quad \tilde{\boldsymbol{v}} = \boldsymbol{N}\hat{\tilde{\boldsymbol{v}}} \quad \text{and} \quad \tilde{\boldsymbol{K}} = \int_{\mathcal{I}} \tilde{\boldsymbol{B}}^T \boldsymbol{C} \, \tilde{\boldsymbol{B}} \; d\tilde{V}, \tag{47}$$

where $\boldsymbol{N}(\boldsymbol{r})$ is the matrix of shape functions in the coordinates of the parameter space. The first terms of the linear form reads

$$\int_{\mathcal{I}} \tilde{\boldsymbol{R}}^T \boldsymbol{\pi}_+ \tilde{\boldsymbol{v}} d\tilde{V} = \hat{\tilde{\boldsymbol{R}}}^T \tilde{\boldsymbol{L}}_R \hat{\tilde{\boldsymbol{v}}}, \tag{48}$$

$$\text{with } \tilde{\boldsymbol{R}} = \boldsymbol{N}\hat{\tilde{\boldsymbol{R}}} \quad \text{and} \quad \tilde{\boldsymbol{L}}_R = \int_{\mathcal{I}} \boldsymbol{N} \, \boldsymbol{N}_{\boldsymbol{+}} d\tilde{V}, \tag{49}$$

where $\boldsymbol{N}_{\boldsymbol{+}}$ is the matrix of hierarchically expanded shape functions. The transformed stress jumps on $\partial\mathcal{I}$ yield the second term of the linear form

$$\int_{\partial\mathcal{I}} \tilde{\boldsymbol{J}}^T \boldsymbol{\pi}_+ \tilde{\boldsymbol{v}} d\tilde{O} = \hat{\tilde{\boldsymbol{J}}}^T \tilde{\boldsymbol{L}}_J \hat{\tilde{\boldsymbol{v}}}, \tag{50}$$

$$\text{with } \tilde{\boldsymbol{J}} = \boldsymbol{N}\hat{\tilde{\boldsymbol{J}}} \quad \text{and} \quad \tilde{\boldsymbol{L}}_J = \int_{\partial\mathcal{I}} \boldsymbol{N} \, \boldsymbol{N}_{\boldsymbol{+}} d\tilde{S}. \tag{51}$$

Note that letters with a hat denote nodal values at finite element nodes. Thus, inserting equs. 46 - 51 into Eq. 45, we get the main equation of error analysis, computed on a unit patch as

$$Max(\det \boldsymbol{\mathcal{J}}_{\Omega_e}) \hat{\tilde{\boldsymbol{e}}}^T \tilde{\boldsymbol{K}} \hat{\tilde{\boldsymbol{v}}} = \hat{\tilde{\boldsymbol{R}}}^T \tilde{\boldsymbol{L}}_R \hat{\tilde{\boldsymbol{v}}} + \hat{\tilde{\boldsymbol{J}}}^T \tilde{\boldsymbol{L}}_J \hat{\tilde{\boldsymbol{v}}}. \tag{52}$$

The corresponding energy norm of the error η_e is

$$\eta_e^2 = Max(\det \boldsymbol{\mathcal{J}}_{\Omega_e}) \hat{\tilde{\boldsymbol{e}}}^T \tilde{\boldsymbol{K}} \hat{\tilde{\boldsymbol{e}}} = \hat{\tilde{\boldsymbol{R}}}^T \tilde{\boldsymbol{L}}_R \hat{\tilde{\boldsymbol{e}}} + \hat{\tilde{\boldsymbol{J}}}^T \tilde{\boldsymbol{L}}_J \hat{\tilde{\boldsymbol{e}}}. \tag{53}$$

Herewith it is possible to derive two different error–estimates, the first oriented to *Johnson's and Hansbo's* [23] proposal using an eigenvalue problem, and the second one proposed here, using the inversion of $\tilde{\boldsymbol{K}}$.

5.3 Solution of a general eigenvalue problem

At first, two base–vector systems $\hat{\tilde{\boldsymbol{g}}}_{Ri}$ and $\hat{\tilde{\boldsymbol{g}}}_{Ji}$ are defined which are orthonormal with respect to the energy norm on patch $\mathcal{I}$, i.e.

$$\hat{\tilde{\boldsymbol{g}}}_{Ri}^T \tilde{\boldsymbol{K}} \hat{\tilde{\boldsymbol{g}}}_{Rj} = \delta_{ij} \quad , \quad \hat{\tilde{\boldsymbol{g}}}_{Ji}^T \tilde{\boldsymbol{K}} \hat{\tilde{\boldsymbol{g}}}_{Jj} = \delta_{ij}, \tag{54}$$

with the conditions for general eigenvalue problems, see Eqs. 48 and 50,

$$\hat{\tilde{\boldsymbol{g}}}_{Ri}^{T}\tilde{\boldsymbol{L}}_{R}\hat{\tilde{\boldsymbol{g}}}_{Ri} - c_{Ri}^{2}\hat{\tilde{\boldsymbol{g}}}_{Ri}^{T}\tilde{\boldsymbol{K}}\hat{\tilde{\boldsymbol{g}}}_{Ri} = 0 \tag{55}$$

$$\hat{\tilde{\boldsymbol{g}}}_{Ji}^{T}\tilde{\boldsymbol{L}}_{J}\hat{\tilde{\boldsymbol{g}}}_{Ji} - c_{Ri}^{2}\hat{\tilde{\boldsymbol{g}}}_{Ji}^{T}\tilde{\boldsymbol{K}}\hat{\tilde{\boldsymbol{g}}}_{Ji} = 0 \,.$$

The displacement error, Eq. 47, can be represented by these base vectors $\hat{\tilde{\boldsymbol{g}}}_{Ri}$ and $\hat{\tilde{\boldsymbol{g}}}_{Ji}$ as

$$\hat{\tilde{\boldsymbol{e}}} = \sum_{i} \tilde{e}_{Ri}\hat{\tilde{\boldsymbol{g}}}_{Ri} \quad \text{with} \quad \tilde{e}_{Ri} = \hat{\tilde{\boldsymbol{e}}}^{T}\tilde{\boldsymbol{K}}\hat{\tilde{\boldsymbol{g}}}_{Ri} \text{ and} \tag{56}$$

$$\hat{\tilde{\boldsymbol{e}}} = \sum_{i} \tilde{e}_{Ji}\hat{\tilde{\boldsymbol{g}}}_{Ji} \quad \text{with} \quad \tilde{e}_{Ji} = \hat{\tilde{\boldsymbol{e}}}^{T}\tilde{\boldsymbol{K}}\hat{\tilde{\boldsymbol{g}}}_{Ji}, \tag{57}$$

with the coefficients $\tilde{e}_{Ri}$ and $\tilde{e}_{Ji}$. Then we get the bilinear form for the unit patch element $\mathcal{I}$ in the parameter space from Eq. 53

$$Max(\det \boldsymbol{\mathcal{J}}_{\Omega_e})\hat{\tilde{\boldsymbol{e}}}^{T}\tilde{\boldsymbol{K}}\hat{\tilde{\boldsymbol{e}}} = \sum_{ij} \tilde{e}_{Ri}\hat{\tilde{\boldsymbol{g}}}_{Ri}^{T}\tilde{\boldsymbol{K}}\hat{\tilde{\boldsymbol{g}}}_{Rj}\tilde{e}_{Rj} = \tag{58}$$

$$= \sum_{i} \tilde{e}_{Ri}^{2} = \sum_{ij} \tilde{e}_{Ji}\hat{\tilde{\boldsymbol{g}}}_{Ji}^{T}\tilde{\boldsymbol{K}}\hat{\tilde{\boldsymbol{g}}}_{Jj}\tilde{e}_{Jj} = \sum_{i} \tilde{e}_{Ji}^{2}.$$

The error norm η_e in the form of Eq. 53 results in

$$\eta_e^2 = Max(\det \boldsymbol{\mathcal{J}}_{\Omega_e})\hat{\tilde{\boldsymbol{e}}}^{T}\tilde{\boldsymbol{K}}\hat{\tilde{\boldsymbol{e}}} = \tag{59}$$

$$= \sum_{i} \hat{\tilde{\boldsymbol{R}}}^{T}\tilde{\boldsymbol{L}}_{R}\underbrace{\hat{\tilde{\boldsymbol{g}}}_{Ri}\tilde{\boldsymbol{e}}_{Ri}}_{\hat{\tilde{\boldsymbol{e}}}} + \sum_{i} \hat{\tilde{\boldsymbol{J}}}^{T}\tilde{\boldsymbol{L}}_{J}\underbrace{\hat{\tilde{\boldsymbol{g}}}_{Ji}\tilde{\boldsymbol{e}}_{Ji}}_{\hat{\tilde{\boldsymbol{e}}}} \,.$$

With the triangle inequality we get the estimate

$$Max(\det \boldsymbol{\mathcal{J}}_{\Omega_e})\hat{\tilde{\boldsymbol{e}}}^{T}\tilde{\boldsymbol{K}}\hat{\tilde{\boldsymbol{e}}} \le \left\{ \sum_{i}[\hat{\tilde{\boldsymbol{R}}}^{T}\tilde{\boldsymbol{L}}_{R}\hat{\tilde{\boldsymbol{g}}}_{Ri}]^2 \underbrace{\sum_{i}[\tilde{\boldsymbol{e}}_{Ri}]^2}_{a(\tilde{\boldsymbol{e}},\tilde{\boldsymbol{e}})_{\mathcal{I}}} + \right. \tag{60}$$

$$\left. \sum_{i}[\hat{\tilde{\boldsymbol{J}}}^{T}\tilde{\boldsymbol{L}}_{J}\hat{\tilde{\boldsymbol{g}}}_{Ji}]^2 \underbrace{\sum_{i}[\tilde{\boldsymbol{e}}_{Ji}]^2}_{a(\tilde{\boldsymbol{e}},\tilde{\boldsymbol{e}})_{\mathcal{I}}} \right\}^{1/2} ,$$

and finally, the improved error estimator computed in the parameter space reads

$$\eta_e^2 = a(\boldsymbol{e},\boldsymbol{e})_{\mathcal{P}_e} \le Max(\det \boldsymbol{\mathcal{J}}_{\Omega_e})a(\tilde{\boldsymbol{e}},\tilde{\boldsymbol{e}})_{\mathcal{I}} \le \tag{61}$$

$$\left\{\sum_i [\hat{\tilde{R}}^T \tilde{L}_R \hat{\tilde{g}}_{Ri}]^2 \quad + \quad \sum_i [\hat{\tilde{J}}^T \tilde{L}_J \hat{\tilde{g}}_{Ji}]^2\right\} Min(\det \mathcal{J}_{\Omega_e})^{-1}.$$

$\hat{\tilde{R}}$ and $\hat{\tilde{J}}$ are transformed nodal residuals and jumps from the real patch $\mathcal{P}_e$ onto the unit patch $\mathcal{I}$. Note: For an upper bound estimation it is necessary to define the factor $\det \mathcal{J}_\Omega$ as a minimum because of the division of this term in Eq. 61.

$$Min(\det \mathcal{J}_{\Omega_e}) = \min_{\boldsymbol{x} \in \Omega_e} \ (\ \det \mathcal{J}(\boldsymbol{x}) \ ) \ . \tag{62}$$

In [23] also the solution of a general eigenvalue problem is treated without introducing the unit patch mapping.

5.4 Inversion of the stiffness matrix of the unit patch and projections

Starting from Eq. 52, the solution exists for each $\hat{\tilde{v}}$ if

$$Max(\det \mathcal{J}_{\Omega_e}) \tilde{K} \hat{\tilde{e}} = \tilde{L}_R^T \hat{\tilde{R}} + \tilde{L}_J^T \hat{\tilde{J}} \tag{63}$$

holds. $\tilde{K}$ is a positive definite matrix (see table 1) and can be inverted as

$$\hat{\tilde{e}} = Min(\det \mathcal{J}_{\Omega_e}) \tilde{K}^{-1} \left\{ \tilde{L}_R^T \hat{\tilde{R}} + \tilde{L}_J^T \hat{\tilde{J}} \right\}. \tag{64}$$

We set $\det \mathcal{J}_\Omega = Min(\det \mathcal{J}_{\Omega_e})$ in order to get an upper bound. The error estimator η in the energy norm finally results in

$$\eta_e^2 = a(\boldsymbol{e}, \boldsymbol{e})_{\mathcal{P}_e} \leq Max(\det \mathcal{J}_{\Omega_e}) a(\tilde{\boldsymbol{e}}, \tilde{\boldsymbol{e}})_{\mathcal{I}} \leq$$
$$\left\{ \hat{\tilde{R}}^T \tilde{L}_R + \hat{\tilde{J}}^T \tilde{L}_J \right\} \tilde{K}^{-1} \left\{ \tilde{L}_R^T \hat{\tilde{R}} + \tilde{L}_J^T \hat{\tilde{J}} \right\} Min(\det \mathcal{J}_{\Omega_e})^{-1}. \tag{65}$$

The estimators Eq. 61 and Eq. 65 look similar. The computational effort is nearly the same. Note that the calculation on the unit patch has only to be performed once. The eigenvalue – method admits a fast approximated estimation of dominating terms whereas the inversion method yields a strict upper bound estimate. Both methods are effective and need only about 10% of the computing time for assembling the global stiffness matrix. Note that the matrices $\tilde{K}$, $\tilde{L}_R$ etc have to be computed and stored only once.

5.5 Directional error estimator

A classical result of the error analysis for linear elastic problems, see e.g. Babuškaand Miller [4] or Johnson and Hansbo [23], is the following estimation

$$\eta_e^2 = a(\boldsymbol{e}, \boldsymbol{e}) = ||\boldsymbol{e}||^2_{E(\Omega_e)} = ||\boldsymbol{u} - \boldsymbol{u}_h||^2_{E(\Omega_e)} = \tag{66}$$

$$= ||c_R h \boldsymbol{R}||^2_{L_2(\Omega_e)} + ||\frac{1}{2} c_J h^{1/2} \boldsymbol{J}||^2_{L_2(\partial\Omega_e)} ,$$

with the residual part of the interior domain, $\boldsymbol{R} = \boldsymbol{b} + \mathrm{div}\boldsymbol{\sigma}(\boldsymbol{u}_h)$ and the projected traction jumps at the boundaries, $\boldsymbol{J}$. But the L_2-norms in the formulas above have averaging and smoothing properties, especially for elements with large length to width ratios (anisotropic elements). A suitable error estimator starts with the equation

$$\eta_e^2 = a(\boldsymbol{e},\boldsymbol{e}) = a(\boldsymbol{e},\boldsymbol{e}_+) = \int_{\Omega_e} \boldsymbol{R}^T \boldsymbol{e}_+ d\boldsymbol{x} + \frac{1}{2}\int_{\partial\Omega_e} \boldsymbol{J}^T \boldsymbol{e}_+ dS \tag{67}$$

with the unknown updated error $\boldsymbol{e}_+ \in \boldsymbol{V}_+ := \boldsymbol{V} \ominus \boldsymbol{V}_h$ of the displacements. The strategy is now to choose the next higher possible degree of polynomials in the approximation space or a suitable refinement for every direction $z \in r,s,t$ of the parameter coordinates. The approximation space is adjusted to the anisotropic refinement (directions of the anisotropic coordinates r,s,t) with $\boldsymbol{N}_+^z(r,s,t)$ denoting the refined shape functions

$$\boldsymbol{e}_+^z = \boldsymbol{N}_+^z(r,s,t)\hat{\boldsymbol{e}}_+^z \in \boldsymbol{V}_+ \subset \boldsymbol{V} \; ; \; z = r,s,t \quad \text{and} \tag{68}$$

$$\boldsymbol{e}^z = \boldsymbol{N}^z(r,s,t)\hat{\boldsymbol{e}}^z \in \boldsymbol{V} \quad . \tag{69}$$

This yields an approximated directional error, which is a sum of the error components over all elements and directions r,s,t

$$\eta_z^2 = \sum_e \left\{ \int_{\Omega_e} \boldsymbol{R}\cdot\boldsymbol{e}_+^z dV + \frac{1}{2}\int_{\partial\Omega_e} \boldsymbol{J}\cdot\boldsymbol{e}_+^z dS \right\} \quad \text{and} \quad \eta_e^2 = \sum_{z=r,s,t} \eta_z^2 . \tag{70}$$

Such, local contributions to the global directional estimator are available on a unit patch which test the three directions with higher p or h-refinement, e.g., $p_+ = 1;2$ or $h_+ = \frac{h}{2};\frac{h}{4}$. These general results are mapped in an energy–invariant form to the real patches in order to control h– and p– adaptivity.

6 Absolute residual error estimates by equilibrated residua and enhanced test functions for simple elements with Neumann boundary conditions (PEM)

The "Posterior Equilibrium Method" (PEM) for quantitative local computation of directional (anisotropic) global or local error estimates is based on an equivalent problem for direct residual error estimation and yields sharper upper bounds and much better effectivity indices than the explicit Residual Error Estimation Method (REM), especially for problems with locking phenomena. Furthermore, this method admits directional estimators for the discretization error as well as effective estimators for the model– error.

PEM starts with the a posteriori computation of improved equilibrated boundary tractions $\overset{\star}{\boldsymbol{t}}_h$ at the element interfaces by local variational problems on element level and then – in the 2nd step – provides improved element displacements by solving again local variational problems with Neumann boundaries, using the extended test- space $\boldsymbol{V}_{h+}$ from which an appropriate norm of the discretization error with strict upper bound can be defined from the difference of the previous global FE-solution $\boldsymbol{u}_h$ and the improved local displacements $\boldsymbol{u}_{h+}$. The local error term then reads

$$\eta_e^2 = a(\boldsymbol{u}_{h+} - \boldsymbol{u}_h, \boldsymbol{u}_{h+} - \boldsymbol{u}_h) = a(\boldsymbol{e}_{h+}, \boldsymbol{e}_{h+}) \ . \tag{71}$$

Different from [25, 26, 8] and [13, 14, 1] the improved boundary tractions $\overset{\star}{\mathbf{t}}_h$ are not determined by using averaged neighbored tractions but by using the energy- equivalent nodel forces $\widehat{\boldsymbol{p}}_{he}(\boldsymbol{u}_h)$ from the preceding global finite element solution as righthand sides of an overdetermined system of linear algebraic equations on element level which is regularized by a leastsquare condition, see Eq. 92 and [18, 29, 42].

6.1 Absolute error estimates by postprocessing with equilibrated residua

We consider the local Neumann problem with the displacement error $\boldsymbol{e} \in \boldsymbol{V}$ such that we get the equilibration condition from the variational form, Eqs. 27, 29 and 31, for element patch P_k with inner node k or likewise for every element e of the patch, Eq. 29,

$$\underbrace{\int_{\Omega_k} \boldsymbol{\sigma}(\boldsymbol{e})^T \boldsymbol{\varepsilon}(\boldsymbol{v}) \, d\boldsymbol{x}}_{a(\boldsymbol{e}, \boldsymbol{v})_k} = \int_{\Omega_k} \boldsymbol{b}^T \boldsymbol{v} \, d\boldsymbol{x} - \underbrace{\int_{\Omega_k} \boldsymbol{\sigma}(\boldsymbol{u}_h)^T \boldsymbol{\varepsilon}(\boldsymbol{v}) \, d\boldsymbol{x}}_{a(\boldsymbol{u}_h, \boldsymbol{v})_k} + \int_{\partial\Omega_k} \overset{\star}{\mathbf{t}}^T \boldsymbol{v} \, d\boldsymbol{s} \quad \forall \boldsymbol{v} \in \boldsymbol{V}_{h+} \tag{72}$$

with at least two further hierarchical test modes (one symmetric and one antisymmetric). The condition of equilibration, Eq. 72 then results in

$$\int_{\Omega_k} \boldsymbol{b}^T \boldsymbol{v} \, d\boldsymbol{x} - \int_{\Omega_k} \boldsymbol{\sigma}_h^T \boldsymbol{\varepsilon}(\boldsymbol{v}) \, d\boldsymbol{x} + \int_{\partial\Omega_k} \overset{\star}{\mathbf{t}}^T \boldsymbol{v} \, d\boldsymbol{s} = 0 \quad \forall \boldsymbol{v} \in \boldsymbol{Z} \tag{73}$$

where $\boldsymbol{Z} \subset \boldsymbol{V}_h$ is a space with rigid body modes, only. From this the equilibrated boundary tractions result as averages of neighbored elements, see [1, 26, 29, 42, 13, 14] and [18, 44, 45],

$$\overset{\star}{\mathbf{t}} = \frac{1}{2}(\boldsymbol{\sigma}_h^+ + \boldsymbol{\sigma}_h^-)^T \boldsymbol{n} + \boldsymbol{g} \quad , \tag{74}$$

The functions $\boldsymbol{g} \in \{\text{linear polynomials}\}$ are necessary for equilibration, i.e. for the solvability of local Neumann problems. This results in the error estimator η_k

$$\eta_k^2 = \int_{\Omega_k} \boldsymbol{\sigma}(\boldsymbol{e})^T \boldsymbol{\varepsilon}(\boldsymbol{e}) \, d\boldsymbol{x} = a(\boldsymbol{e}, \boldsymbol{e})_{\Omega_k} \ . \tag{75}$$

Remark: The equilibrated tractions are not fully consistent to the original local FE-problem with the property

$$\eta_k^2 = \int_{\Omega_k} \boldsymbol{\sigma}(\boldsymbol{e})^T \boldsymbol{\varepsilon}(\boldsymbol{e}) \, d\boldsymbol{x} \neq 0 \quad \text{if} \quad \boldsymbol{e} \in \boldsymbol{V}_h \ . \tag{76}$$

Therefore directional error estimation is possibly not reliable for averaging, Eq. 74, and we therefore introduce an expanded test space for equilibration, namely $\boldsymbol{Z} \to \boldsymbol{V}_h$, see [29, 42]. This leads to a local Neumann problem fulfilling the Galerkin orthogonality condition exactly in the form

$$\eta_k^2 = \int_{\Omega_k} \boldsymbol{\sigma}(\boldsymbol{e})^T \boldsymbol{\varepsilon}(\boldsymbol{e})\, d\boldsymbol{x} = 0 \quad \text{if} \quad \boldsymbol{e} \in \boldsymbol{V}_h \quad \text{and} \quad \boldsymbol{Z} - \boldsymbol{V}_h \,. \tag{77}$$

It should be mentioned that anisotropic error estimation is more favorable for weakly elliptic problems, having in mind strongly anisotropic elasticity tensors.

In the sequel the equivalence of the error estimator based on equilibrated residua and the Babuška- Miller residual error estimator is shown.

After integration by parts of Eq. 72,

$$\begin{aligned} \overbrace{\int_{\Omega_k} \boldsymbol{\sigma}^T(\boldsymbol{e})\boldsymbol{\varepsilon}(\boldsymbol{v})\, d\boldsymbol{x}}^{a(\boldsymbol{e},\boldsymbol{v})_k} &= \int_{\Omega_k} \boldsymbol{b}^T\, \boldsymbol{v} dx + \overbrace{\int_{\Omega_k} \operatorname{div}(\boldsymbol{\sigma}_h)^T \boldsymbol{v} dx - \int_{\partial\Omega_k} (\boldsymbol{\sigma}_h \boldsymbol{n})^T \boldsymbol{v} dx}^{-a(\boldsymbol{u}_h,\boldsymbol{v})_k} \\ &+ \underbrace{\int_{\partial\Omega_k} [\frac{1}{2}(\boldsymbol{\sigma}_h^+ + \boldsymbol{\sigma}_h^-)\boldsymbol{n} + \boldsymbol{g}]\boldsymbol{v} dx}_{\int_{\partial\Omega_k} \overset{\star}{\mathbf{t}}^T \boldsymbol{v} dx}, \end{aligned} \tag{78}$$

we get

$$\int_{\Omega_k} \boldsymbol{\sigma}(\boldsymbol{e})^T \boldsymbol{\varepsilon}(\boldsymbol{v})\, d\boldsymbol{x} = \int_{\Omega_k} \boldsymbol{R}^T \boldsymbol{v}\, d\boldsymbol{x} + \int_{\partial\Omega_k} \boldsymbol{J}^T \boldsymbol{v}\, d\boldsymbol{s} + \int_{\partial\Omega_k} \boldsymbol{g}^T \boldsymbol{v}\, d\boldsymbol{s} \tag{79}$$

with the domain residuum $\boldsymbol{R} := \operatorname{div}(\boldsymbol{\sigma}_h) + \boldsymbol{f}$ and the traction jumps $\boldsymbol{J}$

$$\boldsymbol{J} := \begin{cases} \bar{\boldsymbol{t}} - \boldsymbol{\sigma}(\boldsymbol{u}_h)\boldsymbol{n} & \text{on} \quad \Gamma_t \\ \{-[\boldsymbol{\sigma}(\boldsymbol{u}_h)]^+ - [\boldsymbol{\sigma}(\boldsymbol{u}_h)]^- \boldsymbol{n}\} & \text{on} \quad \Gamma_i \\ \boldsymbol{0} & \text{on} \quad \Gamma_u \end{cases} . \tag{80}$$

at Neumann boundaries and element interfaces.

The consistency condition

$$\int_{\Gamma_i} \boldsymbol{g}^T \boldsymbol{v}\, d\boldsymbol{s} = \int_{\Gamma_i} (\boldsymbol{g}^+ + \boldsymbol{g}^-)^T \boldsymbol{v}\, d\boldsymbol{s} = 0 \,, \tag{81}$$

is fulfilled such that

$$\sum_k \left[\int_{\Omega_k} \boldsymbol{R}^T \boldsymbol{v}\, d\boldsymbol{x} + \int_{\partial\Omega_k} \boldsymbol{J}^T \boldsymbol{v}\, d\boldsymbol{s} + \int_{\partial\Omega_k} \boldsymbol{g}^T \boldsymbol{v}\, d\boldsymbol{s} \right] = \int_{\Omega} \boldsymbol{R}^T \boldsymbol{v}\, d\boldsymbol{x} + \int_{\partial\Omega} \boldsymbol{J}^T \boldsymbol{v}\, d\boldsymbol{s} \tag{82}$$

holds and yield the Babuška– Miller residual error estimator, Eq. 26.

A condition for the equivalence to the Babuška– Miller error estimator is an upper bound for $\boldsymbol{g}$ w.r.t the L_2– norm of the traction jumps $\boldsymbol{J}$, i.e.

$$\int_{\partial\Omega_k} |\boldsymbol{g}|^2 dx = ||\boldsymbol{g}||^2_{L_2(\partial\Omega_k)} \leq c||\boldsymbol{J}||^2_{L_2(\partial\Omega_k)} \quad \text{with} \quad c \in \boldsymbol{R}_+. \tag{83}$$

For proving this choose $\boldsymbol{g}$ such that

$$||\boldsymbol{g}||_{L_2(\partial\Omega_k)} \to \min \;, \tag{84}$$

which yields the same condition for regularization at each element boundary $\partial\Omega_k$ as

$$\sum_k || \overset{\star}{\mathbf{t}} - \boldsymbol{\sigma}(\boldsymbol{u}_h)\boldsymbol{n}||_{L_2(\partial\Omega_k)} \to \min \tag{85}$$

see also Eqs. (103) and (107). We get for each internal boundary

$$\int_{\partial\Omega_k} \left[\left(\overset{\star}{\mathbf{t}} - \boldsymbol{\sigma}_h^+ \boldsymbol{n} \right)^2 + \left(- \overset{\star}{\mathbf{t}} - \boldsymbol{\sigma}_h^- (-\boldsymbol{n}) \right)^2 \right] d\boldsymbol{s} = \tag{86}$$

$$\int_{\partial\Omega_k} \left[\left(\frac{1}{2}(\boldsymbol{\sigma}_h^+ + \boldsymbol{\sigma}_h^-)\boldsymbol{n} + \boldsymbol{g} - \boldsymbol{\sigma}_h^+ \boldsymbol{n} \right)^2 + \left(-\frac{1}{2}(\boldsymbol{\sigma}_h^+ + \boldsymbol{\sigma}_h^-)\boldsymbol{n} - \boldsymbol{g} - \boldsymbol{\sigma}_h^- (-\boldsymbol{n}) \right)^2 \right] d\boldsymbol{s} = \tag{87}$$

$$= \int_{\partial\Omega_k} \left[\left(\frac{1}{2}(-\boldsymbol{\sigma}_h^+ + \boldsymbol{\sigma}_h^-)\boldsymbol{n} + \boldsymbol{g} \right)^2 + \left(\frac{1}{2}(\boldsymbol{\sigma}_h^+ - \boldsymbol{\sigma}_h^-)\boldsymbol{n} + \boldsymbol{g} \right)^2 \right] d\boldsymbol{s} = \tag{88}$$

$$= \int_{\partial\Omega_k} \left[2 \left(\frac{1}{2}(-\boldsymbol{\sigma}_h^+ + \boldsymbol{\sigma}_h^-)\boldsymbol{n} \right)^2 + 2\,(\boldsymbol{g})^2 \right] d\boldsymbol{s} \to \min \quad \rightsquigarrow \tag{89}$$

$$\left(\int_{\partial\Omega_k} (\boldsymbol{g})^2 \, d\boldsymbol{s} \right)^{1/2} = ||\boldsymbol{g}||_{L_2(\partial\Omega_k)} \to \min \tag{90}$$

This proves that the different derivations of equilibrated boundary tractions by Ladevèze & Leguillon [25], Ainsworth & Oden [1] and Brink & Stein [14] are equivalent to the version of Stein, Ohnimus [42] in the isotropic case.

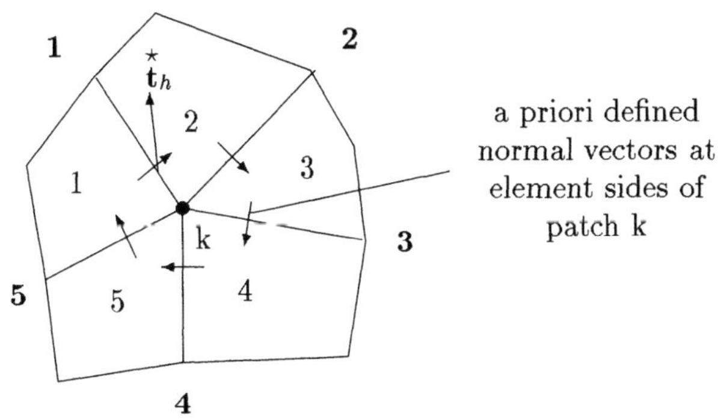

Figure 4: Patch Ω_k for node k; element boundaries are described by side numbers i=1–5. A priori defined normal vectors at element sides i are plotted

6.2 Computation of equilibrated boundary tractions on element level

The improved boundary tractions $\overset{\star}{\mathbf{t}}_h$ are determined at each element surface under the necessary condition of C^0–continuity in normal direction, but not in tangential direction due to the divergence theorem. Such we get the weak equilibrium condition for the local Neumann problem

$$\int_{\partial\Omega_e} \overset{\star}{\mathbf{t}}{}_h^T \; \boldsymbol{v}_h \, d\boldsymbol{s} \;=\; a(\boldsymbol{u}_h, \boldsymbol{v}_h)_{\Omega_e} - \int_{\Omega_e} \boldsymbol{b}^T \; \boldsymbol{v}_h \, d\boldsymbol{x} \qquad \forall\, \boldsymbol{v}_h \in \boldsymbol{V}_h(\Omega_e), \tag{91}$$

and the energy-equivalent nodal forces $\widehat{\boldsymbol{p}}_{he}(\boldsymbol{u}_h)$ from the present finite element analysis – which produce exact equilibrium with the volume loadings $\boldsymbol{b}$ – we get

$$a(\boldsymbol{u}_h, \boldsymbol{v}_h)_{\Omega_e} - \int_{\Omega_e} \boldsymbol{b}^T \; \boldsymbol{v}_h \, d\boldsymbol{x} = \widehat{\boldsymbol{p}}{}_{he}^T \, \widehat{\boldsymbol{v}}_h \;, \tag{92}$$

with nodal values $\widehat{\boldsymbol{v}}_h$ of the test functions.

At first sight a completely coupled system of equations arises from Eqs. 71 and 92 due to the C^0–continuity in normal direction of element surfaces. But due to the fact that the $\overset{\star}{\mathbf{t}}_h$ are not necessarily continuous in tangential direction, they can be computed approximately on patchs – or directly on element level. Matrix notation of Eq. 92 yields

$$\widehat{\boldsymbol{v}}{}_h^T \, \widehat{\boldsymbol{p}}_{he}(\boldsymbol{u}_h) = \widehat{\boldsymbol{v}}{}_h^T \underbrace{\int_{\partial\Omega_e} \boldsymbol{N}_v^T \, \overline{\boldsymbol{N}}_t \, d\boldsymbol{s}}_{\boldsymbol{M}_e = \boldsymbol{I}} \; \widehat{\overset{\star}{\mathbf{t}}}_h \qquad \forall\, \widehat{\boldsymbol{v}}_h \in \widehat{\boldsymbol{V}}_h \, . \tag{93}$$

Herein the shape functions $\mathbf{N}_v$ for the virtual displacements $\boldsymbol{v}_h = \mathbf{N}_v \; \widehat{\boldsymbol{v}}_h \; \in \boldsymbol{V}_h$ are used and not yet defined shape functions $\overline{\boldsymbol{N}}_t$ for the boundary tractions $\overset{\star}{\mathbf{t}}_h = \overline{\boldsymbol{N}}_t \; \widehat{\overset{\star}{\mathbf{t}}}_h \; \in$

$[L_2(\Gamma_e)]^3$. As the orientation of the boundary tractions is dependent on the direction of the normal vector of an element surface, a positive direction has to be defined in advance, figure 4. As the boundary tractions may have tangential jumps along the element edges or surfaces, we can represent matrix $\boldsymbol{M}_e$ as a unit matrix if suitable biorthogonal ansatz–functions $\overline{\boldsymbol{N}}_t$ with respect to $\boldsymbol{N}_v$ are chosen as biorthogonal contravariant base functions $N^p_{\mathcal{L}i}$ with respect to the covariant Lagrangian base functions $\overline{N}^p_{\mathcal{L}j}$, given here for $p=1$ and $p=2$, see [47] for higher p and for integrated Legendre shape functions,

$$p=1 \qquad \begin{array}{ll} N^1_{\mathcal{L}0} = \frac{1}{2}(1-r) & \overline{N}^{10}_{\mathcal{L}} = \frac{1}{2}(1-3r) \\ N^1_{\mathcal{L}1} = \frac{1}{2}(1+r) & \overline{N}^{11}_{\mathcal{L}} = \frac{1}{2}(1+3r)\ , \end{array} \tag{94}$$

$$p=2 \qquad \begin{array}{ll} N^2_{\mathcal{L}0} = -\frac{1}{2}r(1-r) & \overline{N}^{20}_{\mathcal{L}} = \frac{3}{8}(-2-4r+10r^2) \\ N^2_{\mathcal{L}1} = 1-r^2 & \overline{N}^{21}_{\mathcal{L}} = \frac{3}{8}(3-5r^2) \\ N^2_{\mathcal{L}2} = \frac{1}{2}r(1+r) & \overline{N}^{22}_{\mathcal{L}} = \frac{3}{8}(-2+4r+10r^2)\ , \end{array} \tag{95}$$

with

$$\delta^j_i = \int_{-1}^{+1} N^p_{\mathcal{L}_i}(r)\ \overline{N}^{pj}_{\mathcal{L}}(r)\, dr\ . \tag{96}$$

For the 3D-discretization we need 2D-biorthogonal parameterized boundary tractions which follow from the tensor products of the above 1D-covariant base functions. With Lagrangian ansatz function $\boldsymbol{N}_v(r,s)$ we get the covariant shape functions

$$\boldsymbol{N}_v(r,s) \ = \ \boldsymbol{N}_{\mathcal{L}}(r,s) \ = \ \boldsymbol{N}_{\mathcal{L}}(r) \otimes \boldsymbol{N}_{\mathcal{L}}(s)\ , \tag{97}$$

and the related biorthogonal contravariant basis functions $\overline{\boldsymbol{N}}_t(r,s)$

$$\overline{\boldsymbol{N}}_t(r,s) \ = \ \frac{\overline{\boldsymbol{N}}_{\mathcal{L}}(r,s)}{det\left(J(r,s)\right)_{\partial\Omega_e}} \ = \ \frac{\overline{\boldsymbol{N}}_{\mathcal{L}}(r) \otimes \overline{\boldsymbol{N}}_{\mathcal{L}}(s)}{det\left(J(r,s)\right)_{\partial\Omega_e}}\ . \tag{98}$$

After parameterizing the new boundary tractions we can directly find the equilibrium condition for a nodal patch Ω_k, Figure 4, or use Eq. 93, yielding

$$\begin{array}{lll} \text{element } 1: & \widehat{\overset{\star}{t}}_{h1} - \widehat{\overset{\star}{t}}_{h5} & = \widehat{p}_{h1}(\boldsymbol{u}_h) \\ \text{element } 2: & \widehat{\overset{\star}{t}}_{h2} - \widehat{\overset{\star}{t}}_{h1} & = \widehat{p}_{h2}(\boldsymbol{u}_h) \\ \text{element } 3: & \widehat{\overset{\star}{t}}_{h3} - \widehat{\overset{\star}{t}}_{h2} & = \widehat{p}_{h3}(\boldsymbol{u}_h) \\ \text{element } 4: & \widehat{\overset{\star}{t}}_{h4} - \widehat{\overset{\star}{t}}_{h3} & = \widehat{p}_{h4}(\boldsymbol{u}_h) \\ \text{element } 5: & \widehat{\overset{\star}{t}}_{h5} - \widehat{\overset{\star}{t}}_{h4} & = \widehat{p}_{h5}(\boldsymbol{u}_h) \end{array} \tag{99}$$

and find the matrix representation

$$\overbrace{\begin{bmatrix} +1 & & & & -1 \\ -1 & +1 & & & \\ & -1 & +1 & & \\ & & -1 & +1 & \\ & & & -1 & +1 \end{bmatrix}}^{\boldsymbol{T}} \overbrace{\begin{bmatrix} \hat{\overset{\star}{t}}_{h1} \\ \hat{\overset{\star}{t}}_{h2} \\ \hat{\overset{\star}{t}}_{h3} \\ \hat{\overset{\star}{t}}_{h4} \\ \hat{\overset{\star}{t}}_{h5} \end{bmatrix}}^{\hat{\overset{\star}{\mathbf{t}}}_h} = \overbrace{\begin{bmatrix} \hat{p}_{h1}(\boldsymbol{u}_h) \\ \hat{p}_{h2}(\boldsymbol{u}_h) \\ \hat{p}_{h3}(\boldsymbol{u}_h) \\ \hat{p}_{h4}(\boldsymbol{u}_h) \\ \hat{p}_{h5}(\boldsymbol{u}_h) \end{bmatrix}}^{\hat{\boldsymbol{p}}_h(\boldsymbol{u}_h)} . \tag{100}$$

The topology matrix $\boldsymbol{T}$, see Eq. 100, is not regular, such that det $(\boldsymbol{T}^T\ \boldsymbol{T}) = 0$. In 2D–problems there is one zero eigenvalue of $\boldsymbol{T}^T\boldsymbol{T}$, and in 3D–problems there are five zero eigenvalues. $\boldsymbol{T}$ is quasi invertible if the inverse matrix $\tilde{\boldsymbol{T}}^{-1}$ has the same rank deficiency as $\boldsymbol{T}$ itself, i.e. $\boldsymbol{T} \cdot \tilde{\boldsymbol{T}}^{-1} = \tilde{\boldsymbol{I}}$ is a diagonal matrix with zero entries, one for 2D–problems and up to 5 or more for 3D–problems.

To regularize Eq. 100 on element level, two additional conditions are introduced:

The **first additional condition** holds at Neumann boundaries where equilibrium with the equivalent finite element–nodal forces is satisfied explicitly as

$$\hat{\boldsymbol{p}}_h(\bar{\boldsymbol{t}}) = \hat{\overset{\star}{\mathbf{t}}}_h \qquad \text{on} \quad \Gamma_e t \tag{101}$$

with

$$\hat{\boldsymbol{p}}_h(\bar{\boldsymbol{t}}) := \int_{\Gamma_e t} \boldsymbol{N}_v(\boldsymbol{x})\bar{\boldsymbol{t}} d\boldsymbol{s} \,. \tag{102}$$

Herein $\boldsymbol{N}_v(\boldsymbol{x})$ are the shape function of the approximation space.

The **second additional condition** is gained with FE– stresses $\boldsymbol{\sigma}_h(\boldsymbol{u}_h)$ in Ω_e by postulating that the related boundary tractions $\boldsymbol{t}_h(\boldsymbol{u}_h)$ are approximately equal to the improved traction $\overset{\star}{\mathbf{t}}_h$ in a weak sense as

$$\int_{\partial\Omega_e} (\boldsymbol{\sigma}_h(\boldsymbol{u}_h)\ \boldsymbol{n})^T\ \boldsymbol{v}_h d\boldsymbol{s} \simeq \int_{\partial\Omega_e} \overset{\star}{\mathbf{t}}_h^T\ \boldsymbol{v}_h d\boldsymbol{s} \quad \forall\, \boldsymbol{v}_h \in \boldsymbol{V}_h \,, \tag{103}$$

$$\underbrace{\int_{\partial\Omega_e} (\boldsymbol{\sigma}_h(\boldsymbol{u}_h)\ \boldsymbol{n})^T\ \mathbf{N}_v d\boldsymbol{s}\ \hat{\boldsymbol{v}}_h}_{\hat{\boldsymbol{p}}_h(\boldsymbol{\sigma}_h)} \simeq \hat{\overset{\star}{\mathbf{t}}}_h^T \int_{\partial\Omega_e} \frac{\underbrace{\overline{\mathbf{N}}_t^T}_{\overline{\mathbf{N}}_{\mathcal{L}}^T}}{\det\,(J(r,s))_{\partial\Omega_e}} \underbrace{\mathbf{N}_v}_{\mathbf{N}_{\mathcal{L}}} d\boldsymbol{s}\ \hat{\boldsymbol{v}}_h \tag{104}$$

$$\widehat{\boldsymbol{p}}_h(\boldsymbol{\sigma}_h)^T\,\widehat{\boldsymbol{v}}_h \simeq \widehat{\overset{\star}{\mathbf{t}}}{}_h^T \underbrace{\int\limits_{\partial I} \frac{1}{\det\left(J(r,s)\right)_{\partial\Omega_e}} \overline{\mathbf{N}}_{\mathcal{L}}^T\,\mathbf{N}_{\mathcal{L}}\,\det\left(J(r,s)\right)_{\partial\Omega_e} d\boldsymbol{s}}_{\mathbf{I}}\;\widehat{\boldsymbol{v}}_h \tag{105}$$

$$\Rightarrow \quad \widehat{\boldsymbol{p}}_h(\boldsymbol{\sigma}_h) = \widehat{\overset{\star}{\mathbf{t}}}_h\;. \tag{106}$$

Therefore, the following regularization by the least–squares–approximation is used

$$\frac{1}{2}(\widehat{\boldsymbol{p}}_h(\boldsymbol{\sigma}_h) - \widehat{\overset{\star}{\mathbf{t}}}_h)^2 \rightarrow Min \quad . \tag{107}$$

Summarizing the calculation of equilibrated tractions $\overset{\star}{\mathbf{t}}_h$:

$$\begin{array}{lll} \text{Nodal forces} & \boldsymbol{T}\,\widehat{\overset{\star}{\mathbf{t}}}_h - \widehat{\boldsymbol{p}}_h(\boldsymbol{u}_h) = \mathbf{0} & \multirow{2}{*}{\text{exactly fulfilled conditions}} \\ \text{Boundary tractions} & \widehat{\overset{\star}{\mathbf{t}}}_h - \widehat{\boldsymbol{p}}_h(\boldsymbol{t}) = \mathbf{0} & \\ \text{Regularization} & \frac{1}{2}(\widehat{\boldsymbol{p}}_h(\boldsymbol{\sigma}_h) - \widehat{\overset{\star}{\mathbf{t}}}_h)^2 \rightarrow Min & \text{weakly fulfilled conditions} \end{array} \tag{108}$$

6.3 Improved local solutions and approximation of the error using enhanced test functions

Additional variational problems are solved for each element in order to get improved displacements $\boldsymbol{u}_{h+}$ from the new traction $\overset{\star}{\mathbf{t}}_h$, using an extended test space $\boldsymbol{v}_{h+} \in \boldsymbol{V}_{h+} \subset \boldsymbol{V}$ as well as a corresponding trial space $\boldsymbol{u}_{h+} \in \boldsymbol{V}_{h+} \subset \boldsymbol{V}$. The locally improved solution $\boldsymbol{u}_{h+}$ follows as

$$a(\boldsymbol{u}_{h+}, \boldsymbol{v}_{h+})_{\Omega_e} = \int\limits_{\partial\Omega_e} \boldsymbol{v}_{h+}^T\,\overset{\star}{\mathbf{t}}_h\,d\boldsymbol{s} + \int\limits_{\Omega_e} \boldsymbol{b}^T\boldsymbol{v}_{h+}\,d\boldsymbol{x} \quad \forall\,\boldsymbol{v}_{h+} \in \boldsymbol{V}_{h+}\,, \tag{109}$$

where the extended test– and solution– spaces have to be reduced by the rigid body modes. Other proposals [26, 8] fix the edge nodes of each element patch, i.e. use Dirichlet boundaries.

The locally computed **global approximation** error estimator η_{De} follows from the energy norm of the difference $\boldsymbol{u}_{h+} - \boldsymbol{u}_h$ as

$$\eta_{De}^2 = a((\boldsymbol{u}_{h+} - \boldsymbol{u}_h), (\boldsymbol{u}_{h+} - \boldsymbol{u}_h))_{\Omega_e}\;. \tag{110}$$

7 Coupled error-controlled adaptive mechanical modeling and finite element approximation

7.1 Concepts and problems

The concept of model-adaptivity is similar to that for solution adaptivity, chapter 6.3 but uses an hierarchically expanded mathematical model 2, e.g. the expansion from a $2\frac{1}{2}$D-elastic plate theory (model 1) to 3D-elasticity (model 2) which especially concerns the constitutive equations. We want to get the quantitative locally computed approximated error of the model 1 under consideration with respect to the expanded model 2, and, of course, the model error should not be larger than the approximation error. In conjunction with the model expansion we also need dimensional adaptivity of the ansatz-functions in thickness direction, such that we realize an integrated solution– dimension– and model– adaptivity. Special problems arise in case of composites or reinforced concrete plates with anisotropic physical layers which are discretized and adapted by equivalent numerical layers with variable anisotropic hierarchical ansatz– spaces in thickness– and lateral– directions. Material meso– models (regarding fiber lamination or / and geometrical imperfections of fibers) are possible for individual layers followed by proper homogenizations.

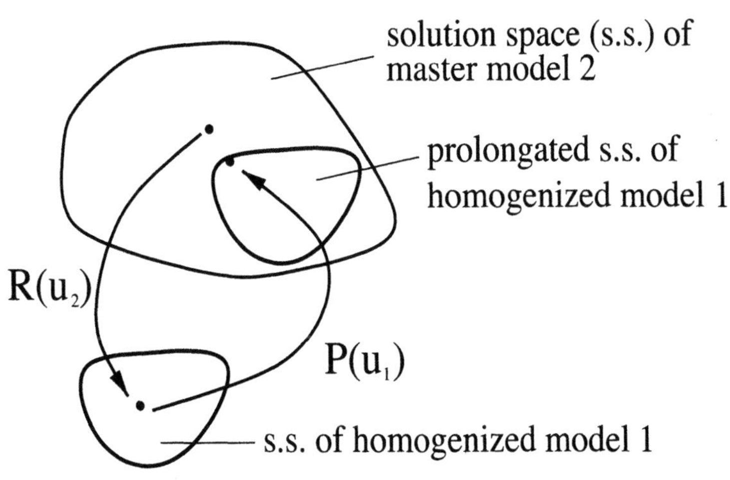

Figure 5: Integrated discretization– dimension– and model– adaptivity by using the restriction operator R and the prolongation operator P

7.2 Model adaptivity based on equilibrated residua and enhanced test functions

We want to compute the improved solution $\boldsymbol{u}_{2h+}$ of the expanded model 2 with the bilinear form

$$a_2(\boldsymbol{u}_{2h+},\boldsymbol{v}_{h+})_{\Omega_e} = \int_{\partial\Omega_e} \boldsymbol{v}_{h+}^T \overset{\star}{\mathbf{t}}_h \, d\boldsymbol{s} + \int_{\Omega_e} \boldsymbol{v}_{h+}^T \boldsymbol{b} \, d\boldsymbol{x} \quad , \quad \forall \, \boldsymbol{v}_{h+} \in \boldsymbol{V}_{2h+} \tag{111}$$

where $\boldsymbol{V}_{2h+}$ is the dimensionally expanded test space and using again the improved equilibrated boundary tractions. It should be remarked that the dimensions of $\boldsymbol{u}_{2h+}$ and 1 $\boldsymbol{u}_{1h+}$ and of the locally extended test spaces have to be equal as well as those of the related strains. A prolongation operator $\mathcal{P}_u$ with the property $\boldsymbol{u}_{1h+} \to \boldsymbol{u}_{2h+} := \mathcal{P}_u \boldsymbol{u}_{1h+}$ has to exist. If the dimensions coincide a priori and the structures of model 1 and 2 are the same, then $\mathcal{P}_u$ is the identity mapping. A further linear mapping (a prolongation $\mathcal{P}_\sigma$) has to be applied for the stresses as $\boldsymbol{\sigma}_{1h+} \to \boldsymbol{\sigma}_{2h+} := \mathcal{P}_\sigma \boldsymbol{\sigma}_{1h+}$ where $\mathcal{P}_\sigma$ maps the dimension of σ_1 into that of σ_2 and contains the constitutiv equation of model 2, Figure 5.

The **local part of the model error estimator** η_{Me} in the energy norm then follows from the difference of improved stresses (by PEM) of the hierarchically expanded model, $\boldsymbol{\sigma}_{2h+}$, and the prolongated stresses $\mathcal{P}_\sigma \boldsymbol{\sigma}_{1h+}$ of model 1 as

$$\eta_{Me}^2 = a_M(\boldsymbol{\sigma}_{2h+} - \mathcal{P}_\sigma \boldsymbol{\sigma}_{1h+}, \boldsymbol{\sigma}_{2h+} - \mathcal{P}_\sigma \boldsymbol{\sigma}_{1h+})_{\Omega_e} \, . \tag{112}$$

If one uses the difference of improved displacements $\boldsymbol{u}_{2h+} - \boldsymbol{u}_{1h+}$ from the expanded model 2 and the reduced model 1, the error estimator usually becomes too large due to deformation constraints of model 2 yielding locking with unrealistic large stresses in boundary layers computed from the displacements of model 1. The mapping $\boldsymbol{u}_{1h+} \to \boldsymbol{u}_{2h+} := \mathcal{P}_u \boldsymbol{u}_{1h+}$ onto the expanded model 2 is not known in general, and the assumption of $\mathcal{P}_u = \boldsymbol{I}$, i.e. the identity mapping, is wrong.

Using the stress for the model error, Eq. 112, the prolongation $\mathcal{P}_\sigma$ of the improved stresses $\boldsymbol{\sigma}_{1h+}$ of model 1 into the solution space of model 2 is identity matrix in case of the 2 1/2 D- Reissner–Mindlin theory as model 1 and the 3D- elasticity theory as model 2, see the related elasticity matrices, Eqs. 8 and 6.

In order to suppress influences on the model error which are only caused by large discretization errors, the model is only adapted if the regional model error (especially in boundary layers) is larger than the related discretization error, i.e. $\eta_D^2 > \eta_M^2$. Furthermore, the model error is only activated after reaching a prescribed tolerance of the approximation error, e.g. $\eta_{D_{rel.}}^2 \leq 5\%$.

We are collecting the following possible model error estimators:

The estimator for the model error expression in the energy norm is

$$\eta_{Mu_e}^2 = ||\boldsymbol{u}_{2h+} - \mathcal{P}_u \boldsymbol{u}_{1h+}||_{E_2(\Omega_e)}^2 . \tag{113}$$

Due to the discussed locking effects (especially in thickness direction), this estimator is not adequate.

Next, we introduce the model error estimator in the energy norm but using the improved stresses of the current and the expanded model directly. Herein, the locking phenomena do not appear because the reduced model (e.g. a thin plate in bending) yields fairly accurate stresses in thickness direction without locking. This error estimator has another type of prolongation operator such that it reads

$$\eta^2_{M\sigma_e} = ||\boldsymbol{\sigma}_{2h+} - \mathcal{P}_{\boldsymbol{\sigma}}\boldsymbol{\sigma}_{1h+}||^2_{E_2(\Omega_e)} \cdot \tag{114}$$

A relative model error in the L_2– norm, taking into account the neglected deformation modes of the reduced model, is given by

$$\zeta^2_{Mu_e} = ||\boldsymbol{u}_{2h+} - \mathcal{P}_u\boldsymbol{u}_{1h+}||^2_{L_2\Omega_e} \tag{115}$$

which is useful for the expansion of 2D–plate and shell models. For a pure displacement method an energy oriented estimator

$$\eta^2_{MEe} = ||\boldsymbol{u}_{2h+}||^2_{E_2(\Omega_e)} - ||\boldsymbol{u}_{1h+}||^2_{E_1(\Omega_e)} \tag{116}$$

can be used, too. But in most investigated cases, this indicator is not sharp because the energies of the two models don't differ very much in case of boundary layers.

The following two simple examples with one trilinear isoparametric cubic element (Q1–element), Figure 6, show that the model error represented in the norm $\eta^2_{M\sigma_e}$ is most effective.

We use a linear elastic, nearly incompressible plane stress state (in the r–s– parameter plane) as the reduced 2D–model and 3D-elasticity as the expanded model.

The first example has the geometry–factor $\frac{\text{length}}{\text{thickness}} = 1$, see figure 6a, and the second one has the ratio $\frac{\text{length}}{\text{thickness}} = 10$, see figure 6b. From an engineering point of view the model error should be smaller in the second example.

Table 2: Model errors, equs. 113–116 for example 1, Figure 6a

Error	$\nu = 0.0$	$\nu = 0.3$	$\nu = 0.499$
η^2_{Mu}	0 %	12.1 %	4160 %
$\eta^2_{M\sigma}$	0 %	0 %	0 %
ζ^2_{Mu}	0 %	8.25 %	19.8 %
η^2_{ME}	0 %	0 %	0 %

As already mentioned, the error estimator η^2_{Mu} is not usefull, see tables 2 and 3. The error estimator $\eta^2_{M\sigma}$ yields reasonable results as well as the error estimator ζ^2_{Mu} in the L_2 norm, and they converge to zero if the element length tends to infinity.

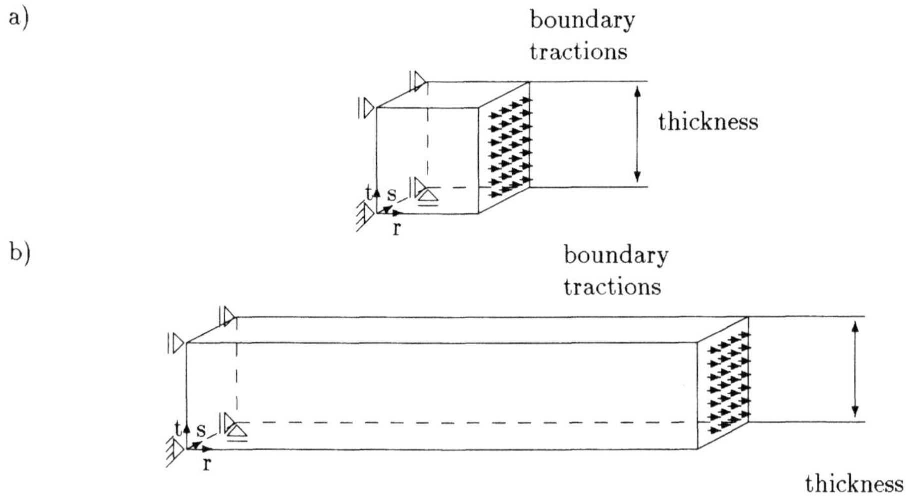

Figure 6: Two simple examples for illustrating the different model error estimators, using a) short element, b) long element

Table 3: Model errors, equs. 113–116 for example 2, Figure 6b

Error	$\nu = 0.0$	$\nu = 0.3$	$\nu = 0.499$
η^2_{Mu}	0 %	12.11 %	4160 %
$\eta^2_{M\sigma}$	0 %	0 %	0 %
ζ^2_{Mu}	0 %	0.09 %	0.25 %
η^2_{ME}	0 %	0 %	0 %

8 Directional error estimator for Prandtl-Reuss elastoplasticity by equilibrated residua and enhanced test functions (PEM)

We consider perfect elastoplasticity (Prandtl-Reuss elastoplasticity) with the convex von Mises yield function at process time $t \in (0, T)$ as the load parameter, see also chapter 10

$$\Phi(\boldsymbol{\sigma}(t)) = |dev\boldsymbol{\sigma}(t)| - \sqrt{\frac{2}{3}}y_0 \leq 0 \tag{117}$$

and Hill's associated flow rule where the 2nd order strainrate tensor $\dot{\boldsymbol{\varepsilon}}^{pl}$ is to be understood as the vector of principal strainrates

$$\dot{\boldsymbol{\varepsilon}}^{pl}(t) = \lambda(t)\ \nabla_\sigma \Phi(t) = \lambda(t)\ \boldsymbol{n}(t)\ , \tag{118}$$

with the out ward positiv normal direction $\boldsymbol{n}(t)$ of the current plastic flow function Φ which is the 2nd order tensor

$$\boldsymbol{n}(t) = \frac{dev\boldsymbol{\sigma}(t)}{|dev\boldsymbol{\sigma}(t)|} . \tag{119}$$

The Kuhn–Tucker conditions of the accompanied variational inequality for loading, unloading and neutral states are

$$\lambda(t) \geq 0 \, , \; \Phi(\boldsymbol{\sigma}(t)) \leq 0 \, , \; \lambda(t)\Phi(\boldsymbol{\sigma}(t)) = 0 \, . \tag{120}$$

To satisfy these conditions we use the additive split of elastic and plastic strain rates with linear kinematic relations

$$\dot{\boldsymbol{\varepsilon}}^{el} :- \boldsymbol{\varepsilon}(\dot{\boldsymbol{u}}(t)) - \dot{\boldsymbol{\varepsilon}}^{pl}(t) = \boldsymbol{\nabla}_s\dot{\boldsymbol{u}}(t) \quad \dot{\boldsymbol{c}}^{pl}(t) \tag{121}$$

with

$$\boldsymbol{\varepsilon}^{pl}(T) = \int_0^T \dot{\boldsymbol{\varepsilon}}^{pl}(\xi)d\xi \, . \tag{122}$$

The volume-specific positiv definit (elastic) strain energy $\Psi(\boldsymbol{\varepsilon}^{el})$ for linear elasticity is given by the quadratic functional

$$\Psi(\boldsymbol{\varepsilon}^{el}) = \frac{1}{2}\boldsymbol{\varepsilon}^{el} : \boldsymbol{C} : \boldsymbol{\varepsilon}^{el} \tag{123}$$

with the symmetric 4th order elasticity tensor $\boldsymbol{C}$, Eq. 6, such that the stresses are the derivatives

$$\boldsymbol{\sigma}(t) = \left.\frac{\partial\Psi(\boldsymbol{\varepsilon}^{el})}{\partial\boldsymbol{\varepsilon}^{el}}\right|_t = \boldsymbol{C} : \boldsymbol{\varepsilon}^{el}(t) = \boldsymbol{C} : (\boldsymbol{\varepsilon}(\boldsymbol{u}(t)) - \boldsymbol{\varepsilon}^{pl}(t)) \, . \tag{124}$$

We assume that the time integration of the flow rule is sufficiently exact (quasi exact). Then we get the Prandtl-Reuss analytical elastoplastic tangent tensor $\boldsymbol{C}^{PR}(t)$ as

$$\dot{\boldsymbol{\sigma}}(t) = \boldsymbol{C} : \dot{\boldsymbol{\varepsilon}}^{el}(t) = \boldsymbol{C} : (\boldsymbol{\varepsilon}(\dot{\boldsymbol{u}}(t)) - \dot{\boldsymbol{\varepsilon}}^{pl}(t)) = \underbrace{(\boldsymbol{C} - 2\mu\boldsymbol{n}(t) \otimes \boldsymbol{n}(t))}_{\boldsymbol{C}^{PR} = \boldsymbol{C}^{PRT}} : \boldsymbol{\varepsilon}(\dot{\boldsymbol{u}}(t)) \, . \tag{125}$$

We solve the given elastoplastic problem approximately by a semidiserete operator split in two steps, first with a spatial finite element discretization $\boldsymbol{u}_h(t)$, yielding the plastic strains $\boldsymbol{\varepsilon}_h^{pl}(t)$, the plastic strain rates $\dot{\boldsymbol{\varepsilon}}_h^{pl}(t)$, the stresses $\boldsymbol{\sigma}_h(t)$, the stress rates $\dot{\boldsymbol{\sigma}}_h(t)$ as well as the flow direction $\boldsymbol{n}_h(t)$ at time t , and secondly by integrating a system of ordinary differential equations from time t to $t + \Delta t$.

Now we consider a two step strategy to get an error estimator for the equilibrium, for the directions of plastic flow and for the plastic strain field. In the first step we assume that the flow direction and the plastic strain field are exact at time t. Then we only have to control the error of the equilibrium

$$div \; \boldsymbol{\sigma}_h(t) + \boldsymbol{b}(t) = \boldsymbol{0} \quad \forall \, t \in (0,T) \tag{126}$$

by the strong residuum, Eq. 21, and we get the weak residuum formulated in rate form for the error $\boldsymbol{e} \in \mathcal{U}$, Eqs. 15, 26, as

$$\int_\Omega \boldsymbol{\sigma}(\dot{\boldsymbol{e}}) : \boldsymbol{\varepsilon}(\boldsymbol{v}) d\boldsymbol{x} = \dot{l}(\boldsymbol{v}) - \int_\Omega \dot{\boldsymbol{\sigma}}_h : \boldsymbol{\varepsilon}(\boldsymbol{v}) d\boldsymbol{x} =: \dot{R}(\boldsymbol{v}) \quad \forall \boldsymbol{v} \in \mathcal{V} \,, \tag{127}$$

with

$$\dot{l}(\boldsymbol{v}) := \int_\Omega \dot{\boldsymbol{b}}(t) \cdot \boldsymbol{v} d\boldsymbol{x} + \int_{\Gamma_t} \dot{\boldsymbol{t}}(t) \cdot \boldsymbol{v} \, d\boldsymbol{s} \,. \tag{128}$$

The test space $\mathcal{V}$ is defined as

$$\mathcal{V} := \left\{ \boldsymbol{v} | \boldsymbol{v} \in (H^1(\Omega); L_2((0,T))) \,, \, \boldsymbol{v}(\boldsymbol{x}, t=0) = \boldsymbol{0} \,, \, \boldsymbol{v}(\boldsymbol{x}, t) = \boldsymbol{0} \text{ on } \Gamma_D \right\} \tag{129}$$

and the solution space $\mathcal{U}$

$$\mathcal{U} := \left\{ \boldsymbol{u} | \boldsymbol{u} \in \mathcal{V} \,, \, \dot{\boldsymbol{u}} \in (H^1(\Omega); L_2((0,T))) \right\} \tag{130}$$

such that we get the important inclusion property

$$\mathcal{U} \subset \mathcal{V} \,. \tag{131}$$

For the semidiscrete approximation method (continuous in time and discrete in space) we introduce the spatial finite test and solution spaces as

$$\mathcal{V}_h := \left\{ \boldsymbol{v}_h | \boldsymbol{v}_h \in (H^1(\Omega); L_2((0,T))) \,, \, \boldsymbol{v}_h(\boldsymbol{x}, t=0) = \boldsymbol{0} \,, \, \boldsymbol{v}_h(\boldsymbol{x}, t) = \boldsymbol{0} \, \text{on} \, \Gamma_D \right\} \tag{132}$$

and

$$\mathcal{U}_h := \left\{ \boldsymbol{u}_h | \boldsymbol{u}_h \in \mathcal{V}_h \,, \, \dot{\boldsymbol{u}}_h \in (H^1(\Omega); L_2((0,T))) \right\} \tag{133}$$

with isoparametric spatial polynomial approximations as shape functions. We get the same inclusion property as for the continuous spaces as

$$\mathcal{U}_h \subset \mathcal{V}_h \,. \tag{134}$$

The residual rates $\dot{\boldsymbol{R}}(\boldsymbol{v})$ have to be equilibrated in the same way as in chapter 6.2, namely

$$\dot{\overset{\star}{\boldsymbol{R}}}(\boldsymbol{v}) := \dot{\boldsymbol{R}}(\boldsymbol{v}) \int_{\Gamma_e} \dot{\overset{\star}{\mathbf{t}}}_h(t) \cdot \boldsymbol{v} \, d\boldsymbol{s} \qquad \forall \boldsymbol{v} \in \mathcal{V}(\Omega_e), \tag{135}$$

with the equilibrium condition

$$\dot{\overset{\star}{\boldsymbol{R}}}(\boldsymbol{v}) := \dot{\boldsymbol{R}}(\boldsymbol{v}) \qquad \forall \boldsymbol{v} \in \mathcal{V}(\Omega) \,. \tag{136}$$

One has to mention that the equilibrated residual $\dot{\overset{\star}{\boldsymbol{R}}}(\boldsymbol{v})$ and the residual $\dot{\boldsymbol{R}}(\boldsymbol{v})$ have to be equal regarding the full test space $\mathcal{V}(\Omega)$, but not regarding locally extended test spaces

$\overset{\star}{\mathcal{V}}(\Omega)$. In order to determine the discrete equilibrated tractions one has to solve the following problem for each element separately

$$\int_{\partial\Omega_e} \dot{\overset{\star}{\mathbf{t}}}_h(t) \cdot \boldsymbol{v}_h \, d\boldsymbol{s} = \int_{\Omega_e} \dot{\boldsymbol{\sigma}}_h(t) : \boldsymbol{\varepsilon}(\boldsymbol{v}) dx - \int_{\Omega_e} \dot{\boldsymbol{b}}(t) \cdot \boldsymbol{v}_h \, d\boldsymbol{x} \qquad \forall\, \boldsymbol{v}_h \in \mathcal{V}_h(\Omega_e)\,, \tag{137}$$

because they have to approximate the correct tractions which solve the local discrete variational problem. The consistency condition, as in Eq. 28, has to be satisfied by the rates of the tractions,

$$\dot{\overset{\star}{\mathbf{t}}}{}_h^{+}(\boldsymbol{x},t) - \dot{\overset{\star}{\mathbf{t}}}{}_h^{-}(\boldsymbol{x},t) = \mathbf{0} \qquad \forall\, \boldsymbol{x} \in \Gamma_i, \forall t \in (0,T)\,. \tag{138}$$

It can be proved that an upper bound error estimator for the current tangential space is gained. To prove this we define the total energy norm $||\boldsymbol{v}||_{\mathcal{E}}$ of the test functions as

$$||\boldsymbol{v}||_{\mathcal{E}}^2 := \int_0^T \int_\Omega \boldsymbol{\varepsilon}(\dot{\boldsymbol{v}}(t)) : \boldsymbol{C} : \boldsymbol{\varepsilon}(\boldsymbol{v}(t)) \, d\boldsymbol{x} \, dt = \int_\Omega \frac{1}{2}\boldsymbol{\varepsilon}(\boldsymbol{v}(T)) : \boldsymbol{C} : \boldsymbol{\varepsilon}(\boldsymbol{v}(T)) \, d\boldsymbol{x} \;\; \forall \boldsymbol{v} \in \mathcal{V}, \tag{139}$$

and we get the error estimator

$$||\boldsymbol{e}||_{\mathcal{E}} := \sup_{\boldsymbol{v}\in \boldsymbol{V}} \frac{\int_0^T \dot{R}(\boldsymbol{v}) dt}{||\boldsymbol{v}||_{\mathcal{E}}} \tag{140}$$

which we want to prove now. Take $\boldsymbol{v} = \boldsymbol{e}$ as a special choice and use Eq. 127 we get the estimate from above

$$\sup_{\boldsymbol{v}\in \boldsymbol{V}} \frac{\int_0^T \dot{R}(\boldsymbol{v}) dt}{||\boldsymbol{v}||_{\mathcal{E}}} \geq \frac{\int_0^T \int_\Omega \boldsymbol{\sigma}(\dot{\boldsymbol{e}}) : \boldsymbol{\varepsilon}(\boldsymbol{e}) dx dt}{||\boldsymbol{e}||_{\mathcal{E}}} = ||\boldsymbol{e}||_{\mathcal{E}} \tag{141}$$

and from below, using the Cauchy-Schwartz inequality

$$\sup_{\boldsymbol{v}\in \boldsymbol{V}} \frac{\int_0^T \dot{R}(\boldsymbol{v}) dt}{||\boldsymbol{v}||_{\mathcal{E}}} = \sup_{\boldsymbol{v}\in \boldsymbol{V}} \frac{\int_0^T \int_\Omega \boldsymbol{\sigma}(\dot{\boldsymbol{e}}) : \boldsymbol{\varepsilon}(\boldsymbol{v}) dx dt}{||\boldsymbol{v}||_{\mathcal{E}}} \leq \sup_{\boldsymbol{v}\in \boldsymbol{V}} \frac{||\boldsymbol{e}||_{\mathcal{E}} ||\boldsymbol{v}||_{\mathcal{E}}}{||\boldsymbol{v}||_{\mathcal{E}}} = ||\boldsymbol{e}||_{\mathcal{E}}\,. \tag{142}$$

Therefor Eq. 140 holds by using Eqs. 141 and 142. This result is only correct for the assumption that the flow direction is exact or frozen within the considered load increment. For local calculation of the error estimator we solve Eq. 127 for each element at any time t separately and get $\dot{\overset{\star}{\boldsymbol{e}}}$

$$\int_{\Omega_e} \boldsymbol{\sigma}(\dot{\overset{\star}{\boldsymbol{e}}}) : \boldsymbol{\varepsilon}(\boldsymbol{v}) d\boldsymbol{x} = \int_{\Omega_e} \boldsymbol{\varepsilon}(\dot{\overset{\star}{\boldsymbol{e}}}) : \boldsymbol{C} : \boldsymbol{\varepsilon}(\boldsymbol{v}) d\boldsymbol{x} = \dot{R}_{\Omega_e}(\boldsymbol{v}) \quad \forall \boldsymbol{v} \in \mathcal{V}(\Omega_e)\;;\; \forall t \in (0,T)\,. \tag{143}$$

By solving this eqution exactly, we get an upper bound error estimator since the space $\mathcal{V}$ is extended to the broken space $\overset{\star}{\mathcal{V}}$ which has additional discontinuous test and solution functions at the element boundaries such that the proof reads

$$
\begin{aligned}
||\boldsymbol{e}||_{\mathcal{E}} &= \sup_{\boldsymbol{v}\in\boldsymbol{V}} \frac{\int_0^T \dot{R}(\boldsymbol{v})dt}{||\boldsymbol{v}||_{\mathcal{E}}} \\
&= \sup_{\boldsymbol{v}\in\boldsymbol{V}} \frac{\int_0^T \dot{\overset{\star}{R}}(\boldsymbol{v})dt}{||\boldsymbol{v}||_{\mathcal{E}}} \\
&\leq \sup_{\boldsymbol{v}\in\overset{\star}{\boldsymbol{V}}} \frac{\int_0^T \dot{\overset{\star}{R}}(\boldsymbol{v})dt}{||\boldsymbol{v}||_{\mathcal{E}}} \\
&= \sup_{\boldsymbol{v}\in\overset{\star}{\boldsymbol{V}}} \frac{\sum_e \int_0^T \int_{\Omega_e} \boldsymbol{\sigma}(\dot{\overset{\star}{\boldsymbol{e}}}) : \boldsymbol{\varepsilon}(\boldsymbol{v}) d\boldsymbol{x} dt}{||\boldsymbol{v}||_{\mathcal{E}}} \\
&= \sup_{\boldsymbol{v}\in\overset{\star}{\boldsymbol{V}}} \frac{\sum_e \int_{\Omega_e} \boldsymbol{\sigma}(\overset{\star}{\boldsymbol{e}}) : \boldsymbol{\varepsilon}(\boldsymbol{v}) d\boldsymbol{x}}{||\boldsymbol{v}||_{\mathcal{E}}} \\
&\leq \sup_{\boldsymbol{v}\in\overset{\star}{\boldsymbol{V}}} \frac{\left(\sum_e \int_{\Omega_e} \boldsymbol{\sigma}(\overset{\star}{\boldsymbol{e}}) : \boldsymbol{\varepsilon}(\boldsymbol{e}) d\boldsymbol{x}\right)^{1/2} \left(\sum_e \int_{\Omega_e} \boldsymbol{\sigma}(\overset{\star}{\boldsymbol{v}}) : \boldsymbol{\varepsilon}(\boldsymbol{v}) d\boldsymbol{x}\right)^{1/2}}{||\boldsymbol{v}||_{\mathcal{E}}} \\
&= \left(\sum_e ||\overset{\star}{\boldsymbol{e}}(T)||^2_{\mathcal{E}(\Omega_e)}\right)^{1/2} \\
&= \left(\sum_e \frac{1}{2} \int_{\Omega_e} \boldsymbol{\varepsilon}(\overset{\star}{\boldsymbol{e}}(T)) : \boldsymbol{C} : \boldsymbol{\varepsilon}(\overset{\star}{\boldsymbol{e}}(T))\, d\boldsymbol{x}\right)^{1/2} =: \left(\sum_e \eta_e^2\right)^{1/2} = \eta\,.
\end{aligned}
\tag{144}
$$

One has to mention that this error estimator is also true in case of pure elasticity. The error holds at time T. If we want to consider the (process) time influence, the so-called accumulation error, then we have to use a norm, different from that in Eq. 139 e.g., namely

$$
||\boldsymbol{v}||^2_{\mathcal{B}} := \int_0^T \int_0^t \int_{\Omega} \boldsymbol{\varepsilon}(\dot{\boldsymbol{v}}(\xi)) : \boldsymbol{C} : \boldsymbol{\varepsilon}(\boldsymbol{v}(\xi))\, d\boldsymbol{x}\, d\xi\, dt = \int_0^T \int_{\Omega} \frac{1}{2}\boldsymbol{\varepsilon}(\boldsymbol{v}(t)) : \boldsymbol{C} : \boldsymbol{\varepsilon}(\boldsymbol{v}(t))\, d\boldsymbol{x}\, dt \;\; \forall \boldsymbol{v} \in \mathcal{V}\,,
\tag{145}
$$

The proof is similar to that for Eq. 144. Our estimator yields an upper global bound because we describe the plastic domain more and more accurate in time, such that the error estimator is asympotically exact.

For the second step we consider an error indicator for the evolution of plastic strain. For this we test the local solution $\overset{\star}{\boldsymbol{e}}$ of the error from Eq. 127 w.r.t. the current elasto-plastic tangent, Eq. 125. The motivation is that we have to control whether the stress rates and the elastic strain rates are orthogonal to the flow direction $\boldsymbol{n}_h(t)$. Therefore, we introduce the plastic strain error indicator in the energy related norm as

$$\eta_{pl}^2 = \| \dot{\overset{\star}{\boldsymbol{e}}} \|_{\mathcal{E}}^{pl\,2} := \sum_e \int_0^T \int_{\Omega_e} \boldsymbol{\varepsilon}(\dot{\overset{\star}{\boldsymbol{e}}}(t)) : (\boldsymbol{C} - \boldsymbol{C}^{PR}(t)) : \boldsymbol{\varepsilon}(\dot{\overset{\star}{\boldsymbol{e}}}(t)) \, d\boldsymbol{x} \, dt \, . \tag{146}$$

We have to integrate this term over the time because the elastoplastic tangent $\boldsymbol{C}^{PR}$ is time dependent.

For directional error estimation and related anisotropic mesh refinements we solve the local varational problem, Eq. 143, separately for each convected coordinate and calculate the error estimators separately for the three spatial directions. Thus, we get three directionally different contributions of the error.

For a guaranteed upper bound error estimator one has to solve the local variational problem

$$a(\overset{\star}{\mathrm{e}}, \boldsymbol{v})_{\Omega_e} = \overset{\star}{\mathcal{R}} (\boldsymbol{v})_{\Omega_e} \quad \forall \boldsymbol{v} \in \mathcal{V}(\Omega_e) \tag{147}$$

for each elemente exactly, but this is not possible for realistic engineering problems. Therefore, we solve the local problem approximately in an expanded approximation space with additional hierarchically expanded polynomial test and trial functions. The polynomial order of the current test space is p, and the local extension is k, such that we get the local polynomial order $p+k$. We define the anisotropic error estimator η_a^{p+k} for the error norm from Eq. 145, including the accumulation of the error in time, as

$$\eta_a^{p+k} := \left((\eta_r^{p+k})^2 + (\eta_s^{p+k})^2 + (\eta_t^{p+k})^2 \right)^{1/2} \tag{148}$$

where r, s, t are the scaled coordinates of the unit cube, and the isotropic error estimator η_i^{p+k} as

$$\eta_i^{p+k} := \eta_{rst}^{p+k} \tag{149}$$

with simultaneous testing in all directions. The effectivity index for the anisotropic error estimator reads

$$\theta_a^{p+k} = \frac{\eta_a^{p+k}}{\|\boldsymbol{e}\|_{\mathcal{B}}} \tag{150}$$

and for the isotropic error estimator , correspondingly

$$\theta_i^{p+k} = \frac{\eta_i^{p+k}}{\|\boldsymbol{e}\|_{\mathcal{B}}} \, . \tag{151}$$

In example Figure 15, section 9.2.2, the effectivity indices for $p = 1$ with different k are given. The relative error is $||\boldsymbol{e}_{\boldsymbol{\sigma}}||_{\mathcal{B}}/||\boldsymbol{\sigma}||_{\mathcal{B}} \approx 10\%$ for the start mesh. The adaptive computation of the effectivity index $\theta = \left(\frac{\eta_r^2 + \eta_s^2 + \eta_t^2}{||\boldsymbol{e}||_{\mathcal{B}}^2}\right)^{1/2}$ is performed by using the isotropic and the anisotropic error estimators, both for $p = 1$ and $k = 2$. The results are:

k	θ_i^{1+k}	θ_a^{1+k}
1	0.928	0.768
2	1.074	0.861
3	1.136	0.892
4	1.156	0.903
5	1.165	0.907

For further results see section 9.2.2.

Different from the error analysis for elastoplastic deformations in this section using equilibrated residua and directionally enhanced test spaces, averaging techniques and element wise higher interpolations are presented for elastoplasticity in section 10, considering all approximation errors in space and time in the frame of primal FEM.

9 Examples for residual error estimators

Several problems are treated with error–controlled adaptive meshes and ansatz–functions. Reliability and effectivity are controlled. The developed error estimators η are computed with the following norms: the global error in the energy norm $a(\boldsymbol{u} - \boldsymbol{u}_h, \boldsymbol{u} - \boldsymbol{u}_h)^{1/2}$, the global error $(\int_\Omega |\boldsymbol{u} - \boldsymbol{u}_h|^2 dx)^{1/2}$ in the L_2–norm, as well as the effectivity index $\theta = \eta/(a(\boldsymbol{u} - \boldsymbol{u}_h, \boldsymbol{u} - \boldsymbol{u}_h))^{1/2}$, where solutions for very fine reference meshes from convergence tests were used instead of analytical solutions.

9.1 Clamped thin bar with isotropic material properties

The treated adaptive methods, are applied to an isotropic elastic cantilever beam in plane strain state ($\varepsilon_{zz} = 0, \gamma_{xz} = 0, \gamma_{yz} = 0$) with a 3D-elastic material equation, Eq. 6. Comparative results are presented in figures 7 and 9. Figure 8 shows adaptive meshes for quadratic isoparametric shape functions, resulting from anisotropic equilibration of residua (PEM). Figure 7b shows the convergence of the potential energy. The locking effect is obvious for linear shape functions (points $\bigcirc$), such that the adaptive process is not efficient. From Figure 7c one can observe that the effectivity index in case for anisotropic error estimators is very good for p=2. As expected, the (explicit) isotropic residual error estimator (by REM) with computed interpolation constants yields a lower bound for Dirichlet boundaries of element patches (points +) and an upper bound for Neumann boundaries of single elements (points $\star$), but the effectivity is worse than by using PEM.

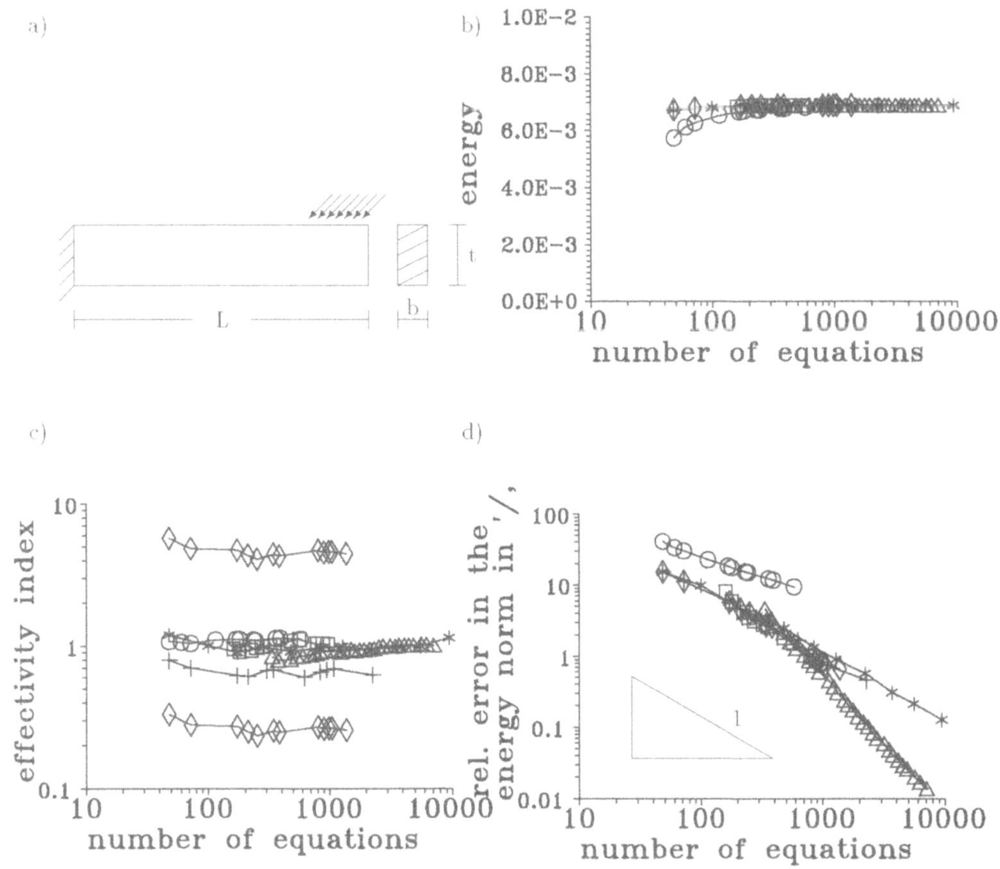

Figure 7: 2D-beam with anisotropic mesh adaptation and isotropic material equation;

$L = 10m$, $t = 2m$, $b = 1m$, $E = 206900MPa$, $\nu = 0.29$, $f = 10KN/m^2$; a) system and system data; b) convergence of the energy; c) effectivity index of the estimated error against the **nearly exact** error in the energy norm; d) relative error in the energy norm in %; ○ : p=1; anisotropic solution adaptivity with PEM, □ : p=2; anisotropic solution adaptivity with PEM, △ : p=3; anisotropic solution adaptivity with PEM, ◇ : p=2; isotropic solution adaptivity with REM with c_u=1.04 for upper bound and c_e=0.06 for lower bound, ⋆ p=2; isotropic residual solution adaptivity with local Neumann boundary, + p=2; isotropic residual solution adaptivity with local Dirichlet boundary

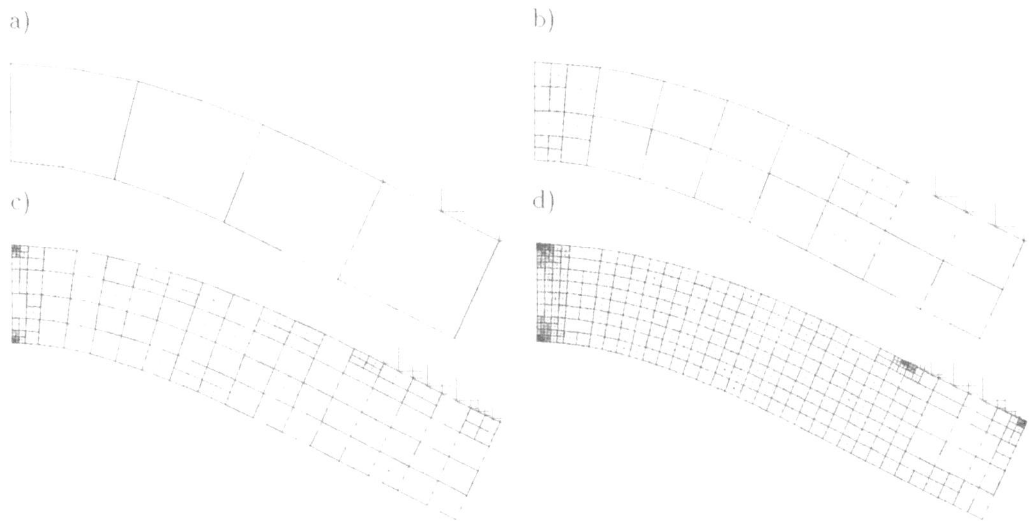

Figure 8: Example of a transversally loaded beam in 2D state of stress, clamped at the left baoundary, with anisotropic mesh adaptation and isotropic material equation; a) startmesh; b) mesh 4; c) mesh 7; d) mesh 10; all for p=2;

Isotropic REM–estimators (points ◇) have similar effects with lower effectivity but again the convergence rate, figure 7d, is similar because in this example, the isotropic refinement strategy (four daughter elements are generated from one quadrangle) is quasi optimal. For the anisotropic PEM with polynomial order higher than 1 (□ and △) the effectivity index and the convergence rate are very good and robust. The interpolation constants for isotropic REM were estimated by using Eq. 23 and solutions from eigenvalue problem with $c_u = 1.04$ for the upper bound and $c_l = 0.06$ for the lower bound, (points ◇). Residual locally computed error estimators (REM) using Dirichlet– or Neumann–boundaries proved to be **reliable** and **effective** in case of **isotropic** estimations. The computed interpolation constants c_l and c_u may estimate the error too small or too large by about one order of magnitude but nevertheless, the estimators control the isotropic mesh adaptivity efficiently.

9.2 Example of a thin cantilever beam with anisotropic material equations

Anisotropic error estimators using equilibrated boundary tractions are used for a thin–walled cantilever beam with transversally isotropic, laterally orthotropic linear elastic material. The system is shown in figure 9a with the material paramters $E_1 = 100\,000$ MPa, $E_2 = 1000$ MPa, $\nu = 0.4$, $G_{12} = 30\,000$ MPa, $G_{13} = 300$ MPa and the fiber orientations $\alpha_{1,2} = 30°; 90° + 30°$. Some adapted meshes are presented in figure 10 a–d for the polynomial order p=2. One can recognize the material anisotropy in the

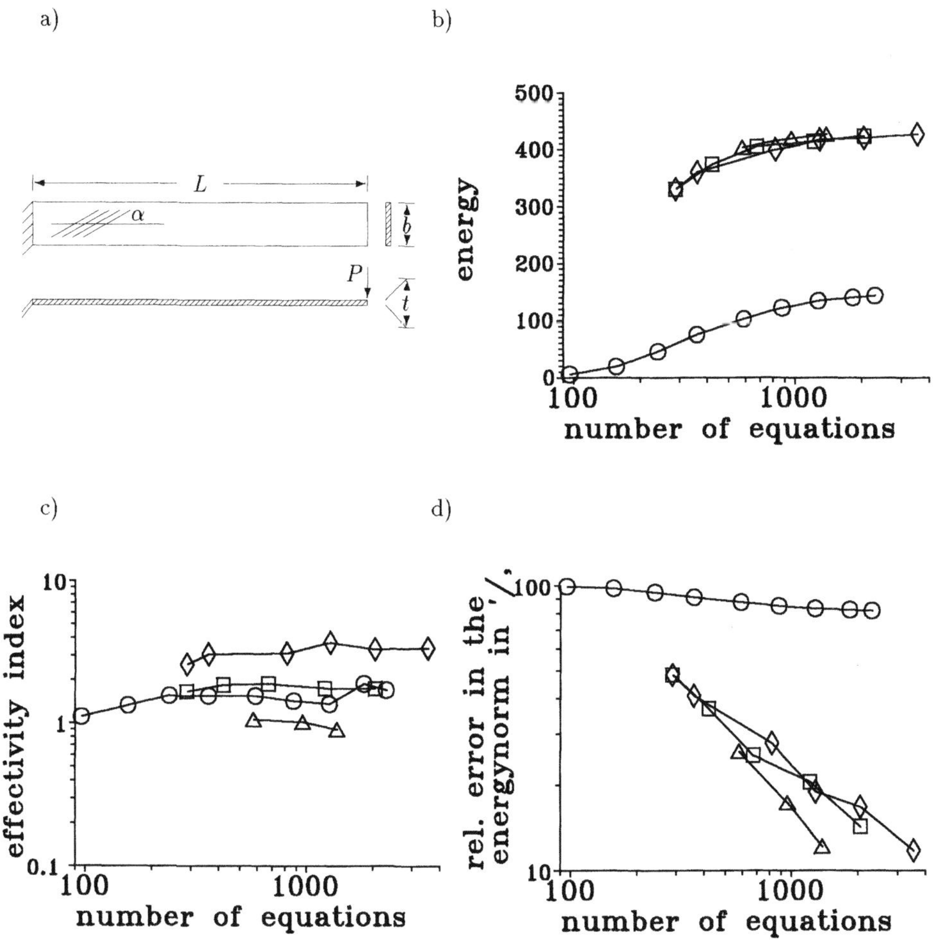

Figure 9: Example of a beam in 2D–state of stress, clamped at the left boundary and transversally loaded by a line load over the thickness with anisotropic mesh adaptation for transversally isotropic, laterally orthotropic material;

$L = 6.4m$, $t = 10cm$, $b = 80cm$, $P = 10KN$; a) system and loading; b) convergence of the energy; c) effectivity index of the estimated error with respect to the nearly exact error in the energy norm; ○ : p=1; anisotropic solution adaptivity with PEM, □ : p=2; anisotropic solution adaptivity with PEM, △ : p=3; anisotropic solution adaptivity with PEM, ◇ : p=2; isotropic solution adaptivity with REM, see section 4.2;

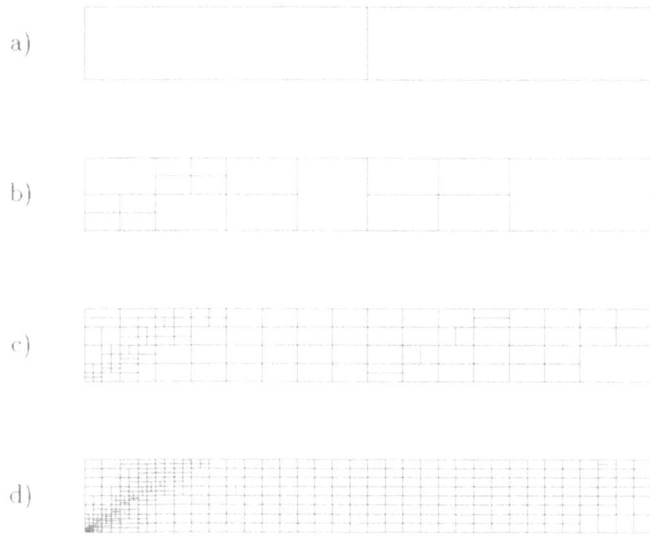

Figure 10: Example of anisotropic mesh adaptation for transversally isotropic, laterally orthotropic material; a) startmesh; b) mesh 4; c) mesh 7; d) mesh 10, all for p=2;

refined meshes. The effectivity index is about 1, and the convergence rates for the PEM–estimators are optimal and demonstrate the reliability and robustness of PEM. One can especially observe that higher polynomials ($p > 2$) produce higher convergence order if sufficient regularity of the solution is present. The relative residual error estimator is not effective due to the large condition number (about 100) of the anisotropic material tensor, Eq. 9. Nevertheless, a good convergence order of the finite element solution is achieved. This is caused by the modified estimation of the interpolation constant for which only isotropic elasticity with $E = 100\,000$ MPa was used. In case that also the transversal Young's modulus is used the interpolation constant has to be multiplied with $E_1/E_2 = 10$. Hence, the isotropic mesh refinements due to isotropic residual error estimators can be efficient without computing the quantitative errors, i.e. the interpolation constants.

9.2.1 Example of integrated solution– and model-adaptivity of a clamped plate in bending

Here, the automatic error–controlled expansion from $2\frac{1}{2}$D-plate theory with elasticity matrix Eq. 8 to 3D-elasticity with elasticity matrix Eq. 6 and also the dimensional ex-

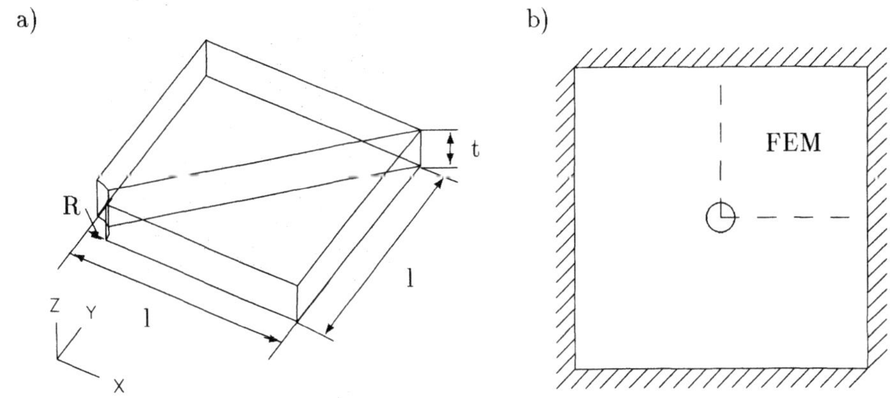

Figure 11: Results of model adaptivity for a trnsversally plate in bending with a hole; calculation with anisotropic solution– and model adaptivity using local Neumann problems and PEM; a) system and systemdata, $\frac{l}{t} = 10; \frac{l}{R} = 10$, l=10m; $E = 206900 MPa$, $\nu = 0.29$, $f = 10 KN/m^2$;

tension of the ansatz–spaces in thickness direction is performed in connection with the discretization error of 2D-theory. System data and constant lateral loading are given in figure 11, and figues 12 – 14 show adapted meshes and models together with isoplots of stresses. The blue coloured domain depicts subdomains treated by 2D-theory, and the red coloured subdomains characterize the use of 3D-theory. In order to avoid constraints (→ locking) in the transition elements of differently modeled subdomains, a special hybrid transition element was developed which averages extensions in thickness direction in integral form. The expanded 3D-elastic domain grows from the clamped boundaries to the inner domain, due to the adaptive process with global error estimators and decreasing error tolerances. Within the disturbed boundary layers at the clamped edge, strong 3D-refinements take place in order to capture the root– singularity, and anisotropic mesh refinements make PEM especially effective. It can be observed that the physical 3D–boundary layer of the 2D-modeled plate affects 3D–model adaptivity automatically. An important result lies in the fact that the size of the 3D-modeled subdomain depends on the size of the admitted error tolerance of the approximation error, i.e. the 3D-subdomain grows with decreasing error tolerance.

9.2.2 Example of a 3D-thick plate with a hole for Prandtl–Reuss elastoplasticity

The adaptive computations are carried out using PEM with isotropic and anisotropic error estimators, regarding equilibrium for the elastic tangents and the error norm with accumulation of the error in time direction, Eq. 145. The geometrical and physical data

Figure 12: Isoplots of τ_{xz} and σ_{xx} for model– adaptivity calculation of an elastic square plate in bending with a hole and transverse loading; results for start mesh

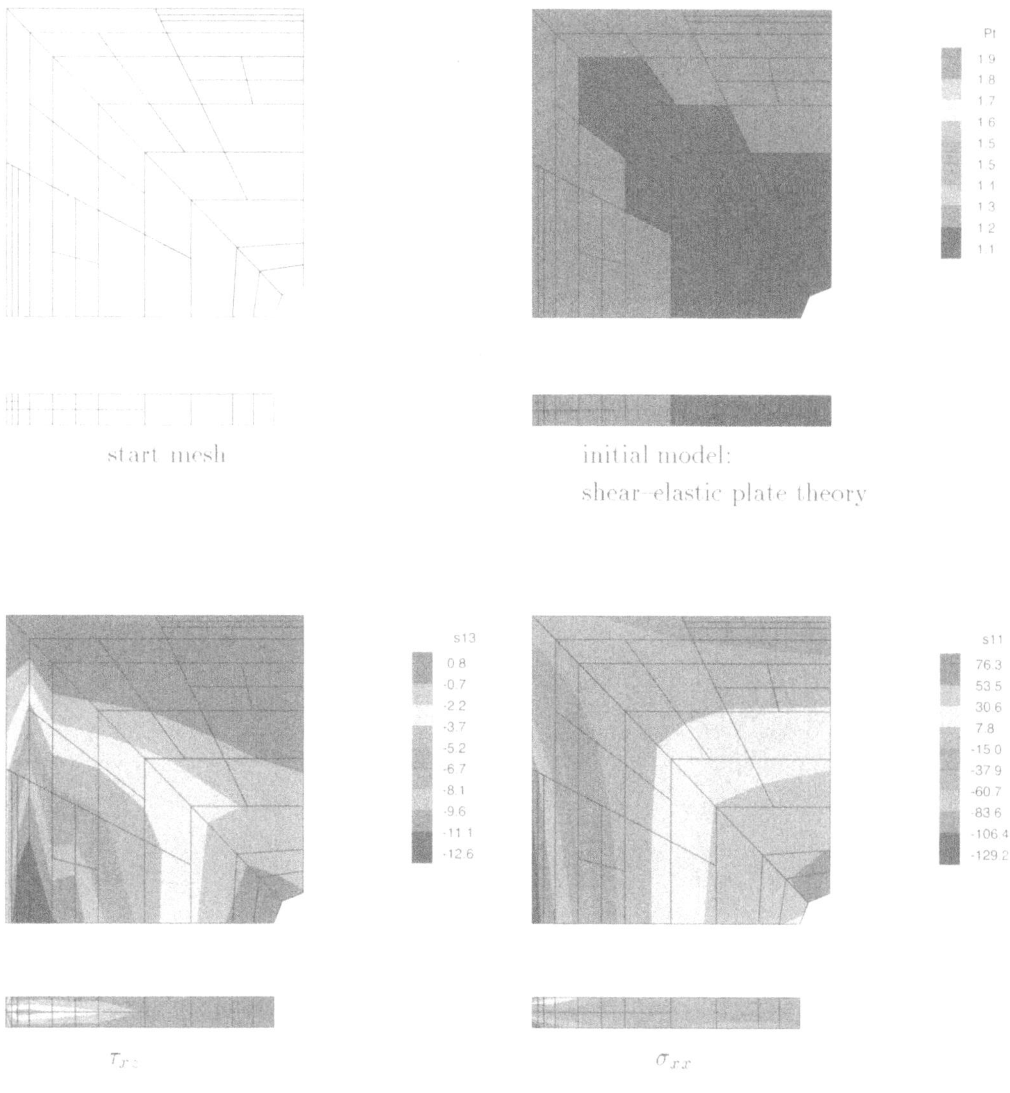

Figure 13: Isoplots of τ_{xz} and σ_{xx} for model- adaptivity calculation of an elastic square plate in bending with a hole and transverse loading; results for mesh 8

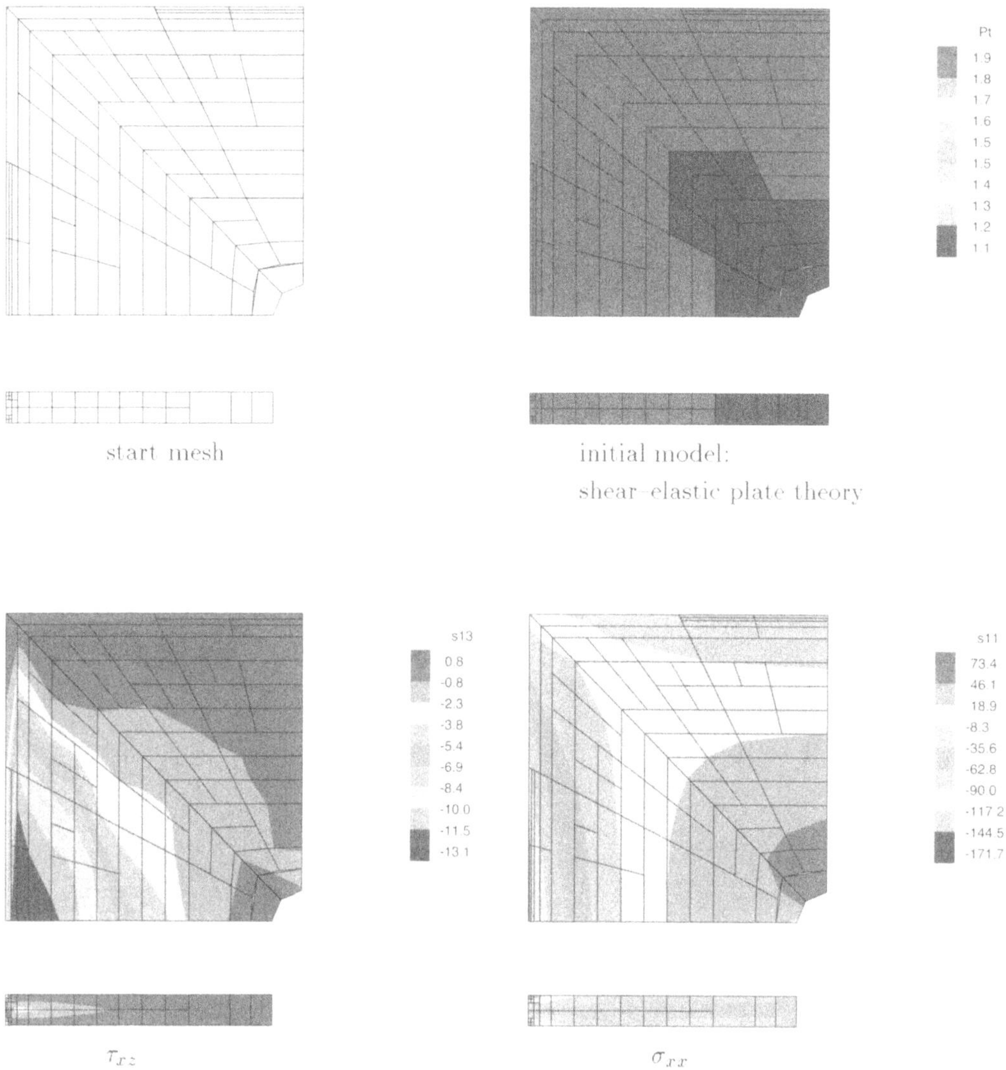

Figure 14: Isoplots of τ_{xz} and σ_{xx} for model- adaptivity calculation of an elastic square plate in bending with a hole and transverse loading; results for mesh 10

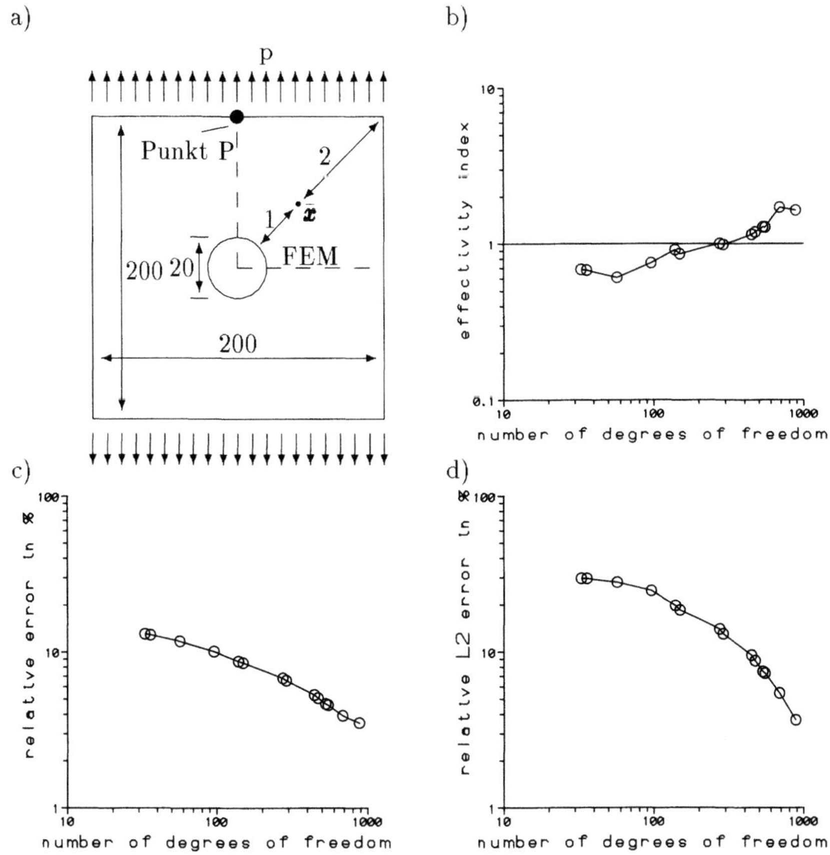

Figure 15: Elasto–plastic 3D– thick plate with a hole; calculated with anisotropic solution adaptivity using PEM; a) system and system data: thickness t = 20 mm; Poisson's ratio $\nu = 0.29$; Youngs's modulus E = 206900 N/mm^2; yield stress $y_0 = 450$ N/mm^2; loading p = 410 N/mm^2, ○ : anisotropic error estimation of the equilibrium in the energy norm; b) effectivity index, c) error in the energy norm, d) error in the L_2 norm

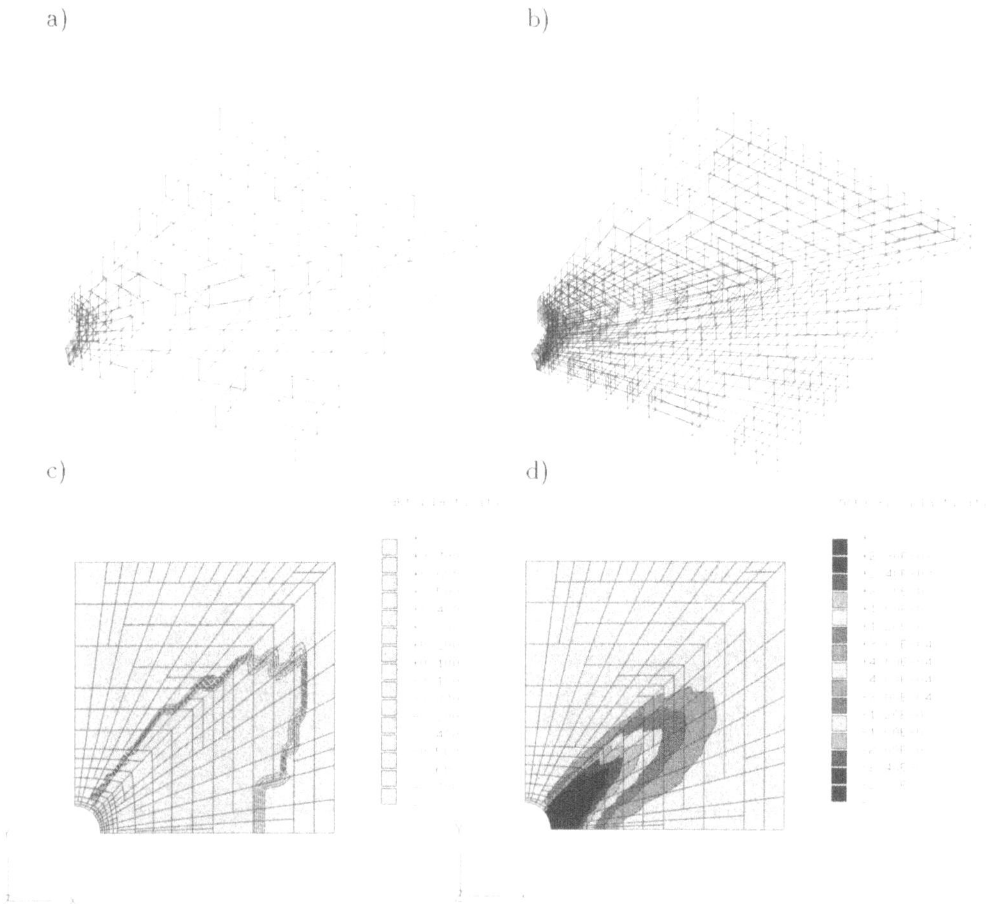

Figure 16: Elastoplastic 3D– thick plate with a hole; a) 10th adapted mesh, b) 30th adapted mesh, c) zone of plastic deformations, d) norm of plastic strains

are given in figure 15a. In Figure 15b-d the effectivity index, the error in the norm $||\boldsymbol{e}||_B$ and the error in the time integrated L_2-norm is given. The convergence rate is in agreement with the a priori prediction. One can see that the effectivity index is smaller than 1 for coarse meshes and tends to an upper error bound, see also [40, 46, 10]. The reference solution was calculated with a stongly adapted mesh (30 steps of adaptation), see Figure 16b; the 10th adapted mesh is shown in Figure 16a. The plastic region and the norm of the plastic strains are shown in Figure 16c and d. We have to mention that the FE-calculation was performed with three symmetry boundaries, such that one can see one half of the thickness and one fourth of the plate.

10 Global averaging and residual error estimates for elastoplastic deformations

10.1 Introduction and theory of elastoplasticity

Error controlled adaptive strategies should be used in order to achieve a prescribed accuracy of the computed solutions at minimum cost. Locally computed global error estimators in the primal form of the finite-element-method for Hencky- and Prandtl-Reuss-elastoplasticity without and with nonlinear isotropic hardening are presented, controlling global errors of equilibrium, the direction of plastic strain rates, the yield condition and the numerical integration of the flow rule. Higher polynomials of order p+2 are used for postprocessing on element level for getting improved stresses and for the transfer of physical data between subsequent meshes by least squares techniques, implementing also additional requirements like plastic incompressibility. All error indicators are tested with the benchmark example of a rectangular plane strain problem with a hole, see also [46].

In the adaptive approximation process, aside from the stepwise optimal choice of the spatial discretization, we have to pay attention to the implicit time-dependency of the deformation process.

Therefore, special attention has to be devoted to adaptive refinement and coarsening algorithms for mesh and load-step sizing. The usual decomposition (operator split) of the space-time problem into FEM in space and FDM in time is used which does not permit a rigorous coupled a posteriori error analysis, especially with respect to the accumulation error in time.

Thus, an heuristic assumption based on the comparison of Prandtl-Reuss elastoplasticity with Hencky quasi-plasticity, namely nonlinear elasticity, is used to split the influence of the error of plastic deformation into spatial and time discretization errors [10, 46].

Three separate locally computed global spatial error indicators are derived from global residual and averaging error measures, namely for equilibrium, plastic strains using the plastic dissipation and the yield condition. Each of them yields a scaled local indicator for mesh refinement. The error within the integration of the flow rule by the backward-Euler scheme is estimated by the maximal change of the normal with respect to the yield surface between two time (load) steps. Additionally, the error by updating and interpolation of the material parameters for each adapted mesh has to be considered, at least by a further iteration step of the weak equilibrium conditions before the next load step.

A new technique considering the accumulation error in time is given in the form of a staggered combination of adaptivity in space and time and compared with the procedure by Gallimard et. al [18], computing the loading history for each spatial FE-mesh with different equal time (load) steps.

Next, the basic equations for elastoplastic deformations with hardening are given pointing out the differences between the Hencky-deformation theory and the Prandtl-Reuss-flow-theory. The choice of an adequate finite element is treated in subsection two. The third subsection deals with algorithmic questions such as the combination of space and time adaptivity during the computation, the data transfer between different meshes, as mentioned above, and the features between global and goal–oriented error estimations. Subsection four presents the transfer of process-depending material parameters between subsequent meshes. The fifth subsection describes the benchmark example used for the numerical tests. Subsection six presents examples for the different sources of errors in elastoplastic deformations as well as computed error estimators.

We consider the solid body $\Omega \subset I\!R^3$ with boundary $\partial\Omega = \Gamma = \Gamma_u \cup \Gamma_t$. The variational form reads with the test function $\boldsymbol{v} \in \boldsymbol{V}$; $\boldsymbol{V} = \{\boldsymbol{v}(\boldsymbol{x}) \in [H_0^1(\Omega)]^3, \Omega \subset I\!R^3, \boldsymbol{v} = \boldsymbol{0}$ on $\Gamma_u\}$, see also section 8,

$$\int_\Omega \boldsymbol{\varepsilon}(\boldsymbol{v}(\boldsymbol{x})) : \boldsymbol{\sigma}(\boldsymbol{u}(\boldsymbol{x}))\, d\Omega = \int_\Omega \boldsymbol{v}(\boldsymbol{x}) \cdot \boldsymbol{f}\, d\Omega + \int_{\Gamma_t} \boldsymbol{v}(\boldsymbol{x}) \cdot \bar{\boldsymbol{t}}\, d\Gamma \tag{152}$$

with mixed boundary conditions for the displacements $\boldsymbol{u} = \boldsymbol{0}$ and $\boldsymbol{v} = \boldsymbol{0}$ on the Dirichlet boundary Γ_u and given surface tractions $\boldsymbol{t} = \boldsymbol{\sigma} \cdot \mathbf{n} = \bar{\boldsymbol{t}}$ on Γ_t. Herein, $\bar{\boldsymbol{t}}$ are prescribed tractions and $\boldsymbol{f}$ are body forces. $\boldsymbol{\sigma}$ and $\boldsymbol{\varepsilon}$ denote the macro stress tensor and macro strain tensor, respectively and are square integrable.

Among the theories of elastoplastic deforming materials we consider first the deformation theory or Hencky–plasticity [21] and the flow theory by Prandtl-Reuss [33]. The latter has the advantage that it can not only describe changes of the material parameters during the loading process but also elastic unloading with remaining plastic deformations because of the existence of an 'internal time' which is represented by means of the loading history. Consequently, Hencky–plasticity is often referred to as 'non-linear elasticity' with a yield condition.

The main difference between the two theories is the determination of the final state of strains which is given for the Hencky-plasticity by

$$\boldsymbol{\varepsilon}^p = c \cdot \operatorname{dev}\boldsymbol{\sigma} \tag{153}$$

with an arbitrary constant c instead of the rate equation

$$\dot{\boldsymbol{\varepsilon}}^p = \dot{\gamma} \operatorname{dev}\boldsymbol{\sigma} \tag{154}$$

for Prandtl-Reuss plasticity. With the introduction of an associated flow rule with respect to the yield function, the difference gets even more apparent, as the normals with respect to the 'plastic flow' surfaces are constant throughout the load history for the Hencky-deformation theory, namely

$$\frac{\boldsymbol{\sigma}_i}{||\mathrm{div}\boldsymbol{\sigma}_i||} = \frac{\boldsymbol{\sigma}_{i+1}}{||\mathrm{div}\boldsymbol{\sigma}_{i+1}||}. \tag{155}$$

The definitions and basic relations of Prandtl-Reuss elastoplasticity for small strains with nonlinear isotropic hardening are presented in Table 4 in a canonical form.

Linear kinematics	$\boldsymbol{\varepsilon}^e := \boldsymbol{\varepsilon} - \boldsymbol{\varepsilon}^p$
Internal variable for isotropic hardening (micro strain)	α
Free energy function	$\Psi \ = \hat{\Psi}(\boldsymbol{\varepsilon}^e, \alpha)$
Macro stress	$\boldsymbol{\sigma} \ = \partial_{\boldsymbol{\varepsilon}^e} \hat{\Psi}$
Micro stress	$\beta \ = -\partial_\alpha \hat{\Psi}$
Dissipation	$\mathcal{D}^p = \boldsymbol{\sigma} : \dot{\boldsymbol{\varepsilon}}^p - \beta \cdot \dot{\alpha} \geq 0$
Yield function	$\Phi \ = \hat{\Phi}(\boldsymbol{\sigma}, \beta)$
Flow rule	$\dot{\boldsymbol{\varepsilon}}^p \ = \dot{\gamma}\, \partial_{\boldsymbol{\sigma}} \hat{\Phi}$
Evolution of α	$\dot{\alpha} \ = -\dot{\gamma}\, \partial_\beta \hat{\Phi}$
Loading and unloading (Kuhn-Tucker conditions)	$\dot{\gamma} \geq 0, \quad \hat{\Phi} \leq 0, \quad \dot{\gamma} \cdot \hat{\Phi} = 0$

Table 4: The basic equations for Prandtl-Reuss elastoplasticity $\forall\, \boldsymbol{x} \in \Omega \subset \mathbb{R}^3$

Herein the free energy function $\Psi = \hat{\Psi}(\boldsymbol{\varepsilon}^e, \alpha)$ can be split into a macroscopic part of isotropic elastic deformations

$$\hat{\Psi}_{Macro}(\boldsymbol{\varepsilon}^e) = \frac{1}{2}\, \kappa\, (\mathrm{tr}\, \boldsymbol{\varepsilon}^e)^2 + \mu\, \mathrm{tr}\, (\boldsymbol{\varepsilon}^e)^2 \tag{156}$$

with bulk modulus κ and shear modulus μ, and a microscopic part

$$\hat{\Psi}_{Micro}(\alpha) = \frac{1}{2}\, h\, \alpha^2 + (y_\infty - y_\circ) \left(\alpha + \frac{1}{\omega}\, (\exp(-\omega\alpha) - 1) \right) \tag{157}$$

which contains combined linear and exponential isotropic hardening with stress saturation.

The yield function $\Phi = \hat{\Phi}(\boldsymbol{\sigma}, \beta)$ can also be split into a macro- and a micro-part

$$\Phi = \hat{\Phi}(\boldsymbol{\sigma}, \beta) = ||\mathrm{dev}\boldsymbol{\sigma}|| - \sqrt{\frac{2}{3}}\, \hat{y}(\beta) \tag{158}$$

with the macroscopic stress tensor $\boldsymbol{\sigma}$ and the microscopic stress β related to isotropic hardening. The saturation law for the yield stress which combines the initial yield stress with linear and exponential hardening reads

$$\hat{y}(\beta) = y_\circ + h\, \alpha + (y_\infty - y_\circ)\, (1 - \exp(-\omega\, \alpha)). \tag{159}$$

This material model has 6 parameters and one internal variable, namely α . The volume-specific dissipation function for the macro- and the micro-terms

$$\mathcal{D}^p = \boldsymbol{\sigma} : \dot{\boldsymbol{\varepsilon}}^p - \beta \cdot \dot{\alpha} \tag{160}$$

measures the rate of energy dissipation due to inelastic deformation. The isotropic hardening variable α is related to the squareroot of the dislocation density in case of ductile metal materials.

10.2 Chosen 2D finite elements

We consider four finite elements for elastoplasticity: the bilinear quadrilateral Q1-element, the Q1P0-element by Nagtegaal, Parks & Rice [27] and the enhanced Q1E5-element by Simo and Rifai [38]. Furthermore we tested the dual mixed triangular BDM-element based on the Brezzi, Douglas & Marini spaces, which was extended to elastoplasticity by Schröder, Klaas, Stein & Miehe [24, 36, 37].

These four elements were compared by the computing times and the convergence behavior of the displacements u_x at node 2, see benchmark system Figure 21a,b (stretched plate with a hole in plane strain state), see also [46].

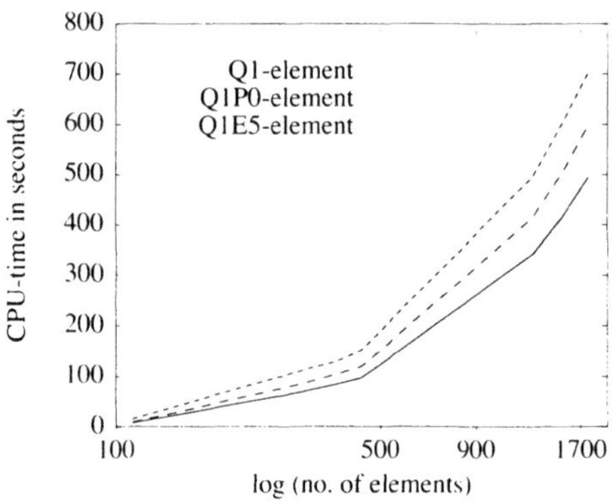

Figure 17: CPU times for different elements and meshes for benchmark system, figures 21a, 22b

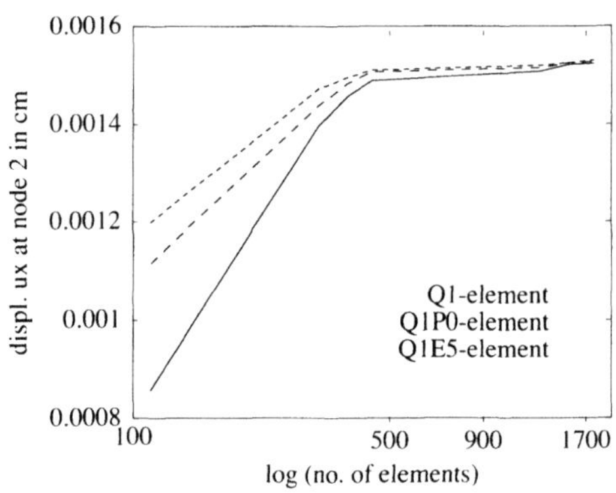

Figure 18: Displacement u_x at node 2 for different elements at load parameter $\lambda = 4.5$

Due to the physical constraint of plastic incompressibility, linear displacement-based element formulations are unstable. Elements with mixed Ansatz-spaces are superior, as the Q1P0-element with an additional constant pressure d.o.f. per element, or the Q1E5-element with enhanced strain modes.

Figure 17 depicts the CPU times of three different element formulations for several meshes with growing number of elements, while Figure 18 shows the convergence of the displacements u_x at node 2.

We favorized the Q1P0-element as it needs less computing time than the enhanced Q1E5-element.

The BDM-element yields better convergence than the Q1P0-element but at large computing time. A mesh of 64 BDM-elements gave comparable results with about 4000 Q1-elements but took approximately the same computing time.

10.3 Strategies for adaptive FEM in space and time

10.3.1 Two methods for including coupled space-time errors

Several adaptive strategies are available for time-dependent material laws in the frame of an operator split with FEM in space and FDM in time.

One of them is the algorithm used by Gallimard, Ladevèze & Pelle [18], depicted in figure 19, which uses a constant mesh and a constant time step for one whole loading path. Based on two error indicators for space and time discretization, a new mesh and time step scheme is generated, and the loading process is repeated such that this algorithm is only adaptive at the end of each re-computation of the loading path.

Its advantage is the simple structure as it needs neither a transfer of variables and computed data between subsequent meshes nor the capability of 'restarting' the computation process from a given time in load history. Its big disadvantage is the re-compute of the whole load path several times which needs high computational effort.

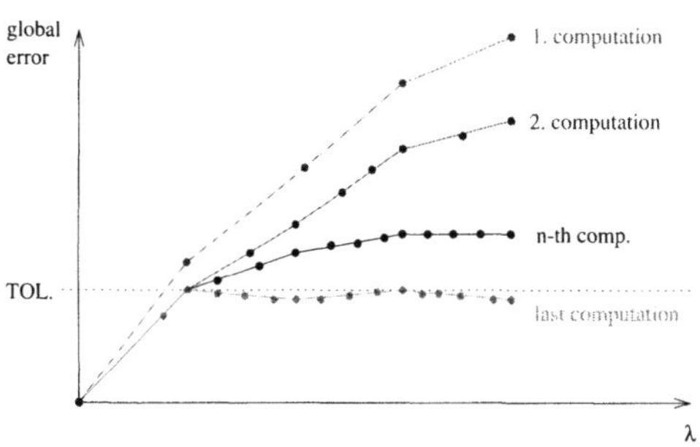

Figure 19: The sequential algorithm by Gallimard, Ladevèze & Pelle

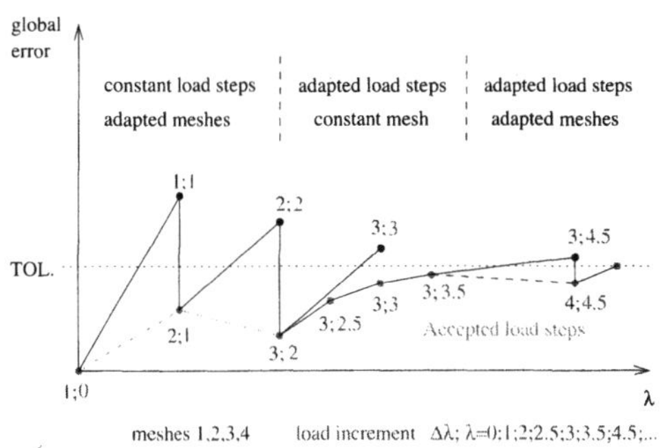

Figure 20: The algorithm by Barthold, Schmidt & Stein for staggered space-time adaptivity

In contrast to this, the algorithm by Barthold, Schmidt & Stein [10, 35, 46], see figure 20, estimates the error in space and time after each applied load increment. Consequently, only those load increments with a total error of $\eta_{total} > TOL.$ are to be re-computed with a refined mesh and/or time step size. If the time- dependent error is too big in comparison with the error in space a back-stepping in time is applied in order to consider the accumulation effect. The advantages are high flexibility as well as low computing time. The disadvantage is the complicated algorithm as it needs the transfer of computed data between two given meshes. It should be pointed out that adaptivity of time discretization is especially important for un- and reloading processes like cyclic loading.

10.3.2 Global or local error indicators

For years it was a standard procedure to estimate global errors by element-wise computation, e.g. the error of equilibrium by means of the energy norm. The goal there is to obtain equidistributed error contributions throughout the system i.e. to get the same quality of the solution in each point of a system.

Since a few years, goal-oriented error estimators, e.g. local error estimators and indicators were developed by e.g. Ainsworth [2], Fish and Markolefas [17], Babuškaand Strouboulis [7], Rannacher and Suttmeier [34], Ohnimus, Stein and Walhorn [30] and Stein and Rüter [44] which search for an optimal adaptive process with respect to highest accuracy of a local quantity, a function of it or a functional, e.g. the J-integral in fracture mechanics.

Shortcomings of this strategy arise for non-linear consitutitve equations, because the region of maximum stresses can vary during the loading process. Therefore, the adaptive process would have to adapt to these conditions. Another crucial point is the so-called pollution error, which represents the influence of the global error onto the local error and reflects St. Venants principle for strong elliptic problems. The pollution error needs a global solution, using e.g. Green's functions. In total, goal-oriented error estimation is of growing importance.

It may be sufficient for complex systems with an easy load history to compute a solution of guaranteed quality for few given physical properties at only some distinguished points, instead of equilibrating the error throughout the whole domain, leading to very fine meshes nearly everywhere.

As a consequence, the choice of the optimal strategy for error estimation and error control depends on the goal of the computation and the complexity of the problem, as well as on the number of loading cases.

10.3.3 Refinement strategy

A given element e is refined if it contributes an error of e.g. $\geq$70% of the maximum error,

$$\eta_e \geq 0.7 \max_{E=1,n_{el}} \eta_E,$$

and it is to be coarsened if e.g.

$$\eta_e \leq 0.2 \max_{E=1,n_{el}} \eta_E.$$

With these rules a refinement of all those elements is assured which are critical for the overall quality of the solution.

Though the time discretization error is computed for each element separately, the time-step length is changed globally due to the operator split of FEM in space and FDM in time. The time-step-length is refined if the L_2-norm of each element contribution is higher than a given tolerance. Time-step-coarsening is performed for an L_2-norm of the element contributions less than

$$\eta_{time} < 2^{-n}\Delta t \cdot \mathrm{TOL.},$$

with n being the number of performed time-step-refinements. Especially for loading- and unloading processes the time-integration error must be controlled.

10.4 Mesh transfer of computed data

The transfer of computed data between subsequent hierarchical meshes can be done by a simple interpolation with Gauss-point based additional shape functions $\mathbf{M}(\xi, \eta)$ as proposed by Ortiz/Quigley in [31].

A better local approximation for element patches is possible with the so-called super-convergent-patch- recovery (SPR) technique by Zienkiewicz and Zhu [48], (polynomial smoothing) which is based on the fact that smoothed solutions have a higher rate of convergence in the numerical integration points than elsewhere if the regularity of the true solution is sufficiently high. From these points, an improved solution is computed by means of a least squares approximation

$$F(a) = \sum_{i=1}^{n_{sp}} (\boldsymbol{\sigma}_h - \boldsymbol{\sigma}_*)^2 = \sum_{i=1}^{n_{sp}} (\boldsymbol{\sigma}_h - \mathbf{Pa})^2 \tag{161}$$

$$\frac{dF}{d\mathbf{a}} = 0 \quad \longrightarrow \quad \sum_{i=1}^{n_{sp}} \underbrace{\mathbf{P}^T\mathbf{P}}_{\mathbf{M}}\mathbf{a} = \sum_{i=1}^{n_{sp}} \mathbf{P}^T\boldsymbol{\sigma}_h \rightarrow \mathrm{Min}, \tag{162}$$

with the polynomial vector $\mathbf{P} = [1, x, y, xy]$ for the bilinear 4-node-element, the vector $\mathbf{a}$ of unknowns in Gauss-points and the number n_{sp} of super-convergent points on the patch. The property of super convergence is restricted to limited distortion of the elements, e.g. in the frame of isoparametric finite elements.

As the transfer algorithm uses a patch of elements, it delivers a higher quality interpolation of the computed variables than the simple element-wise one.

In addition, the performance of SPR can be further enhanced by the introduction of secondary conditions $\tilde{\mathbf{P}}$ to the minimal condition of equation (161) where one possible secondary condition is e.g. the preservation of equilibrium, $\mathrm{div}\boldsymbol{\sigma} \stackrel{!}{=} 0$, or of plastic incompressibility, i.e. $tr\boldsymbol{\varepsilon}_{pl} = 0$.

10.4.1 Data of the benchmark system

We have chosen a plane system of a rectangular domain with a circular hole under plane strain constraint for the numerical simulations of the different element formulations and different error estimators and indicators. This benchmark example was also used by the German research project 'Adaptive finite-element-methods in applied mechanics' [46].

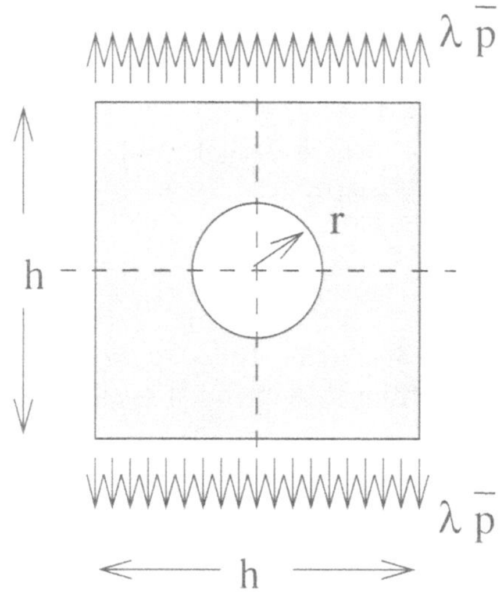

Figure 21a: Benchmark system: Stretched plane strain square with a hole, h=200, r=10

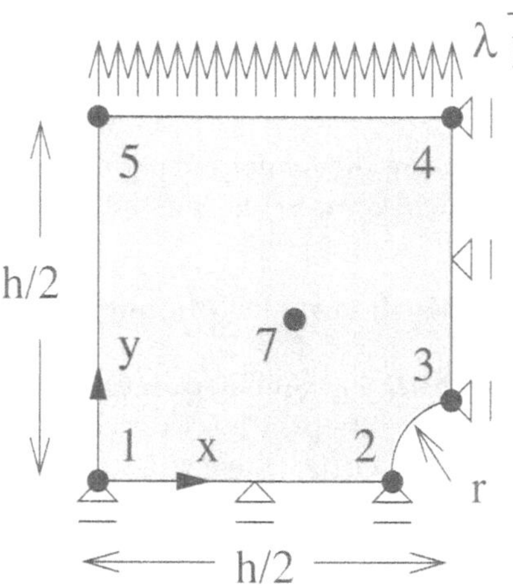

Figure 21b: Upper left quarter, using symmetry conditions, with sample points, see also Figure 22b

One to four material parameters are sufficient for perfect plasticity, while mixed linear and nonlinear isotropic hardening with saturation needs 5 to 7 parameters, table 5. They were determined from tension tests with mild steel specimen by least squares techniques.

1.	Young's modulus	E =206899.94 N/mm^2
2.	Poisson's ratio	ν = 0.29
3.	Bulk modulus	K =164206.00 N/mm^2
4.	Initial yield stress	y_0 = 450.00 N/mm^2
5.	Saturation stress	y_∞ = 750.00 N/mm^2
6.	Linear hardening	h = 129.00 N/mm^2
7.	Hardening exponent	ω = 16.93

Table 5: Material parameters

From a priori knowledge of the type of stress concentrations the adaptive computation starts with a sequence of graded meshes, ranging from 64 elements with 25 nodes to 8192 elements with 8385 nodes used for the examples. Additionally, a reference solution with 24200 elements and 24642 nodes, resulting in 49062 degrees of freedom was considered. The results of these computations are given in the subsequent sections. In order to obtain comparable results for all meshes and material laws, some selected points and data were chosen, table 7, see also Figure 21 b.

In the forthcoming figures 22–27, the spatial distributions of the different error indicators are shown for a load factor of $\lambda = 4.5$ where the critical load for perfect plasticity

Mesh No.	number of elements	nodes	d.o.f.
1	16	25	40
2	64	81	144
3	256	289	544
4	1024	1089	2112
5	4096	4225	8320
6	8192	8385	16640
reference	24200	24642	49062

Table 6: Data of the different starting-meshes for adaptivity, figure 22b

Physical quantity		node No.
Displacement	u_x	2
Displacement	u_y	4
Displacement	u_x	5
Stress	σ_{yy}	2
Stress	σ_{xx}	7
Integral	$\int_4^5 \|u_y\|\, dx$	4–5

Table 7: Selected points and data of system, Figure 21a,b, for computation

is $\lambda_{crit} = 4.66$, a $\lambda_{crit} = 7.89$ holds for isotropic hardening according to Eq. (150) and material parameters in Table 5. The load from $\lambda = 0$ to 4.5 was applied in 18 load steps which renders the implicit time discretization error to be negligible against the errors due to spatial discretization. All computations hold for hardening material, table 5.

10.5 Global error estimates and results for mesh adaptivity

10.5.1 Spatial discretization errors

As elastoplastic deformations are history dependent, the displacement fields $\boldsymbol{u}(\boldsymbol{x},t)$ determine the actual stress and strains states for all points $\boldsymbol{x} \in \Omega$ at all times $0 \leq t \leq T$. Thus, the spatial discretization error $\boldsymbol{e}_u$ measures the difference between the exact displacement field $\boldsymbol{u}$ and its finite-dimensional finite element approximation $\boldsymbol{u}_h$, $\boldsymbol{e} = \boldsymbol{u} - \boldsymbol{u}_h$.

A physically meaningful error measure is achieved by considering the point-wise incremental total stress power $\dot{\mathcal{P}}$, where $\dot{\gamma} \geq 0$ assures the fulfillment of the yield condition $\Phi \leq 0$

$$\dot{\mathcal{P}} := \boldsymbol{\sigma} : \dot{\boldsymbol{\varepsilon}} + \dot{\gamma}\Phi = \underbrace{\boldsymbol{\sigma} : \dot{\boldsymbol{\varepsilon}}^e + \beta\dot{\alpha}}_{\dot{\Psi}} + \underbrace{\boldsymbol{\sigma} : \dot{\boldsymbol{\varepsilon}}^p - \beta\dot{\alpha}}_{\mathcal{D}^p} + \dot{\gamma}\Phi = \dot{\Psi} + \mathcal{D}^p + \dot{\gamma}\Phi. \tag{163}$$

All terms have the physical meaning of power, i.e. work per unit time. Thus, the stress power $\dot{\mathcal{P}}$ within elastoplasticity plays the role of the potential energy in elasticity. Therefore, a natural choice of a spatial error norm is based on $\dot{\mathcal{P}}$, measuring the increase of the spatial discretization error at each time t.

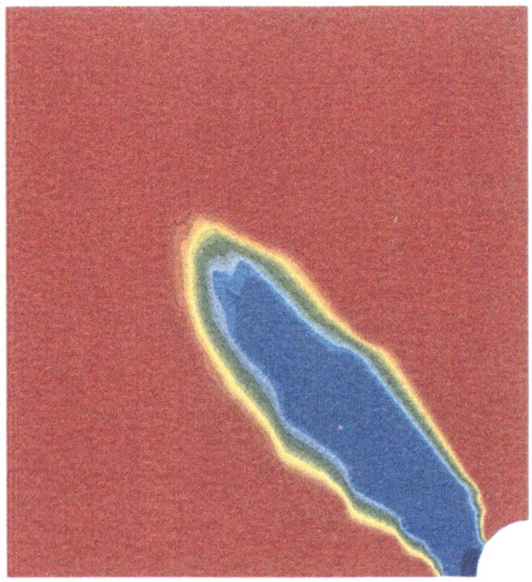

Figure 22a: v. Mises stresses at $\lambda = 4.5$ for hardening material

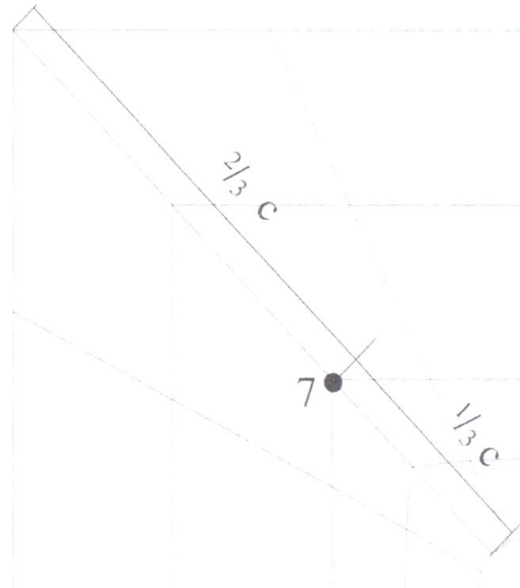

Figure 22b: Coarsest graded mesh with 16 elements a=61.953, b=28.047, c=131.421

Time integration over the interval of interest $[0, t]$ with $t_0 = 0$ and $t = t_{n+1}$ yields

$$\int_0^t \dot{\mathcal{P}} d\tau = \int_0^t \left(\dot{\Psi} + \mathcal{D}^p + \dot{\gamma}\Phi\right) d\tau \tag{164a}$$

$$= \Psi + \int_0^t (\mathcal{D}^p + \dot{\gamma}\Phi)\, d\tau = \Psi + \sum_{i=0}^{n} \int_{t_i}^{t_{i+1}} (\mathcal{D}^p + \dot{\gamma}\Phi)\, d\tau, \tag{164b}$$

with $\dot{\mathcal{P}} = \dot{\Psi}$ in the limit case of pure elasticity.

The occurring errors in the primal FEM can be characterized as follows, see Eqns. 164a, 164b:

1. The error of equilibrium related to the elastic strains and stresses:

$$\eta_{eq}^2 = ||(\boldsymbol{\sigma}^e - \boldsymbol{\sigma}_h^e) : (\boldsymbol{\varepsilon}^e - \boldsymbol{\varepsilon}_h^e)||^2_{L_2(\Omega)} + ||(\alpha - \alpha_h)H(\alpha - \alpha_h)||^2_{L_2(\Omega)} \tag{165}$$

2. The error increment of the plastic strain rates, related to plastic dissipation, multiplied by the related stresses in order to get the same dimension for $\eta_{\dot{\boldsymbol{\varepsilon}}^p}$ and η_{eq}:

$$\eta_{\dot{\boldsymbol{\varepsilon}}^p}^2 = ||\boldsymbol{\sigma} : (\dot{\boldsymbol{\varepsilon}}^p - \dot{\boldsymbol{\varepsilon}}_h^p)||^2_{L_2(\Omega)} + ||\beta(\dot{\alpha} - \dot{\alpha}_h)||^2_{L_2(\Omega)} \tag{166}$$

3. The error increment in the Kuhn-Tucker conditions for loading and unloading:

$$\eta_{KT} = \int_t \int_\Omega |\dot{\gamma}_h \Phi_h|^2 \, d\Omega \, dt. \tag{167}$$

10.5.2 Spatial error indicators

The global error of the equilibrium conditions can also be controlled by means of the residual error estimator η_{eq} by Babuška, Rheinboldt & Miller [6] and Johnson, Hansbo [23]. It reads

$$\eta_{eq}^2 = C_R h^2 ||\mathbf{R}_h||^2_{L_2(\Omega)} + C_j h ||\mathbf{J}_h||^2_{L_2(\Gamma)}, \qquad [\eta_{eq}] = \sqrt{\mathrm{Nm}} \tag{168}$$

with the residuals $\mathbf{R}_h = \mathrm{div}\ \boldsymbol{\sigma}_h - \boldsymbol{f}_h$, the jumps of the tractions on inter-element boundaries $\boldsymbol{J}_h$ and the interpolations constants C_R and C_j.

Figure 23a shows the spatial distribution of η_{eq} for the load $\lambda = 4.5$.

Figure 23a: η_{eq} for a load of $\lambda = 4.5$

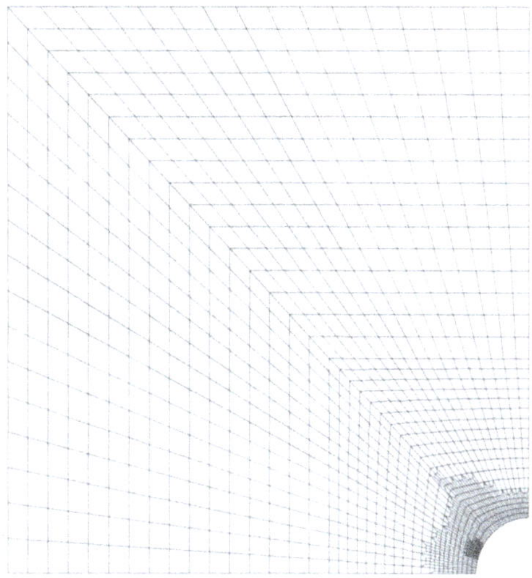

Figure 23b: The resulting mesh for η_{eq} with 1168 elements and 1242 nodes

Figures 23a, 23b illustrate that the areas of high errors of equilibrium are concentrated at the edge of the hole close to the singularity. It is obvious that this error indicator is not sufficient to describe the effects of elastoplastic deformations.

The error increment of the plastic strain rates $\dot{\boldsymbol{\varepsilon}}^p$ and their directions at the and of a load-step can be estimated by equation (166) as

$$\eta^2_{\dot{\boldsymbol{\varepsilon}}^p} = ||\boldsymbol{\sigma} : (\dot{\boldsymbol{\varepsilon}}^p - \dot{\boldsymbol{\varepsilon}}^p_h)||^2_{L_2(\Omega)} + ||\beta \cdot (\dot{\alpha} - \dot{\alpha}_h)||^2_{L_2(\Omega)}, \qquad [\eta_{\dot{\boldsymbol{\varepsilon}}^p}] = \sqrt{\mathrm{Nm}}, \tag{169}$$

introduced in [10]. As the exact solutions $\boldsymbol{\sigma}$, $\dot{\boldsymbol{\varepsilon}}^p$, β and $\dot{\alpha}$ are unknown, we use averaged discretized solutions $\boldsymbol{\sigma}_*$, $\dot{\boldsymbol{\varepsilon}}^p_*$, β_* and $\dot{\alpha}_*$, gained by L_2-projections according to [47]. Thus $\eta^2_{\dot{\boldsymbol{\varepsilon}}^p}$ reads

$$\eta^2_{\dot{\boldsymbol{\varepsilon}}^p} \approx ||\boldsymbol{\sigma}_* : (\dot{\boldsymbol{\varepsilon}}^p_* - \dot{\boldsymbol{\varepsilon}}^p_h)||^2_{L_2(\Omega)} + ||\beta_* \cdot (\dot{\alpha}_* - \dot{\alpha}_h)||^2_{L_2(\Omega)} \tag{170}$$

Figure 24a: $\eta_{\dot{\varepsilon}^p}$ at a load of $\lambda = 4.5$

Figure 24b: The refined mesh for $\eta_{\dot{\varepsilon}^p}$ with 1285 elements, 1368 nodes

Figures 24a, 24b show the distribution of $\eta_{\dot{\varepsilon}^p}$ at $\lambda = 4.5$ and the related mesh.

The error indicator by Perič and Owen [32]

$$\eta_{PO} \quad = \quad ||(\boldsymbol{\sigma} - \boldsymbol{\sigma}_h) : (\dot{\boldsymbol{\varepsilon}}^p - \dot{\boldsymbol{\varepsilon}}^p_h)||^2_{L_2(\Omega)} \tag{171}$$

$$\eta_{PO} \quad \approx \quad ||(\boldsymbol{\sigma}_* - \boldsymbol{\sigma}_h) : (\dot{\boldsymbol{\varepsilon}}^p_* - \dot{\boldsymbol{\varepsilon}}^p_h)||^2_{L_2(\Omega)}, \tag{172}$$

combines the error of equilibrium within the term $\boldsymbol{\sigma} - \boldsymbol{\sigma}_h$ with the error of plastic strain rates in form of the Z^2-indicator using averaging. This results in a concentration of the mesh refinement in the vicinity of the hole as depicted in Figure 25a,b. Thus, the mesh is refined closer to the hole. It is open whether the combined indicator, Eq. 172, has asymptotic properties for both the equilibrium and the plastic strains.

A third error occurs in the yield function Φ and is controlled by the the error indicator of the Kuhn-Tucker conditions for loading and unloading

$$\dot{\gamma} \geq 0, \qquad \Phi \leq 0, \qquad \dot{\gamma}\Phi = 0. \tag{173}$$

The related global error indicator η_{KT} by Barthold, Schmidt and Stein [10] reads with (167)

$$\eta^2_{KT} = \int_t \int_\Omega |\dot{\gamma}\Phi - \dot{\gamma}_h \Phi_h|^2 \, d\Omega \, dt = \int_t \int_\Omega |\dot{\gamma}_h \Phi_h|^2 \, d\Omega \, dt; \qquad [\eta_{KT}] = \sqrt{\text{Nm}}. \tag{174}$$

The yield condition is fulfilled at all quadrature points but not identically elsewhere in the elements. The idea for η_{KT} is consequently an extrapolation of the given Gauss-point values to integration points of higher polynomial order as shown in Figure 26.

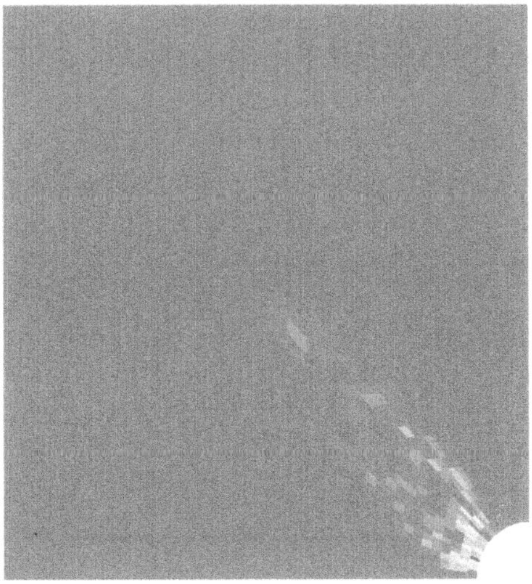

Figure 25a: η_{PO} at a load of $\lambda = 4.5$

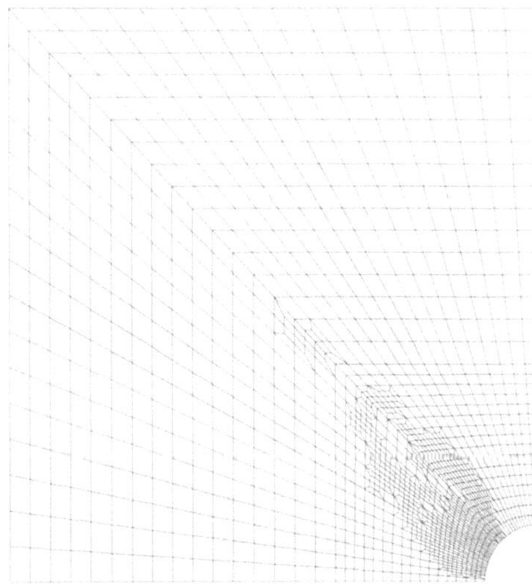

Figure 25b: The refined mesh for η_{PO} with 1324 elements, 1398 nodes

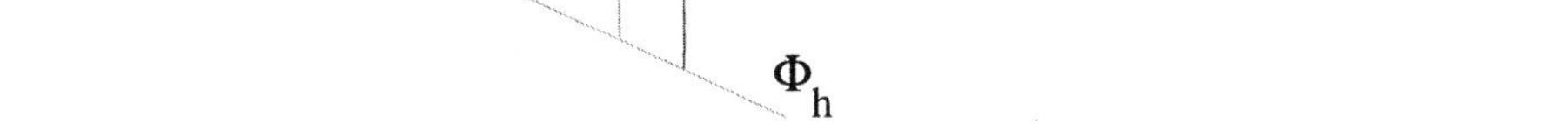

Figure 26: Interpolation of Gauss-point values by numerical integration points of a higher polynomial order

Figures 26a and 26b show the error distribution of η_{KT} at a load of $\lambda = 4.5$ and the resulting mesh refinement. This estimator yields mesh refinement at the transition zone between the elastic and the plastic parts of the domain, i.e. in those regions in which the elements are not fully plastic.

10.5.3 Estimation of time discretization errors

Using implicit numerical integration schemes such as here the backward Euler rule, the time discretization error occurs for each time interval $[t_i; t_{i+1}]$.

A physically motivated time discretization error estimate can be derived by the com-

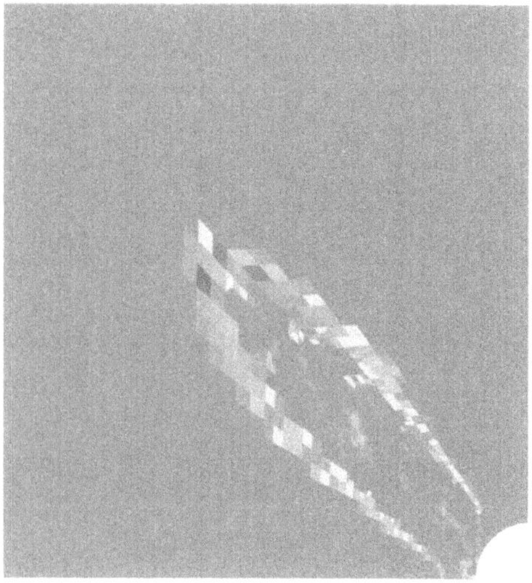

Figure 27a: η_{KT} at a load $\lambda = 4.5$

Figure 27b: Refined mesh for η_{KT}

parison of Hencky- and Prandtl-Reuss-elastoplasticity theory. In the latter theory, the plastic strain increments $\dot{\boldsymbol{\varepsilon}}^p$ (instead of the plastic strains $\boldsymbol{\varepsilon}^p$) are related to the stress deviator in the flow rule. Therefore the Hencky-quasiplasticity is a non-linear elastic material-law with an additional yield condition without any time-discretization errors.

The basic idea consists in detecting points within the plastic domain with maximal deviation of Prandtl-Reuss-plasticity from Hencky-plasticity.

The error estimator η_{time} in the numerical integration of the flow rule is such yields

$$\eta_{time} := ||\boldsymbol{\sigma}||_{L_2(\Omega)} \left\| \int_t \dot{\boldsymbol{\varepsilon}}^p \, dt - \Delta\boldsymbol{\varepsilon}^p \right\|_{L_2(\Omega)} \tag{175}$$

with the scaling factor $||\boldsymbol{\sigma}||_{L_2(\Omega)}$ which is introduced in order to characterize the influence on the stress power. The physical dimension of the square of this error indicator is — as for the others — the dimension of work Nm.

For numerical computations we furthermore assume in the frame of operator split, that the spatial discretization does not affect the time discretization error significantly, i.e. the unknowns in the definition above can be replaced by $\boldsymbol{\sigma}_h, \dot{\boldsymbol{\varepsilon}}_h^p$ and $\Delta\boldsymbol{\varepsilon}_h^p$. But the coupling of both errors, especially the accumulation of the time-dependent error, is considered by the staggered refinements in time and space as outlined in section 10.3.1 and in fig. 20.

This error estimator can then be bounded for each time step by values at the end of the time step

$$\left\| \int_{t_i}^{t_{i+1}} \dot{\boldsymbol{\varepsilon}}_h^p \, dt - \Delta\boldsymbol{\varepsilon}_h^p \, dt \right\|_{L_2(\Omega)} = \left\| \int_{t_i}^{t_{i+1}} \dot{\gamma}\mathbf{n}(t) \, dt - \Delta t \gamma_{i+1}\mathbf{n}_{i+1} \right\|_{L_2(\Omega)} \tag{176}$$

$$= \left\| \int_{t_i}^{t_{i+1}} [\dot{\gamma}\mathbf{n}(t) - \gamma_{i+1}\mathbf{n}_{i+1}]\, dt \right\|_{L_2(\Omega)} \tag{177}$$

$$\leq \int_{t_i}^{t_{i+1}} ||\dot{\gamma}\mathbf{n}(t) - \gamma_{i+1}\mathbf{n}_{i+1}||_{L_2(\Omega)}\, dt \tag{178}$$

$$\leq \Delta t \max_{t\in[t_i;t_{i+1}]} ||\dot{\gamma}\mathbf{n}(t) - \gamma_{i+1}\mathbf{n}_{i+1}||_{L_2(\Omega)} \tag{179}$$

where $\mathbf{n} = \mathrm{dev}\boldsymbol{\sigma}/||\mathrm{dev}\boldsymbol{\sigma}||$ denotes the outer normal tensor to the yield surface.

The assumptions

- The difference between the scaled directions $\mathbf{n}(t)$ and $\mathbf{n}_{i+1}$ in the current time step is maximal for $t = t_i$.
- The Lagrangian multiplier $\gamma(t)$ is chosen as γ_{i+1} as the worst case.

are introduced in order to obtain the computable bound

$$\eta_{time} = \gamma_{i+1}||\mathbf{n}_i - \mathbf{n}_{i+1}||\Delta t. \tag{180}$$

Figures 27a,b show the spatial distribution of η_{time} at different loads. It is significant for this error indicator that the regions with beginning plasticity contribute high values to η_{time} whereas already plastified sub-domains add only low values. Therefore, a mesh coarsening in fully plastic regions can be controlled with this indicator. This option was implemented in the computer code.

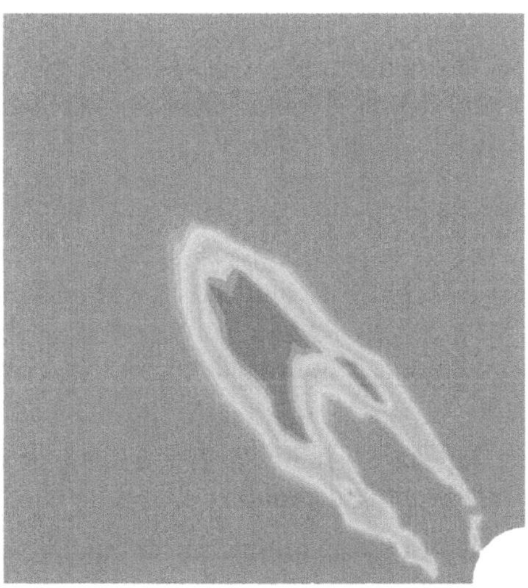

Figure 28a: η_{time} at $\lambda = 4.5$

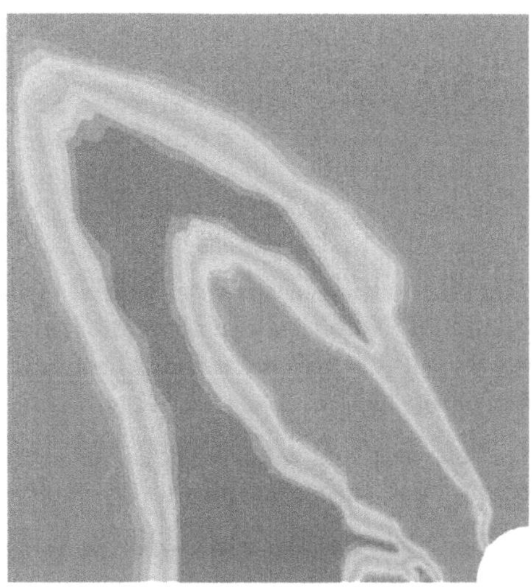

Figure 28b: η_{time} at $\lambda = 4.75$

Figure 28 shows the importance of adaptivity in time for the stresses σ_{xx} at point 7, given in figure 22b. There, interior unloading occurs with decaying stresses which are very sensitive with respect to adaptivity. Mesh No. 3, given in Figure 31b, was the start mesh for two computations with different load discretizations of 10 and 25 load steps, respectively. The differences of results are obvious for load factors $\lambda > 4.0$ where the maximum stresses differ by 10 %.

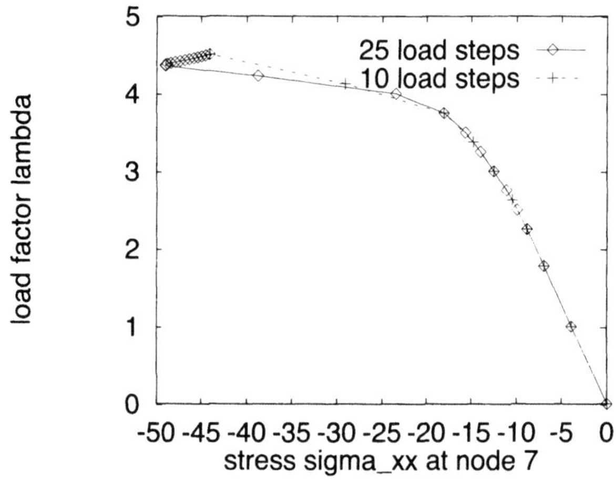

Figure 29: Load history with different numbers of load steps, influence on the sensitive stress σ_{xx} at node 7

10.5.4 Total approximation error estimate

All the three error indicators in space as well as the indicator in time, described above are computed separately and scaled in their natural norms. A given element e is refined if it contributes an error which is above 70% of the maximum refine if

$$\eta_e \geq 0.7 \max_{E=1,n_{el}} \eta_E$$

The distribution of the total spatial error is shown in figure 29 and the related mesh in Figure 29. Note that the computation of $\eta_{all} := (\eta_{eq}^2 + \eta_{\varepsilon^p}^2 + \eta_{KT}^2)^{1/2}$ is possible as the squares of all three error indicators have the same physical dimension of work Nm, see Eq. (163).

This resulting refinement is concentrated at the edge of the hole and in the plastic domain and takes care for all physical error of the discretized problem.

The figures of the spatial distributions of the different error estimators and indicators as well as the related meshes shown in section 9.4 demonstrate that it is not sufficient to apply error-controlled adaptivity for elastic deformations only as they are not capable for describing all physical effects of the plastic flow theory. A true estimation of the global total error can only be achieved by a combination of all related error measures.

Figure 30a: All spatial error indicators at $\lambda = 4.5$; % $\eta_{all} := (\eta^2_{eq} + \eta^2_{\dot{\varepsilon}^p} + \eta^2_{KT})^{1/2}$

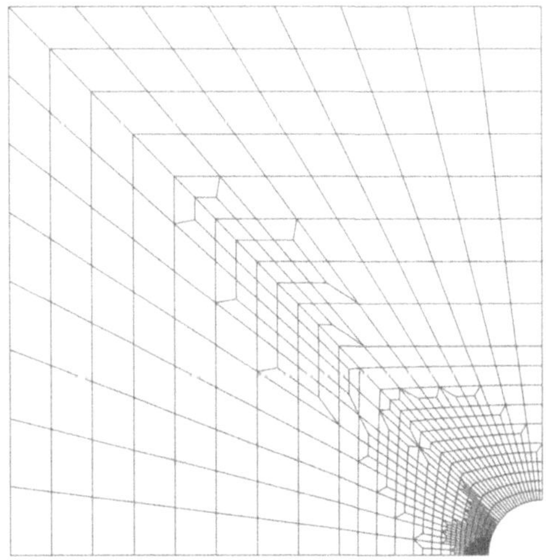

Figure 30b: Refined mesh for all spatial error indicators η_{all}

The analysis described above is based on the assumption that spatial and time discretization errors are decoupled. A consistent estimation of the time discretization error should be possible by a Galerkin method in space and time.

10.6 Further numerical results

The following numerical examples were performed for the benchmarking system of a rectangular domain with a circular hole under plane strain as described in section 10.5.. The results were computed for three material models:

1. linear elastic material
2. Hencky-type non-linear elastic materials
3. elastoplastic material with Hill's flow theory and nonlinear isotropic hardening,

each case for a sequence of graded meshes first, usually followed by error controlled adaptive meshes. Meshes 2, 3 and 4 are shown in figures 30a- c, mesh No. 1 see Figure 22b. It can be observed that the combination of graded meshes for a specific problem, followed by error–controlled adaptive meshes may be more efficient than to start the adaptive process with a homogeneous mesh, see also section 10.5.4.

Figure 31a: Graded mesh No. 2 with 64 elements

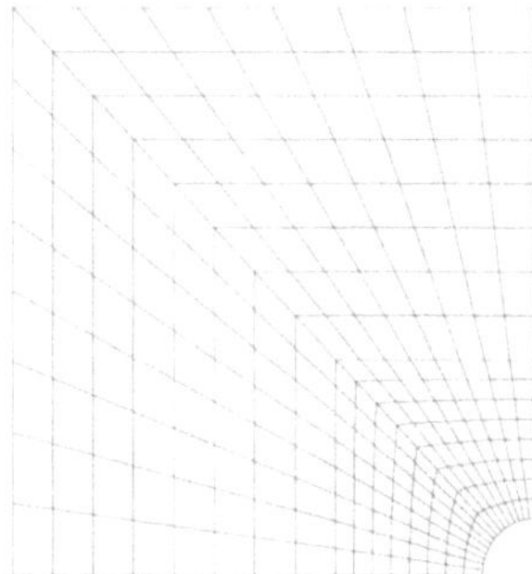

Figure 31b: Graded mesh No. 3 with 256 elements

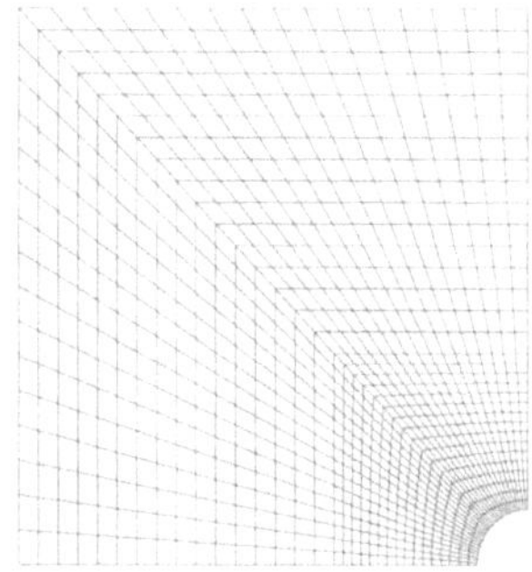

Figure 31c: Graded mesh No. 4 with 1024 elements

10.6.1 Elastic Material

The quoted mesh number 2, Figure 31a, was refined in 8 steps using the equilibrium error estimator by Babuška- Miller yielding 800 elements with 1680 degrees of freedom. The results of the adaptive run are shown in figures 31a-f in comparison with the results of the sequence of graded meshes. The locations of the selected points for results are given in figure 21b

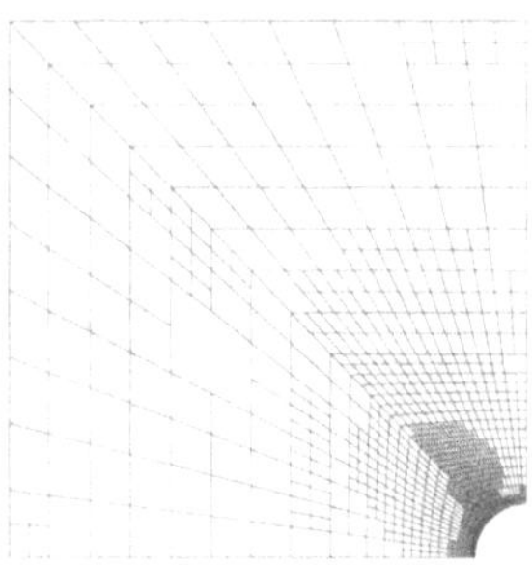

Figure 32a: Adapted mesh for elastic deformations at load factor $\lambda = 4.5$

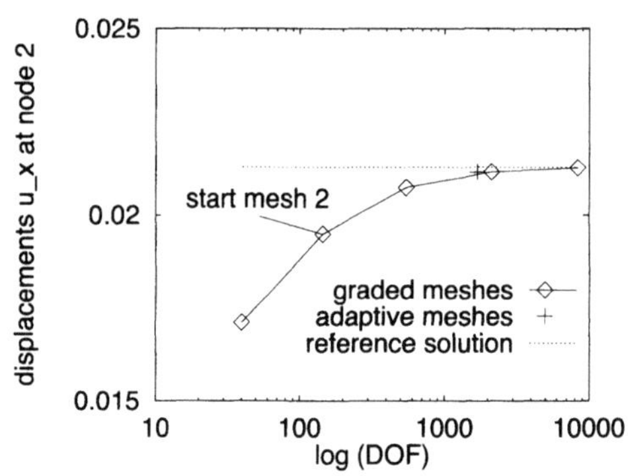

Figure 32b: Displacements $\mathbf{u}_x$ of node 2, figure 21b

The good convergence of the adaptive elastic computations for all selected quantities is obvious. The asymptotic values were computed with the reference mesh of 24200 elements.

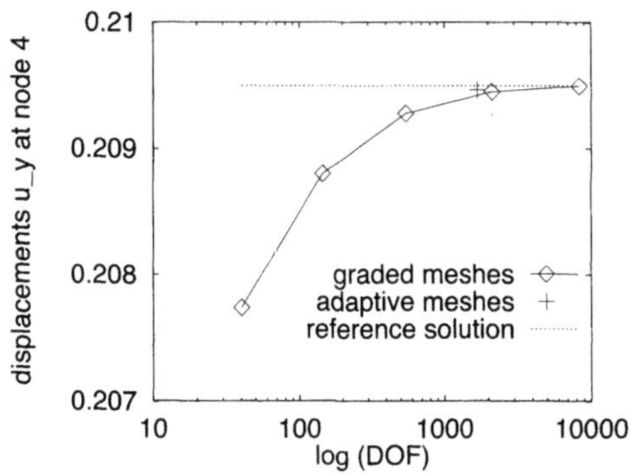

Figure 32c: Displacements $\mathbf{u}_y$ of node 4, figure 21b

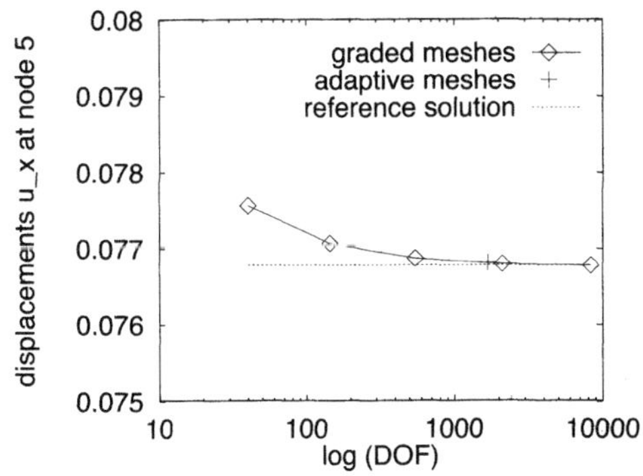

Figure 32d: Displacements $\mathbf{u}_x$ of node 5, figure 21b

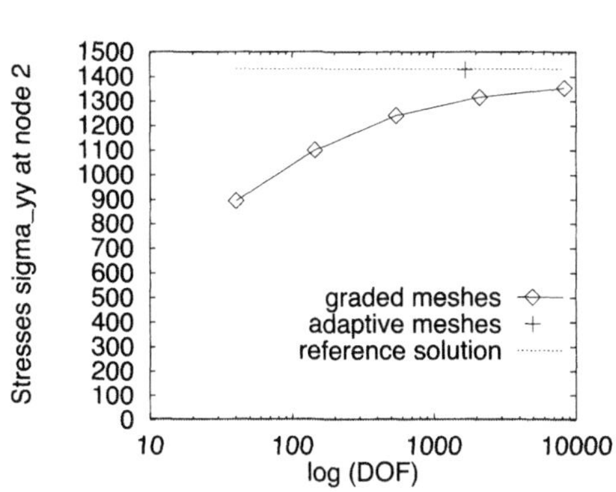

Figure 32e: Stresses $\boldsymbol{\sigma}_{yy}$ of node 2, figure 21b

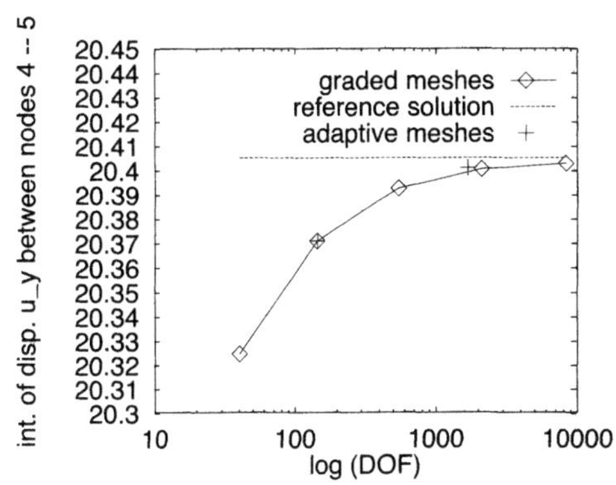

Figure 32f: $\int_4^5 |\mathbf{u}_y|\, dx$

10.6.2 Hencky-type non-linear elasticity

Mesh number 2, Figure 30a, was refined in 10 steps using the equilibrium criterion by Babuškaand Miller, yielding 875 elements with 1816 degrees of freedom. The results of the adaptive computations are shown in the following figures 33a-d in comparison with the results of the sequence of graded meshes.

The displacements in figures 32b-d show lower efficiency of the adaptive method than in the elastic case.

While the wanted asymptotic convergence is attained easily in the elastic case, it is harder to reach it for the Hencky-plasticity in spite of the use of about 10,000 unknowns.

As the related figures for Prandtl-Reuss elastoplasticity are qualitatively the same, they are not shown here.

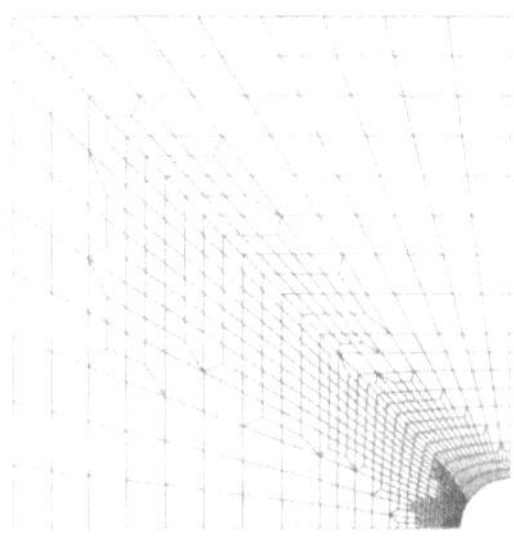

Figure 33a: Adapted mesh at $\lambda = 4.5$ for Hencky-plasticity

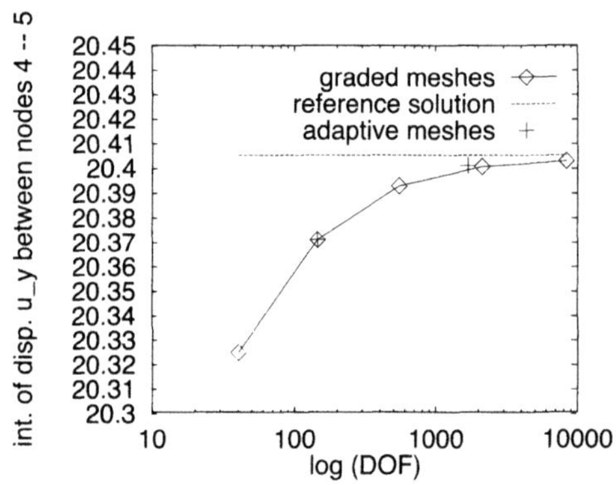

Figure 33b: $\int_4^5 |\mathbf{u}_y|\, dx$ at $\lambda = 4.5$ for Hencky-plasticity

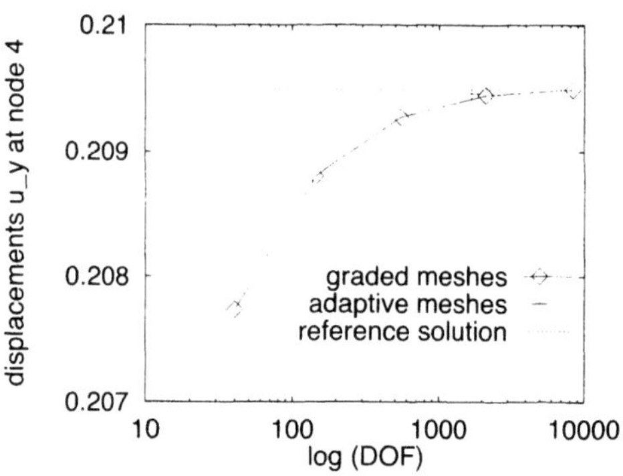

Figure 33c: Displacements $\mathbf{u}_y$ of node 4 at $\lambda = 4.5$ for Hencky-plasticity

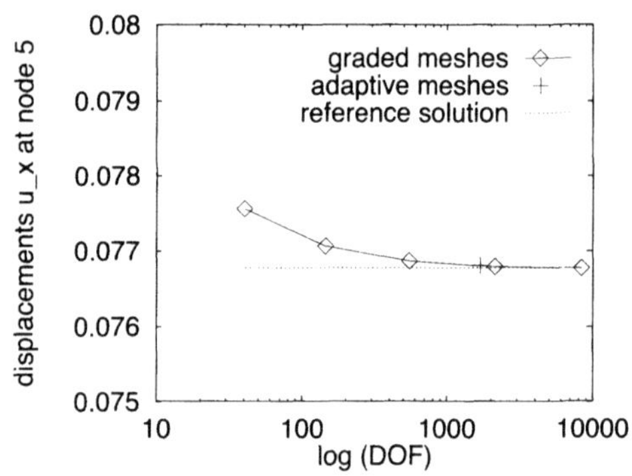

Figure 33d: Displacements $\mathbf{u}_x$ of node 5 at $\lambda = 4.5$ for Hencky-plasticity

10.6.3 Prandtl-Reuss-Elastoplasticity with mixed hardening

The main goal of the computations, see figures 34a-c, was to test the numerical efficiency of the adaptive refinements controlled by the different error estimators and indicators described in section 10.5.1.

In the figures 35a-c below, the convergence rates for the three parts of spatial error are shown for a sequence of uniformly refined graded meshes from 256 (mesh no. 3) to 4096 elements (mesh no. 5) for a load of $\lambda = 4.5$. Adaptive computations were performed from start meshes number 2 and 4. Figure 35a presents the error of equilibrium η_{eq}; Figure 35b depicts the evolution of the error $\eta_{\dot{\boldsymbol{\varepsilon}}^p}$ of the plastic strain rates, and figure 35c shows the evolution of the error η_{KT} of the Kuhn-Tucker-conditions. A higher rate of convergence is obtained for all the three adaptive computations.

Finally the convergence of the combined spatial error indicators was considered. Figure 35d shows again that the convergence rate is higher for the adaptive computation starting from the stassered mesh number 4. In the figures 34a-f, an adaptive computation starting

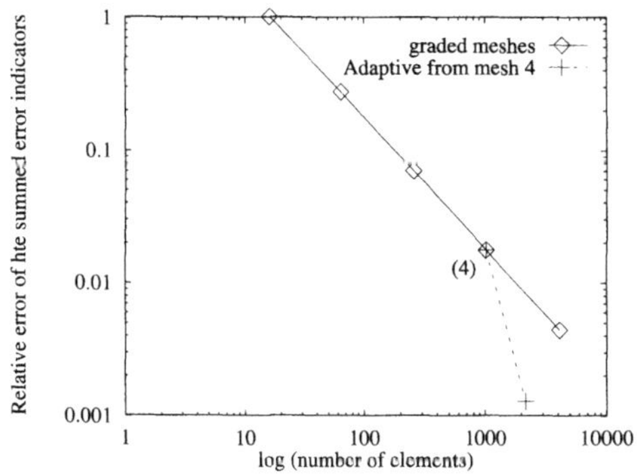

Figure 34a: Convergence due to η_{eq}

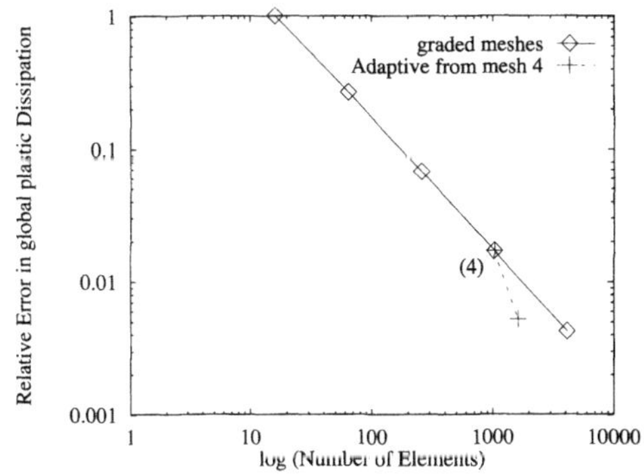

Figure 34b: Convergence due to $\eta_{\dot{\varepsilon}^p}$

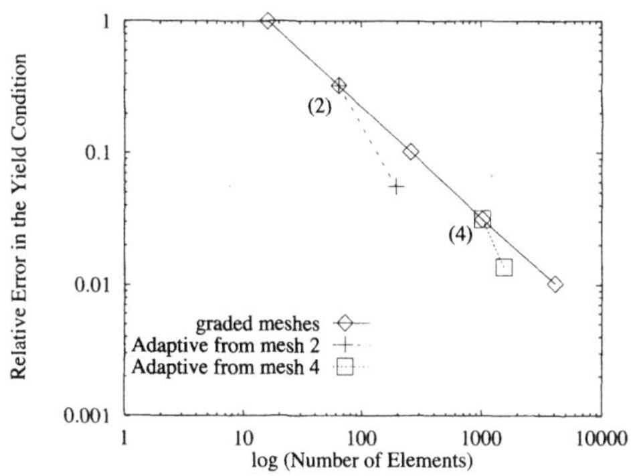

Figure 34c: Convergence due to η_{KT}

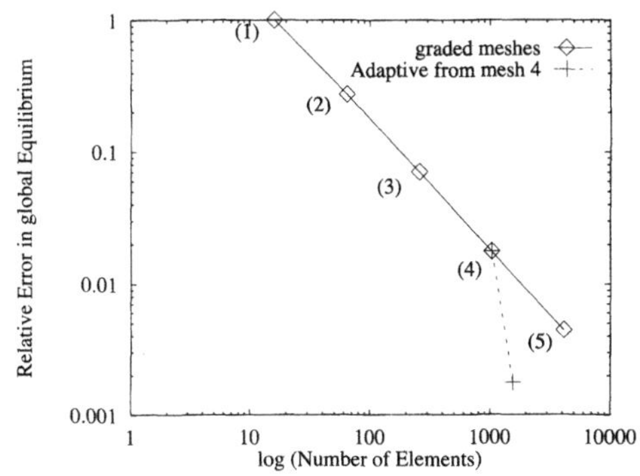

Figure 34d: Convergence due to the total spatial error indicator

from mesh 1, yielding finally 802 elments with 1674 degrees of freedom, is shown in comparison to the solutions of the sequence of graded meshes. Mesh No. 1 with 16 elements remains elastic throughout the whole loading process.

11 Node regular h-,d- and p-mesh refinement of 3D- and 2D-problems including transition elements

11.1 Node regular refinement of hexahedron meshes

In many problems of structural mechanics we have to analyze thin-walled and at least partially orthogonal geometries. These can be discretized with hexahedron meshes much better than with tetraedron meshes. Furthermore the following arguments vote for a node-regular refinement:

Figure 35a: Adapted mesh at $\lambda = 4.5$ for Prandtl-Reuss elastoplasticity

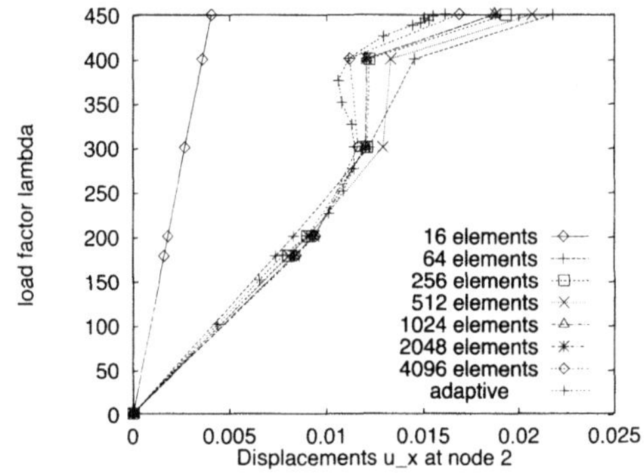

Figure 35b: Displacements $\mathbf{u}_x$ of node 2 over λ for Prandtl-Reuss elastoplasticity

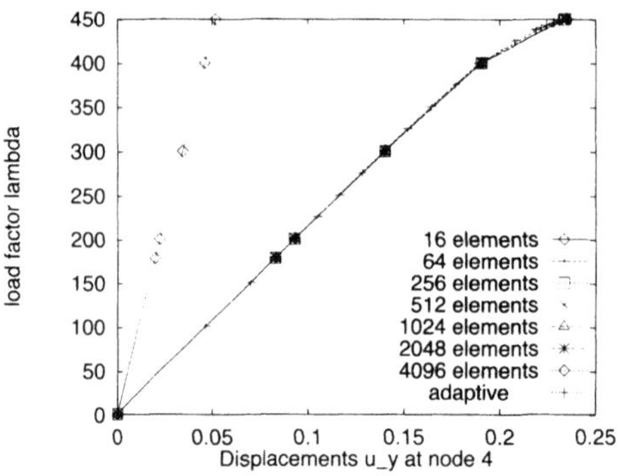

Figure 35c: Displacements $\mathbf{u}_y$ of node 4 over λ for Prandtl-Reuss elastoplasticity

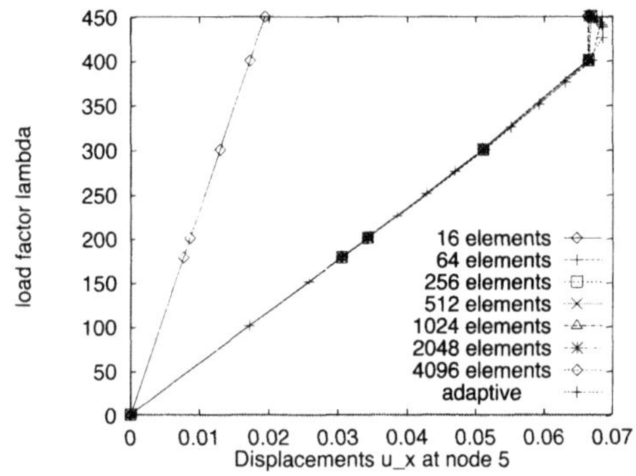

Figure 35d: Displacements $\mathbf{u}_x$ of node 5 over λ for Prandtl-Reuss elastoplasticity

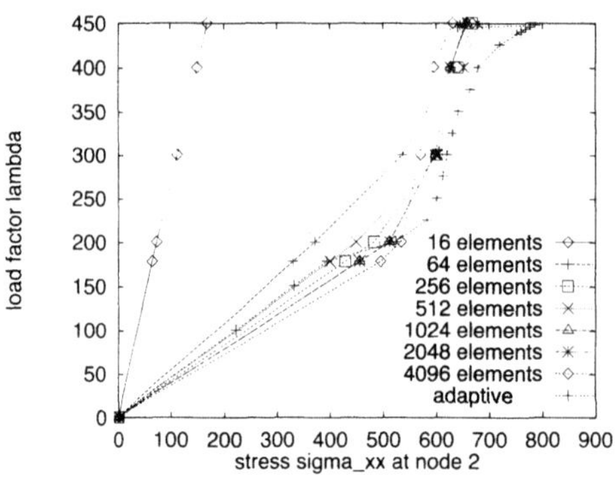

Figure 35e: Stresses $\boldsymbol{\sigma}_{yy}$ of node 2 over λ for Prandtl-Reuss elastoplasticity

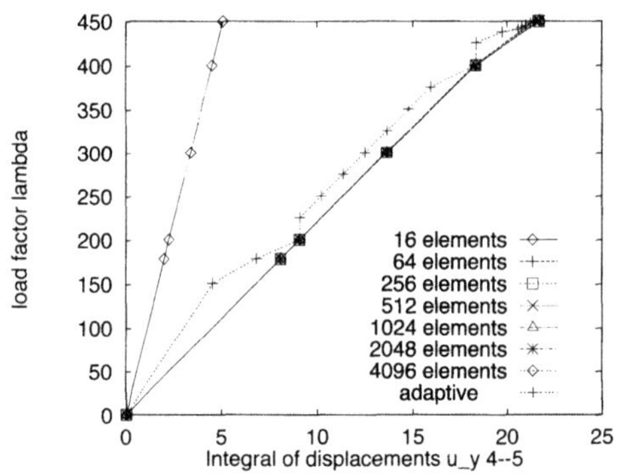

Figure 35f: $\int_4^5 |\mathbf{u}_y|\, dx$ over λ for Prandtl-Reuss elastoplasticity

- The resulting system of equations can be given to the solver without further preparation like adding a penalty term or condensating the hanging nodes.

- The meshes are also suitable for mixed element formulations for which in the case of hanging nodes the stability might not be ensured.

- The node- edge and face patches needed for local error estimators can be defined without any case distinction.

Unfortunately the node–regular local refinement of hexahedron meshes is much more complicated than the refinement of tetraedron- or 2-dimensional quadrilateral meshes.

First we want to present the underlying data structures. The topology of a finite element mesh is composed of the topological components nodes, edges, faces, regions and their relations to each other.

Figure 35 describes the relations between the classes used for the topological and geometrical description of a finite element mesh. For a better overall view, the representation of mesh is separated into topology and geometry. The h-refinement algorithm only needs the topological view.

In addition to the 'horizontal' relations, e.g. the inclusion of an edge in a face, the 'vertical' relations of the basic elements are of interest concerning the adaptive mesh refinement. These 'vertical' relations are the so-called 'parent-child' relations between a coarse-grid-element and the elements resulting from the refinement of this element. This relation is needed to compute the transfer operators for the multilevel preconditioning. In Figure 36 these entities and relations are shown.

The refinement algorithm is formulated in pure topological terms. Further specifications concerning the type of elements are not used here. Of course, the 'constructors' of the specific numerical elements are called by the 'virtual function call' mechanism during the refinement of the element concerned. The coordinates of the new nodes are computed subsequently. Since each new node is the midpoint of a topological element (i.e. edge, face or region), its coordinates can be computed by using the geometric description of the regarding element.

Besides the representation of topology and geometry the following components are needed to define the regular h-adaption:

Transition elements: during local mesh refinement, shown in Figure 37, a transition between refined and non-refined areas has to be found. This is not possible in one step for node-regular elements of cubic topology (all regular elements are hexahedrons). Therefore two transition elements, shown in Figure 38, are necessary:

The transition element in Figure 38b contains the partial region ABCDE with a volume percentage of 3% which is a degenerated hexahedron – a so-called trieder – arising by collapsing of four corners into one. Three hyperbolic faces confine this trieder, see Figure 39. These trieders only appear in the inner of the considered cube without surfaces at the outer boundary. They disappear in the case of further refinement.

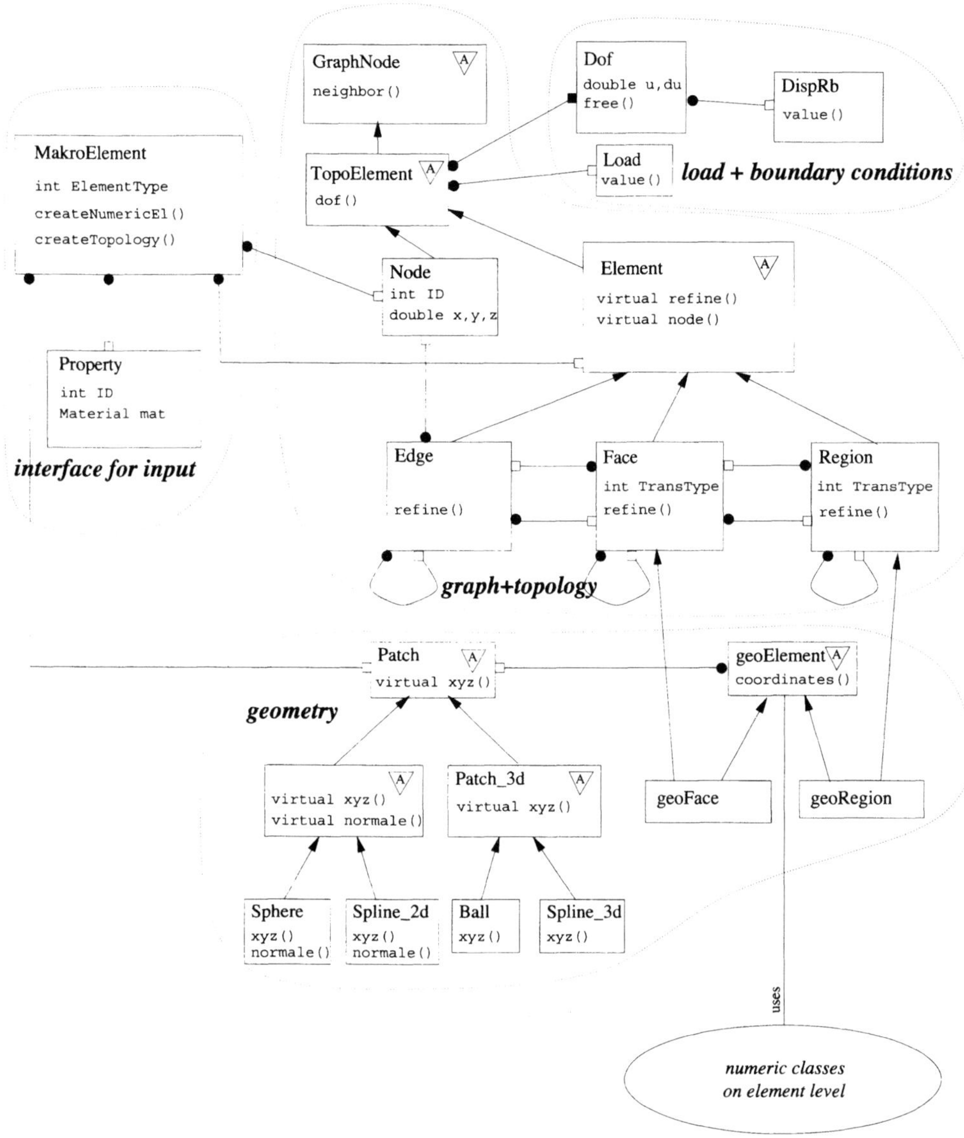

Figure 36: Hierarchy of the classes used for the topological and geometrical description of a finite element mesh

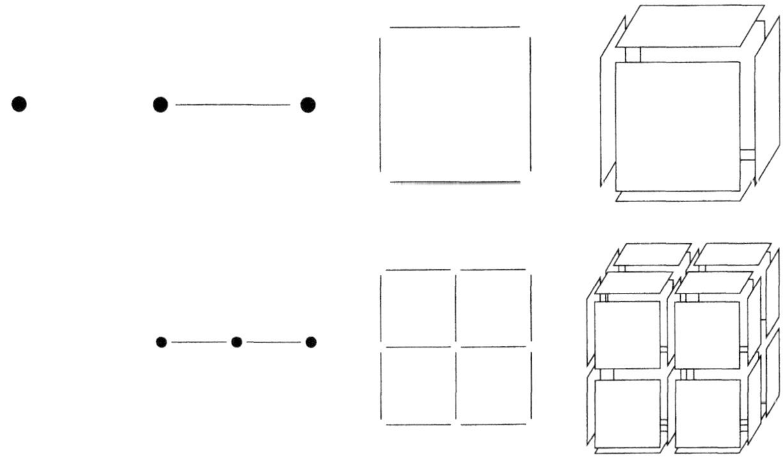

Figure 37: Horizontal and vertical relations between the basic elements of a finite element mesh

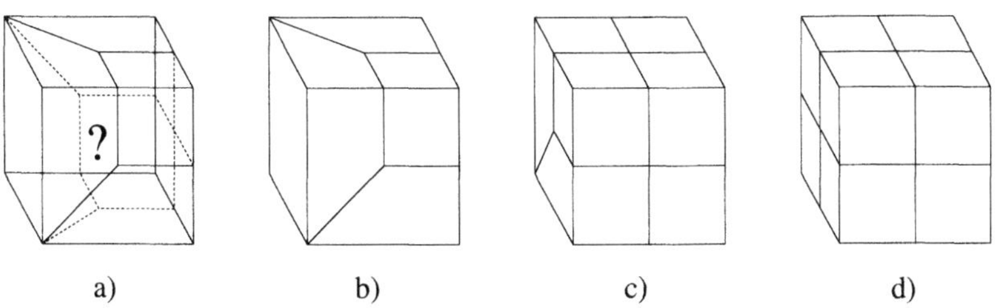

Figure 38: Transition from coarse to refined faces

a) Needed transition from the left coarse face to the refined right face; not possible in one step,
b) - d) Transition from the left coarse face in b) to the refined left face in d) using the intermediate step in c)

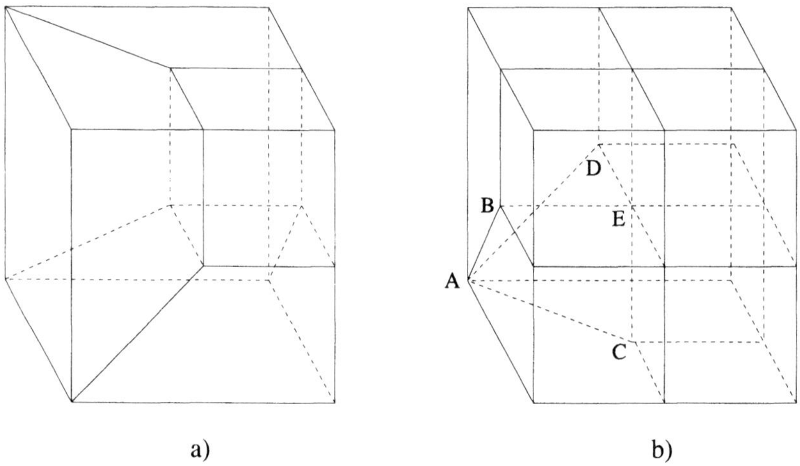

Figure 39: Transition elements. The points ABCDE in b) are the corners of a trieder, also shown in Figure 40

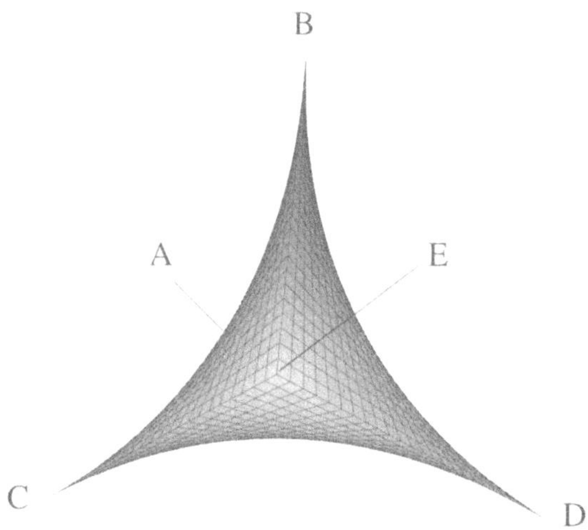

Figure 40: The trieder as a degenerated hexahedron

Due to the strongly disturbed geometry, the integration over this region has to be done with one higher integration order. The trieder can be parametrized by

$$F : \begin{cases} [0,1]^3 \longrightarrow trieder \\ (\xi, \eta, \theta) \longrightarrow (x = \xi - (1-\xi)\eta\theta, y = \eta - (1-\eta)\xi\theta, z = \theta - (1-\theta)\xi\eta) \end{cases}$$

Of course, this mapping is bijective in the interior of the unit cube. It has the Jacobian $det J_F = (1-\xi)(1-\eta)(1-\theta)(1+\xi+\eta+\theta)$ in which each parameter arises only quadratically. Therefore, applying the next higher Gaussian integration formula yields the same integration order as such applied to the undisturbed hexahedron.

Furthermore, integrals over this region can be computed analytically according to

$$\int_{trieder} f(x,y,z)dxdydz = \int_0^1 \int_x^1 \int_{xy}^{x/y} f(x,y,z) + f(y,x,z)dzdydx \quad .$$

For polynomials the integration formula reads

$$\int_{trieder} x^k y^l z^m dxdydz = \frac{2}{(2+k+l)(2+k+m)(2+l+m)}$$

which gives the possibility to derive adapted Gaussian integration rules for the trieder.

Refinement rules: these rules define the manner in which an element of a particular refinement status reacts on the refinement of one of its faces which is shown on the left hand side of Figure 40a) and 40b) with dotted lines. The new state of refinement is shown for each case in Figure 40a) and 40b) on the right hand side with a solid line. The dot marks the node coloring of a hexahedron in each case.

Figure 41 shows the region of adapted elements when one element in the center of the block is refined. In this case all outer faces are unchanged; the block contains 80 hexahedrons. The rather large amount of elements for local isotropic refinement might be seen as a disadvantage compared with node- irregular refinements, but usually h-adaptivity is necessary in larger inner regions or at boundaries but not pointwise.

Overall refinement algorithm by recursion: the recursive structure of the refinement algorithm can be verbally described as follows:
Each entity (edge, face or region) to be refined adapts itself according to an internal set of refinement-rules and calls all including and included entities(i.e. all entities which are in horizontal relation) in order to adapt themselves to the new state of refinement. If a called entity is stored on another processor, the recursive call is substituted by a message to the processor concerned.
This recursion guarantees the node regularity of the adapted mesh. Our implementation of this algorithm consists of about 700 C++ lines plus 300 lines for the included 2D-refinement.

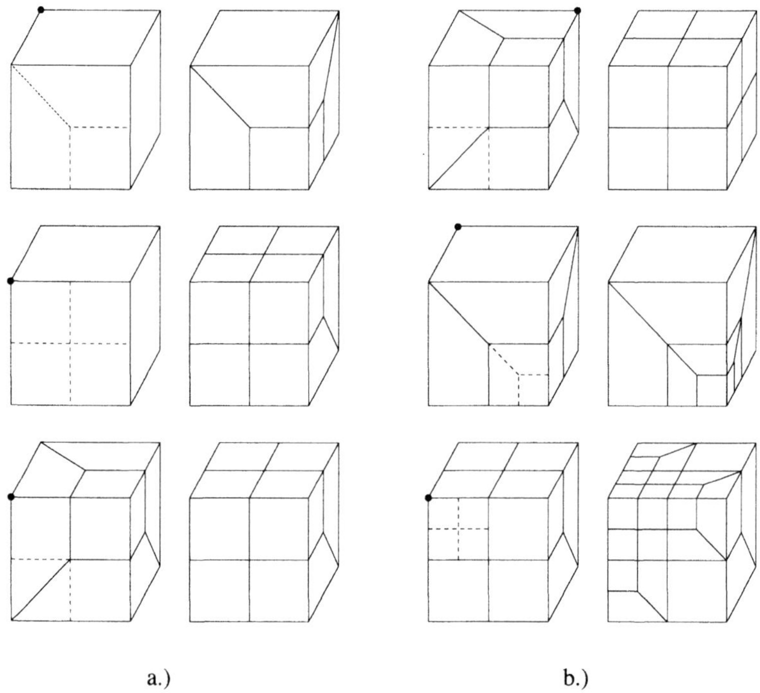

Figure 41: Refinement rules for the hexahedron, elementary rules in a.) and some more complex rules in b.)

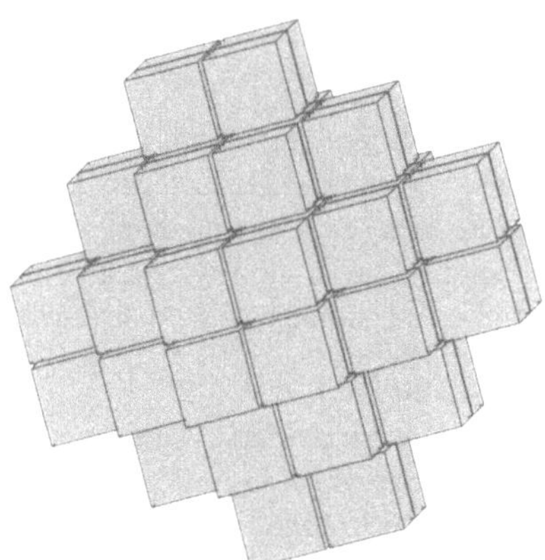

Figure 42: Adapted region for isotropic refinement of one cube-element in the center of the block yielding 80 refined elements

Figure 42 shows the mesh, which is obtained by subsequent refining four times the left front corner element, starting with a regular mesh of $2 \times 2 \times 2$ elements. The element in the upper back right corner remains unchanged.

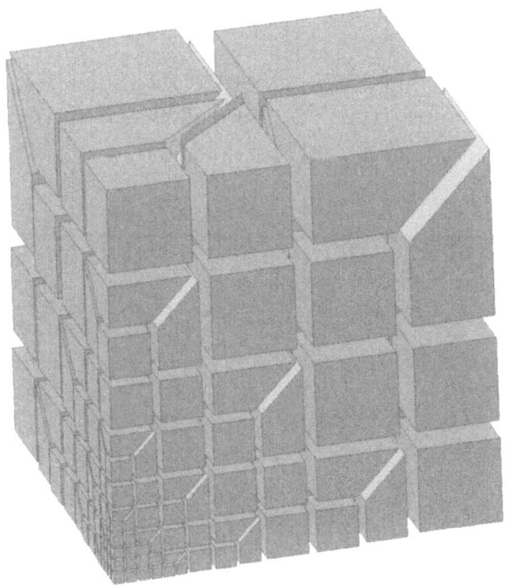

Figure 43: For times in the lower left front area refined mesh

11.2 Transition from 2D- to 3D-discretization

Three topologically distinct elements are used to perform the transition from two dimensional to three dimensional areas of discretizations, see figure 43. Together with an hybrid shell and a 3D-continuum element, they permit the combination of these models in order to get good three-dimensional stress results in crucial areas without the necessity of using 3D-continuum elements for thin- walled partial structures. These transition elements have nodes with three translational degrees of freedom at one side as well as shell nodes with five degrees of freedom (3 displacements and 2 rotations) of the cross-section at the opposite side. They couple the continuum with the shell model in a purely kinematical way, see [3].

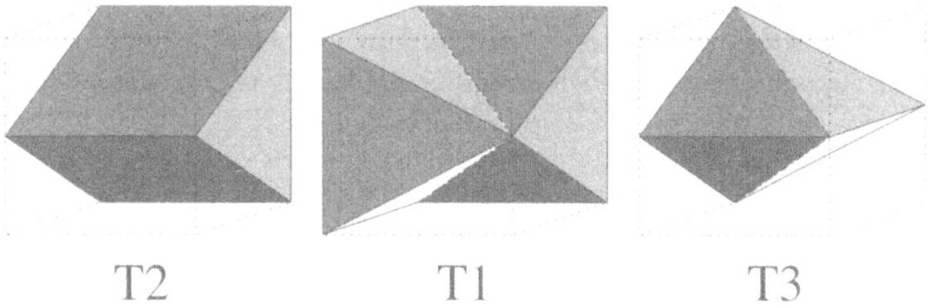

Figure 44: The three transition elements, needed for dimension adaptivity, with two shell and four continuum nodes (type T2), one shell and six continuum nodes (type T1) and three shell and 2 continuum nodes (type T3)

11.2.1 Hierarchical higher order Ansatzfunctions for hexahedrons

Given are the p-orders $\mathrm{grd}_i(\mathrm{i}=0 \ldots 11, \mathrm{grd}_i > 0)$ of the ansatz on the edges of a hexahedron, see Figure 44. The edge ansatz order may be chosen in an arbitrary way.

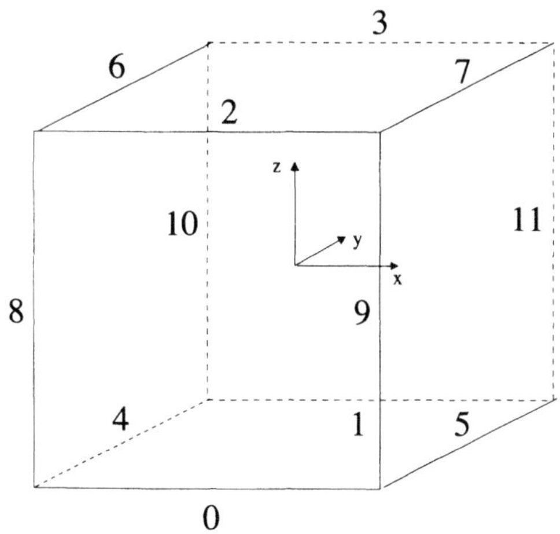

Figure 45: Numbering of the edges of an hexahedron

The resulting set of the corresponding ansatzfunctions consists of

- trilinear node functions,
- edgefunctions:

$(1-y)\ (1-z)\ P_i(x)/4$	for	$1 < i \leq grd_0$,	$(1+y)\ (1-z)\ P_i(x)/4$	for	$1 < i \leq grd_1$,
$(1+y)\ (1+z)\ P_i(x)/4$	for	$1 < i \leq grd_2$,	$(1-y)\ (1+z)\ P_i(x)/4$	for	$1 < i \leq grd_3$,
$(1-x)\ (1-z)\ P_i(y)/4$	for	$1 < i \leq grd_4$,	$(1+x)\ (1-z)\ P_i(y)/4$	for	$1 < i \leq grd_5$,
$(1+x)\ (1+z)\ P_i(y)/4$	for	$1 < i \leq grd_6$,	$(1-x)\ (1+z)\ P_i(y)/4$	for	$1 < i \leq grd_7$,
$(1-x)\ (1-y)\ P_i(z)/4$	for	$1 < i \leq grd_8$,	$(1+x)\ (1-y)\ P_i(z)/4$	for	$1 < i \leq grd_9$,
$(1+x)\ (1+y)\ P_i(z)/4$	for	$1 < i \leq grd_{10}$,	$(1-x)\ (1+y)\ P_i(z)/4$	for	$1 < i \leq grd_{11}$

- facefunctions:

$(1-x)\ P_i(y)\ P_j(z)$	for	$1 < i \leq \max(grd_4, grd_6)$,	and	$1 < j \leq \max(grd_8, grd_{10})$,
$(1+x)\ P_i(y)\ P_j(z)$	for	$1 < i \leq \max(grd_5, grd_7)$,	and	$1 < j \leq \max(grd_9, grd_{11})$,
$(1-y)\ P_i(x)\ P_j(z)$	for	$1 < i \leq \max(grd_0, grd_2)$,	and	$1 < j \leq \max(grd_8, grd_9)$,
$(1+y)\ P_i(x)\ P_j(z)$	for	$1 < i \leq \max(grd_1, grd_3)$,	and	$1 < j \leq \max(grd_{10}, grd_{11})$,
$(1-z)\ P_i(x)\ P_j(y)$	for	$1 < i \leq \max(grd_0, grd_1)$,	and	$1 < j \leq \max(grd_4, grd_5)$,
$(1+z)\ P_i(x)\ P_j(y)$	for	$1 < i \leq \max(grd_2, grd_3)$,	and	$1 < j \leq \max(grd_6, grd_7)$

- bubblefunctions

$P_i(x)P_j(y)P_k(z)$

$1 < i \leq \max(grd_0, grd_1, grd_2, grd_3)$,

$1 < i \leq \max(grd_4, grd_5, grd_6, grd_7)$,

$1 < i \leq \max(grd_8, grd_9, grd_{10}, grd_{11})$.

P_i are the integrated Legendre polynomials, with the wellknown properties:

$P_i(-1) = P_i(1) = 0$, $\int_{-1}^{1} P_i' P_j' dx = \delta_{ij}$.

This hierarchical ansatz has two advantages concerning iterative solving compared with the corresponding Lagrange polynomials, namely that the transfer operators for multilevel preconditioning become simple and that the condition number of the resulting system is much smaller.

A refinement algorithm for hexahedron meshes preserving node-regularity was implemented in the FE-program ParaFEP for parallel processing. In one of the two used kinds of transition elements degenerated hexahedrons arise with a small volume percentage. The integration over these elements has to be done in an adequate way to keep the integration order of regular elements.

12 The hierarchically graded multilevel method

12.1 Outline of the method

The p-method is very powerfull to achieve good approximations of the solution of PDE's, especially if this solution is smooth enough, i.e if the regularity is big enough. Otherwise

– due to load jumps or boundary layers – disturbed areas require many anisotropic h-refinements to yield a good approximation of the discretization. Therefore, local h-adaptivity is adequate here. Let us collect some observations concerning the h- and p-method:

- Using high p-order yields a better approximation for the regular part of the solution.
- High p-order avoids locking phenomena.
- At least the isotropic h-refinement effects a cubic complexity for 3D-problems measured by the number of refinement levels.
- An element of ansatz-order p requires a computational effort of order of at least p^6!

One possible consequence of these facts is the strategy:

- Use the highest p-order only on the coarsest grid, to cover the smooth solution part.
- Use h-refinement in areas with irregular solution parts with a small ansatz-order (e.g. $p_J = 1$ or 2) independently of the order on the coarser grid, i.e. no elements of a high order are generated.

12.2 The hierarchically graded multilevel-ansatz

This leads to a generalization of the hp-method which could be called a hierarchically graded multilevel-ansatz method, see Figure 45.

Hierarchically graded means here:

> Ansatz order p is graded in the hierarchy from coarse to fine grid in a decreasing manner.
>
> To avoid linear dependencies of the ansatz-functions, the ansatz on the coarser grids has to be reduced by the polynomials which are already covered by the ansatz spaces defined on finer levels. This reduction will be called furtheron *filtering*.

This leads to the **definition**:

Let $\mathcal{M}_j$ (j=0..J) be an hierarchical sequence of h-refined meshes, $p_0 \geq p_1 \geq .. \geq p_J \geq 1$ *a decreasing sequence of natural numbers and* $\mathbf{P}(j, p)$ *the continuous polynomials of maximal order* p, *defined on level i element-wise . Then the hierarchically graded multilevel ansatz HGM(*$\mathcal{M}^j$, p_j *(j=0. . J)) consists of the union of the level-wise defined ansatz spaces*

$$\bigcup \{\mathbf{P}(j, p_j) \setminus \mathbf{P}(j, p_{j+1}) : j=0..J-1\} \cup \mathbf{P}(J, p_J).$$

Properties of the resulting ansatz spaces are:

- HGM($\mathcal{M}_j$, p_j (j=0...J)) spans a subspace of the ansatz space of order p_0 on the finest level J.
- The span of HGM($\mathcal{M}_j$, p_j (j=0...J)) includes the ansatz spaces of order p_j on level j (j= 0.. J).

Therefore the approximation quality of HGM($\mathcal{M}_j$, p_j (j=0...J)) is worse than the one of the space mentioned in the first item but better than those of the spaces mentioned in the second item.

Remark: This definition can be generalized in the way, that the series p_j (j=0...J) may be given macro-element-wise, i.e. for elements in $\mathcal{M}_0$, and for each direction independently.

12.2.1 Comparison with the overlay technique of E. Rank

The total fine grid can be seen as an overlay of each of the coarser grids.

The necessary Dirichlet-condition on $\Gamma_{overlay}$ in the overlay technique corresponds to the filtering in the HGM-method.

The main goal of the overlay technique and the HGM-ansatz is:

> Combine the power of the p-method with strong local h-refinement for local irregularities or for other local phenomena.

12.3 Aspects of implementation

The following couplings between the ansatz functions have to be integrated numerically on each level $\mathcal{M}_j$:

- the couplings between the ansatz functions assigned to this level (of maximal order p_j) with integration order $2p_j$,
- the coupling with higher p-ansatz functions p_k on coarser element levels $\mathcal{M}_k$ (k¡j) with integration order $p_j + p_k$.

The required elementary steps of adaptivity are:

- Increasing the ansatz order of elements in $\mathcal{M}_0$ by one.
- „Pushing " of the ansatz order of elements in $\mathcal{M}_j$ onto their child elements (in $\mathcal{M}_{j+1}$) .

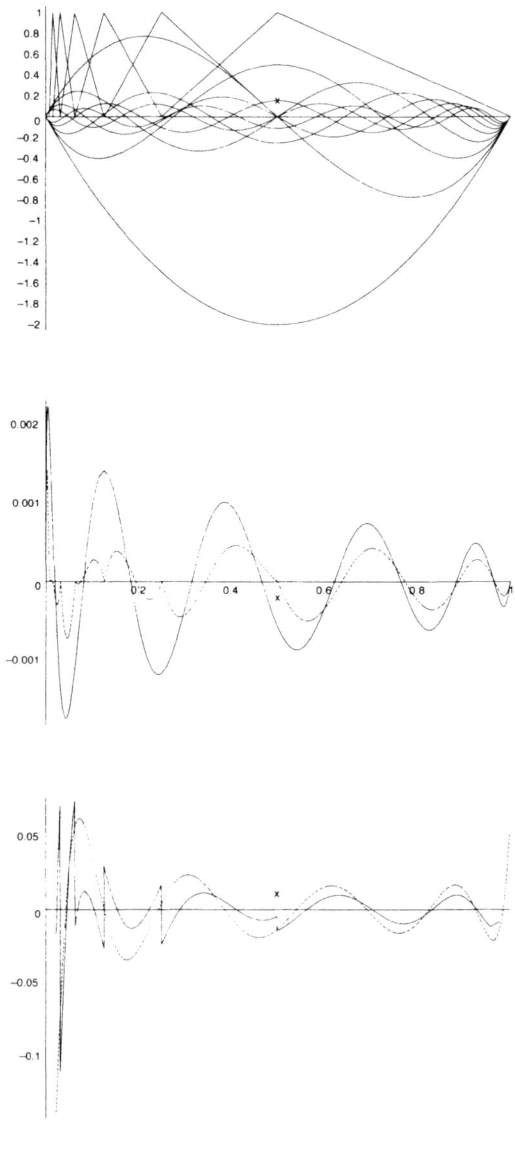

a.)

b.)

c.)

Figure 46: HGM-ansatz with $p_0=9$ and $p_j = 1$ (j=1..8) in a.), u-u_p (solid line) and u- u_{hgm}(dotted line) in b.) as well as $\frac{d}{dx}$(u-u_p) (dotted line) and $\frac{d}{dx}$(u-u_{hgm}) (solid line) in c.).

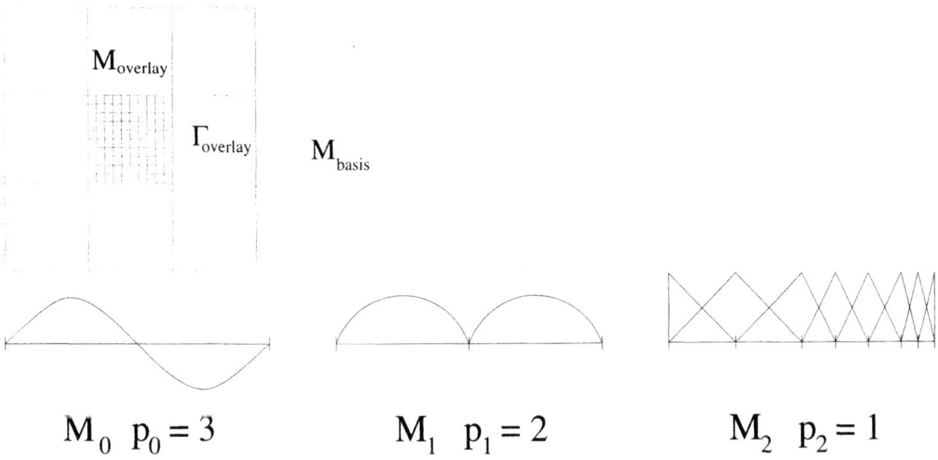

Figure 47: An overlay- and an HGM-discretization

- h-refinement of some elements of the finest level $\mathcal{M}_J$. The new child elements get the ansatz order 1.

Each HGM-discretization can be achieved by successive application of these steps. On the other hand each of the generated discretization in such a way fulfills the necessary condition:

$$element_1 \subset element_2 \Longrightarrow order\ of\ element_1 < order\ of\ element_2. \quad (181)$$

For each of these elementary steps of adaptivity another error indicator can and should be used.

12.4 Results and problems

Figure 46b shows the HGM-ansatz with $p_0 = 9$ and $p_j = 1$ (j=1..8). A one–dimensional Laplace problem with the analytical solution $u(x) = (1\text{-}x)^{\frac{4}{5}}\sin(\pi x)$ (with a singularity in x=0) has been solved using an ansatz of order p=9 and with HGM-ansatz, see Figure 46a. Figure 46b shows the approximation errors using these ansatz functions, and in Figure 46c the first derivation of these errors is presented. Especially near the singularity the HGM-ansatz yields a noticeably better FE-approximation.

1. Using node-regular mesh-adaptivity, the meshes are, due to transition elements, not really hierarchical. Therefore the filtering effects of the HGM-ansatz do not cover the complete polynomial ansatz spaces of order p_j on coarser grids $\mathcal{M}_j$ for

j¡J. This leads to the loss of the monoton convergence (in the energy norm) of h-adaption. This effect can be avoided by using the trial space $\mathbf{u}_0$ + span ¡HGM$(\mathcal{M}_j$, p_j $(j=0..J))$, where $\mathbf{u}_0$ is the solution of the last discretization.

2. Let us consider the ansatz HGM$((\mathcal{M}_0,\mathcal{M}_1$, (2,1)) in the one-dimensional case and the Laplace-problem as underlying PDE. The error $\mathbf{u}-\mathbf{u}_0$ (with $\mathbf{u}_0$ as the solution using the quadratic ansatz on $\mathcal{M}_0$) is well approximated in areas of a smooth solution with the second integrated Legendre-Polynom. Therefore this error vanishes near the midpoint of each element. This fact has as the consequence, that in these areas (by adding linear functions on the refined mesh $\mathcal{M}_1$ elementwise, i.e. using HGM$(\mathcal{M}_0,\mathcal{M}_1$, (2,1)), will not significantly improve the approximation quality of the discretization but strongly worsen the condition number of the resulting system of equations. Therefore we need a sharp error indicator for h-adaptivity.

13 Final remarks

The main issue of this part of the Udine-lectures was the developement of methods for getting a posteriori quantitative (implicit), locally computed anisotropic residual error estimators by equilibration of residua with higher test functions, called Posterior Equilibrium Method (PEM). It is gained from a mathematically equivalent substitute for the relative (explicit) residual error estimation (REM) and is based on the posterior computation of improved boundary tractions on element level by local equilibration of the residua of the previous FE-analysis.

The approximation error estimator follows from a further local variational problem for getting improved displacements from the improved boundary tractions as righthand sides and using extended testfunctions ($p_+ = p+1, p+2$ or $h_+ = h/2, h/4$) of hierarchical test spaces.

Furthermore, effective error estimators for elastoplastic deformations can be derived from PEM, using tangential representations. Again anisotropic estimators are superior against isotropic ones in case of singularities and layer zones with 1D– character.

An important feature of PEM is the integration of model– and dimension– adaptivity into the error controlled adaptive discretization process which again seems only to be effective with PEM because of locking problems which arise with REM at the transition from a lower to an expanded mathematical model. PEM was also extended for different types of nonlinearities, see [45, 46].

It was shown that the primal FEM for elastoplasticity of Prandtl-Reuss-type (FEM in space and FDM in time) yields essentially three spatial discretization errors, namely for equilibrium, the Kuhn-Tucker conditions (yield condition and loading conditions) and of the direction of the plastic strain rates which can be interpreted as a control of the plastic dissipation condition. At the time there is not yet a common mathematical basis for these three errors due to the complicated trace-spaces in which the elastoplastic solutions are located.

For this task two types of error estimators were presented, namely implicit residual estimators (PEM) and estimators with averaging techniques and implicit postprocessing using higher polynomials lolcally called SPRM. Dual and dual hybrid FE-functionals, like with BDM-approximation spans as applied in [12] for elasticity and in [36, 37] for elastoplasticity, shall be investigated further-on. Moreover local and more general goal-oriented error analysis by using dual arguments for the far-field solution of goal-oriented error estimators is in strong progress.

In the concept of operator split (FEM in space, FDM in time) of time (or load-history) dependent problems the coupling of space-time discretization errors is possible in a crude way using local time step control with time increments for the whole spatial system. Our staggered or pilgrim step scheme for space-time error control, section 10, is promising, especially for cyclic loading processes. A true coupled space-time error analysis is only possible for FEM in space *and* time which is attractive for parallel computing by, e.g., splitting a time slab into p subproblems, p being the number of processors, [15, 22].

For the parallel adaptive finite element code ParaFEP we presented a brief description of the refinement algorithm for hexahedron meshes preserving node-regularity. In one of the two used types of transition elements, degenerated hexahedrons, so-called trihedrons, arise with a small volume percentage and vanish in the next step. The higher integration order over these elements has to be applied in an adequate way to keep the overall integration order of accuracy.

Summarizing, one can state that absolute, i.e. quantitative, error estimators – especially permitting anisotropic mesh refinements– of the type of equilibration of residua provide a quality jump for solution- and integrated model-adaptivity. They yield quasi optimal mesh refinements and rather sharp estimators of the absolute errors as well as effective computations with scalar and parallel computers for large scaled engineering structures. Apart from a averaging techniques these error estimators provide upper bounds.

References

[1] M. Ainsworth and J.T. Oden. A procedure for a posteriori error estimation for h-p finite element methods. *Computer Methods in Applied Mechanics and Engineering*, 101:73-96, 1992.

[2] M. Ainsworth and J.T. Oden. A posteriori error estimation in finite element analysis. Mathematics and computer science technical reports, University of Leicester, 1995.

[3] N. Apel. Formulierung und Implementierung eines Schalenelementes mit endlichen Rotationen unter C++, sowie eines Übergangselementes zu 3d-Kontinuumselementen mit hierarchischen Ansatzfunktionen und beliebiger p-Ordnung in Hinblick auf parallele FE-Berechnungsverfahren. *diploma thesis, Hanover, Nov. 1998*, 1998.

[4] I. Babuška. Some aspects of the h and hp version of the finite element method. *Numerical Methods in Engineering Theory and Applications*, 1987.

[5] I. Babuška and W.C. Rheinboldt. Error estimates for adaptive finite element computations. *SIAM Journal on Numerical Analysis*, 15:736–754, 1978.

[6] I. Babuška and A. Miller. A-posteriori error estimates and adaptive techniques for the finite element method. Technical report, Institute for Physical Science and Technology, University of Maryland, 1981.

[7] I. Babuška, T. Strouboulis, S.K. Gangaraj, and C.S. Upadhyay. A posteriori estimation and adaptive control of the pollution error in the h-version of the finite element method. *International Journal for Numerical Methods in Engineering*, 38:4207–4235, 1995.

[8] R. E. Bank and A. Weiser. Some a posteriori error estimators for elliptic partial differential equations. *Mathematics of Computation*, 44:283–301, 1985.

[9] F.-J. Barthold, M. Schmidt, and E. Stein. Error estimation and mesh adaptivity for elasto-plastic deformations. *Computational Plasticity, Proceedings of the fifth International Conference on Computational Plasticity COMPLAS V, CIMNE, 10th–17th March 1997, Barcelona, Spain, edited by D.R.J. Owen, E. Onate and E. Hinton,* Part 1:597–602, 1997.

[10] F.-J. Barthold, M. Schmidt and E. Stein. Error indicators and mesh refinements for finite elements computations of elastoplastic deformations. *Computational Mechanics*, 22:225–238, 1998.

[11] F.-J. Barthold and E. Stein. Error estimation and mesh adaptivity for elasto-plastic deformations. *ZAMM: Proc. ICIAM/GAMM*, 76 Sup. 1:159–162, 1996.

[12] D. Braess, O. Klaas, R. Niekamp, E. Stein, and F. Wobschal. Error indicators for mixed finite elements in 2-dimensional linear elasticity. *Computer Methods in Applied Mechanics and Engineering*, 127:345–356, 1995.

[13] U. Brink. Adaptive gemischte finite Elemente in der linearen finiten Elastostatik und deren Kopplung mit Randelementen. *Institut für Baumechanik und Numerische Mechanik, Dissertation Universität Hannover IBNM-Bericht*, F98/1, 1997.

[14] U. Brink and E. Stein. A-posteriori error estimation in large-strain elasticity using equilibrated local Neumann problems. *Computer Methods in Applied Mechanics and Engineering*, 161(1-2):77–101, 1998.

[15] M. Buch, A. Idesman, R. Niekamp, and E. Stein. Finite elements in space and time for parallel computing of viscoelastic deformations. *Computational Mechanics*, 24:386–395, 1999.

[16] H. Bufler and E. Stein. Zur Plattenberechnung mittels finiter Elemente. *Ing. Archiv*, 39:248–260, 1970.

[17] J. Fish and S. Markolefas. Adaptive global-local refinement strategy based on the interior error estimates of the h-method. *International Journal for Numerical Methods in Engineering*, 37:827–838, 1994.

[18] L. Gallimard, P. Ladevèze, and J. P. Pelle. Error estimation and adaptivity in elastoplasticity. *International Journal for Numerical Methods in Engineering*, 39:189–217, 1996.

[19] W. Hackbusch. Multi–grid method and applications. *Springer Series in Computational Mathematics, Springer–Verlag, Berlin Heidelberg New York Tokyo*, 4, 1985.

[20] W. Hackbusch. Integralgleichungen, Theorie und Numerik. *Teubner Studienbücher, Mathematik, ISBN 3-519-02370-9*, 1989

[21] H. Hencky. Zur Theorie plastischer Deformationen und der hierdurch im Material hervorgerufenen Nebenspannungen. In *Proceedings of the 1st International Congress on Applied Mechanics, Delft*, 1924.

[22] A. Idesman, R. Niekamp, and E. Stein. Continuous and discontinous galerkin method with finite elements in space and time for parallel computing of viscoelastic deformation. *Computer Methods in Applied Mechanics and Engineering*, 190:1049–1063, 2000.

[23] C. Johnson and P. Hansbo. Adaptive finite element methods in computational mechanics. *Computer Methods in Applied Mechanics and Engineering*, 101:143–181, 1992.

[24] O. Klaas, J. Schröder, E. Stein, and C. Miehe. A regularized mixed element for plane elasticity. implementation and performance of the BDM element. IBNM-Bericht 93 / 6, Institut für Baumechanik und Numerische Mechanik, Universität Hannover, 1993.

[25] P. Ladevèze and D. Leguillon. Error estimate procedure in the finite element method and applications. *SIAM J. Num. Anal.*, 20:485–509, 1983.

[26] P. Ladevèze and E.A.W. Maunder. A general methodology for recovering equilibration finite element tractions and stress fields for plate and solid elements. *in First international workshop on Trefftz Method, 30. May - 1. June, 1996, Cracov, Poland, Archives of Mechanics*, pages 34–35, 1995.

[27] J.C. Nagtegaal, D.M. Parks, and J.R. Rice. On numerically accurate finite element solutions in the fully plastic range. *Computer Methods in Applied Mechanics and Engineering*, 4:153–177, 1974.

[28] R. Niekamp, , and E. Stein. An object-oriented approach for parallel two- and three-dimensional adaptive finite element computations. *Computers and Structures*, pages 317–328, 2002.

[29] S. Ohnimus. *Theorie und Numerik dimensions- und modelladaptiver Finite-Elemente-Methoden von Flächentragwerken, Dissertationsschrift.* Berichte des Instituts für Baumechanik und Numerische Mechanik der Universität Hannover, F96/6,1996.

[30] S. Ohnimus, E. Stein, and E. Walhorn. Local error estimates of FEM for displacement and stresses in linear elasticity by solving local Neumann problems. *International Journal for Numerical Methods in Engineering* 52:727-746, 2001.

[31] M. Ortiz and J. J. Quigley. Adaptive mesh refinement in strain localization problems. *Computer Methods in Applied Mechanics and Engineering*, 90:781–804, 1991.

[32] D. Perić, J. Yu, and D.R.J. Owen. On error estimates and adaptivity in elastoplastic solids: Applications to the numerical simulation of strain localization in classical and cosserat continua. *International Journal for Numerical Methods in Engineering*, 37:1351–1379, 1994.

[33] R. Prandtl. Spannungsverteilung in plastischen Körpern. In *Proceedings of the 1st International Congress on Applied Mechanics, Delft*, 1924.

[34] R. Rannacher and F.-T. Suttmeier. A posteriori error control in finite element methods via duality techniques: Application to perfect plasticity. Technical report, Preprints of SFB 350, Universität Heidelberg, 1997.

[35] M. Schmidt. Fehlerschätzung und adaptive Lastschrittsteuerung bei elastoplastischen Problemen. Master's thesis, Institut für Baumechanik und Numerische Mechanik, 1996.

[36] J. Schröder, O. Klaas, E. Stein, and C. Miehe. A physically nonlinear dual mixed finite element formulation. Technical report, Institut für Baumechanik und Numerische Mechanik, Universität Hannover, 1995.

[37] J. Schröder, O. Klaas, E. Stein, and C. Miehe. A physically nonlinear dual mixed finite element formulation. *Computer Methods in Applied Mechanics and Engineering*, 144:77–92, 1997.

[38] J.C. Simo and M.S. Rifai. A class of mixed assumed strain methods and the method of incompatible modes. *International Journal for Numerical Methods in Engineering*, 29:1595 – 1693, 1980.

[39] E. Stein and R. Ahmad. An equilibrium method for stress calculation using finite element displacement models. *Computer Methods in Applied Mechanics and Engineering*, 10:175–198, 1977.

[40] E. Stein, F.-J. Barthold, S. Ohnimus, and M. Schmidt. Adaptive finite elements in elastoplasticity with mechanical error indicators and Neumann–type estimators. *Advances in Adaptive FEM, Cachan, 17.-19. Sep. 97, Proceedings edited by P. Ladevèze and T.J. Oden in Elsevier Science B.V.*, ISBN 0-08-043327-8, 1997.

[41] E. Stein and S. Ohnimus. Dimensional adaptivity in linear elasticity with hierarchical test–spaces for h– and p– refinement processes. *Engineering with Computers*, 12:107–119, 1996.

[42] E. Stein and S. Ohnimus. Coupled model– and solution–adaptivity in the finite–element method. *Computer Methods in Applied Mechanics and Engineering*, 150:327–350, 1997.

[43] E. Stein and S. Ohnimus. Equilibrium method for postprocessing and error estimation in the finite elemente method. *in First international workshop on Trefftz Method, 30. May - 1. June, 1996, Cracov, Poland, Computer Assisted Mechanics and Engineering Sciences*, 4:645–666, 1997.

[44] E. Stein and M. Rüter. Finite element methods for elasticity with error-controlled discretization and model adaptivity. *E. Stein, R. de Borst and T.J.R. Hughes (editors), Encyclopedia of Computational Mechanics. John Wiley & Sons, Chichester*, 2004.

[45] E. Stein, M. Rüter and S. Ohnimus. Adaptive finite element analysis and modeling of solids and structures—findings, problems and trends. *International Journal for Numerical Methods in Engineering*, in print, 2004.

[46] E. Stein. (Edt.) Error - controlled adaptive finite elements in solid mechanics. John Wiley & Sons Ltd. 2002

[47] O. C. Zienkiewicz and J. Z. Zhu. A simple error estimator and adaptive procedure for practical engineering analysis. *International Journal for Numerical Methods in Engineering*, 24:337–357, 1987.

[48] O. C. Zienkiewicz and J. Z. Zhu. The superconvergent patch recovery and adaptive finite element refinement. *Computer Methods in Applied Mechanics and Engineering*, 101:207–224, 1992.

Three Lectures on Error Estimation and Adaptivity

Peter Hansbo

Department of Solid Mechanics, Chalmers University of Technology, Sweden
E-mail: `hansbo@solid.chalmers.se`

Keywords: Adaptivity, plasticity, ordinary differential equations, mesh generation

Summary. In the first Section, we consider global control of the discretization error in time and space–time using finite element methods in both space and time. The approach we use gives the opportunity for adaptive control of general functionals based on residual–type *a posteriori* error estimates. In the second Section we apply the general theory to model problems in elasto–plasticity. We consider both the Hencky total deformation theory and the Prandtl-Reuss quasi-static theory, including load step control. Finally, in the third Section, we review some standard techniques for generating meshes, and for handling local mesh refinement.

Contents

1 Error Estimators and Adaptivity in Time and Space–Time

1.1 Time Adaptivity for Ordinary Differential Equations

Growth of Perturbations in an Ordinary Differential Equation. We will first consider the case of ordinary differential equations for initial–value problems. To this end, we introduce the (nonlinear) Ordinary Differential Equation (ODE)

$$\frac{dy}{dt} = f(t, y(t)), \quad y(0) = y_0. \tag{1}$$

Solving this problem numerically means to find a sequence of numbers y_n approximating $y(t)$ at discrete times t_n, $y_n \approx y(t_n)$. We can then view any y_n as a perturbed version of the corresponding $y(t_n)$ and we must—if we want to nail down the error in the numerics—consider the question of how perturbations grow *in the continuous problem*. The reason for this is as follows: at each time–step we introduce an error. We can compute rather precisely how large this error is, but this will only yield information about the *local* error contribution. After the time–step has been performed, we now have two problems, one with exact initial data and one with perturbed initial data. Information about the different solution trajectories in the sequel tells us how the error is *transported*, which is crucial if we want to estimate the error accurately.

Thus, consider a perturbed initial value $\tilde{y}(0) = y_0 + \delta_0$, leading to a perturbed solution, $\tilde{y} = y + \delta$, which satisfies

$$\frac{d\tilde{y}}{dt} = f(t, \tilde{y}(t)), \quad \tilde{y}(0) = y_0 + \delta_0,$$

and thus

$$\frac{d\delta}{dt} = f(t, y(t) + \delta(t)) - f(x, y(t)),\ \delta(0) = \delta_0.$$

This problem can be written, on $I = (0, t)$, as

$$\frac{d\delta}{dt} = \delta \left.\frac{\partial f}{\partial y}\right|_{y+\xi\delta}, \quad \delta(0) = \delta_0.$$

for some $\xi \in (0, 1)$. Thus it is the sign of $\partial f/\partial y$ that determines the growth of perturbations. As a special case we may consider the linear case $f(t, y(t)) := \lambda\, y(t) + g(t)$, where λ is a complex number and g is a given function of time. This leads to $\partial f/\partial y = \lambda$, i.e., the perturbation equation is an autonomous version of the original equation. The error transport properties are then precisely the same as in the original problem:

- Re $\lambda < 0$ means the error is damped exponentially,
- Re $\lambda = 0$ means the error is transported without damping,
- Re $\lambda > 0$ means the error is blown up exponentially.

It should be clear that this kind of information abut the error transport properties is crucial.

In the classical error control for ODE:s, see, e.g., Hairer et al. (1992), the concept of error transport is used as follows. Consider taking one step of size $2\,k$ followed by two steps of size k in a time-stepping scheme of (global) order k^p, i.e.

$$|y_n - y(t_n)| \leq Ck^p$$

for all n. Assume that the error at time $t = t_0$ is zero, and that we want to control the error $e_2 = |y_2 - y(t_2)|$ after the two small time-steps. We know that $e_1 \leq C_1 k^{p+1}$ (local truncation error), and consequently e_2 is the sum of two parts: local truncation error plus the growth of e_1, obtained from

$$\frac{de}{dt} = e\,\frac{\partial f}{\partial y}\bigg|_{y(t_1)+\xi e}, \quad e(0) = e_1$$

or

$$\frac{de}{dt} \leq Ce, \; e(0) = e_1$$

with $C = \max|\partial f/\partial y|$ in the neighborhood of y_1. Thus $e \leq (I + Ck + O(k^2))e_1$. The total error is obtained by adding the local truncation error to the transported error:

$$\begin{aligned} e_2 &\leq C_1 k^{p+1} + (I + Ck + O(k^2))C_1 k^{p+1} \\ &\leq 2C_1 k^{p+1} + O(k^{p+2}) \end{aligned}$$

and for the solution Y after one large step, we have

$$|y(2\,k) - Y| \leq C_1(2\,k)^{p+1}$$

and

$$|y_2 - Y| \leq C_1(2\,k)^{p+1} - 2C_1 k^{p+1}$$

so that

$$\frac{|y(2\,k) - y_2|}{|y_2 - Y|} \approx \frac{2C_1 k^{p+1}}{C_1 2^{p+1} k^{p+1} - 2C_1 k^{p+1}} = \frac{1}{2^p - 1}$$

The error for the short time-steps can thus be estimated by

$$|y(2\,k) - y_2| \approx \frac{|y_2 - Y|}{2^p - 1}.$$

Unfortunately, this is still a *local* estimate: the argument is based on the assumption of there being no error at time $t = t_0$; so the errors from earlier time–steps are not accounted for. The error controls in actual use in general ODE packages are much more refined versions of a similar approach, often using two methods of different order. Nevertheless, they are usually still of local type, and in general fail to give global error control.

Consider next the following *weak* approach: instead of directly solving the ODE, we seek an admissible (i.e., weakly differentiable and fulfilling the initial condition) solution y such that

$$\int_0^T \left(\frac{dy}{dt} - f(t, y(t))\right) v\,dt = 0 \tag{2}$$

for all admissible (weakly differentiable, zero at time $t = 0$) functions v. Clearly, any solution to (1) will also be a solution to (2). We now want to control perturbation growth for $\delta = \tilde{y} - y$. The perturbation equation can now alternatively be though of as finding δ such that

$$\int_0^T \left(\frac{d\delta}{dt} - (f(t, y(t) + \delta(t)) - f(x, y(t))) \right) v \, dt = 0$$

for all admissible v, or, equivalently, finding δ such that

$$\int_0^T \left(\frac{d\delta}{dt} - A\,\delta \right) v \, dt = 0 \tag{3}$$

for all admissible v, where

$$A = \int_0^1 \frac{\partial f}{\partial y}(t, \xi\tilde{y} + (1-\xi)y) \, d\xi.$$

Note that this is an *exact* linearization: with $\hat{y}(\xi) = \xi\tilde{y} + (1-\xi)y$, we find that

$$\int_0^1 \frac{d}{d\xi} f(t, \hat{y}(\xi)) \, d\xi = f(t, \tilde{y}) - f(t, y)$$

and

$$\int_0^1 \frac{d}{d\xi} f(t, \hat{y}) \, d\xi = \int_0^1 \frac{\partial}{\partial y} f(t, \hat{y}) \frac{d\hat{y}}{d\xi} \, d\xi = \int_0^1 \frac{\partial}{\partial y} f(t, \hat{y}) \, d\xi \, (\tilde{y} - y) = A\,(\tilde{y} - y)$$

so that $A\,(\tilde{y} - y) = f(t, \tilde{y}) - f(t, y)$. We now introduce the auxiliary *dual* problem

$$-\frac{dz}{dt} - A^{\mathrm{T}} z = 0 \quad \text{from } t = T \text{ to } t = 0$$

where $z(T) = 1$ and A^{T} is the adjoint operator to A, defined by

$$\int_0^T (A\,\delta)\, z \, dt = \int_0^T \delta\,(A^{\mathrm{T}} z) \, dt$$

(e.g., in the case δ is a vector, A^{T} is the transpose of A). Then

$$0 = \int_0^T (\tilde{y} - y) \left(-\frac{dz}{dt} - A^{\mathrm{T}} z \right) dt = \int_0^T \left(\frac{d\delta}{dt} z - \delta\, A^{\mathrm{T}} z \right) dt - [\delta\, z]_0^T ,$$

and we find that

$$\delta(T) = \delta(0) z(0) + \int_0^T \left(\frac{d\delta}{dt} - A\delta \right) z \, dt.$$

Now, the second term on the right-hand side is equal to zero by the perturbation growth equation (3) and thus $\delta(T) = \delta(0)z(0)$, so $z(0)$ directly gives information about how perturbations grow (are transported). In the case of a numerical solution, however, we can think of perturbations as being produced at each timestep due to discretization because we cannot expect a numerically computed $\tilde{y}$ to fulfill (1), i.e.,

$$\frac{d\tilde{y}}{dt} - f(t, \tilde{y}(t)) = R(t) \neq 0.$$

Then

$$\int_0^T \left(\frac{d\delta}{dt} - A\delta \right) z\, dt = \int_0^T \left(\frac{d\tilde{y}}{dt} - \frac{dy}{dt} - (f(t, \tilde{y}(t)) - f(t, y(t))) \right) z\, dt = \int_0^T R\, z\, dt,$$

and we have

$$\delta(T) = \delta(0)z(0) + \int_0^T R\, z\, dt. \tag{4}$$

The perturbation size at final time depends on the *production* (handled by the residual R) and *transport* (handled by the dual solution z) of perturbations.

The weak formulation used to obtain a perturbation growth equation indicates the usefulness of a Finite Element (FE) approach. However, it is not possible to just apply the finite element method in a straightforward way: this would result in a globally coupled system of equations which would not be competitive compared to finite difference methods. In the following Section we will consider two different FE methods that lead to time–stepping schemes without global coupling. For more information concerning these FE methods, we refer to Eriksson et al. (1996), Estep and French (1994), and Estep (1995).

Continuous Finite Element Methods for Initial Value Problems. We consider the time interval $I = (0, T]$ subdivided into time intervals $I_n = (t_{n-1}, t_n]$. On each such time interval, we shall seek a polynomial approximation Y of $y(t)$. We write the space of all polynomials of degree less or equal to q on I_n as $\mathcal{P}q(I_n)$, and we will seek a time–continuous Y such that $Y|_{I_n} \in \mathcal{P}^q(I_n)$. Thus, given $Y_{n-1} := Y(t_{n-1})$, we can find $Y|_{I_n} \in \mathcal{P}^q(I_n)$ by requiring that

$$\int_{t_{n-1}}^{t_n} \left(\frac{dY}{dt} - f(t, Y) \right) v\, dt = 0 \quad \forall v \in \mathcal{P}^{q-1}(I_n). \tag{5}$$

Note that the polynomial Y has $q+1$ unknowns, and that one is removed by requiring continuity at $t = t_{n-1}$; the remaining q unknowns are found by requiring (5) to hold for the q linearly independent polynomials making up the basis for $\mathcal{P}^{q-1}(I_n)$, e.g., $v = 1, t, t^2, \ldots$

As an example, consider $q = 1$, whence

$$Y_n - Y_{n-1} = \int_{t_{n-1}}^{t_n} f(t, Y)\, dt \approx \begin{cases} \dfrac{k_n}{2}\left(f(t_{n-1}, Y_{n-1}) + f(t_n, Y_n)\right) \text{ (trapezoidal rule)}, \\ k_n f\left((t_{n-1} + t_n)/2, (Y_{n-1} + Y_n)/2\right) \text{ (midpoint rule)}. \end{cases}$$

Thus, by quadrature, we can regain some classical implicit methods for ODE:s. Note that these are second order schemes; generally it can be proven that the continuous method with polynomials of degree q are of global order $2\,q$ (in the finite difference sense).

We shall now establish an *a posteriori* error estimate for the continuous method. Thus, we consider a dual problem of the type

$$-\frac{dz}{dt} - A^{\mathrm{T}} z = J_1 \quad \text{from } t = T \text{ to } t = 0, \quad z(T) = J_2, \tag{6}$$

where A^{T} is the adjoint of

$$A := \int_0^1 \frac{\partial f}{\partial y}(t, \xi y + (1-\xi)Y)\, d\xi, \tag{7}$$

and where J_1 and J_2 can be chosen freely for different types of error control. For example, defining the error $e := y - Y$, we may choose

- Control of the Euclidean norm of $e(T)$: $J_1 = 0$, $J_2 = e(T)/|e(T)|$.
- Root-mean-square control of $e(t)$:

$$J_1 = \frac{e(t)}{\left(\int_0^T e^2(t)\, dt\right)^{1/2}}, \quad J_2 = 0.$$

Note that these error controls require an approximation of the error, which must come from solutions on two different meshes. However, it is not the *size* of the error, merely the *direction*, that is used as data for the dual problem.

Now, using the fact that $e(0) = 0$, we have that

$$\begin{aligned} \int_0^T e\, J_1\, dt &= \int_0^T e \left(-\frac{dz}{dt} - A^{\mathrm{T}} z\right) dt \\ &= \int_0^T \left(\frac{de}{dt} - A\, e\right) z\, dt - e(T)\, J_2, \end{aligned}$$

so that

$$\begin{aligned} \int_0^T e\, J_1\, dt + e(T)\, J_2 &= -\int_0^T \left(\frac{dY}{dt} - f(t,Y)\right) z\, dt \\ &= \sum_n \int_{I_n} \left(\frac{dY}{dt} - f(t,Y)\right) (\pi_{q-1} z - z)\, dt, \end{aligned} \tag{8}$$

where the last step follows from (5) with $\pi_{q-1} z$ a projection of z onto the polynomial space $\mathcal{P}^{q-1}(I_n)$. Comparing with the perturbation growth equation (4) we note that the fact that the FE method is a projection method allows us to subtract from z its projection (onto $\mathcal{P}^{q-1}(I_n)$ in this case), which makes the estimate much sharper.

We emphasize that the different terms in the last integral in (8) signify different things: the residual $dY/dt - f(t,Y)$ is the local error production in each time-step, and the term $\pi_{q-1} z - z$ points to the relative importance of this residual when it comes to the error we wish to estimate.

Unfortunately, we do not have access to the continuous solution z of the dual problem, so one might (correctly) argue that this is not a fully *a posteriori* estimate. Basically, we still have two possibilities to use this estimate for time-step selection and/or error control:

1. To use an interpolation estimate, i.e., to say that, on I_n, $|z-\pi_{q-1}z| \leq Ck_n^q \max_{I_n} |D^q z|$, where $D^q z$ is the q:th derivative of z. We also need to get hold of the interpolation constant C, which inherently means that we consider a worst case scenario. We must then choose between using analytical estimates of $D^q z$ and C (if available), which is safe but leads to overestimation of the error, or trying to evaluate these quantities numerically (which is less safe, but more practical).
2. To estimate z directly by an auxiliary computation. In this case we may for example compute z using the same method we use for the forward problem, since then the computed $Z \approx z$ will be in $\mathcal{P}^q(I_n)$, and it makes sense to talk of $Z - \pi_{q-1}Z$ as an approximation of $z - \pi_{q-1}z$. (This would not be the case for $Z - \pi_q Z \equiv 0$).

The general experience is that the second approach is the best when it comes to estimating the error; the first approach requires taking absolute values in the integral in (8) which tends to over-estimate the error.

A practical algorithm for estimating the error and changing the time–step size could be:

1. Compute the actual problem on two consecutive meshes if an estimate of $e(t)$ is needed.
2. Choose a mesh on which to solve the dual problem with the chosen J_2 as initial data and J_1 as forcing term.
3. Use the computed solution Z of the dual problem in the term $\pi_{q-1}Z - Z$, or use an interpolation estimate.
4. Redistribute the timestep sizes according to where the estimate predicts a large error contribution.
5. Stop when the estimated error lies below your chosen tolerance.

In comparison with traditional local error estimates this is a very expensive approach, which should only be used when it is important to have a good estimate of the error. Simplified forms, where knowledge of the physical problem is taken into account, are currently being investigated.

Discontinuous Finite Element Methods for Initial Value Problems. The continuous Galerkin method discussed in the previous Section is non-dissipative and may not perform well in cases where data are rough (e.g. for rough initial data). We would also like a space-time finite element method with better smoothing properties, and there is one: the Discontinuous Galerkin (DG) method. The DG method can be defined as follows: on each time-step, select a polynomial space $P^q(I_n)$ and seek $Y \in P^q(I_n)$ such that

$$\int_{t_n}^{t_{n+1}} \left(\frac{dY}{dt} - f(t,Y) \right) v\,dt + [Y_n]\,v_n^+ = 0$$

for all $v \in P^q(I_n)$, where $[Y_n] = Y_n^+ - Y_n^-$ and

$$v_n^{\pm} = \lim_{s\to 0+} v(t_n \pm s).$$

As an example, consider the case $q = 0$, i.e., a piecewise constant approximation on each time step. We then obtain, with the notation $Y_n = Y_n^+$, $Y_{n+1} = Y_n^-$,

$$Y_{n+1} = Y_n + \int_{t_n}^{t_{n+1}} f(t, Y_{n+1})\, dt$$

which, with the quadrature choice

$$\int_{t_n}^{t_{n+1}} f(t, Y_{n+1})\, dt \approx k_n f(t_{n+1}, Y_{n+1})$$

becomes the implicit Euler method. (It should be pointed out that this is a really lousy first order quadrature scheme. While it does not destroy the first order convergence of the method, one should consider instead midpoint quadrature which does not cost more in terms of computing time.) The class of methods defined by the DG method is dissipative and, using a polynomial approximation of degree q, the accuracy is $2q + 1$ (in the finite difference sense).

The a posteriori error estimation follow the same track as in the time-continuous case. Using again the dual problem (6), followed by integration by parts on each subinteval, we end up with

$$e(T)\, J_2 + \int_0^T e\, J_1\, dt = \sum_n \left(\int_{t_n}^{t_{n+1}} \left(\frac{de}{dt} - Ae \right) \cdot z\, dt + [e_n] \cdot z_n^+ \right).$$

For the dual problem, we have used the linearization (7).

Then it follows that

$$\begin{aligned} e(T)\, J_2 + \int_0^T e\, J_1\, dt &= \sum_n \int_{t_n}^{t_{n+1}} \left(\frac{d(y - Y)}{dt} - A(y - Y) \right) \cdot z\, dt + \sum_n [y_n - Y_n] \cdot z_n^+ \\ &= -\sum_n \int_{t_n}^{t_{n+1}} \left(\frac{dY}{dt} - f(t, Y) \right) \cdot z\, dt - \sum_n [Y_n] \cdot z_n^+ \end{aligned}$$

Using the FE equation, we find the error representation formula

$$e(T)\, J_2 + \int_0^T e\, J_1\, dt = \sum_n \int_{t_n}^{t_{n+1}} \left(\frac{dY}{dt} - f(t, Y) \right) \cdot (\pi_q z - z)\, dt + \sum_n [Y_n] \cdot (\pi_q z_n^+ - z_n^+)$$

where $\pi_q z$ is a piecewise polynomial in $P^q(I_n)$.

When trying to estimate the difference $\pi_q z - z$ we now we have the additional difficulty that we must use a higher order approximation Z of z in order to make the approximation $\pi_q z - z \approx \pi_q Z - Z$, since if $Z \in P^q(I_n)$ then $\pi_q Z - Z \equiv 0$. One obvious way around this problem is to use interpolation based on two time-steps for the determination of Z, so that $Z \in P^{q+1}(I_n)$. The general observation when using this type of strategy is that the more we invest in solving the dual problem, the better we are able to predict the error.

Example 1. We consider the problem of solving the ordinary differential equation

$$\frac{dy}{dt} = \sin(t/A) - A\,y, \quad y(0) = 1, \tag{9}$$

which has the exact solution

$$y(t) = \frac{1 + A + A^4 - A\,e^{At}\cos(t/A) + A^3\,e^{At}\sin(t/A)}{(1 + A^4)\,e^{At}}.$$

We choose $A = 2$, and we wish to compute this problem for $0 \le t \le 2$. The exact solution is plotted in Figure 1. We seek to control the error at final time, $e = |y(2) - Y(2)|$. Since this problem is damped, we expect the discretization errors at early times to affect the error less than the discretization errors at later times. A plot of the dual solution for a given final error is also given in Figure 1.

For the approximate solution of (9), we use the time-discontinuous finite element method. In Figure 2, we show the a typical time step distribution and the corresponding approximate solution, based on a simple adaptive scheme dividing those elements with the largest contribution to the total error (according to a fixed fraction of the total number of elements). We note in particular, that the approximate solution is far off at early times, which does not affect the accuracy at final time due to the damping properties in (9).

1.2 Adaptivity in Space–Time for Partial Differential Equations

Residual Estimates in Space. We shall next discuss how the time-adaptivity can be combined with spatial adaptivity for complete control of the discretization error. We will focus on residual-type error estimates in space and time. To this end, we first consider a linear abstract stationary problem of the type

$$Au = f \text{ in } \Omega, \quad u = 0 \text{ on } \partial\Omega.$$

To obtain an *a posteriori* error estimate, we proceed as in the time-dependent case. We assume that we are given the variational problem of finding $u \in V$ such that

$$\int_\Omega Au\,v\,d\Omega = \int_\Omega f\,v\,d\Omega$$

for all $v \in V$, and formulate its finite element counterpart: find $U \in V_h \subset V$ such that

$$\int_\Omega AU\,v\,d\Omega = \int_\Omega f\,v\,d\Omega$$

for all $v \in V_h$. Next, we introduce the dual problem of finding $\phi \in V$ such that

$$\int_\Omega A^*\phi\,v\,d\Omega = \int_\Omega J\,v\,d\Omega$$

for a given J and for all $v \in V$. Here, the adjoint operator A^* is defined by

$$\int_\Omega A^*v\,w\,d\Omega = \int_\Omega v\,Aw\,d\Omega$$

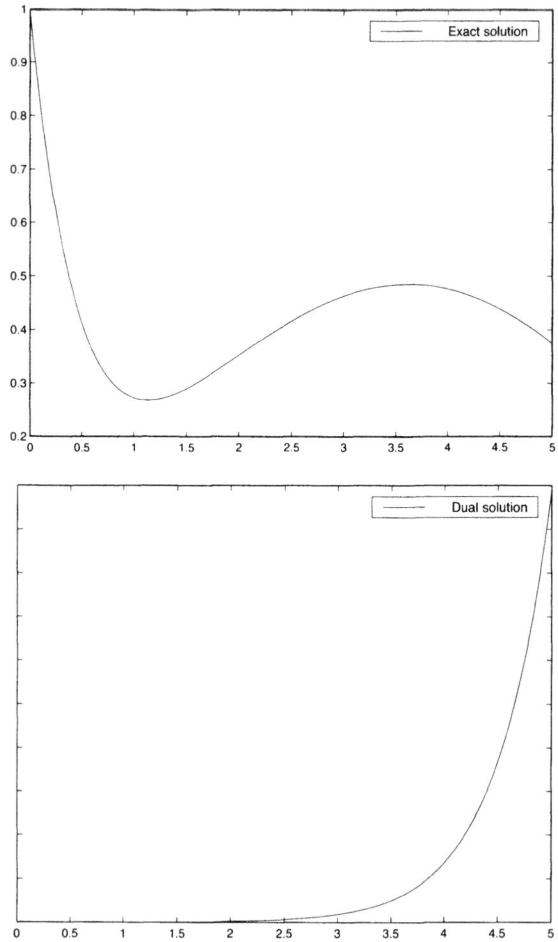

Figure 1. Exact solution to (9) (top), and a typical dual solution (bottom).

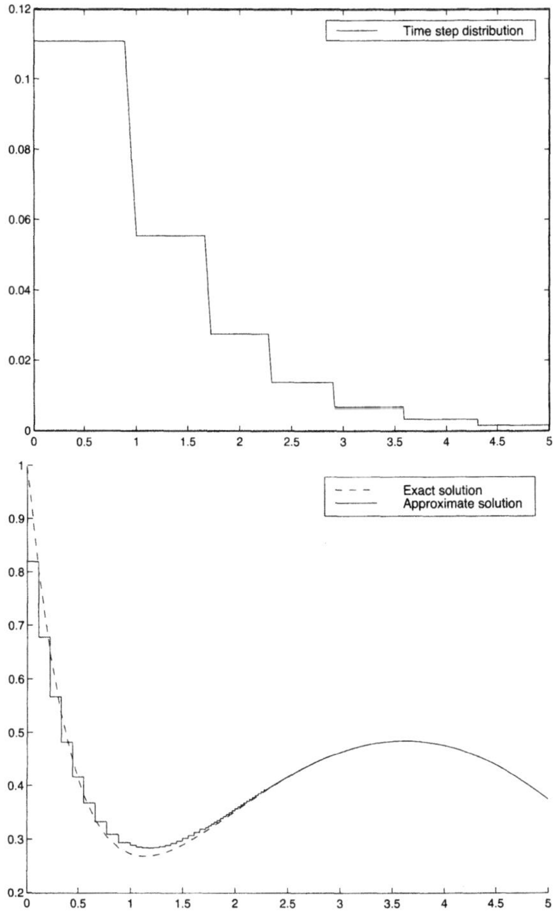

Figure 2. Time step distribution (top), and approximate solution (bottom).

for $w \in V^*$, the space dual to V. The Galerkin orthogonality with respect to V_h can be formulated as

$$\int_\Omega A\,(u-U)\; v\, d\Omega = 0$$

for all $v \in V_h$. By definition, we have that

$$\int_\Omega (u-U)\; J\, d\Omega = \int_\Omega (u-U)\; A^* \phi\, d\Omega$$

or

$$\int_\Omega (u-U)\; J\, d\Omega = \int_\Omega (Au - AU)\; \phi\, d\Omega = \int_\Omega (f - AU)\; \phi\, d\Omega,$$

and, using the orthogonality, we find that

$$\int_\Omega (u-U)\; J\, d\Omega = \int_\Omega (f - AU)\; (\phi - \pi_h \phi)\, d\Omega, \tag{10}$$

which is our error representation formula.

We can now define an adaptive strategy for the approximate solution of our abstract variational problem.

1. Choose J for suitable control of the error.
2. Solve the dual problem (approximately) for ϕ.
3. Evaluate (approximately) the difference $\phi - \pi_h \phi$ and the residual. To evaluate the error, simply compute the integral on the right-hand side or (10).
4. Refine those elements where the product of the residual and interpolation error is large.

Important questions are how to choose J, how accurately ϕ has to be evaluated and how to compute the weights $\phi - \pi_h \phi$. The answer is that this depends on the problem type and on the desired level of certainty. A simple strategy would be to solve the dual problem on the same mesh as the original, use $J = e \approx U_h - U_H$, where U_h and U_H is the solution on two consecutive meshes, followed by projecting ϕ patchwise to estimate $\phi - \pi_h \phi$. Experience shows that this is sufficient to obtain error estimates in the range of a factor 1.5–2 of the actual value for a large class of problems, cf. Rannacher (1999) and references therein.

Example 2. We take the problem of linear elasticity as a concrete example of how to proceed. Navier's equations can be formulated as follows: find $\boldsymbol{u}$ such that

$$-\nabla \cdot \boldsymbol{\sigma}(\boldsymbol{u}) = \boldsymbol{f} \quad \text{in } \Omega \subset \mathbf{R}^2,$$

$$\boldsymbol{\sigma}(\boldsymbol{u}) = \boldsymbol{S} : \boldsymbol{\varepsilon}(\boldsymbol{u}),$$

$$(\boldsymbol{\varepsilon}(\boldsymbol{u}))_{ij} = \frac{1}{2}\left(\frac{\partial u_i}{\partial x_j} + \frac{\partial u_j}{\partial x_i}\right).$$

with boundary conditions

$$\boldsymbol{u} = 0 \quad \text{on } \partial\Omega_D, \quad \boldsymbol{\sigma}(\boldsymbol{u}) \cdot \mathbf{n} = \boldsymbol{g} \quad \text{on } \partial\Omega_N.$$

A variational formulation of Navier's equations leads to the problem of finding $\boldsymbol{u} \in V = \left[H_0^1(\Omega)\right]^2$ such that

$$\int_\Omega \boldsymbol{\sigma}(u) : \boldsymbol{\varepsilon}(\boldsymbol{v})\, d\Omega = \int_\Omega \boldsymbol{f} \cdot \boldsymbol{v}\, d\Omega + \int_{\partial\Omega_N} \boldsymbol{g} \cdot \boldsymbol{v}\, ds$$

for all $\boldsymbol{v} \in V$, where we have used the notation

$$\boldsymbol{\sigma} : \boldsymbol{\varepsilon} = \sum_i \sum_j \sigma_{ij} \varepsilon_{ij}.$$

The FEM-version of this problem reads: find $\boldsymbol{U} \in \boldsymbol{V}_h$ such that

$$\int_\Omega \boldsymbol{\sigma}(\boldsymbol{U}) : \boldsymbol{\varepsilon}(\boldsymbol{v})\, d\Omega = \int_\Omega \boldsymbol{f} \cdot \boldsymbol{v}\, d\Omega + \int_{\partial\Omega_N} \boldsymbol{g} \cdot \boldsymbol{v}\, ds$$

for all $\boldsymbol{v} \in V_h$. Here, Galerkin orthogonality takes the form

$$\int_\Omega (\boldsymbol{\varepsilon}(\boldsymbol{u}) - \boldsymbol{\varepsilon}(\boldsymbol{U})) : \boldsymbol{S} : \boldsymbol{\varepsilon}(\boldsymbol{v})\, d\Omega = 0$$

for all $\boldsymbol{v} \in V_h$.

Next, we introduce the dual problem of finding $\boldsymbol{z}$ such that

$$-\nabla \cdot \boldsymbol{\sigma}(\boldsymbol{z}) = \boldsymbol{J}_1 \quad \text{in } \Omega,$$

$$\boldsymbol{z} = 0 \quad \text{on } \partial\Omega_D, \quad \boldsymbol{\sigma}(\boldsymbol{z}) \cdot \mathbf{n} = \boldsymbol{J}_2 \quad \text{on } \partial\Omega_N.$$

leading to the variational problem of finding $\boldsymbol{z} \in V$ such that

$$\int_\Omega \boldsymbol{\sigma}(\boldsymbol{z}) : \boldsymbol{\varepsilon}(\boldsymbol{v})\, d\Omega = \int_\Omega \boldsymbol{J}_1 \cdot \boldsymbol{v}\, d\Omega + \int_{\partial\Omega_N} \boldsymbol{J}_2 \cdot \boldsymbol{v}\, ds.$$

With $\boldsymbol{e} = \boldsymbol{u} - \boldsymbol{U}$ we thus find that

$$\begin{aligned}\int_\Omega \boldsymbol{J}_1 \cdot \boldsymbol{e}\, d\Omega + \int_{\partial\Omega_N} \boldsymbol{J}_2 \cdot \boldsymbol{e}\, ds &= \int_\Omega \boldsymbol{\sigma}(\boldsymbol{z}) : \boldsymbol{\varepsilon}(\boldsymbol{e})\, d\Omega = \int_\Omega \boldsymbol{\varepsilon}(\boldsymbol{z}) : \boldsymbol{S} : \boldsymbol{\varepsilon}(\boldsymbol{e})\, d\Omega \\ &= \int_\Omega \boldsymbol{\varepsilon}(\boldsymbol{z} - \pi_h \boldsymbol{z}) : (\boldsymbol{\sigma}(\boldsymbol{u}) - \boldsymbol{\sigma}(\boldsymbol{U}))\, d\Omega.\end{aligned}$$

Integration by parts yields

$$\begin{aligned}\int_\Omega \boldsymbol{J}_1 \cdot \boldsymbol{e}\, d\Omega + \int_{\partial\Omega_N} \boldsymbol{J}_2 \cdot \boldsymbol{e}\, ds = &\sum_K \int_K (\boldsymbol{z} - \pi_h \boldsymbol{z}) \cdot (\boldsymbol{f} + \nabla \cdot \boldsymbol{\sigma}(\boldsymbol{U}))\, dx \\ &- \sum_K \int_{\partial K} \frac{1}{2} (\boldsymbol{z} - \pi_h \boldsymbol{z}) \cdot [\boldsymbol{\sigma}(\boldsymbol{U}) \cdot \mathbf{n}]\, ds\end{aligned} \tag{11}$$

where K is a finite element with boundary ∂K and $[\cdot]$ denotes jump in the indicated quantity across the boundary of K. The jump term is modified on $\partial\Omega_N$ to $\boldsymbol{\sigma}(\boldsymbol{U}) \cdot \mathbf{n} - \boldsymbol{g}$.

Equation (11) is the error representation formula, used to evaluate the error. In order to proceed and get local estimates (purely for assessing which elements are to be refined), we use Cauchy inequalities to find

$$\int_\Omega \boldsymbol{J}_1 \cdot \boldsymbol{e}\, d\Omega + \int_{\partial\Omega_N} \boldsymbol{J}_2 \cdot \boldsymbol{e}\, ds \leq \sum_K \|\boldsymbol{z} - \pi_h \boldsymbol{z}\|_{L_2(K)} \|\boldsymbol{f} + \nabla \cdot \boldsymbol{\sigma}(\boldsymbol{U}))\|_{L_2(K)}$$
$$+ \sum_K \int_{\partial K} \frac{1}{2} \|\boldsymbol{z} - \pi_h \boldsymbol{z}\|_{L_2(\partial K)} \| \left[\boldsymbol{\sigma}(\boldsymbol{U}) \cdot \mathbf{n}\right] \|_{L_2(\partial K)}.$$

The total error depends on the local error (the residual) and the transported error through $\boldsymbol{z} - \pi_h \boldsymbol{z}$; this is no different from the time-dependent case. However, there is now a special case. If we want to to control the error in energy norm,

$$\|\boldsymbol{e}\|_E = \left(\int_\Omega \boldsymbol{e} : \boldsymbol{S} : \boldsymbol{e}\, d\Omega \right)^{1/2},$$

we find that we should use $\boldsymbol{J}_2 = 0$ and $\boldsymbol{J}_1 = -\nabla \cdot \boldsymbol{\sigma}(\boldsymbol{e})$, in which case $\boldsymbol{z} = \boldsymbol{e}$ and

$$\int_\Omega \boldsymbol{e} : \boldsymbol{S} : \boldsymbol{e}\, d\Omega \leq \sum_K \|\boldsymbol{e} - \pi_h \boldsymbol{e}\|_{L_2(K)} \|\boldsymbol{f} + \nabla \cdot \boldsymbol{\sigma}(\boldsymbol{U}))\|_{L_2(K)}$$
$$+ \sum_K \int_{\partial K} \frac{1}{2} \|\boldsymbol{e} - \pi_h \boldsymbol{e}\|_{L_2(\partial K)} \| \left[\boldsymbol{\sigma}(\boldsymbol{U}) \cdot \mathbf{n}\right] \|_{L_2(\partial K)}.$$

We can then use the following interpolation estimates in the energy norm:

$$\|\boldsymbol{e} - \pi_h \boldsymbol{e}\|_{L_2(K)} \leq C_1^i h_K \|\boldsymbol{e}\|_{E,\tilde{K}}$$

and

$$\|\boldsymbol{e} - \pi_h \boldsymbol{e}\|_{L_2(\partial K)} \leq C_2^i h_K^{1/2} \|\boldsymbol{e}\|_{E,\tilde{K}}$$

(for technical reasons, we must use $\|\boldsymbol{e}\|_{E,\tilde{K}}$, the enery norm on $\tilde{K}$, where $\tilde{K}$ is the union of K and its neighbors) so that

$$\|\boldsymbol{e}\|_E \leq \sum_K C_1^i h_K \|\boldsymbol{f} + \nabla \cdot \boldsymbol{\sigma}(\boldsymbol{U}))\|_{L_2(K)} + \sum_K C_2^i h_K^{1/2} \| \left[\boldsymbol{\sigma}(\boldsymbol{U}) \cdot \mathbf{n}\right] \|_{L_2(\partial K)}.$$

This is a classical estimate, see, e.g., Kelly et al. (1984). Here, there is no need to solve a dual problem, which means that there is *no error transport in the energy norm.* This fact is deceptive, and has led to an extreme emphasis on the energy norm, which is unfortunate in view of

1. the limited amount of information contained in this norm;
2. the fact that many problem classes do not possess an energy functional;
3. that it has implanted the seductive but erroneous idea that error control can always be based on local estimates.

Error Estimates in Space and Time. Finally, we come to the full problem of controlling the error in space and time. In principle, it is just a straightforward application of the general approach, so we will just give an example of the combination of the time-discontinuous Galerkin method in combination with a standard finite element method in space. This idea was first suggested by Eriksson and Johnson (1991). For an example of time-continuous Galerkin in combination with finite elements in space, see Bangerth and Rannacher (1999).

Example 3. Consider the problem of time-dependent heat conduction:

$$\frac{\partial u}{\partial t} - \nabla^2 u = f \quad \text{in } \Omega,\ 0 < t < T$$

with $u = 0$ on $\partial\Omega$ and $u(\mathbf{x}, 0) = u_0$.

On each time level $t_n < t < t_{n+1}$, we have a spatial mesh and a polynomial approximation in time defining the discrete FEM-space V_h^n. We note in particular that the time-discontinuous Galerkin method allows for non-matching meshes between consecutive time-levels by the built-in projection represented by the jump term. This is nice, because the refinement zone must be able to move to capture local phenomena in space-time. In the time-continuous Galerkin method, on the other hand, we must use an intermediate projection step which requires separate analysis. In principle, one a finite number of projection steps per unit time if the order of the method is to be maintained.

The space-time discretization yields the time-stepping scheme

$$\int_{t_n}^{t_{n+1}} \int_{\Omega} \left(\frac{\partial U}{\partial t} v + \nabla U \cdot \nabla v \right) d\Omega dt + \int_{\Omega} [U_n] v_n^+ \, d\Omega = \int_{t_n}^{t_{n+1}} \int_{\Omega} f \, v \, d\Omega \, dt$$

for all v in V_h^n. Introduce a dual problem

$$-\frac{\partial z}{\partial t} - \nabla^2 z = 0 \quad \text{in } \Omega,$$

solved backwards in time, with $z = 0$ on $\partial\Omega$ and

$$z(\cdot, T) = \frac{u(\cdot, T) - U(\cdot, T)}{\|u(\cdot, T) - U(\cdot, T)\|_{L_2(\Omega)}}.$$

In analogy with the previous cases, we have

$$\begin{aligned} \|u(\cdot, T) - U(\cdot, T)\|_{L_2(\Omega)} = & \sum_n \sum_K \int_{t_n}^{t_{n+1}} \int_K \left(f + \nabla^2 U - \frac{\partial U}{\partial t} \right) (z - \pi_{hk} z) \, dx dt \\ & - \sum_n \sum_K \int_K [U_n](z_n^+ - \pi_{hk} z_n^+) \, dx \\ & - \frac{1}{2} \sum_n \sum_K \int_{\partial K} [\mathbf{n} \cdot \nabla U](z - \pi_{hk} z) \, dx dt. \end{aligned}$$

where $\pi_{hk} z$ belongs to V_h^n. This representation has exactly the same structure as previously: the total error depends on the the local error and the effect of error transport.

It should be emphasized at this point that there has been very little work done on total control of the discretization error in time and space simultaneously. Notable exception are the works by Sandboge (1998) and Bangerth and Rannacher (1999). Since the dual problem in general has to be solved over the whole of space-time, it is very important to cut down on the computational work. For instance, in many cases we need to have the forward solution on the whole of space-time in order to compute the dual problem. It is crucial that this can be done on a coarser reference mesh lest the problem size will be intractable. It may turn out that, for some problems, the duality-based approach will indeed be intractable with today's computing resources, in which case simpler tricks must be used. Still, the duality based approach is a kind of super model, giving insights into what kinds of shortcuts and dirty tricks will have a chance of producing at least nominally reliable results. In any case, it is more than likely that error control and adaptivity in the future will be more concerned with control of norms other than the energy norm, and with control of errors in other quantities than norms (drag and lift in fluid mechanics, the J-integral in crack mechanics, etc.). The approach suggested in this Section is one alternative, but it does not preclude other possibilities.

2 Error Estimates and Adaptivity for Elasto–Plasticity

In this Section, we give an overview of residual–based *a posteriori* error estimates and adaptivity for model problems in elasto–plasticity. We consider both the Hencky total deformation theory and the Prandtl-Reuss quasi-static theory, including load step control.

2.1 Model Problems in Ideal Plasticity

The simplest model for plasticity is the *Hencky* (or total deformation) model. In standard engineering notation, using the strain ε and the stress $\boldsymbol{\sigma}$, this model may be formulated as follows: for a given body load $\boldsymbol{f}$ (and given boundary conditions), find the displacement field $\boldsymbol{u}$ such that

$$\begin{aligned}
-\nabla\cdot\boldsymbol{\sigma} &= \boldsymbol{f} && \text{(equilibrium)}\\
\boldsymbol{\sigma} &= \boldsymbol{S}:\varepsilon^e && \text{(constitutive equation)}\\
\varepsilon &= \tfrac{1}{2}\left(\nabla\boldsymbol{u}+\nabla\boldsymbol{u}^T\right) && \text{(small strain)}\\
\varepsilon &= \varepsilon^e+\boldsymbol{q} && \text{(additive decomposition)}\\
\boldsymbol{q} &= \boldsymbol{g}(\boldsymbol{\sigma}) && \text{(constitutive equation)}
\end{aligned}$$

where $\boldsymbol{S}$ is the tensor of elastic moduli, ε^e is the elastic strain, $\boldsymbol{q}$ is the (visco-)plastic strain, and $\boldsymbol{g}$ is the overstress function (to be defined). The Hencky model is suitable for proportional loading only. In case of loading and unloading, one must instead choose the *Prandtl-Reuss* (or quasi-static) formulation of plasticity. To this end, we introduce a (fictitious) time t and seek $\boldsymbol{u}$ such that

$$
\begin{aligned}
&-\nabla \cdot \boldsymbol{\sigma}(t,\cdot) = \boldsymbol{f}(t,\cdot) && \text{(equilibrium)} \\
&\dot{\boldsymbol{\sigma}} = \boldsymbol{S} : \dot{\boldsymbol{\varepsilon}}^e && \text{(constitutive equation)} \\
&\dot{\boldsymbol{\varepsilon}} = \tfrac{1}{2}\left(\nabla \dot{\boldsymbol{u}} + \nabla \dot{\boldsymbol{u}}^T\right) && \text{(small strain)} \\
&\dot{\boldsymbol{\varepsilon}} = \dot{\boldsymbol{\varepsilon}}^e + \dot{\boldsymbol{q}} && \text{(additive decomposition)} \\
&\dot{\boldsymbol{q}} = \boldsymbol{g}(\boldsymbol{\sigma}) && \text{(constitutive equation)}
\end{aligned}
$$

Here, a superscribed dot means rate with respect to time.

The Principle of Maximum Plastic Dissipation. We shall consider constitutive models that can be described by a convex elastic domain E in stress space defined by a *yield function* ϕ:

$$E = \{\boldsymbol{\sigma} : \quad \phi(\boldsymbol{\sigma}) \leq 0\}.$$

Thus, in the inviscid limit, a constraint on the actual stress will be that $\boldsymbol{\sigma} \in E$. An alternative way of formulating the yield condition in the convex case is the well known *Maximum Plastic Dissipation* (MPD) property: the actual stress $\boldsymbol{\sigma} \in E$ is the one, among all admissible stresses $\boldsymbol{\tau} \in E$, leading to maximum plastic dissipation D, where

$$D \equiv \boldsymbol{\sigma} : (\dot{\boldsymbol{\varepsilon}} - \dot{\boldsymbol{\varepsilon}^e}) \geq 0.$$

Note that MPD is particular to convex yield functions.

We may write MPD as

$$\boldsymbol{\sigma} : (\dot{\boldsymbol{\varepsilon}} - \dot{\boldsymbol{\varepsilon}}^e) \geq \boldsymbol{\tau} : (\dot{\boldsymbol{\varepsilon}} - \dot{\boldsymbol{\varepsilon}}^e) \quad \forall \boldsymbol{\tau} \in E$$

or

$$(\boldsymbol{\sigma} - \boldsymbol{\tau}) : \dot{\boldsymbol{q}} \geq 0 \quad \forall \boldsymbol{\tau} \in E.$$

From a geometrical analogy we see, in Figure 3, the relation between MPD and the standard engineering view that the plastic strain rate is proportional to the gradient of the yield function, the *associated flow rule*,

$$\dot{\boldsymbol{q}} = \lambda \nabla_\sigma \phi(\boldsymbol{\sigma})|_{\phi=0},$$

where λ is the plastic multiplier. In the case of viscoplasticity, the strain rate is instead given by:

$$\dot{\boldsymbol{q}} = \frac{1}{\gamma}\eta(\phi(\boldsymbol{\sigma}))\nabla_\sigma \phi(\boldsymbol{\sigma}),$$

where η is the *overstress* function, fulfilling either $\eta = 0$ or $\eta' > 0$, and γ is known as the *relaxation time*. One well known example is the Duvaut–Lions model of viscoplasticity, cf. Duvaut and Lions (1976):

$$\dot{\boldsymbol{q}} := \frac{1}{\gamma}\phi(\boldsymbol{\sigma})\nabla_\sigma \phi(\boldsymbol{\sigma})$$

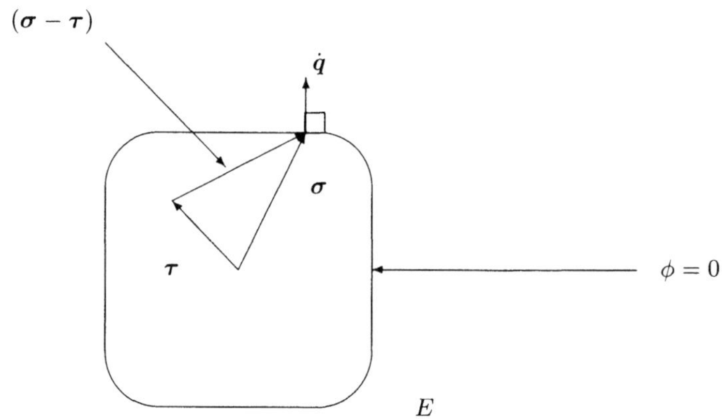

Figure 3. Convex yield function

or

$$\dot{\boldsymbol{q}} := \frac{1}{\gamma}\left(\boldsymbol{\sigma}^D - \sigma_Y \frac{\boldsymbol{\sigma}^D}{|\boldsymbol{\sigma}^D|}\right) \text{ if } |\boldsymbol{\sigma}^D| > \sigma_Y.$$

Two direct consequences of MPD will be of use in the sequel:

1. Monotonicity:

$$(\boldsymbol{\sigma}_A - \boldsymbol{\sigma}_B) : (\dot{\boldsymbol{q}}_A - \dot{\boldsymbol{q}}_B) = (\boldsymbol{\sigma}_A - \boldsymbol{\sigma}_B) : \dot{\boldsymbol{q}}_A + (\boldsymbol{\sigma}_B - \boldsymbol{\sigma}_A) : \dot{\boldsymbol{q}}_B \geq 0.$$

2. Contraction: assume that $(\dot{\boldsymbol{u}}_B, \boldsymbol{\sigma}_B, \dot{\boldsymbol{q}}_B)$ constitutes a perturbation of $(\dot{\boldsymbol{u}}_A, \boldsymbol{\sigma}_A, \dot{\boldsymbol{q}}_A)$. We have that

$$\boldsymbol{S}^{-1} : (\dot{\boldsymbol{\sigma}}_A - \dot{\boldsymbol{\sigma}}_B) + \dot{\boldsymbol{q}}_A - \dot{\boldsymbol{q}}_B = \boldsymbol{\varepsilon}(\dot{\boldsymbol{u}}_A) - \boldsymbol{\varepsilon}(\dot{\boldsymbol{u}}_B)$$

so that, with the *complementary energy* functional

$$a(\boldsymbol{\sigma}, \boldsymbol{\tau}) := \int_\Omega \boldsymbol{\tau} : \boldsymbol{S}^{-1} : \boldsymbol{\sigma}\, d\Omega, \tag{12}$$

we have

$$a(\dot{\boldsymbol{\sigma}}_A - \dot{\boldsymbol{\sigma}}_B, \boldsymbol{\sigma}_A - \boldsymbol{\sigma}_B) \leq 0$$

by MPD and equilibrium. This relation implies that

$$\frac{d}{dt}\|\boldsymbol{\sigma}_A - \boldsymbol{\sigma}_B\|_a^2 \leq 0 \tag{13}$$

where

$$\|\boldsymbol{\sigma}\|_a^2 := a(\boldsymbol{\sigma}, \boldsymbol{\sigma}).$$

We note in particular that the contraction property means that perturbations of stresses do not grow; note also that this quality holds pointwise. The same properties will be shared by materials governed by viscoplastic constitutive relations based on a convex yield function. Though the proof is not as straightforward as in the case of plasticity, it still basically hinges on the convexity of the yield function and the associated flow rule, cf. Simo (1991).

2.2 Hencky Plasticity

We shall first be concerned with the simpler case of the Hencky theory, where we do not have to worry about load step control. Hencky plasticity is described by the stress-strain relation

$$\varepsilon^{e}(\boldsymbol{u}) = \boldsymbol{S}^{-1} : \boldsymbol{\sigma}$$

and the additive split

$$\boldsymbol{\varepsilon} = \boldsymbol{\varepsilon}^{e} + \boldsymbol{q}.$$

Combined they give the Hencky equations: find $\boldsymbol{\sigma} \in W = \{\boldsymbol{\tau} : \tau_{ij} \in L_2(\Omega)\}$ and $\boldsymbol{u} \in V$ such that

$$a(\boldsymbol{\sigma}, \boldsymbol{\tau}) + (\boldsymbol{q}, \boldsymbol{\tau}) - (\boldsymbol{\varepsilon}(\boldsymbol{u}), \boldsymbol{\tau}) = 0 \quad \forall \boldsymbol{\tau} \in W$$

(where $a(\boldsymbol{\sigma}, \boldsymbol{\tau})$ is the complementary energy functional defined by (12)) and, from equilibrium,

$$(\boldsymbol{\sigma}, \boldsymbol{\varepsilon}(\boldsymbol{v})) - (\boldsymbol{f}, \boldsymbol{v}) = 0 \quad \forall \boldsymbol{v} \in V.$$

Here

$$(\boldsymbol{\sigma}, \boldsymbol{\tau}) = \int_{\Omega} \boldsymbol{\sigma} : \boldsymbol{\tau} \, \mathrm{d}\Omega,$$

and, for simplicity, we take $V \subset [H^1(\Omega)]^3$. Strictly speaking, this is not possible to do, since the Hencky equations allow for slip lines in the solution, which means that the strains may only be defined as measures. This will not concern the present discussion, but a detailed discussion of this difficulty is given by Temam (1985).

Finite Element Formulation. For simplicity, we shall consider only the case of a piecewise linear finite element approximation in space. In particular, this choice avoids the issue of whether or not a mixed method with independent approximations for displacements and stresses is preferable to a method based on displacements only (in which case the constitutive equations are taken to hold in the Gauss–points). For the constant-strain element, the element is either fully plastic or fully elastic, which means that there is no difference between a mixed method (with a piecewise constant approximation of the stresses) and a pure displacement method. We thus consider $V_h \subset V$ to be the constant strain FE space, and we use the following FEM formulation: find $(\boldsymbol{\sigma}_h, \boldsymbol{u}_h)|_{\Omega \times I_n} \in W \times V_h$ such that

$$a\left(\boldsymbol{\sigma}_h, \boldsymbol{\tau}\right) + \left(\boldsymbol{q}_h, \boldsymbol{\tau}\right) = \left(\boldsymbol{\varepsilon}\left(\boldsymbol{u}_h\right), \boldsymbol{\tau}\right) \tag{14}$$

for all $\boldsymbol{\tau} \in W$,

$$\left(\boldsymbol{\sigma}_h, \boldsymbol{\varepsilon}\left(\boldsymbol{v}\right)\right) = \left(\boldsymbol{f}, \boldsymbol{v}\right) \tag{15}$$

for all $\boldsymbol{v} \in V_h$. Here, Equation (14) can be thought of as holding pointwise in the element; it is formulated as a weak statement for convenience only. We also need the constitutive relation coupling $\boldsymbol{q}$ to $\boldsymbol{\sigma}$. However, the following discussion will only hinge on the convexity properties of this relation, and we will leave until later a more detailed discussion on the effect of it.

Error Estimation in Complementary Energy for Hencky Plasticity. To obtain a first error estimate in complementary energy norm for the Hencky problem, we subtract the finite element problem from the continuous problem to obtain

$$a(\boldsymbol{\sigma} - \boldsymbol{\sigma}_h, \boldsymbol{\tau}) + (\boldsymbol{q} - \boldsymbol{q}_h, \boldsymbol{\tau}) = (\varepsilon(\boldsymbol{u}) - \varepsilon(\boldsymbol{u}_h), \boldsymbol{\tau}) \quad \forall \boldsymbol{\tau} \in W.$$

We define $\|\boldsymbol{\sigma}\|_a^2 := a(\boldsymbol{\sigma}, \boldsymbol{\sigma})$ and $\boldsymbol{e}_u := \boldsymbol{u} - \boldsymbol{u}_h$, and set $\boldsymbol{\tau} = \boldsymbol{\sigma} - \boldsymbol{\sigma}_h$ to find that

$$\|\boldsymbol{\sigma} - \boldsymbol{\sigma}_h\|_a^2 = (\varepsilon(\boldsymbol{u} - \boldsymbol{u}_h), \boldsymbol{\sigma} - \boldsymbol{\sigma}_h) - (\boldsymbol{q} - \boldsymbol{q}_h, \boldsymbol{\sigma} - \boldsymbol{\sigma}_h) \leq (\varepsilon(\boldsymbol{e}_u), \boldsymbol{\sigma} - \boldsymbol{\sigma}_h),$$

where the last step follows from monotonicity.

Using the equilibrium equation, we note that

$$(\boldsymbol{\sigma} - \boldsymbol{\sigma}_h, \varepsilon(\boldsymbol{v})) = 0 \quad \forall \boldsymbol{v} \in V_h,$$

so that

$$\|\boldsymbol{\sigma} - \boldsymbol{\sigma}_h\|_a^2 \leq (\varepsilon(\boldsymbol{e}_u - \pi_h \boldsymbol{e}_u), \boldsymbol{\sigma} - \boldsymbol{\sigma}_h),$$

where π_h denotes an interpolant or projection onto the mesh. Since

$$\boldsymbol{e}_u - \pi_h \boldsymbol{e}_u = \boldsymbol{u} - \boldsymbol{u}_h - \pi_h \boldsymbol{u} + \pi_h \boldsymbol{u}_h = \boldsymbol{u} - \pi_h \boldsymbol{u},$$

this can be written

$$\|\boldsymbol{\sigma} - \boldsymbol{\sigma}_h\|_a^2 \leq (\varepsilon(\boldsymbol{u} - \pi_h \boldsymbol{u}), \boldsymbol{\sigma} - \boldsymbol{\sigma}_h). \tag{16}$$

Now, we cannot estimate $\varepsilon(\boldsymbol{u} - \pi_h \boldsymbol{u})$ in terms of complementary energy, as we would do in the case of linear elasticity. However, by integration by parts in (16) we find that

$$\begin{aligned} \|\boldsymbol{\sigma} - \boldsymbol{\sigma}_h\|_a^2 \leq & \textstyle\sum_K \int_K (\boldsymbol{f} + \nabla \cdot \boldsymbol{\sigma}_h) \cdot (\boldsymbol{u} - \pi_h \boldsymbol{u}) \ \mathrm{d}\Omega \\ & -\tfrac{1}{2} \textstyle\sum_K \int_{\partial K} \mathbf{n}_K \cdot [\boldsymbol{\sigma}_h] \cdot (\boldsymbol{u} - \pi_h \boldsymbol{u}) \ ds, \end{aligned} \tag{17}$$

where $[\boldsymbol{\sigma}_h]$ denotes the jump in stress between elements and $\mathbf{n}_K$ is the outward unit normal vector to the element. The estimate (17) can thus be used to get an estimate of the error. To obtain an elementwise estimate to be used for mesh refinement purposes, we use Cauchy's inequality to obtain

$$\|\boldsymbol{\sigma} - \boldsymbol{\sigma}_h\|_a^2 \leq \sum_K \omega_K \rho_K, \tag{18}$$

where, introducing the size h_K of element K,

$$\omega_K = \max \begin{cases} h_K^{-2} \|\boldsymbol{u} - \pi_h \boldsymbol{u}\|_{L_2(K)} \\ h_K^{-3/2} \|\boldsymbol{u} - \pi_h \boldsymbol{u}\|_{L_2(\partial K)} \end{cases}$$

and

$$\rho_K = h_K^2 \|\boldsymbol{f} + \nabla \cdot \boldsymbol{\sigma}_h\|_{L_2(K)} + \frac{1}{2} h_K^{3/2} \|\mathbf{n}_K \cdot [\boldsymbol{\sigma}_h]\|_{L_2(\partial K)}.$$

We approximate the weights ω_K by

$$\omega_K \approx C_K^i \|D_h^2 \boldsymbol{u}_h\|_{L_2(K)}$$

where $C_K^i \approx 1$ is an interpolation constant, and where $D_h^2 \boldsymbol{u}_h$ denotes a discrete approximation of the maximum second order derivative of $\boldsymbol{u}_h$. The estimate (18) is not very sharp due to the fact that Cauchy's inequality has been used on each element. Thus it is better to use directly (17) when trying to assess the size of the error, and to use (18) as a refinement marker only.

The estimates (17) and (18) give information about the error in the stresses in complementary norm only. If we wish to control other quantities, such as functionals of the error (e.g., error in traction) or other norms (e.g., maximum displacement error), we must use the general theory advocated by Eriksson et al. (1995), first suggested and developed for plasticity by Rannacher and Suttmeier (1998).

As an example of the type of refinement we obtain using the estimate (18), we give a plane strain elasto-plastic example. The problem is displacement driven, clamped at the top and compressed. The problem is symmetric, and we are computing on the top left quarter of the physical domain. In Figure 4, we give the adapted mesh and the corresponding plastic zone. Note the marked refinement on the transition zone between elastic and plastic zones; this is a line of non-smoothness in the stresses.

A General Approach to Nonlinear Problems. Assume that we have a nonlinear variational problem: find $u \in V$ such that

$$(A(u), v) = (f, v) \quad \forall v \in V,$$

and a FEM counterpart: find $u_h \in V_h$ such that

$$(A(u_h), v) = (f, v) \quad \forall v \in V_h.$$

We then have the following Galerkin orthogonality property

$$(A(u) - A(u_h), v) = 0 \quad \forall v \in V_h,$$

or

$$(B \cdot (u - u_h), v) = 0 \quad \forall v \in V_h,$$

where

$$B := \int_0^1 A'(\xi u + (1 - \xi) u_h)\, d\xi.$$

Note that B constitutes an *exact* linearization, since

$$\int_0^1 \frac{d}{d\xi} A(\xi u + (1 - \xi) u_h)\, d\xi = A(u) - A(u_h)$$

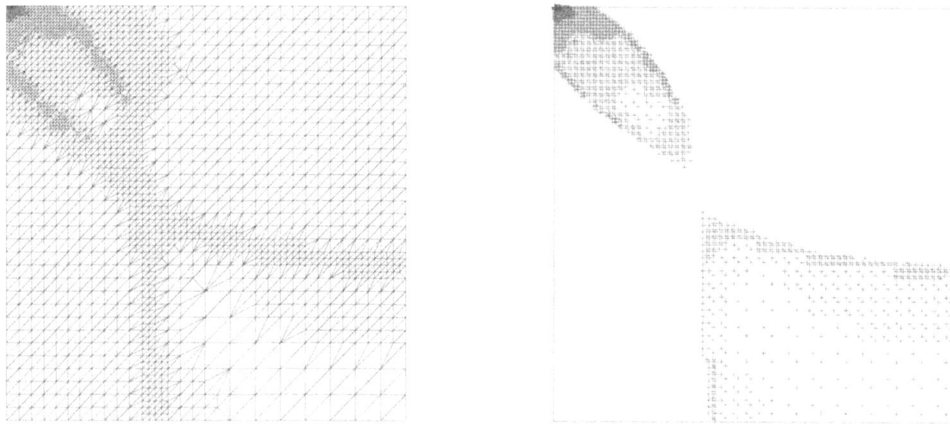

Figure 4. Refined mesh and plastic zone.

and

$$\int_0^1 \frac{d}{d\xi} A(\xi u + (1-\xi)u_h)\, d\xi = B \cdot (u - u_h).$$

However, since the exact solution is not known in general, the exact linearization is intractable. In practice, one must accept a linearization error an approximate B by

$$B \approx A'(u_h).$$

Next, we must define a linearized dual continuous problem as follows: find φ such that

$$B^{\mathrm{T}} \cdot \varphi = J,$$

where J can be chosen freely, and where B^{T} is the adjoint of B, defined by

$$(B^{\mathrm{T}} \cdot u, v) = (u, B \cdot v)$$

for all $u \in V$ and $v \in V$. Then

$$(u - u_h, J) = (u - u_h, B^T \cdot \varphi) = (B \cdot (u - u_h), \varphi) =$$

$$= (A(u) - A(u_h), \varphi) = (f - A(u_h), \varphi - \pi_h \varphi).$$

To obtain an error estimate involving a quantity of interest, J is chosen as some functional of the error. As an example, we may take $J = u - u_h$, in which case (by Cauchy's inequality)

$$\|u - u_h\|^2_{L_2(\Omega)} \leq \sum_K \|f - A(u_h)\|_{L_2(K)} \|\varphi - \pi_h \varphi\|_{L_2(K)}$$

where K denotes an element in the mesh. This is an estimate of the same type as (18), this time with weights $\rho_K := h_K^2 \|f - A(u_h)\|_{L_2(K)}$ and $\omega_K := h_K^{-2} \|\varphi - \pi_h \varphi\|_{L_2(K)}$. Thus, ρ_K will always signify the *local error contribution*, whereas ω_K will depend on the type of error control, and signifies *error transport*, as discussed in Section 1.1.

Note that the right-hand side J of the dual problem may not be known *a priori*. Thus, J has to be estimated, e.g., by solving on two consecutive meshes. In conclusion, the duality approach entails:

- (Approximate) linearization
- Solution of dual problem (possibly with estimated right-hand side)
- Estimation of $\varphi - \pi_h \varphi$

Though many questions remain unanswered (how good is the linearization; how well need J be approximated, etc.), this approach has been shown to give a good hold of the error in the works by Rannacher and Suttmeier (1998, 1999). It should be noted that the linearization process is given for free if the plasticity problem is solved using a Newton method; in this case B is nothing but the Newton matrix, and the dual problem is computed using one more Newton iteration. Thus, the duality approach is particularly attractive in the nonlinear case. If a Newton method is not used, however, the full duality based adaptivity requires additional programming effort, and the complementary energy norm control may be considered a simplified version (where $\boldsymbol{\varphi} = \boldsymbol{u}$) which can be implemented without much effort. Thus, complementary energy norm control of the stresses plays the same role in plasticity as energy norm control in linear elasticity: it is considerably simpler and considerably much more limited.

Estimates Based on Stress Projection. The basic idea behind stress projection methods, introduced for adaptive purposes by Zienkiewicz and Zhu (1987), is to find a (hopefully) improved stress field $\boldsymbol{\sigma}^*$ and use the triangle inequality

$$\|\boldsymbol{\sigma} - \boldsymbol{\sigma}_h\|_a \leq \|\boldsymbol{\sigma} - \boldsymbol{\sigma}^*\|_a + \|\boldsymbol{\sigma}^* - \boldsymbol{\sigma}_h\|_a$$

followed by the basic assumption

$$\|\boldsymbol{\sigma} - \boldsymbol{\sigma}^*\|_a << \|\boldsymbol{\sigma} - \boldsymbol{\sigma}_h\|_a$$

formalized for instance as

$$\|\boldsymbol{\sigma}-\boldsymbol{\sigma}^*\|_a = O(h^{k+1}) \quad \text{if} \quad \|\boldsymbol{\sigma}-\boldsymbol{\sigma}_h\|_a = O(h^k).$$

It is not so simple to actually find such a stress field $\boldsymbol{\sigma}^*$ on general unstructured meshes, and in practice one usually has to settle for less strict arguments. Nevertheless, this approach is popular in commercial codes due to its simplicity, and due to the fact that stress projections are already performed for visualization purposes. In practice, one often uses L_2–projections of the discrete stress components onto the same mesh used to compute the displacements. The lack of strict proofs notwithstanding, this approach works well for smooth solutions.

In the case of plasticity, we recall

$$\|\boldsymbol{\sigma}-\boldsymbol{\sigma}_h\|_a^2 \leq (\boldsymbol{\sigma}-\boldsymbol{\sigma}_h, \boldsymbol{\varepsilon}(\boldsymbol{u}-\boldsymbol{u}_h)) \leq \|\boldsymbol{\sigma}-\boldsymbol{\sigma}_h\|_a \|\boldsymbol{\varepsilon}(\boldsymbol{u})-\boldsymbol{\varepsilon}(\boldsymbol{u}_h)\|_{a^{-1}}$$

where

$$\|\boldsymbol{\varepsilon}\|_{a^{-1}} = \int_\Omega \boldsymbol{\varepsilon} : \boldsymbol{S} : \boldsymbol{\varepsilon}\, dx$$

We can now smooth the computed strain field to obtain $\boldsymbol{\varepsilon}^*$ and use

$$\|\boldsymbol{\sigma}-\boldsymbol{\sigma}_h\|_a \lesssim \|\boldsymbol{\varepsilon}^*-\boldsymbol{\varepsilon}(\boldsymbol{u}_h)\|_{a^{-1}}.$$

The simplicity of this approach is hard to beat! However, the lack of smoothness of the strains casts doubt on the reliability of stress projection approaches in plasticity. Note:

$$\|\boldsymbol{\sigma}-\boldsymbol{\sigma}_h\|_a^2 \leq (\boldsymbol{\sigma}-\boldsymbol{\sigma}_h, \boldsymbol{\varepsilon}(\boldsymbol{u})-\boldsymbol{\varepsilon}^*+\boldsymbol{\varepsilon}^*-\boldsymbol{\varepsilon}(\boldsymbol{u}_h)) \leq$$

$$\leq \|\boldsymbol{\sigma}-\boldsymbol{\sigma}_h\|_a \left(\|\boldsymbol{\varepsilon}(\boldsymbol{u})-\boldsymbol{\varepsilon}^*\|_{a^{-1}} + \|\boldsymbol{\varepsilon}^*-\boldsymbol{\varepsilon}(\boldsymbol{u}_h)\|_{a^{-1}}\right)$$

and we must hope that

$$\|\boldsymbol{\varepsilon}(\boldsymbol{u})-\boldsymbol{\varepsilon}^*\|_{a^{-1}} << \|\boldsymbol{\varepsilon}-\boldsymbol{\varepsilon}(\boldsymbol{u}_h)\|_{a^{-1}}$$

which cannot be the case for non-smooth strains.

Another approach to stress smoothing is to make it residual based. We consider the case of a *continuous* smoothed stress field $\boldsymbol{\sigma}^*$ and use the relation

$$\|\boldsymbol{\sigma}-\boldsymbol{\sigma}_h\|_a^2 \leq (\boldsymbol{\sigma}-\boldsymbol{\sigma}_h+\boldsymbol{\sigma}^*-\boldsymbol{\sigma}^*, \boldsymbol{\varepsilon}(\boldsymbol{u}-\pi_h\boldsymbol{u}))$$

together with integration by parts, which leads to

$$\|\boldsymbol{\sigma}-\boldsymbol{\sigma}_h\|_a^2 \leq \sum_K \int_K (\boldsymbol{f}+\nabla\cdot\boldsymbol{\sigma}^*)\cdot(\boldsymbol{u}-\pi_h\boldsymbol{u})\ \mathrm{d}\Omega + \sum_K \int_K (\boldsymbol{\sigma}^*-\boldsymbol{\sigma}_h) : \boldsymbol{\varepsilon}(\boldsymbol{u}-\pi_h\boldsymbol{u})\ \mathrm{d}\Omega.$$

Say the first term on the right-hand side is small; then

$$\|\boldsymbol{\sigma}-\boldsymbol{\sigma}_h\|_a^2 \leq \sum_K \omega_K \rho_K,$$

where $\omega_K = \|\boldsymbol{\sigma}^* - \boldsymbol{\sigma}_h\|_{a,K}$ and

$$\rho_K = \left(\int_K \boldsymbol{\varepsilon}(\boldsymbol{u} - \pi_h \boldsymbol{u}) : \boldsymbol{S} : \boldsymbol{\varepsilon}(\boldsymbol{u} - \pi_h \boldsymbol{u}) \right)^{1/2}.$$

It seems more likely to be possible to find a $\boldsymbol{\sigma}^*$ close to equilibrium than to find a superconvergent strain field, at least in the case of non-smooth strains.

Remark 1. In fact, stress smoothing can be combined with the duality approach, as discussed by Cirak and Ramm (2000). Basically, this is not so surprising: the error, as viewed in the duality approach, is obtained as a product of the local discretization error (which now is computed by stress smoothing) and the dual solution. We do not have to use the residual to express the discretization error.

2.3 Load Step Control for Prandtl-Reuss.

Since the Prandtl-Reuss problem is time-dependent, we begin by recalling the problem of solving the ordinary differential equation

$$\frac{dy}{dt} = f(t, y(t)), \quad y(0) = y_0.$$

For adaptivity, the main question is: how do perturbations of y grow?

As discussed in Section 1.1, the sign of $\partial f/\partial y$ determines the growth of perturbations. Damped problems are more suited for local step control, simply because perturbations do not persist, and will not seriously affect the error at final time. We are now considering (visco)plasticity, which we know is a dissipative (damped) problem since by Equation (13)

$$\frac{d}{dt}\|\boldsymbol{\sigma} - \tilde{\boldsymbol{\sigma}}\|_a^2 \leq 0,$$

i.e., perturbation of the stresses will not grow. Thus, we will begin by suggesting a local load step control method, and then take a closer look at the problem of global error control.

To obtain a local error control, we first split the total error at time $t = t_n$ using a fully discrete stress $\boldsymbol{\sigma}_h$ and a time-discrete, spatially continuous stress $\boldsymbol{\sigma}_*$. By the triangle inequality we have

$$\|\boldsymbol{\sigma}(t_n) - \boldsymbol{\sigma}_h^n\|_a \leq \|\boldsymbol{\sigma}(t_n) - \boldsymbol{\sigma}_*^n\|_a + \|\boldsymbol{\sigma}_*^n - \boldsymbol{\sigma}_h^n\|_a. \tag{19}$$

Next, we apply the backward Euler method to the spatially continuous problem, which requires solving

$$\boldsymbol{S}^{-1} : \left(\boldsymbol{\sigma}_*^n - \boldsymbol{\sigma}_*^{n-1}\right) + k_n\left(\dot{\boldsymbol{q}}_*^n - \boldsymbol{\varepsilon}(\dot{\boldsymbol{u}}_*^n)\right) = 0. \tag{20}$$

Inserting the actual stress field into (20), we obtain the following representation of the truncation error

$$\boldsymbol{S}^{-1} : (\boldsymbol{\sigma}(t_n) - \boldsymbol{\sigma}(t_{n-1})) + k_n(\dot{\boldsymbol{q}}(t_n) - \boldsymbol{\varepsilon}(\dot{\boldsymbol{u}}(t_n))) = -\int_{I_n} (t - t_n)\, \boldsymbol{S}^{-1} : \ddot{\boldsymbol{\sigma}}\, dt \neq 0, \tag{21}$$

cf. Rannacher and Suttmeier (1999).

If we now subtract (20) from (21), multiply by $\boldsymbol{\sigma}_*^n - \boldsymbol{\sigma}(t_n)$, use MPD and equilibrium at $t = t_n$, we obtain

$$\|\boldsymbol{\sigma}_*^n - \boldsymbol{\sigma}(t_n)\|_a \leq \|\boldsymbol{\sigma}_*^{n-1} - \boldsymbol{\sigma}(t_{n-1})\|_a + C_\sigma k_n^2$$

where $C_\sigma = \| \max_{t\in I_n} \ddot{\boldsymbol{\sigma}}(t)\|_a$. This fact motivates the assumption of existence of a constant C such that, if $\boldsymbol{\sigma}_*^{n-1} = \boldsymbol{\sigma}(t_{n-1})$,

$$\boldsymbol{\sigma}_*^n = \boldsymbol{\sigma}(t_n) + Ck_n^2 + \text{higher order terms}$$

Now for the temporal error "control", we can use a simple approach due to Richardson (cf. Hairer et al. (1992)):

1. Assume $\boldsymbol{\sigma}_*^{n-1} = \boldsymbol{\sigma}(t_{n-1})$ (*local* error control).
2. Take first one step from $t = t_{n-1}$ to $t = t_n$ of size $2\,k_n$ to compute $\hat{\boldsymbol{\sigma}}_*^n$, leading to
$$\hat{\boldsymbol{\sigma}}_*^n = \boldsymbol{\sigma}(t_n) + C(2\,k_n)^2 + \text{HOT}$$
3. Take two steps from $t = t_{n-1}$ to $t = t_n$, each of size k_n to compute $\boldsymbol{\sigma}_*^n$, leading to
$$\boldsymbol{\sigma}_*^n = \boldsymbol{\sigma}(t_n) + 2\,Ck_n^2 + \text{HOT}$$
since the error due to the perturbation from the first step does not grow (MPD yields dissipation).
4. Eliminate Ck_n^2 to arrive at
$$\|\boldsymbol{\sigma}_*^n - \boldsymbol{\sigma}(t_n)\|_a \approx \|\boldsymbol{\sigma}_*^n - \hat{\boldsymbol{\sigma}}_*^n\|_a$$

Note that this approach is very close to solving local problems in spatial adaptivity. The unanswered question is, as always when using local error control, how is the error transported? Solving local problems in space is generally used only for energy norm control where we know that errors are local, but in the present case the error will depend on the perturbation growth.

To continue with the spatial error control, of $\|\boldsymbol{\sigma}_*^n - \boldsymbol{\sigma}_h^n\|_a$, we have

$$\boldsymbol{S}^{-1} : \left(\boldsymbol{\sigma}_*^n - \boldsymbol{\sigma}_*^{n-1}\right) + k_n\left(\dot{\boldsymbol{q}}_*^n - \boldsymbol{\varepsilon}(\dot{\boldsymbol{u}}_*^n)\right) = 0$$

and also

$$\boldsymbol{S}^{-1} : \left(\boldsymbol{\sigma}_h^n - \boldsymbol{\sigma}_h^{n-1}\right) + k_n\left(\dot{\boldsymbol{q}}_h^n - \boldsymbol{\varepsilon}(\dot{\boldsymbol{u}}_h^n)\right) = 0.$$

Subtract, multiply by $\boldsymbol{\sigma}_h^n - \boldsymbol{\sigma}_*^n$ and use MPD:

$$a\left(\boldsymbol{\sigma}_h^n - \boldsymbol{\sigma}_h^{n-1} - \boldsymbol{\sigma}_*^n + \boldsymbol{\sigma}_*^{n-1}, \boldsymbol{\sigma}_h^n - \boldsymbol{\sigma}_*^n\right) \leq k_n\left(\boldsymbol{\varepsilon}\left(\dot{\boldsymbol{u}}_h^n - \dot{\boldsymbol{u}}_*^n\right), \boldsymbol{\sigma}_h^n - \boldsymbol{\sigma}_*^n\right).$$

Note that

$$a\left(\boldsymbol{\sigma}_h^n - \boldsymbol{\sigma}_h^{n-1} - \boldsymbol{\sigma}_*^n + \boldsymbol{\sigma}_*^{n-1}, \boldsymbol{\sigma}_h^n - \boldsymbol{\sigma}_*^n\right) =$$

$$= \|\boldsymbol{\sigma}_h^n - \boldsymbol{\sigma}_*^n\|_a^2 - a\left(\boldsymbol{\sigma}_h^{n-1} - \boldsymbol{\sigma}_*^{n-1}, \boldsymbol{\sigma}_h^n - \boldsymbol{\sigma}_*^n\right) \geq$$

$$\geq \frac{1}{2}\|\boldsymbol{\sigma}_h^n - \boldsymbol{\sigma}_*^n\|_a^2 - \frac{1}{2}\|\boldsymbol{\sigma}_h^{n-1} - \boldsymbol{\sigma}_*^{n-1}\|_a^2.$$

Using Galerkin orthogonality at $t = t_n$, we finally find that

$$\|\boldsymbol{\sigma}_h^n - \boldsymbol{\sigma}_*^n\|_a^2 \leq \|\boldsymbol{\sigma}_h^{n-1} - \boldsymbol{\sigma}_*^{n-1}\|_a^2 + 2k_n \left(\boldsymbol{\sigma}_*^n - \boldsymbol{\sigma}_h^n, \varepsilon\left(\dot{\boldsymbol{u}}_*^n - \pi_h \dot{\boldsymbol{u}}_*^n\right)\right),$$

which means that we now have control of both of the terms on the right-hand side of (19) since $-\nabla \cdot \boldsymbol{\sigma}_*^n = \boldsymbol{f}(t_n, \cdot)$.

The local error approach will give a crude way of determining the load step size, but it will not give us a good hold on the global error. Simply summing up the local error contribution would result in a severe overestimation of the error since perturbations are damped, and an error at early times may have no impact at the error at final time at all. An alternative approach would be to use finite elements in space and time, which gives the possibility of a more detailed estimation of the error. In the following we shall discuss the appearance of the linearization occuring in the duality based approach. For simplicity, we stick to load step control only, but the complete control of the error in space and time follow along the same lines. For some initial results regarding the full space-time adaptivity problem in this context, see Rannacher and Suttmeier (1999).

Load Step Control Only. Again, the constant strain elements allows for the obvious way to interpret the stresses as independent variables. The Prandtl-Reuss equations can thus be written: find $\dot{\boldsymbol{\sigma}}$ and $\boldsymbol{v}$ such that

$$\boldsymbol{S}^{-1} : \dot{\boldsymbol{\sigma}} + \dot{\boldsymbol{q}}(\boldsymbol{\sigma}) = \varepsilon(\boldsymbol{v})$$

and

$$(\boldsymbol{\sigma}, \varepsilon(\boldsymbol{w})) = (\boldsymbol{f}(t), \boldsymbol{w}) \quad \forall \boldsymbol{w} \in V, \quad t \in I,$$

with $\boldsymbol{\sigma} = \boldsymbol{\sigma}_0$ at $t = 0$; $I = (0, T)$. We note in particular that the second equation is linear.

Remark 2. An alternative formulation arises by considering the plastic strains $\boldsymbol{q}$ and displacement rate $\boldsymbol{v}$ as independent variables:

$$\dot{\boldsymbol{q}} + \boldsymbol{S}^{-1} : \dot{\boldsymbol{\sigma}}(\boldsymbol{q}, \varepsilon(\boldsymbol{v})) = \varepsilon(\boldsymbol{v})$$

and

$$(\boldsymbol{\sigma}(\boldsymbol{q}, \varepsilon(\boldsymbol{v})), \varepsilon(\boldsymbol{w})) = (\boldsymbol{f}(t), \boldsymbol{w}) \quad \forall \boldsymbol{w} \in V, \quad t \in I,$$

with both equations nonlinear.

We note that there are conflicting interests in the second formulation:

1. plastic strains may be ill behaved;
2. plastic strains are used in models of damage and cracking.

In engineering, the advantage of 2 often outweighs the disadvantage of 1; the engineer usually knows when it is safe to use $\boldsymbol{q}$ and $\boldsymbol{v}$ as unknowns. Here, we shall stick to $\boldsymbol{\sigma}$ and $\boldsymbol{v}$ as unknowns; the difference in formulating an error estimate is not very large since we follow a general approach.

For our purpose, we need a finite element method also in time, and we shall choose the time-discontinuous Galerkin FEM intoduced in Section 1.1, with a piecewise constant approximation in time on each subinterval $I_n = (t_{n-1}, t_n)$ of length $k_n = t_n - t_{n-1}$. There is a good reason for this choice: as a time-stepping method it is closely related to the backward Euler method which is often used for plasticity. We set $\boldsymbol{\sigma}_h^n = \boldsymbol{\sigma}_h(t_n^-)$ and $\boldsymbol{\sigma}_h^{n-1} = \boldsymbol{\sigma}_h(t_{n-1}^+)$, and formulate the following FEM: find $(\boldsymbol{\sigma}_h^n, \boldsymbol{v}_h^n)|_{\Omega \times I_n} \in W \times V_h^n$ such that for $n = 1, 2, \ldots$,

$$a\left([\boldsymbol{\sigma}_h^n], \boldsymbol{\tau}^n\right) + k_n\left(\dot{\boldsymbol{q}}(\boldsymbol{\sigma}^n), \boldsymbol{\tau}^n\right) = k_n\left(\boldsymbol{\varepsilon}\left(\boldsymbol{v}_h^n\right), \boldsymbol{\tau}^n\right)$$

for all $\boldsymbol{\tau} \in W$, $t \in I_n$,

$$\left(\boldsymbol{\sigma}_h^n, \boldsymbol{\varepsilon}\left(\boldsymbol{w}\right)\right) = (\hat{\boldsymbol{f}}, \boldsymbol{w})$$

for all $\boldsymbol{w} \in V_h^n$, $t \in I_n$, where $[\boldsymbol{\sigma}_h^n] = \boldsymbol{\sigma}_h^n - \boldsymbol{\sigma}_h^{n-1}$, $\boldsymbol{\sigma}_h^0 = \boldsymbol{\sigma}_0$, and

$$\hat{\boldsymbol{f}} = \frac{1}{k_n}\int_{t_{n-1}}^{t_n} \boldsymbol{f}(\cdot, t)\, dt.$$

This is precisely the standard backward Euler scheme (on a fixed mesh), apart from the treatment of $\boldsymbol{f}$.

If we assume that we are solving on a fixed spatial mesh for all times, we can write the spatially discrete, time-continuous, problem on matrix form as

$$\begin{bmatrix} \mathbf{K} & \mathbf{0} \\ \mathbf{0} & \mathbf{0} \end{bmatrix} \begin{bmatrix} \dot{\mathbf{s}} \\ \dot{\mathbf{u}} \end{bmatrix} + \begin{bmatrix} \mathbf{C}(\mathbf{s}) + \mathbf{G}\mathbf{u} \\ \mathbf{G}^{\mathrm{T}}\mathbf{s} \end{bmatrix} = \begin{bmatrix} \mathbf{0} \\ \mathbf{f} \end{bmatrix}$$

with $\mathbf{s}$ and $\mathbf{u}$ vectors containing elemental stresses and nodal displacement rates. This set of ODE:s is of the form

$$\mathbf{A}\dot{\mathbf{x}} + \mathbf{B}(\mathbf{x}) = \mathbf{g} \quad (+\ \mathrm{IC}). \tag{22}$$

We note that $\mathbf{x}^{\mathrm{T}}\mathbf{A}\mathbf{x} = \|\boldsymbol{\sigma}\|_a^2$, which is the norm for which we will attempt error control.

The discontinuous Galerkin method applied to the problem (22) reads: find $\mathbf{x}_h \in P^q(I_n)$ such that for all $\mathbf{v} \in P^q(I_n)$

$$\mathbf{v}_+^{\mathrm{T}}[\mathbf{A}\mathbf{x}_h^n] + \int_{I_n} \mathbf{v}^{\mathrm{T}}\left(\mathbf{A}\dot{\mathbf{x}}_h + \mathbf{B}(\mathbf{x}_h) - \mathbf{g}\right) dt = 0,$$

where $P^q(I_n)$ is the space of polynomials on I_n of degree q or less.

Due to the fact that $\mathbf{B}'$ is symmetric, the approximate linearized dual problem can be written:

$$-\mathbf{A}\dot{\mathbf{y}} + \mathbf{B}'(\mathbf{x}_h)\mathbf{y} = \mathbf{0}$$

with intial condition

$$\mathbf{y}(T) = \mathbf{e}(T) := \mathbf{x}(T) - \mathbf{x}_h(T).$$

We have

$$\|\boldsymbol{\sigma}(T)-\boldsymbol{\sigma}_h(T)\|_a^2 = \mathbf{e}(T)^{\mathrm{T}}\mathbf{A}\mathbf{e}(T) =$$

$$[\text{use the dual problem}] = \mathbf{e}(T)^{\mathrm{T}}\mathbf{A}\mathbf{e}(T) + \sum_{n=1}^{N}\int_{I_n}\mathbf{e}^{\mathrm{T}}\left(-\mathbf{A}\dot{\mathbf{y}}+\mathbf{B}'(\mathbf{x}_h)\mathbf{y}\right)dt =$$

$$[\text{use the symmetry of } \mathbf{B}'] = \sum_{n=1}^{N}\int_{I_n}\mathbf{y}^{\mathrm{T}}\left(\mathbf{A}\dot{\mathbf{e}}+\mathbf{B}'(\mathbf{x}_h)\mathbf{e}\right)dt + \sum_{n=0}^{N-1}\mathbf{y}_+^{\mathrm{T}}[\mathbf{e}^n] \approx$$

$$[\text{use approximate linearization}] \approx \sum_{n=1}^{N}\int_{I_n}\mathbf{y}^{\mathrm{T}}\left(\mathbf{g}-\mathbf{A}\dot{\mathbf{x}}_h-\mathbf{B}(\mathbf{x}_h)\right)dt + \sum_{n=0}^{N-1}\mathbf{y}_+^{\mathrm{T}}[\mathbf{x}_h^n].$$

Finally, invoke the orthogonality property:

$$\|\boldsymbol{\sigma}(T)-\boldsymbol{\sigma}_h(T)\|_a^2 \approx \sum_{n=0}^{N-1}(\mathbf{y}_+-\tilde{\mathbf{y}}_+)^{\mathrm{T}}[\mathbf{x}_h^n] +$$

$$+\sum_{n=1}^{N}\int_{I_n}(\mathbf{y}-\tilde{\mathbf{y}})^{\mathrm{T}}\left(\mathbf{g}-\mathbf{A}\dot{\mathbf{x}}_h-\mathbf{B}(\mathbf{x}_h)\right)dt,$$

where $\tilde{\mathbf{y}} \in P^k(I_n)$.

Dual Problem for Plasticity. We will end the discussion on adaptivity for plasticity by giving more precise information regarding the approximate linearization, showing that there is a particular difficulty in that we cannot in fact get hold of the linearization error. Returning to the Prandtl-Reuss formulation, we thus need to solve a dual problem, i.e., we want to find a dual stress $\boldsymbol{\Sigma}$ and a dual displacement rate $\boldsymbol{z}$ so that

$$-\boldsymbol{S}^{-1} : \dot{\boldsymbol{\Sigma}} + \boldsymbol{B}(\boldsymbol{\sigma},\boldsymbol{\sigma}_h) : \boldsymbol{\Sigma} = \varepsilon(\boldsymbol{z})$$

and

$$\nabla \cdot \boldsymbol{\Sigma} = 0$$

with $\boldsymbol{B}$ the Jacobian of $\dot{\boldsymbol{q}}(\boldsymbol{\sigma})$. This dual problem should then be solved backwards in time with initial data $\boldsymbol{\Sigma}(T) = \boldsymbol{\sigma}(T) - \boldsymbol{\sigma}_h(T)$, $\boldsymbol{z}(T) = 0$. This means that we must evaluate the linearized tensor

$$\boldsymbol{B}(\boldsymbol{\sigma},\boldsymbol{\sigma}_h) := \int_0^1 \frac{\partial\dot{\boldsymbol{q}}}{\partial\boldsymbol{\sigma}}(\xi\,\boldsymbol{\sigma}+(1-\xi)\boldsymbol{\sigma}_h^n)\,d\xi.$$

In the case of Mises–Duvaut-Lions visco-plasticity, we can write down the visco-plastic strain rate as

$$\dot{\boldsymbol{q}} = \frac{1}{\gamma}\left(\boldsymbol{\sigma}^D - \sigma_Y\frac{\boldsymbol{\sigma}^D}{|\boldsymbol{\sigma}^D|}\right) \text{ if } |\boldsymbol{\sigma}^D| > \sigma_Y,$$

where $\gamma \to 0$ is the limit case of plasticity. To perform the differentiation is then a purely technical problem; we obtain

$$\frac{\partial \dot{\boldsymbol{q}}}{\partial \boldsymbol{\sigma}} = \frac{1}{\gamma\,|\boldsymbol{\sigma}^D|}\left((|\boldsymbol{\sigma}^D| - \sigma_Y)\,\boldsymbol{I}^D + \frac{\sigma_Y}{|\boldsymbol{\sigma}^D|^2}\boldsymbol{\sigma}^D \otimes \boldsymbol{\sigma}^D\right)$$

where $\boldsymbol{I}^D$ is such that $\boldsymbol{I}^D : \boldsymbol{\tau} = \boldsymbol{\tau}^D$. We are now concerned with what happens as $\gamma \to 0$. For the first term on the right-hand side, we note that

$$0 \leq \lim_{\gamma \to 0} \frac{|\boldsymbol{\sigma}^D| - \sigma_Y}{\gamma} = \lambda;$$

λ, the plastic multiplier, is finite. This term is thus unproblematic. The second term on the right-hand side,

$$\frac{\sigma_Y}{\gamma\,|\boldsymbol{\sigma}^D|^3}\boldsymbol{\sigma}^D \otimes \boldsymbol{\sigma}^D$$

will not be finite, however. Looking at the differential equation, we note that the effect of this term is to kill the components of the dual solution in the $\boldsymbol{\sigma}^D$–direction. This is quite natural, since the true stress will not change in this direction due to the yield condition; there can be no error in this direction. On the other hand, we do not know this direction, all we have is the numerically computed stresses, $\boldsymbol{\sigma}_h$, not the actual $\boldsymbol{\sigma}$. It would be unwise to just kill of components of $\boldsymbol{\Sigma}$ according to $\boldsymbol{\sigma}_h$. Thus the effect of the second term is too good to be trusted. What to do? Well, tentative numerical experiments show that it is possible to simply discard the problematic term, in which case the dual problem reduces to finding $\boldsymbol{\Sigma}$ and $\boldsymbol{z}$ such that

$$-\boldsymbol{S}^{-1} : \dot{\boldsymbol{\Sigma}} + \frac{\lambda_h}{|\boldsymbol{\sigma}_h^D|}\boldsymbol{\Sigma}^D = \boldsymbol{\varepsilon}(\boldsymbol{z}),$$

$$\nabla \cdot \boldsymbol{\Sigma} = 0.$$

In the case of visco-plasticity, γ is small and the effect of the linearization error will also be smaller. The answer to the question of how to treat the linearization will depend on numerical evidence, and we are currently investigating this issue.

3 Techniques of Mesh Generation and Adaptivity

In this Section, we give some standard techniques for mesh refinement and for mesh generation. The discussion is not exhaustive in any way, in particular with respect to mesh generation—that would require a book of its own. The purpose is rather to introduce the reader to some key concepts.

3.1 Adaptive Control

Any *a posteriori* error estimator is supposed to give information about the local meshsize. The question is: given a residual estimate, how is the mesh to be refined? This question is twofold: first we have to turn the error estimate into a machine for predicting the local meshsize, and then we face the purely technical question of how to implement mesh refinement. We will address these two aspects of adaptivity in the following.

Basically, residual–based error estimates take the form

$$\|u - U\| \leq \sum_K \rho_K$$

or, when using a dual approach,

$$\|u - U\| \leq \sum_k \omega_K \rho_K$$

where ρ_K symbolizes the equilibrium error on element K plus derivative jumps between element K and its neighbors, and ω_K is a weight obtained by solving a dual problem. We thus have a local *error indicator* $\eta_K = \omega_K \rho_K$ and a global, $\eta = \sum_K \eta_K$.

The goal to have $\|u - U\|_\Omega \leq \text{TOL}$, where TOL is a user-defined tolerance. Clearly, it natural to refine where η_K is large! Two typical strategies are:

1. Equidistribution: the contribution from each element should be of the same size, i.e.,
$$\eta \approx N_{\text{ele}} \eta_K \leq \text{TOL},$$
and thus $\eta_K \leq \text{TOL}/N_{\text{ele}}$, where N_{ele} is the (current) number of elements.
2. Refine a fixed percentage of the elements according to the size of η_K.

Here it should be noted that equidistribution gives the possibility of evaluating a new meshsize (cf. Example 4), which a fixed percentage strategy does not. More refined strategies for optimizing the local meshsize are given by, e.g., Rannacher (1999).

Example 4. Assume that we are working with triangles in 2D and that we want to control the energy error. For simplicity, we look at only the internal residual so that

$$\eta_K = \left(\int_K h_K^2 |\boldsymbol{f} - \nabla \cdot \boldsymbol{\sigma}(U)|^2 \, dxdy \right)^{1/2} \leq \frac{\text{TOL}}{N_{\text{ele}}}$$

leading to

$$\frac{1}{2} h_K^4 \leq \frac{\text{TOL}^2}{|\boldsymbol{f} - \nabla \cdot \boldsymbol{\sigma}(U)|^2 N_{\text{ele}}^2}$$

or, equivalently,

$$h_K \leq \left(\frac{\text{TOL}\sqrt{2}}{|\boldsymbol{f} - \nabla \cdot \boldsymbol{\sigma}(U)| N_{\text{ele}}} \right)^{1/2}.$$

This requires knowledge of the final number of elements, which may be estimated without changing the mesh, in the following way:

1. Use $N_{\text{ele}}^{\text{old}}$ to compute local meshsizes
2. Use the local meshsizes to estimate the new number of elements, $N_{\text{ele}}^{\text{new}}$
3. Check whether $N_{\text{ele}}^{\text{new}} \approx N_{\text{ele}}^{\text{old}}$. If not, set $N_{\text{ele}}^{\text{old}} = N_{\text{ele}}^{\text{new}}$ and go to 1.
4. Given a final $N_{\text{ele}}^{\text{new}}$ the new mesh can be constructed.

Since the solution will change when the mesh is refined, we cannot take too many steps in this process.

3.2 Refinement of Elements

In this Section, we will discuss some of the more common ways to implement mesh refinement. The basic problem is *closure*, i.e., the difficulty of obtaining geometrically conforming meshes. As we shall see, this problem is particularly difficult when using quadrilateral and hexahedral elements.

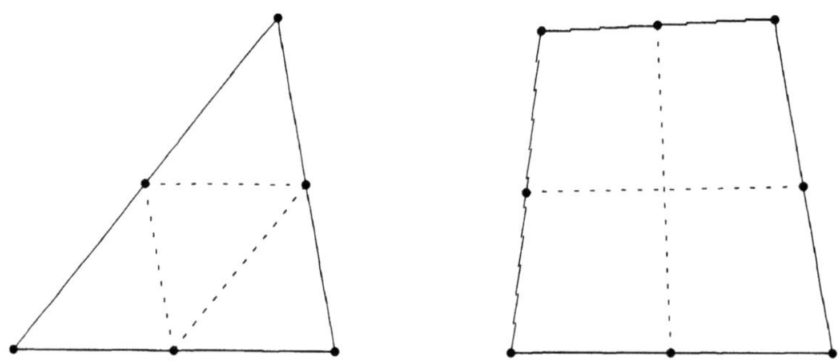

Figure 5. Regular subdivision of the triangle and the quad.

I: Regular Subdivision Regular subdivision means dividing the elements into "children", see Figure 5. In this case we obtain "transition elements" on the edges of the transition between refined and untouched elements, see Figure 6. Thus, we face the question of how to deal with the *hanging nodes* on the edges on the transition elements adjacent to the refined elements. In the case of triangular (or tetrahedral) elements, this situation is usually handled by making a temporary triangulation of the transition element, cf. Bank and Sherman (1981). The reason for not keeping the triangulation on the transition elements is that it may lead to severe degradation of the maximum angle in the mesh and thereby affect the approximation properties of the FEM.

In the case of quadrilateral elements, several approaches have been suggested:

1. To use higher approximations on the side of the transition element with a hanging node. For conformity, this must also affect the approximation on the adjacent elements, which makes this alternative less attractive implementationwise.
2. To divide the transition element into triangular elements. This has the drawback of requiring two sets of basis functions and element matrices of different sizes. Furthermore, this approach is not really feasible in three dimensions, since we will then also need Egyptian pyramids (with a rectangular base), for which a basic difficulty in defining shape functions arises.

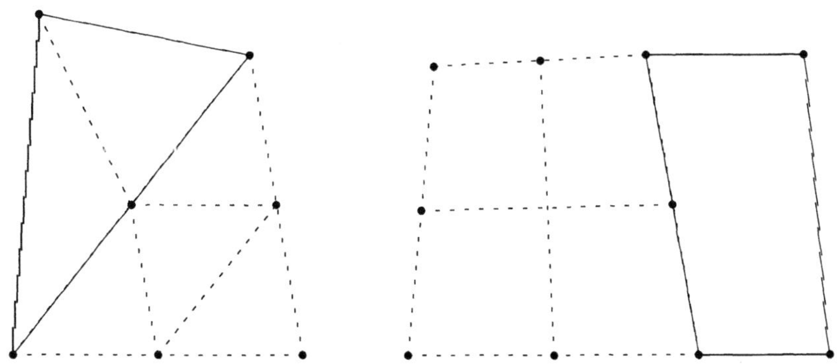

Figure 6. Transition elements for the triangle and the quad.

3. Restricting the approximation at the surplus node so that the approximation becomes conforming, cf. Example 5 below. This means that the transition element itself is handled as an ordinary element; it is the elements adjacent to it that require special handling. One advantage is that there is only one set of basis functions; a disadvantage is that the refinement may be too local—the transition elements are refined elements.
4. Closing the mesh by use of quadrilaterals, cf. Stein, Rust, and Ohnimus (1992). A typical mesh resulting from closing is given in Figure 7. A basic difficulty with this approach is that the closing process not only involves the adjacent elements, but also elements further away. The main advantage is that all elements can be of the same type, the main disadvantage is that the implementation is cumbersome with many special cases to consider.

In summary:

- Regular subdivision is the only approach used for quadrilaterals, and very common with triangles.
- Transition elements are invariably handled by splitting for triangles. For quadrilaterals, the technique of restricting the hanging nodes is the simplest to implement, and thus the most common.
- In three dimensions, regular subdivision on tetrahedral elements is more cumbersome, since we do not know in advance the number of elements that are affected by a hanging node (and thus should be split); there is a difference between an element side and an element edge in 3D, but not in 2D. Furthermore, it is not possible to divide a tetrahedron into similar elements, and special care has to be taken to avoid deterioration of the maximum angle. For a particular implementation, see Löhner and Baum (1992). For hexahedral elements, we have the same problem of not knowing in advance the number of elements affected by hanging nodes on edges. Both the hanging node technique and the closure technique can be extended, but with additional programming effort.

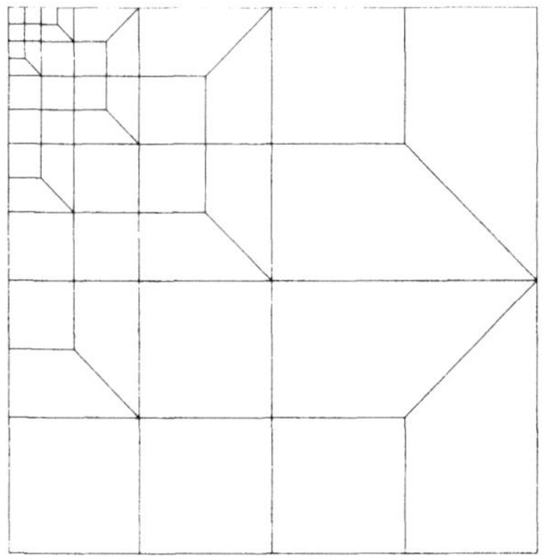

Figure 7. Typical mesh obtained by closure.

Example 5. Consider prescribing the surplus node of element T in Figure 8. For a scalar unknown, we then have that $u_j = (u_k + u_m)/2$, and we may write

$$\tilde{\mathbf{u}} \stackrel{\text{def}}{=} \mathbf{P}\mathbf{u}$$

on element K, leading to

$$\begin{bmatrix} \tilde{u}_i \\ \tilde{u}_j \\ \tilde{u}_k \\ \tilde{u}_l \end{bmatrix} = \begin{bmatrix} 1 & 0 & 0 & 0 \\ 0 & 1/2 & 1/2 & 0 \\ 0 & 0 & 1 & 0 \\ 0 & 0 & 0 & 1 \end{bmatrix} \begin{bmatrix} u_i \\ u_m \\ u_k \\ u_l \end{bmatrix}$$

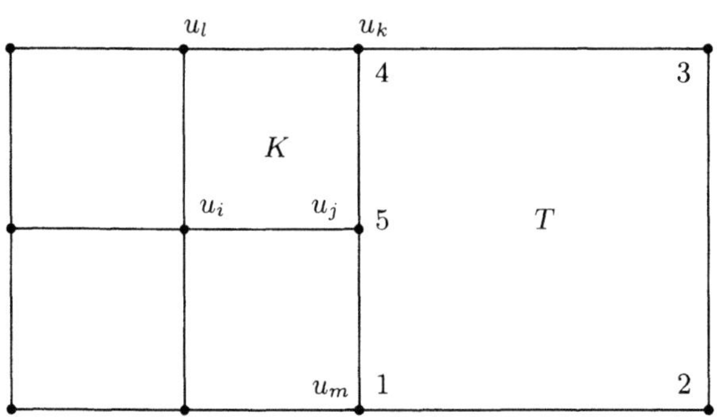

Figure 8. Transition element for the quad.

An element matrix on K must be transformed to take into account the prescribed variable u_j. Symmetry is maintained if

$$\tilde{\mathbf{S}}_K = \mathbf{P}^{\mathrm{T}}\mathbf{S}_K\mathbf{P}$$

and

$$\tilde{\mathbf{f}}_K = \mathbf{P}^{\mathrm{T}}\mathbf{f}_K.$$

$\tilde{\mathbf{S}}_K$ and $\tilde{\mathbf{f}}_K$ are assembled to the global system of equations, which then does not contain u_j.

II: Edge Bisection (Rivara) Methods This approach, pioneered by Rivara (1984), is an attractive alternative to regular subdivision for triangular and tetrahedral elements. The main idea is to divide edges rather than elements, and to select the refinement edges in such a way so as to avoid increasing the maximum angle during refinement. For a given two-dimensional triangular mesh $\mathfrak{T}$ we can write an algorithm as follows:

1. For each element K, mark the longest edge. Call this edge the refinement edge.
2. Divide by connecting the midpoint of the refinement edge with the opposing node. This defines new refinement edges, see Figure 9.
3. Bisect each $K \in \mathfrak{T}$ to obtain a possibly nonconforming mesh.
4. Collect all elements with a hanging node in a set $\mathfrak{S}$.
5. If $\mathfrak{S}$ is not empty, let $\mathfrak{T} = \mathfrak{S}$ and go to 3.

In three dimensions, the algorithm is slightly more involved, but simpler to implement than the regular subdivision technique, see, e.g., Bänsch (1991). Unrefinement can also be handled with this approach (as with the regular subdivision approach).

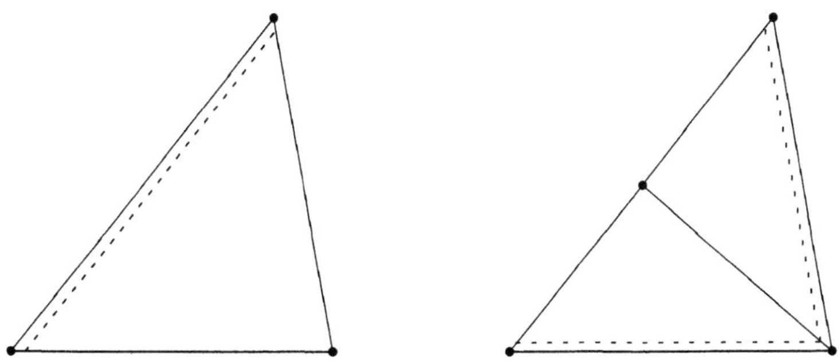

Figure 9. Refinement edges before and after bisection.

3.3 Mesh Generation

In this section we discuss the problem of generating meshes. We shall focus on the generation of unstructured triangular (and tetrahedral) meshes, but also give a simple recipe for generating unstructured quadrilateral meshes. The generation of unstructured hexahedral meshes is considerably more difficult; we will not address this problem, but refer to, e.g., Sheffer et al. (1999).

Structured Grids The generation of structured grids is crucial for finite difference approximations and some finite volume methods, but can of course be used also in combination with finite element methods. The basic idea in structured mesh generation is to generate the grid on a simple domain which is topologically equivalent to the physical domain, followed by mapping the simple domain to the physical domain; the node positions will then be given by this map. (For finite differences, the differential equations must then also be mapped.) This approach is of course severely limited; usually it is extended by using 'blocks' of structured grids that are patched together, either with overlaps or with special treatment of the boundaries in between blocks.

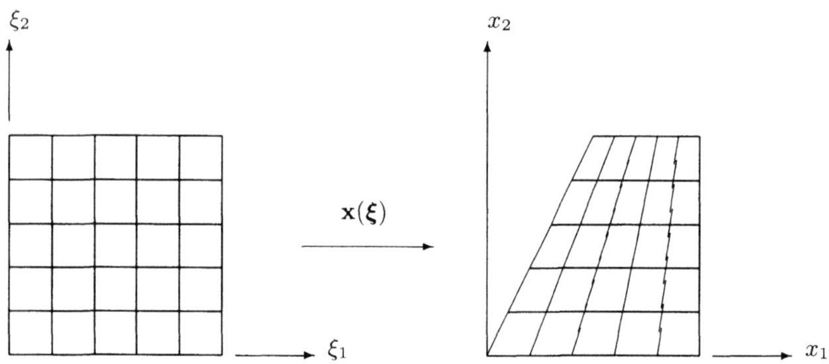

Figure 10. Map defining a mesh.

The map is usually not given explicitly (though there are exceptions); typically, it is defined as the solution of a partial differential equation. This equation is then solved approximately using a discretization of the simple domain, and with the physical location of the boundary nodes as boundary conditions. Examples of partial differential equations used are:

- The Laplacian (very weak control of element size).
- Equations of elasticity (somewhat better control of size).
- Nonlinear equations of elasticity (smoother meshes).

– An equation explicitly defined in order to gain control of orthogonality of meshlines, meshsize etc.

For more information, see Knupp and Steinberg (1993).

Example 6. Winslow's method, given by Winslow (1967), based on the inverse Laplace equation. Consider a bounded domain in the computational $\mathbf{x}$–plane. We may think of the mesh lines in the x_1– and x_2–directions as iso–lines to the solution of Laplacian problems, one for each coordinate direction. Each coordinate line may be thought of as a scalar function solved separately from the other. We may identify the mesh lines in the $\mathbf{x}$–plane with the coordinate lines in the original, simple, domain, which we consider situated in the $\boldsymbol{\xi}$–plane.

We wish to solve

$$\Delta_{\mathbf{x}}\boldsymbol{\xi} = 0$$

or

$$\frac{\partial^2 \boldsymbol{\xi}}{\partial x_1^2} + \frac{\partial^2 \boldsymbol{\xi}}{\partial x_2^2} = 0 \tag{23}$$

with the gridline spacing on the boundary as boundary conditions.

To solve, we have to invert (23):

$$\begin{aligned} \alpha\frac{\partial^2 x_1}{\partial \xi_1^2} - 2\beta\frac{\partial^2 x_1}{\partial \xi_1 \partial \xi_2} + \gamma\frac{\partial^2 x_1}{\partial \xi_2^2} &= 0, \\ \alpha\frac{\partial^2 x_2}{\partial \xi_1^2} - 2\beta\frac{\partial^2 x_2}{\partial \xi_1 \partial \xi_2} + \gamma\frac{\partial^2 x_2}{\partial \xi_2^2} &= 0, \end{aligned} \tag{24}$$

where

$$\begin{aligned} \alpha &= \left(\frac{\partial x_1}{\partial \xi_2}\right)^2 + \left(\frac{\partial x_2}{\partial \xi_2}\right)^2, \\ \beta &= \frac{\partial x_1}{\partial \xi_1}\frac{\partial x_1}{\partial \xi_2} + \frac{\partial x_2}{\partial \xi_1}\frac{\partial x_2}{\partial \xi_2}, \\ \gamma &= \left(\frac{\partial x_1}{\partial \xi_1}\right)^2 + \left(\frac{\partial x_2}{\partial \xi_1}\right)^2. \end{aligned}$$

Note that for equilateral elements, $\alpha = \gamma$ and $\beta = 0$, and we have

$$\Delta_{\boldsymbol{\xi}}\mathbf{x} = 0 \tag{25}$$

Unfortunately, this simple approach may lead to folded meshes in the case of concave boundaries.

A very simple method for the approximate solution of (25), given already by Winslow (1967), is

$$\mathbf{x}_i^{\text{new}} = \frac{1}{N}\sum_{j=1}^{N}\mathbf{x}_j \tag{26}$$

where N is the number of nodes connected to node i.

Unstructured Triangular Grids. We will confine the exposition to the two most widely used methods:

- *Delaunay triangulation*, which is based on modification of an existing (simple) mesh by point insertion
- *Advancing front method*, which creates the mesh as new points are inserted.

The *Delaunay* algorithm is a method from computational geometry for triangulating the convex hull of a set of points in the plane. A good point of the Delaunay triangulation, from the point of view of its approximation properties in combination with finite elements, is that it maximizes the minimum angle compared to any other triangulation of the same set of points. When using the Delaunay approach as a basis for mesh generation, the basic difficulty lies in the triangulation of non-convex domains; the problem is to keep the boundary segments as element sides, which the basic Delaunay algorithm will not do.

A Delaunay mesh generator presupposes the existence of a grid and a set of points which are to make up the new mesh. We thus assume that we are given a a discretization of the boundary into a number of (connected) points and an initial mesh made up of two triangles making up a quadrilateral that covers the domain to be meshed. Next, we wish to insert a new nodal point into the mesh, and produce a new valid mesh such that the angles in the mesh do not deteriorate. The basic tool for achieving this is the *circle condition*, which can be defined as follows. The three points of any element K define a circle. If any node of any other element in the mesh falls within this *circumbscribing circle*, then the edge separating K and its neighbor should be swapped. In practice the circle condition is used in a slightly different way by use of the Bowyer-Watson algorithm (after Bowyer (1981) and Watson (1981)). When a new point is to be inserted into the mesh, we first identify all existing elements whose circumcircle contains the new point. This is done by first locating the element containing the point, followed by checking the neighbor of this element, and then the neighbors of these elements etc., until we no longer find elements whose circumcircle contains the point. (Using a neighbor list, this approach will be much faster than the naive checking of *all* of the elements in the mesh.) The elements that have the point within their circumcircle are then removed from the mesh, resulting in a convex hole in the mesh (the convexity of this domain was proven by Watson (1981)). Finally, the nodes on the edge of the convex domain are joined to the new node to create a complete mesh, see Figures 11 and 12. Special care has to be exercised in order not to swap boundary segments, cf. Chew (1989). This approach produces a mesh subdivision of the original quadrilateral (given a set of points), and the elements falling outside the computational domain are removed in a postprocessing step.

Unfortunately the "maximum minimum angle" quality is lost in three dimensions; the circumshpere condition has to be supplemented by angle checks. Nevertheless, this is the (by far) fastest method of triangulating a given set of points also in 3D. Unfortunately, we cannot always determine the set of points to be triangulated in advance. Therefore, the Delaunay algorithm is sometimes used in combination with the advancing front approach, to be described next, which has a natural way of computing the "optimal" point placements. For further information, consult, e.g., Frey, Borouchaki, and George (1998), Baker (1989).

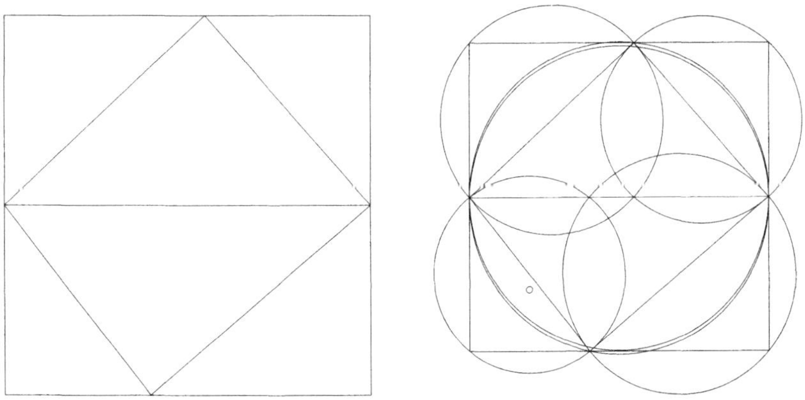

Figure 11. Original mesh, mesh with circumcircles and inserted point.

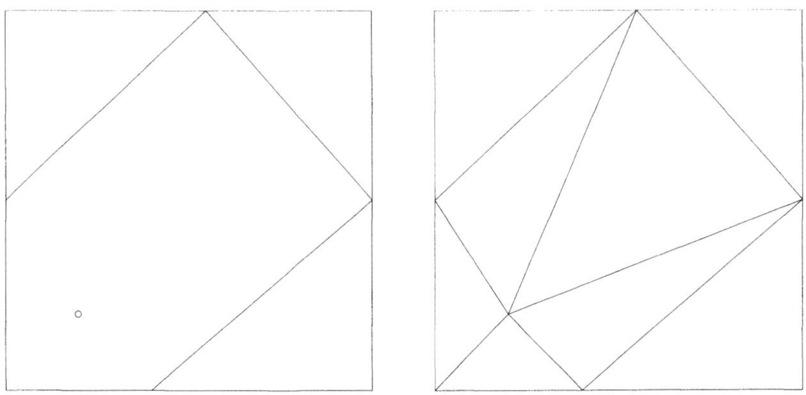

Figure 12. Convex patch and new mesh.

Remark 3. A neat trick for computing the centroid and radius of a circle, given three points on its circumference is the following. A circle is described by the equation $(x - x_m)^2 + (y - y_m)^2 = r^2$, where (x_m, y_m) is the centroid and r is the radius. Thus, we can try to minimize the distance between the line $a + b\,x + c\,x$ and $f(x, y)$, where we define $a = r^2 - x_m^2 - y_m^2$, $b = 2\,x_m$, $c = 2\,y_m$, and $f = x^2 + y^2$. By simple fitting (which, incidentally, can be generalized to a least-squares approach) we can thus solve a *linear* 3×3–system to compute (a, b, c) based on the three points (x_i, y_i) on the circumference. Given (a, b, c), the triple (x_m, y_m, r) is directly computable.

The *advancing front mesh generation technique*, unlike Delaunay triangulation, is based on local construction of elements; no global mesh will exist until the process is completed. Instead, there is a "front" of elements growing into the parts of the domain which have not been meshed, which gives an immediate control over the mesh characteristics: when a new element is to be constructed, the mesh characteristics can be fulfilled exactly, as long as the element does not extend into the existing front.

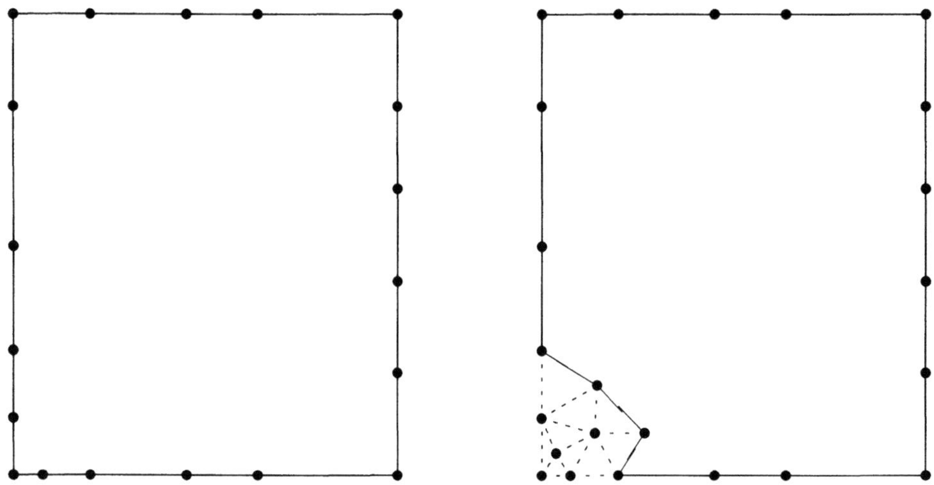

Figure 13. Initial front and front at a later stage

We start out by subdividing the given boundary into segments. These segments constitute the initial front, and they have to be constructed from some representation of the mesh characteristics, usually stored in one of two different ways:

1. On a 'background grid' supplied by the user. Here the mesh characteristics are stored in the nodes of the background grid, and the interpolation is performed using the standard FE basis functions.
2. Using 'distance weighted interpolation'. The user simply provides a set of points in which the mesh characteristics are given, and the interpolation is based on the distance from the given points.

The two node numbers on each segment are ordered so that the interior of the domain is always on one side ("left") of the segments and the exterior on the other. The segments are stored in a queue, e.g., in a heap, so that one always selects the smallest segment. Once a segment has been selected, an "ideal" point is located, i.e., a point which tries to make an element which fulfills the given local element characteristics. The ideal point is stored in a list in which one also puts the points in a neighborhood of the ideal point. For a fast implementation the geometrical search involved in locating neighborhood points necessitates data structures such as trees and linked lists.

Next, we select the point which together with the two points on the segment will form the new triangle. The points in the list are ordered, so that old points are placed before the ideal point if they are "sufficiently close" to it. Finally, one selects the first point in the list which gives rise to a valid element, meaning an element which does not contain any of the old points (and thus does not cross the front). The first description of this approach was given by Perarie et al. (1987). The approach easily extends into 3D, cf. Perarie, Peiró,and Morgan (1992) or Möller and Hansbo (1995).

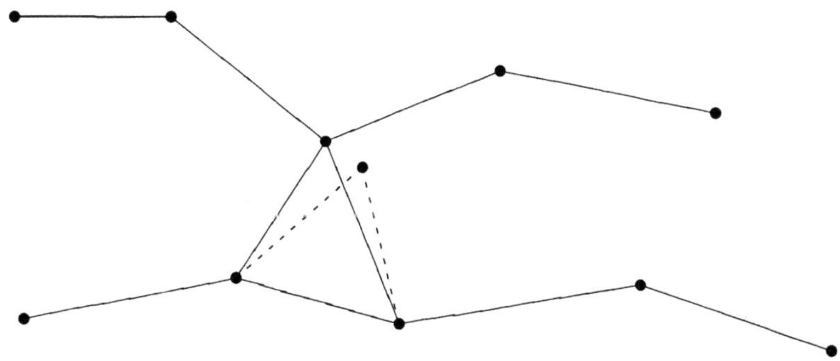

Figure 14. Choosing an alternative point

Unstructured Quadrilateral Grids. We shall confine the discussion to just one simple approach. Given a triangular mesh, one can connect each triangle with its neighbor in such a way that the best possible quadrilateral (for instance with respect to the angles of the quadrilateral) will be constructed from each pair of triangles (checking all the neighbors of a given triangle). The problem is that this approach will lead to stray triangles remaining in the mesh, and these have to be handled in some way. A simple approach circumventing this problem was given by Rank, Schweingruber, and Sommer (1993): First, we create a triangular mesh of double the desired meshsize. Next, we join triangles into quadrilaterals using some angle criterion. We are now left with a set of "good" quadrilaterals and some stray triangles. Finally, each quadrilateral is divided into three new quads, and all triangles are divided by joining the midpoint of the triangle with the midpoints of the sides of the triangle, resulting in four new quads. An example of converting a triangular mesh into a quadrilateral mesh is given in Figure 15.

3.4 Mesh Smoothing

The quadrilateral mesh in Figure 15 is not very good with respect to the maximum angle, and typically this is the case with meshes produced by an automatic mesh generator (even Delaunay, at least unless we have a very good algorithm for point placement). In order to improve the quality of the mesh, one therefore usually resorts to some kind of mesh smoothing. The simplest possible approach is Laplacian smoothing, i.e., the simplified Winslow's method (26). The effect of this approach can be seen in Figure 16. The same

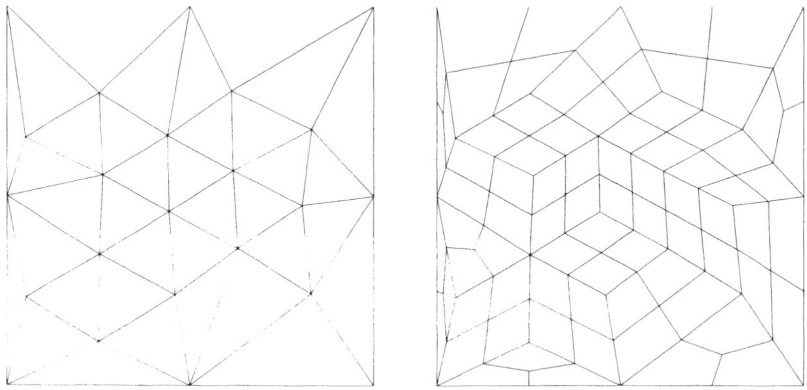

Figure 15. Triangular mesh converted into quadrilaterals.

drawbacks to Laplacian smoothing that were discussed in the context of mesh generation also apply to mesh smoothing: possible folding and bad control of the local meshsize and element stretch properties. A more refined approach to Laplacian smoothing that gives control of element stretch was presented by Hansbo (1995), and an approach based on the inverse Laplace equation (thus avoiding folding) was given by Knupp (1999).

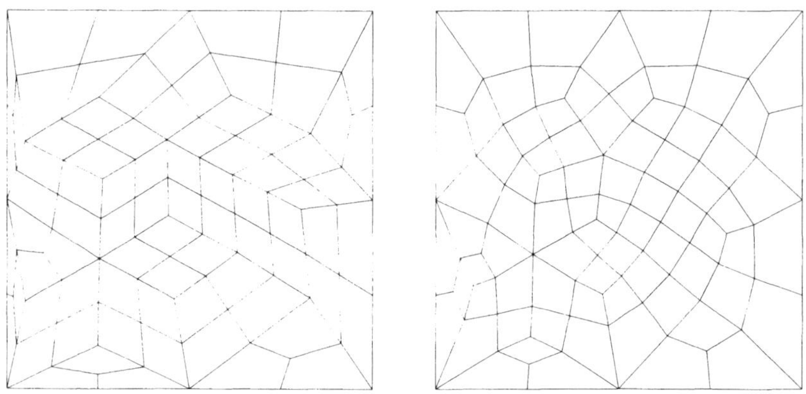

Figure 16. Quadrilateral mesh before and after Laplacian smoothing.

Another related approach to mesh smoothing is the *spring system analogy*, cf. Löhner et al. (1986). This approach is based on viewing the mesh as being composed of compressible bars. Adjacent nodes i and j are thus considered to be joined by a spring of stiffness c_{ij}, exerting a force $\boldsymbol{f}_{ij} = c_{ij}(\mathbf{x}_j - \mathbf{x}_i)$ between the nodes. If the spring stiffness is equal to the inverse of the desired local element size, then the loads will be in balance when the grid is correctly sized on average. By assembling the loads $\boldsymbol{f}_{ij}$ to a global load vector and solving (using viscous relaxation), a smooth and correctly sized grid is obtained. To see the relation to Laplacian smoothing, consider the equilibrium equation for equal-sized elements:

$$\sum_{i=1}^{n} \boldsymbol{f}_{iN} = \sum_{i=1}^{n} (\mathbf{x}_N - \mathbf{x}_i) = 0,$$

which is identical to (26).

The spring system analogy is easy to extend to a mesh consisting of stretched elements (as one would like to have in a boundary layer, for example). It suffers from the same folding tendencies as Laplacian smoothing, however.

References

Bangerth, W. and Rannacher, R. (1999). Finite element approximation of the acoustic wave equation: error control and mesh adaptation. *East-West J. Numer. Math.* 7:263–282.

Baker, T. J. (1989). Automatic mesh generation for complex 3-dimensional regions using a constrained Delaunay triangulation. *Engineering with Computers* 5:161–175.

Bank, R. E. and Sherman, A. H. (1981). An adaptive, multi-level method for elliptic boundary value problems. *Computing* 26:91–105.

Bänsch, E. (1991). An adaptive finite-element strategy for the three-dimensional time-dependent Navier-Stokes equations. *J. Comput. Appl. Math.* 36:3–28.

Becker, R. and Rannacher, R. (1996). A feed-back approach to error control in finite element methods: basic analysis and examples. *East-West J. Numer. Math.* 4:237–264.

Bowyer, A. (1981). Computing Dirichlet tesselations. *The Computer Journal* 24:162–166.

Chew, L. P. (1989). Constrained Delaunay triangulations. *Algorithmica*, 47:97–108.

Cirak, F. and Ramm, E. (2000). A posteriori error estimation and adaptivity for elastoplasticity using the reciprocal theorem. *Int. J. Numer. Meth. Engrg.*, 47:379–393.

Duvaut, G. and Lions, J. L. (1976). *Inequalities in Mechanics and Physics.* Berlin: Springer.

Eriksson, K., Estep, D., Hansbo, P., and Johnson, C. (1995). Introduction to adaptive methods for partial differential equations. *Acta Numerica* 105–158.

Eriksson, K., Estep, D., Hansbo, P., and Johnson, C. (1996). *Computational Differential Equations.* Lund: Studentlitteratur.

Eriksson, K. and Johnson, C. (1991). Adaptive finite element methods for parabolic problems. I: A linear model problem. *SIAM J. Numer. Anal.* 28:43–77.

Estep, D. (1995). A posteriori error bounds and global error control for approximations of ordinary differential equations. *SIAM J. Numer. Anal.* 32:1–48.

Estep, D. and French, D. (1994). Global error control for the continuous Galerkin finite element method for ordinary differential equations . *RAIRO M^2 AN* 28:815–852.

Frey P. J., Borouchaki H., and George P. L. (1998). 3D Delaunay mesh generation coupled with an advancing-front approach. *Comput. Methods Appl. Mech. Engrg.* 157:115–131.

Hansbo, P. (1995). Generalized Laplacian smoothing of unstructured grids. *Commun. Numer. Methods Engrg* 11:455–464.

Hairer, E., Nørsett, S. P., and Wanner, G. (1992). *Solving Ordinary Differential Equations I: Nonstiff Problems.* Berlin: Springer Verlag, 2nd edition.

Johnson, C. and Hansbo, P. (1992). Adaptive finite element methods in computational mechanics. *Comput. Methods Appl. Mech. Engrg.* 101:143–181.

Johnson, C. (1990). Adaptive finite element methods diffusion and convection problems. *Comput. Methods Appl. Mech. Engrg.* 82:301–322.

Kelly, D. W., Gago, R., Zienkiewicz, O. C., and Babuška, I. (1984). A posteriori error analysis and adaptive processes in the finite element method. Part I: Error analysis. *Int. J. Numer. Meth. Engrg.* 19:1593–1619.

Knupp, P. (1999). Winslow smoothing on two-dimensional unstructured meshes. *Engineering with Computers* 15:263–268.

Knupp, P. and Steinberg, S. (1992). *Fundamentals of Grid Generation.* Boca Raton: CRC Press.

Löhner, R. and Baum, J. D. (1992). Adaptive *h*-refinement on 3D unstructured grids for transient problems. *Int. J. Numer. Methods Fluids* 14:1407–1419.

Löhner, R., Morgan, K., and Zienkiewicz, O. C. (1986). Adaptive grid refinement for the compressible Euler equations. In Babuška et al., eds., *Accuracy Estimates and Adaptive Refinements in Finite Element Computations.* New York: Wiley, 281–297.

Möller, P. and Hansbo, P. (1995). On advancing front mesh generation in three dimensions. *Int. J. Numer. Methods Engrg* 38:3551–3569.

Peraire, J., Vahdati, M., Morgan, K., and Zienkiewicz, O. C. (1987). Adaptive remeshing for compressible flow computations. *J. Comput. Phys.* 72:449–466.

Peraire, J., Peiró, J., and Morgan, K. (1992). Adaptive remeshing for three-dimensional compressible flow computations. *J. Comput. Phys.* 103:269–285.

Rank, E., Schweingruber, M., and Sommer, M. (1993). Adaptive mesh generation and transformation of triangular to quadrilateral meshes. *Commun. Numer. Methods Engrg.* 9:121–129.

Rannacher, R. (1999). Error control in finite element computations : An introduction to error estimation and mesh-size adaptation. In Bulgak, H. and Zenger, C., eds., *Error Control and Adapticity in Scientific Computing.* Kluwer Academic. 247–278.

Rannacher, R. and Suttmeier, F.-T. (1998). A posteriori error control in finite element methods via duality techniques: Application to perfect plasticity. *Comput. Mech.*, 21:123–133.

Rannacher, R. and Suttmeier, F.-T. (1999). A posteriori error estimation and mesh adaptivity for finite element models in elasto-plasticity. *Comput. Methods Appl. Mech. Engrg.*, 176:333–361.

Rivara, M. C. (1984). Algorithms for refining triangular grids suitable for adaptive and multigrid techniques. *Int. J. Numer. Methods Engrg.* 20:745–765.

Sandboge, R. (1998). Quantitative error control for finite element methods for one-dimensional compressible flow. *SIAM J. Numer. Anal.*, 35:2014–2034.

Sheffer A., Etzion M., Rappoport A., and Bercovier M. (1999). Hexahedral mesh generation using the embedded Voronoi graph. *Engineering with Computers*, 15:248–262.

Simo, J. C. (1991). Nonlinear stability of the time-discrete variational problem of evolution in nonlinear heat conduction, plasticity and viscoplasticity. *Comput. Methods Appl. Mech. Engrg.*, 88:111–131.

Stein, E., Rust, W., and Ohnimus, S. (1992). *h*- and *d*-adaptive FE methods for two-dimensional structural problems including post-buckling of shells. *Comput. Methods Appl. Mech. Engrg.*, 101:315–354.

Temam, R. (1985). *Mathematical Problems in Plasticity.* Paris: Gauthier-Villars.

Watson, D. F. (1981). Computing the n-dimensional Delaunay tesselation with application to Voronoi polytypes. *The Computer Journal* 24:167–171.

Winslow, A. M. (1967). Numerical solution of the quasilinear Poisson equation in a nonuniform triangle mesh. *J. Comput. Phys.*, 2:149–172.

Zienkiewicz, O. C. and Zhu, J. Z. (1987). A simple error estimator and adaptive procedure for practical engineering analysis. *Int. J. Numer. Methods Engrg.*, 24:337–357.

Hierarchic Modelling in Elasticity by generalized p- and hp-FEM

C. Schwab
Seminar für Angewandte Mathematik
ETH Zürich, HG G 58.1
8092 Zürich, Switzerland

Lecture Notes for the course given during the summer school

"Adaptive FEM for linear and nonlinear Structural Mechanics"

at the Centre International des Sciences Mechaniques (C.I.S.M.), Italy, Oct. 4-8, 1999.

Contents

0 Hierarchic Modeling in Elasticity

This C.I.S.M. course and these notes are about error control in computational mechanics. In contrast to the other lectures in the course, the notes will address the p- and hp-version FEM and the error control with these methods, as well as the modeling error control. The intended audience are applied mathematicians and computational engineers who want to obtain a quick, nontechnical overview over the p-version FEM in solids as well as on the issue of modeling error control.

Error control has two main ingredients: a) error indication/estimation and b) a strategy for error reduction when it is deemed too large. Since the notion of error is central here, we outline here the main error types arising in computational mechanics, thereby also fixing terms and definitions that shall be used in the sequel.

In the background of any simulation of a physical system is what we will call **state-of-the-art model**. It consists of "first principles" partial differential equations, typically certain conservation resp. balance laws of continuum mechanics. These are completed by empirical data that must be measured and hence is uncertain: mainly constitutive equations and parameters, but also boundary data and, more and more importantly, also the domain occupied by the body; the state-of-the-art model is typically 3-d, nonlinear (geometric and material), and transient. For specific applications and the lack of human/computer/time ressources, the state-of-the-art model is often simplified to a working model by one or more of the following:

a) physical assumptions,
b) linearizations,
c) asymptotic analysis,
d) modal expansion and energy projection.

The resulting **working model** is then approximated by a numerical model with a finite number of unknowns to be determined that can be solved on a computer. This approximation is called **discretization** of the working model. We refer to Figure 0.1 for illustration.

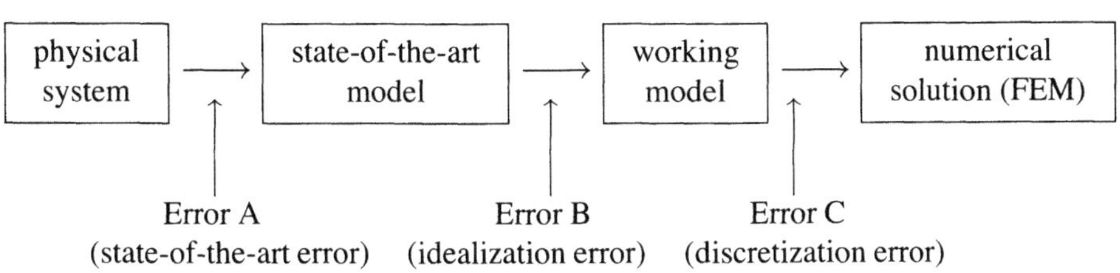

Figure 0.1. Mathematical models and errors

(Clearly, this rather crude distinction of different errors could be refined in various ways, e.g. Error C consists of the actual discretization error as well as e.g. round-off error, numerical integration error etc.).

Adaptive FEM address control and reduction of the discretization error, Error C; control and reduction of type B errors is today often entirely based on the experience of experts, even though systematic, quantitative control of type B errors is often feasible and has been implemented in the commercial FE-code STRESSCHECK [46].

What to do if the errors are too large? A powerful concept is the notion of **hierarchic models** that allows to systematically reduce the errors by adopting "higher" B and C members from a hierarchy of models resp. discretizations. Correspondingly, terms such as "hierarchic elements", "hierarchic modeling" are used with increasing frequency in the literature. Let us give precise descriptions of the meaning of these notions as we shall use them.

- **Hierarchic Models** are ordered sequences of working models. Their exact solutions constitute a converging sequence of functions in the norms appropriate for the objectives of the analysis. In the limit, the solution of the highest model, preferably the state-of-the-art model, is obtained.

- **Hierarchic Finite Element spaces** are a sequence of finite dimensional spaces S_i of increasing dimensions N_i with $S_i \subset S_{i+1}$, $N_i < N_{i+1}$. They are a means to obtain converging sequences of numerical solutions, the limit being the exact solution of a working model. The S_i are characterized by the mesh, the polynomial degree of the elements and by the mapping functions.

- **Hierarchic Finite Element shapefunctions** have some useful properties in implementations, such as preconditioning, error estimation and re-use of stiffness matrices. Hierarchic shapefunctions and bases are **not** required for hierarchic FE spaces S_i.

The outline of these notes is as follows:

We begin by introducing the 5 basic hp-elements that are used for all problem classes (solids, plates, beans, shells etc). This is already the first major difference to the usual FEM - there are no specialized and often very large element libraries for plate and shell problems required in the context of hp-FEM.

Rather, 5 p-element families (triangle-/quad-/hex-/prism-/tetrahedral families) will be sufficient to discretize **all** problems of linear solid mechanics successfully. Here successful means that converged FE solutions with a-posteriori quantitative error control are achievable with these elements. What does "converged" mean in this context? It refers to a negligible discretization error between the numerical solution and the exact solution of a given **mathematical model**. It is given in the form of a variational **principle of virtual work** which is the point of departure for the hp-FEM. In the first part of these notes, we present the hp-FEM for linearized elasticity. The second part presents generalizations: geometric and material nonlinearities, and their formulations.

Next, we discuss the control of the **modelling error**, the error inherent in the mathematical model adopted as basis for the numerical scheme. Examples constitute the choice of a two-dimensional plate - or shell theory in place of three-d elasticity, i.e. the adoption of a lower dimensional working model upon which computations are based. Converged numerical solutions can not reduce the **dimension-reduction** idealization error inherent in the two dimensional plate/shell theory. We present a) **modeling error estimators** for certain plate and shell models and b) **hierarchies of plate and shell models** which allow a continuous transition between dimensionally reduced plate- and shell models and three dimensional elasticity.

Differential equations modelling plate and shell problems are, due to the presence of a small parameter, the **thickness of the domain**, often singularly perturbed. In the zero-thickness limit, constraints are imposed on the solutions. For small, positive thickness of plates and shells, the solutions exhibit **boundary layers** and low order FEM are prone to **locking effects**. High order hp-type FEM are either free of locking or their practical performance shows very low sensitivity to it, without special "tricks", as e.g. reduced integration etc.

Particular attention will be paid to **sandwich or graded materials**, where the material properties depend on the transverse coordinate of the plate/ shell. We describe the derivation and implementation of a hierarchy of shell working models which allow to resolve the interlamina stresses with a number of degrees of freedom independent of the number of layers in the sandwich material. The shell models are derived by semidiscretization in the transverse direction with non-polynomial, laminate-adapted shapefunctions and Galerkin-Projection (rather than asymptotic analysis). The **Galerkin-orthogonality** of the discretization as well as of the modeling errors is the basis for a-posteriori error estimates, either residual or duality-based ones [1, 2]. Similar concepts for homogenization problems arising from lattice and cellular materials are under development [26, 27, 28].

The bibliography contains entries that are not referenced but which constitute supplementary material related to the topics of the course. As a rule, no proofs are given in these notes, but we refer, where appropriate, to the mathematical literature where such proofs can be found.

Notation: Throughout, we denote by $\Omega \subset \mathbb{R}^d$ the domain occupied by the elastic body undergoing deformation, $d = 2, 3$ is the spatial dimension, and by $\Gamma = \partial\Omega$ its boundary. By $L^2(\Omega)$ we denote the square integrable function $u : \Omega \to \mathbb{R}$ for which

$$\|u\| := \Big(\int_\Omega |u|^2 \, dx \Big)^{\frac{1}{2}} < \infty ,$$

and by $H^k(\Omega)$, $k \in \mathbb{N}$, the Sobolev space of functions with weak derivatives up to order k in $L^2(\Omega)$. Spaces of vector functions $\underline{u} : \Omega \to \mathbb{R}^m$ are denoted by $L^2(\Omega)^m$, $H^k(\Omega)^m$ etc. We define further on $H^k(\Omega)$ the seminorms

$$|u|_k := \Big(\sum_{|\alpha|=k} \|D^\alpha u\|^2 \Big)^{\frac{1}{2}}, \quad k = 0, 1, 2, \ldots$$

and the norms

$$\|u\|_k := \Big(\sum_{i=0}^{k} |u|_i^2 \Big)^{\frac{1}{2}} .$$

Acknowledgement: In writing the present notes, the author benefitted much from the support of ESRD Inc., in particular from the support of Dr. Ricardo Actis and Prof. Barna Szabo. Their viewpoint on quality assurance in FEA strongly influenced these notes and their help in preparing the course was invaluable. The notes also benefitted much from the stimulating atmosphere at C.I.S.M.

1 hp-Finite Element Methods in Elasticity

We present variational formulations of the elasticity problem, and hp-FE discretization techniques. Particular emphasis is placed on the efficient implementations. Error control by p-extension and extrapolation is described.

1.1 Governing Equations of linearized elastostatics

By $\Omega \subset \mathbb{R}^d$, $d = 2, 3$, we denote an elastic body. We assume Ω to be bounded, connected and its boundary $\Gamma = \partial\Omega$ to be piecewise analytic and globally Lipschitz. By $\boldsymbol{\sigma}, \boldsymbol{\epsilon} : \Omega \to \mathbb{R}^{d\times d}_{\text{sym}}$ we denote the infinitesimal stress and strain fields, respectively. The body Ω deforms subject to body forces $\underline{f} : \Omega \to \mathbb{R}^d$ and surface tractions $\underline{t} : \Gamma_T \to \mathbb{R}^d$ on the part $\Gamma_T \subseteq \Gamma$ of the surface, whereas on $\Gamma_D := \Gamma \backslash \overline{\Gamma}_T$ the displacements $\underline{u} : \overline{\Omega} \to \mathbb{R}^d$ are prescribed to be $\underline{u} = \underline{g}$.

The governing equations in small displacement theory are:

Equilibrium:

$$\operatorname{div} \boldsymbol{\sigma} + \underline{f} = \underline{0} \text{ in } \Omega\,, \tag{1.1.1}$$

Constitutive Equations (Hooke's law):

$$\boldsymbol{\sigma}(x) = \boldsymbol{C}(x)\,\boldsymbol{\epsilon}(x) \text{ at } x \in \Omega\,, \tag{1.1.2}$$

and **boundary conditions:**

Dirichlet- or **displacement boundary conditions:**

$$\underline{u} = \underline{g} \text{ on } \Gamma_D\,. \tag{1.1.3}$$

Neumann- or **Traction boundary conditions:**

$$\boldsymbol{\sigma} \cdot \underline{n} = \underline{t} \text{ on } \Gamma_T\,. \tag{1.1.4}$$

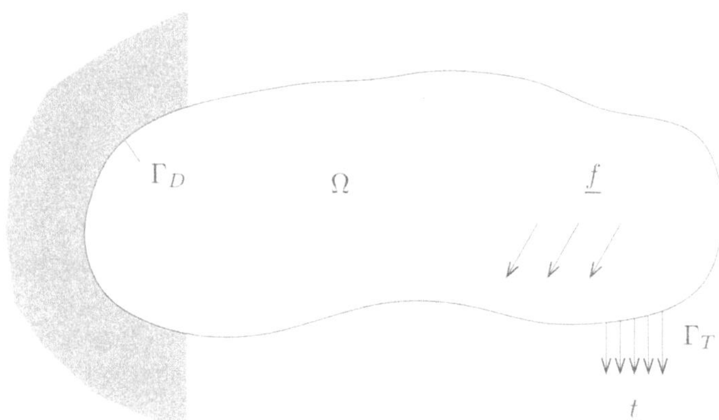

Figure 1.1. Elastic body and the notation

Here $\underline{n}$ is the exterior unit normal vector to Γ, see Figure 1.1, and the strain tensor ϵ is the symmetric gradient of the displacement field $\underline{u}$:

$$\epsilon[\underline{u}] = \frac{1}{2}\left(\nabla \underline{u} + (\nabla \underline{u})^\top\right). \tag{1.1.5}$$

We omit the argument $[\underline{u}]$ if it is clear from the context. The tensors $\boldsymbol{\sigma}$ and ϵ are symmetric and given by $d(d+1)/2$ independent coefficients σ_{ij}, ϵ_{ij}. The material is **homogeneous**, if the compliance tensor $\boldsymbol{C}$ in (1.1.2) is independent of $\underline{x} \in \Omega$. In components, (1.1.2) has the form

$$\sigma_{ij} = C_{ijk\ell}\, \epsilon_{k\ell}, \quad i,j,k,\ell \in \{1,\dots,d\}\,. \tag{1.1.6}$$

The compliance tensor has the symmetries

$$C_{ijk\ell} = C_{jik\ell} = C_{ij\ell k} = C_{k\ell ij}\,, \tag{1.1.7}$$

and contains in dimension $d = 3$ exactly 21 independent constants. If the body is homogeneous and isotropic, (1.1.2) takes the form

$$\sigma_{ij} = \lambda(\mathrm{div}\underline{u})\delta_{ij} + 2\mu\, \epsilon_{ij}[\underline{u}] \tag{1.1.8}$$

where λ and μ are the Lamé-constants. We have

$$2\mu = E/(1+\nu), \quad \lambda = E\nu/((1+\nu)(1-2\nu)) \tag{1.1.9}$$

where $E > 0$ is the Young modulus and ν with $-1 < \nu < 1/2$ the Poisson-ratio. The compliance tensor $\boldsymbol{C}$ is coercive in the sense that there is $\alpha_{\min} > 0$ such that

$$e_{ij}\, C_{ijk\ell}\, e_{k\ell} \geq \alpha_{\min}\, e_{ij}\, e_{ij} \tag{1.1.10}$$

for all symmetric matrices $\boldsymbol{e} = \boldsymbol{e}^\top$ of size d.

1.2 Variational Formulation

The point of departure for any FEM is a **variational formulation** of (1.1.1) - (1.1.3). There are many possible variational formulations. All involve a) the notion of **weak solution**, where the derivatives of the displacement field $\underline{u}$ are taken in the sense of distributions and b) a variational principle. Here, we shall adopt the so-called (primal) **principle of virtual work in the displacements**: find $\underline{u} \in E_g(\Omega)$ such that

$$B(\underline{u}, \underline{v}) = L(\underline{v}) \quad \forall \underline{v} \in E_0(\Omega)\,. \tag{1.2.1}$$

Here the virtual displacement field $\underline{v}$ is referred to as test function, $L(\underline{v})$ on the right hand side is the loading given by (summation over repeated indices)

$$L(\underline{v}) = \int_\Omega f_i\, v_i\, dx + \int_{\Gamma_T} t_i\, v_i\, ds\,, \tag{1.2.2}$$

and $B(\underline{u}, \underline{v})$ is given by

$$B(\underline{u}, \underline{v}) = \int_\Omega \sigma_{ij}[\underline{u}]\, \epsilon_{ij}[\underline{v}]\, dx \stackrel{(1.1.6)}{=} \int_\Omega \epsilon_{ij}[\underline{v}]\, C_{ijk\ell}(x)\, \epsilon_{k\ell}[\underline{u}]\, dx\ . \tag{1.2.3}$$

The set $E(\Omega)$ consists of **kinematically admissible displacements of finite strain energy**, i.e.

$$E_g(\Omega) = \{\underline{u} \in L^2(\Omega)^d \mid B(\underline{u}, \underline{u}) < \infty,\ \underline{u}|_{\Gamma_D} = \underline{g} \text{ on } \Gamma_D\}\ . \tag{1.2.4}$$

If the body is free, $\Gamma_D = \emptyset$ and we write simply $E(\Omega)$. The strain energy $B(\underline{u}, \underline{u})$ is a convenient and natural quantity in which the size of a displacement $\underline{u}$ can be measured. We associate with $\underline{u} \in E_{\underline{g}}(\Omega)$ the **energy-norm** of $\underline{u}$:

$$\|\underline{u}\|_{E(\Omega)} := (B(\underline{u}, \underline{u}))^{\frac{1}{2}},\ \underline{u} \in E_{\underline{g}}(\Omega)\ . \tag{1.2.5}$$

Evidently, for every $\underline{u}, \underline{v} \in E(\Omega)$,

$$\begin{aligned} B(\underline{u}, \underline{v}) &\le \|\underline{u}\|_{E(\Omega)} \|\underline{v}\|_{E(\Omega)} \\ &= \max_{\substack{x\in\overline{\Omega} \\ i,j,k,\ell}} |C_{ijk\ell}(x)|\ \|\boldsymbol{\epsilon}[\underline{u}]\|\ \|\boldsymbol{\epsilon}[\underline{v}]\| \ \le \alpha_{\max} |\underline{u}|_1\, |v|_1 \end{aligned} \tag{1.2.6}$$

so $E(\Omega) \subset H^1(\Omega)^d$. Viceversa, for every $\underline{u} \in E(\Omega)$ we have the **Korn inequality** [34]

$$\|\boldsymbol{\epsilon}[\underline{u}]\| + \|\underline{u}\| \ge C(\Omega)\ \|\underline{u}\|_1 \quad \forall \underline{u} \in E(\Omega)\ , \tag{1.2.7}$$

so $H^1(\Omega)^d \subset E(\Omega)$ and hence $H^1(\Omega)^d = E(\Omega)$.

The **loadings** $\underline{f} : \Omega \to \mathbb{R}^d$, $\underline{t} : \Gamma_T \to \mathbb{R}^d$ **are admissible**, if there exists $C(\underline{f}, \underline{t}) > 0$ such that

$$|L(\underline{v})| \le C(\underline{f}, \underline{t})\ \|\underline{v}\|_{E(\Omega)} \quad \forall \underline{v} \in E(\Omega)\ . \tag{1.2.8}$$

We denote the smallest such C (for given $\underline{f}, \underline{t}$) by $\|L\|_*$. Not all loadings are admissible: $\underline{f} \in L^2(\Omega)^d, \underline{t} \in L^2(\Gamma_T)^d$ are admissible, but point loadings are not.

We address the **well-posedness** of (1.2.1).

Case 1: Displacement Boundary Conditions

We assume that

$$|\Gamma_D| := \int_{\Gamma_D} ds > 0\ , \tag{1.2.9}$$

i.e. the body Ω is fixed along Γ_D and the prescribed displacement $\underline{g}$ is such that

$$\text{there is } \underline{G} \in E(\Omega) \text{ with } \underline{G}|_{\Gamma_D} = \underline{g}\ , \tag{1.2.10}$$

($\underline{G} \in H^1(\Omega)^d$ is sometimes called **lifting of** $\underline{g}$).

Then we may split $\underline{u} = \underline{u}_0 + \underline{G}$ with $\underline{u}_0 \in E_0(\Omega)$ and (1.2.1) reads: find $u_0 \in E_0(\Omega)$ such that

$$B(\underline{u}_0, \underline{v}) = L(\underline{v}) - B(\underline{G}, \underline{v}) \quad \forall \underline{v} \in E_0(\Omega)\ . \tag{1.2.11}$$

For (1.2.9) we have **Korn's first inequality**

$$\|\epsilon[\underline{u}]\| \geq C_K(\Omega, \Gamma_D)\, \|\underline{u}\|_1 \quad \forall \underline{u} \in E_0(\Omega)\,, \tag{1.2.12}$$

and with (1.1.10) follows the **coercivity of** $B(\cdot,\cdot)$:

$$B(\underline{u},\underline{u}) \geq \alpha_{\min} C_K\, \|\underline{u}\|_1^2 \quad \forall \underline{u} \in E_0(\Omega)\,. \tag{1.2.13}$$

By the Lax-Milgram Lemma, for every admissible loading $L(\underline{v})$ there exists a finite energy displacement $\underline{u} = \underline{u}_0 + \underline{G} \in E_g(\Omega)$ solving (1.2.11) which is unique [1].

Case 2: Free body, $\Gamma_T = \Gamma, \Gamma_D = \emptyset$

If the body is free, it can undergo displacements $\underline{u}$ with zero energy, the so-called **rigid body motions**. Define the **rigid body translations**

$$\mathcal{R}_0 = \{\underline{u} : \Omega \to \mathbb{R}^d \,|\, \underline{u} = \text{const.}\}\,, \tag{1.2.14}$$

and the **rigid body rotations**

$$\mathcal{R}_1 = \{\underline{u} : \Omega \to \mathbb{R}^d \,|\, \underline{u} = \boldsymbol{A}\underline{x} \;\text{ for some }\; \boldsymbol{A} = -\boldsymbol{A}^T\}\,, \tag{1.2.15}$$

and set $\mathcal{R} = \mathcal{R}_0 + \mathcal{R}_1$. Then dim $\mathcal{R}_0 = d$, dim $\mathcal{R}_1 = d(d-1)/2$ and

$$\|\epsilon[\underline{u}]\| = 0 \iff \underline{u} \in \mathcal{R}\,. \tag{1.2.16}$$

Hence (1.2.13) cannot hold for all $\underline{u} \in E(\Omega)$ and $\mathcal{R}$ must be removed from $E(\Omega)$. To this end, we set

$$E^{\perp}(\Omega) = \{\underline{u} \in E(\Omega) \,|\, \int_\Omega \underline{u} \cdot \underline{r}\, dx = 0 \quad \forall \underline{r} \in \mathcal{R}\}\,. \tag{1.2.17}$$

Then the **second Korn inequality**

$$\|\epsilon[\underline{u}]\| \geq C_K(\Omega)\|\underline{u}\|_1 \quad \underline{u} \in E^{\perp}(\Omega) \tag{1.2.18}$$

ensures the coercivity (1.2.13) on $E^{\perp}(\Omega)$.

Let $L(\cdot)$ be an admissible loading. Then from (1.2.1) and (1.2.16) with $\underline{v} = \underline{r} \in E(\Omega)$:

$$\forall \underline{r} \in \mathcal{R}: \; |L(\underline{r})| = |B(\underline{u},\underline{r})| \leq \alpha_{\max}\|\underline{u}\|_{E(\Omega)}\|\underline{r}\|_{E(\Omega)} = 0 \tag{1.2.19}$$

i.e. **the loading $L(\cdot)$ must vanish on $\mathcal{R}$** which is the case if and only if **the forces $\underline{f}$ and $\underline{t}$ are in static equilibrium**.

Case 3: Partially free body

[1] for if there were $\underline{u}_1, \underline{u}_2 \in E_g(\Omega)$ solving (1.2.11), $\underline{u}^\delta := \underline{u}_1 - \underline{u}_2 \in E_0(\Omega)$ and $\forall \underline{v} \in E_0(\Omega)$: $B(\underline{u}^1, \underline{v}) = L(\underline{v})$ $\forall \underline{v} \in E_0(\Omega)$: $B(\underline{u}^2, \underline{v}) = L(\underline{v}) \Longrightarrow \forall \underline{v} \in E_0(\Omega)$: $B(\underline{u}^\delta, \underline{v}) = 0$. Pick $\underline{v} = \underline{u}^\delta$ to infer $\|\underline{u}^\delta\|^2_{E(\Omega)} = B(\underline{u}^\delta, \underline{u}^\delta) = 0$, i.e. $\underline{u}^\delta$ has zero energy, i.e. $\|\epsilon[\underline{u}^\delta]\| = 0$. By (1.2.12), $\underline{u}^\delta = \underline{0}$.

Here, only some components of $\underline{u}$ are prescribed on Γ_D, but the set $\mathcal{N} = \mathcal{R} \cap E_0(\Omega) \supsetneq \{\underline{0}\}$. In this case, $L(\cdot)$ must satisfy (1.2.19) with $\mathcal{R}$ replaced by $\mathcal{N}$.

We conclude with the observation that the displacement field $\underline{u}$ of (1.2.1) is the minimizer of the **total potential energy**

$$\Pi(\underline{u}) = \frac{1}{2}\, B(\underline{u}, \underline{u}) - L(\underline{u}) \tag{1.2.20}$$

over $E_g(\Omega)$ resp. over $E^{\perp}(\Omega)$. (1.2.1) are the first order necessary conditions of a minimum for Π:

$$\forall \underline{v} \in E_0(\Omega): \quad 0 \stackrel{!}{=} \frac{d}{d\varepsilon}\, \Pi(\underline{u} + \varepsilon \underline{v})\big|_{\varepsilon=0} = B(\underline{u}, \underline{v}) - L(\underline{v})\,. \tag{1.2.21}$$

1.3 hp-FE Discretization

1.3.1 Basic Notions

In (conforming) FEM for (1.2.1), we minimize the total potential energy Π in (1.2.20) over a finite element subspace $S \subset E(\Omega)$, of finite dimension $N = \dim S < \infty$. Since $\Pi(\cdot)$ is a quadratic functional, there exists a minimizer $\underline{u}_S \in S$ with

$$\Pi(\underline{u}_S) = \min_{\underline{v} \in S} \{\Pi(\underline{v})\} \tag{1.3.1}$$

which satisfies, by (1.2.21),

$$\underline{u}_S \in S: \ B(\underline{u}_S, \underline{v}) = L(\underline{v}) \quad \forall \underline{v} \in S\,. \tag{1.3.2}$$

(1.3.2) is a **discretization** of (1.2.1) and

$$\underline{e}_S = \underline{u} - \underline{u}_S \tag{1.3.3}$$

the **discretization error** (see Figure 0.1). It satisfies the **Galerkin-orthogonality**

$$B(\underline{e}_S, \underline{v}) = 0 \quad \forall \underline{v} \in S\,. \tag{1.3.4}$$

Since $N = \dim(S) < \infty$, there are basis functions $\{\underline{\varphi}_i(x) : i = 1, \ldots, N\}$ such that every FE-solution $\underline{u}_S \in S$ can be written as

$$\underline{u}_S(x) = \sum_{j=1}^{N} \alpha_j\, \underline{\varphi}_j(x) \tag{1.3.5}$$

for certain coefficients α_j. In p- and generalized FEM, the α_j are not necessarily nodal values of $\underline{u}_S(x)$ and are therefore referred to as **amplitudes** of the FE-solution. Once the amplitudes are known, nodal solution values can however readily be calculated.

Remark 1.3.1.

- for any FE-space $S \subset E(\Omega)$, (infinitely) many bases $\{\underline{\varphi}_i\}$ exist,

- the basis functions $\underline{\varphi}_i(x)$ are related to, but generally different from the FE shapefunctions,
- once a basis of S has been selected, the representation (1.3.5) is unique, and (1.3.2) is equivalent to the linear system of size N

$$\boldsymbol{A}\underline{\alpha} = \underline{L} \tag{1.3.6}$$

where $\boldsymbol{A} = \{B(\underline{\varphi}_i, \underline{\varphi}_j) : 1 \leq i, j \leq N\}$ is the **stiffness-matrix** and $\underline{L} = \{L(\varphi_i) : 1 \leq i \leq N\} \in \mathbb{R}^N$ the **load-vector**.

- Sparsity and Conditioning of $\boldsymbol{A}$ depend on the choice of basis, but $\underline{u}_S$ is (in the absence of quadrature and roundoff errors) independent of it.

From the Galerkin-orthogonality (1.3.4) we have

$$\|\underline{e}_S\|_{E(\Omega)} = \|\underline{u} - \underline{u}_S\|_{E(\Omega)} \leq \|\underline{u} - \underline{v}\|_{E(\Omega)} \quad \forall \underline{v} \in S\,, \tag{1.3.7}$$

since, for any $\underline{v} \in S$,

$$\begin{aligned} \|\underline{u} - \underline{v}\|^2_{E(\Omega)} &= \|\underline{u} - \underline{u}_S + \underline{u}_S - \underline{v}\|^2_{E(\Omega)} \\ &= \|\underline{u} - \underline{u}_S\|^2_{E(\Omega)} + \|\underline{u}_S - \underline{v}\|^2_{E(\Omega)} + 2B(\underline{e}_S, \underline{u}_S - \underline{v}) \\ &\overset{(1.3.4)}{=} \|\underline{e}_S\|^2_{E(\Omega)} + \|\underline{u}_S - \underline{v}\|^2_{E(\Omega)} \geq \|\underline{e}_S\|^2_{E(\Omega)} \end{aligned}$$

with equality if and only if $\underline{u}_S = \underline{v}$, i.e. in **energy norm, the FE solution (1.3.2) is the best approximation of $\underline{u}$ from the FE-space S** and the discretization error is minimal in energy norm. This is a distinct feature of the primal FEM, and it is not true for, e.g., mixed FEM. The discretization error in energy norm is closely related to the root mean square error in the stresses.

By far the most important FE-spaces S are continuous, piecewise polynomials of degree p. In hp-FEM, the degree p could be variable and very high. Efficient implementation is necessary to avoid excessive CPU-cost and described in the following. The issue of discretization error control will be addressed in Section 1.4 below. Non-polynomials FE-spaces are necessary for composites and described in Sections 3.4 and 5 below.

1.3.2 hp-FE spaces

The hp-FE spaces are characterized by the FE-mesh $\mathcal{T} = \{K\}$ consisting of elements K, by a distribution $\underline{p} = \{p_K : K \in \mathcal{T}\}$ of polynomial degrees on this mesh and by the element mappings $\underline{F} = \{F_K : K \in \mathcal{T}\}$, whereby each $K \in \mathcal{T}$ is an image of the master element $\widehat{K}$ under the element maps F_K:

$$K = F_K(\widehat{K}) \quad \forall K \in \mathcal{T}\,. \tag{1.3.8}$$

The elements K are open, disjoint and form a partition of Ω, i.e.

$$K \neq K' \Longrightarrow K \cap K' = \emptyset, \qquad \bigcup_{K \in \mathcal{T}} \overline{K} = \overline{\Omega}\,. \tag{1.3.9}$$

In two dimensions, $\widehat{K}$ is either the unit square $\widehat{Q} = (0,1)^2$ or the unit triangle $\widehat{T} = \{(\xi_1, \xi_2) : 0 < \xi_1 < 1, 0 \leq \xi_2 < \xi_1\}$, in three dimensions we have hexahedra, prisms and tetrahedra. On $\widehat{K}$, spaces $S_p(\widehat{K})$ of polynomials of degree p are defined as follows:

$$\mathcal{Q}_p(\widehat{K}) = \text{span}\{\xi_1^{\alpha_1} \dots \xi_d^{\alpha_d} : 0 \leq \alpha_i \leq p, \ i = 1, \dots, d\}, \tag{1.3.10}$$

$$\mathcal{P}_p(\widehat{K}) = \text{span}\{\xi_1^{\alpha_1} \dots \xi_d^{\alpha_d} : 0 \leq \alpha_1 + \alpha_2 + \dots + \alpha_d \leq p\}, \tag{1.3.11}$$

the spaces of tensor product resp. of polynomials of total degree p ($\mathcal{Q}_p$ is used on quad and hex elements, $\mathcal{P}_p$ is used on tetrahedra and on triangles). The C^0-conforming hp-FE spaces are then defined by

$$S = S^{\underline{p}}(\Omega, \mathcal{T}, \underline{F}) = \{u \in C^0(\overline{\Omega}) : u\big|_K \circ F_K \in S_{p_K}(\widehat{K})\} \subset H^1(\Omega). \tag{1.3.12}$$

Remark 1.3.2. We emphasize that the element maps F_K are completely independent of the FE shapefunctions. If they are also from $S_{p_K}(\widehat{K})$, we have isoparametric elements, if F_K is affine, we have affine elements. Other choices, such as non-polynomial blending maps (see, e.g. [20, 48]), are also possible; they allow exact geometry representation.

We emphasize that the polynomial degrees p_K may vary from element to element and, indeed, even within an element. Let us illustrate this for $\widehat{K} = (0,1)^2$. With each quadrilateral element $K \in \mathcal{T}$ we associate then a polynomial degree vector

$$p_K = (p_K^1, \dots, p_K^5), \tag{1.3.13}$$

where p_K^i, $1 \leq i \leq 4$, represent the polynomial degree on side $\hat{a}_{i+4}$ of K and $p_K^5 = (p_K^{\text{horiz.}}, p_K^{\text{vert.}})$ represents the (possibly anisotropic) horizontal and vertical approximation orders, respectively (see Figure 1.2 for definitions of $\hat{a}_{i+4}$).

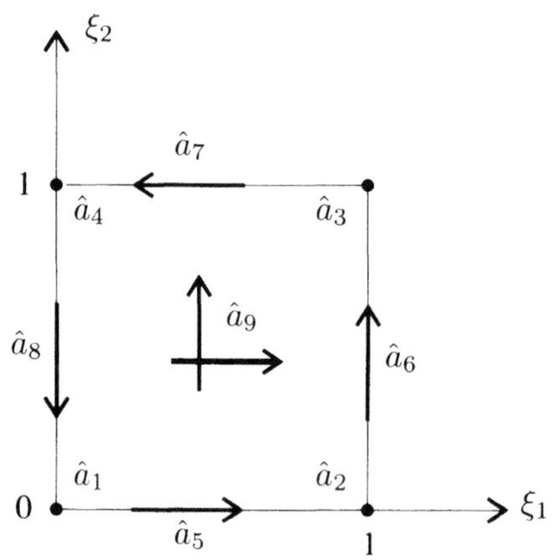

Figure 1.2. Quadrilateral master element $\widehat{K}$ of order $p = (p^1, p^2, p^3, p^4, p^5)$

To ensure interelement compatibility, we require [17]

$$p_K^{\text{horiz.}} \geq \max(p_K^1, p_K^3),\ p_K^{\text{vert.}} \geq \max(p_K^2, p_K^4) \quad \forall K \in \mathcal{T}\,. \tag{1.3.14}$$

This anisotropy in the elemental polynomial degrees is very advantageous for thin-solid elements, as we shall see in Section 2.

If all polynomial degrees p_K are identical, $p_K = p\ \forall K$, we write simply $S^p(\Omega, \mathcal{T}, \underline{F})$ instead of (1.3.12). If $\Omega, \mathcal{T}, \underline{F}$ are clear from the context, we omit them. We note

- In our notion of hp-FEM, the variational formulation, the elemental polynomial degree and the element topology must be separate.
- Conventional element libraries are not designed for concurrent control of errors B and C. Elements like C3D20RHT-20 node, triquadratic displacement, trilinear temperature, hybrid, linear pressure, reduced integration - make systematic quality control impossible.

1.3.3 Elemental shapefunctions

As stated in the introduction, many choices for element shapefunctions are usually possible. All that is required is that they span $S_p(\widehat{K})$, the polynomial space on the master element. Choice of the elemental shapefunctions is guided mainly by a) conditioning and by b) numerical integration. For illustration, we present two frequently used sets of basisfunctions for the unit square $\widehat{K} = (0,1)^2$.

i) Hierarchic shapefunctions of degree p in $\widehat{K}$.
Denote by $L_i(\xi)$ the orthonormal (in $L^2(0,1)$) Legendre polynomials on $(0,1)$. Then the 1-d hierarchical shapefunctions are given by

$$h_1(\xi) = 1 - \xi,\ h_2(\xi) = \xi,\ h_i(\xi) = L_i(\xi) - L_{i-2}(\xi),\ i = 3, \dots\,. \tag{1.3.15}$$

Note that for $i \geq 3$, $h_i(\xi) = c_i \int_0^\xi L_{i-1}(t)\,dt$, and that a FE solution of the form

$$u_{\text{FE}}(\xi) = \sum_{i=1}^{p+1} \alpha_i h_i(\xi)$$

has degrees of freedom α_i which correspond, if $i \geq 3$, not to nodal solution values anymore, while still $u_{\text{FE}}(0) = \alpha_1$, $u_{\text{FE}}(1) = \alpha_2$, of course.

Hierarchic shapefunctions in two dimensions are obtained by tensor products of (1.3.15): for $\xi = (\xi_1, \xi_2) \in \widehat{K}$:

$$N_{ij}(\xi) = h_i(\xi_1)\, h_j(\xi_2),\ \ 1 \leq i, j \leq p+1\,. \tag{1.3.16}$$

Hierarchic shapefunctions have the advantage that **for an increase in the elemental polynomial degree** p_K, the "old" elemental stiffness matrix / the previous elemental load vector can be "recycled", if, for example, profile Cholesky solvers are used, the factorization may be updated from level ℓ to level $\ell + 1$ once the new parts of the global stiffness matrix are calculated simply by resuming Cholesky.

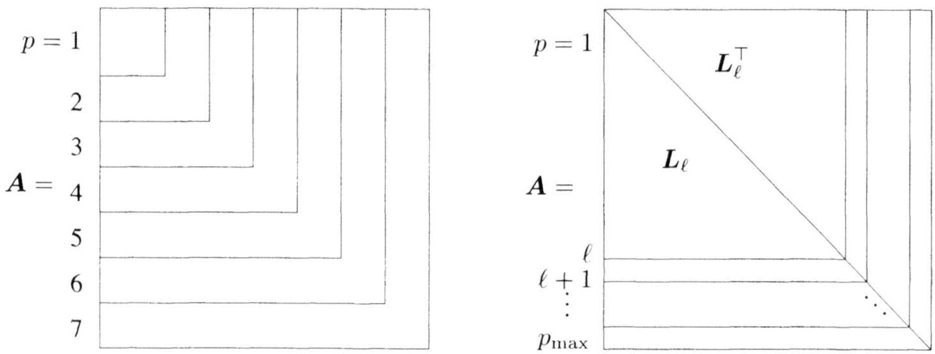

Figure 1.3. Hierarchic structure of global stiffness matrices if hierarchic shapefunctions (1.3.16) and hierarchic Cholesky decomposition are used

Note also that the shapefunctions (1.3.15) are orthogonal in the sense that $h_i'(\xi)$ are orthogonal in $L^2(0,1)$ if $i \geq 3$, and that the 1-d mass-matrix of the h_i is pentadiagonal (see, e.g. [42], Chap. 3).

ii) Non-hierarchic nodal shapefunctions.

For a degree $p \geq 1$ and a set $\mathcal{N} = \{0 = x_0 < x_1 < \ldots . x_p = 1\}$ of $p+1$ nodes in (0,1), define the i-th Lagrange interpolation polynomial of degree p on $\mathcal{N}$

$$\ell_i^{(p)}(\xi, \mathcal{N}) = \prod_{\substack{j=0 \\ j \neq i}}^{p} (\xi - x_i)/(x_i - x_j)\,. \tag{1.3.17}$$

Clearly, for any $\mathcal{N}$, the $\ell_i^{(p)}(\xi, \mathcal{N})$ span the polynomials of degree p on (0,1) and they satisfy

$$\ell_i^{(p)}(x_j, \mathcal{N}) = \delta_{ij} \quad \forall x_j \in \mathcal{N}\,. \tag{1.3.18}$$

whence in the FE-solution of the form

$$u_{\mathrm{FE}}(\xi) = \sum_{i=0}^{p} \alpha_i\, \ell_i^{(p)}(\xi, \mathcal{N})$$

the α_i are the nodal solution values: $u_{\mathrm{FE}}(x_i) = \alpha_i$, for $x_i \in \mathcal{N}$.

It remains to specify the set $\mathcal{N}$. It is advantageous to select nonequidistant sets $\mathcal{N}$. In spectral methods, $\mathcal{N} = \mathcal{N}^p$ is the set of Gauss-Lobatto points, the zeros of $\xi(1-\xi)\, L_i'(\xi)$ in $[0,1]$. This leads to reasonably well-conditioned stiffness and mass-matrices (see, e.g., [9, 35, 36]); under no circumstances, equidistant $\mathcal{N}$ should be used for $p \geq 5$, as they will lead with (1.3.17) to extremely ill-conditioned elemental stiffness/mass matrices). In two and three dimensions, once more tensor products of the 1-d functions (1.3.17) will be employed, e.g.

$$N_{ij}(\xi) := \ell_i^{(p)}(\xi_1, \mathcal{N})\, \ell_j^{(p)}(\xi_2, \mathcal{N}), \quad \xi = (\xi_1, \xi_2) \in \hat{K}\,. \tag{1.3.19}$$

The use of tensor product element shapefunctions is crucial for efficient calculations of elemental stiffness- and mass-matrices, especially in three dimensions and for $p \geq 5$.

1.3.4 Element stiffness matrix computations

The entries A_{ij} of the global stiffness matrix $\boldsymbol{A}$ in (1.3.6) are sums of element contributions, since by (1.3.10)

$$A_{ij} = B(\underline{\varphi}_i, \underline{\varphi}_j) = \sum_{K \in \mathcal{T}} B_K(\underline{\varphi}_i, \underline{\varphi}_j) . \tag{1.3.20}$$

We assume that equal FE spaces for the components of $\underline{u}_S$ are used, i.e.

$$S = S^{\underline{p}}(\Omega, \mathcal{T}, \underline{F})^d, \;\; d = 2, 3, \tag{1.3.21}$$

whence

$$\underline{u}_S\big|_K \circ F_K \in S_{p_K}(\widehat{K})^d = \operatorname{span}\{\underline{N}_I\}_{I \in \mathcal{S}} \tag{1.3.22}$$

where the elemental shapefunctions $\underline{N}_I$ are a basis for $S_{p_K}(\widehat{K})^d$ and I ranges in some index set S; generally I is a unique function of i and K in (1.3.20) and we write $I = I(i, K)$, resp. $i = i(I, K)$. This index transformation is essential for the assembly of $\boldsymbol{A}$ from the **element stiffness matrices**

$$\boldsymbol{A}_K = \{\widehat{B}_K(\underline{N}_I, \underline{N}_J)\}_{I,J \in \mathcal{S}} \tag{1.3.23}$$

where $\widehat{B}_K$ is the pullback of B_K to local coordinates $\xi \in \widehat{K}$; $\widehat{B}_K(N_I, N_J)$ is a combination of terms of the form

$$\widehat{A}_K^{\alpha\beta rs} := \int_{\widehat{K}} \widehat{A}_{rs}^{\alpha\beta}(\xi)\, \hat{\partial}_r\, \varphi_\alpha\, \hat{\partial}_s\, \psi_\beta\, d\xi \;\; \alpha, \beta, r, s \in \{1, \dots, d\} \tag{1.3.24}$$

where $\widehat{A}_{rs}^{\alpha\beta}(\xi)$ are (in general non-polynomial) coefficient functions of the elemental Jacobian $J_K = \det(\underline{F}'_K(\xi))$ and of partials of $\underline{F}_K$, and $\varphi_\alpha(\xi), \psi_\beta(\xi) \in S_{p_K}(\widehat{K})$.

The expression (1.3.24) cannot be evaluated exactly (unless $C_{ijk\ell}|_K$ = const. and $\underline{F}_K$ is affine) and quadrature is needed. We illustrate the sum-factorization and quadrature for dimension $d = 2$ and for quadrilateral elements, i.e. $\widehat{K} = (0, 1)^2$. Let $\mathcal{GL}^q := \{x_i^q\}_{i=0}^q$ denote the $q + 1$ Gauss-Lobatto points in $(0, 1)$, i.e. the roots of $\xi(1 - \xi)L'_q(\xi)$ where L_q is the Legendre-Polynomial of degree q on $(0, 1)$. Then, for every q, there exist quadrature weights $w_k^q > 0$ such that

$$GL_q(u) := \sum_{k=0}^{q} w_k^q\, u(x_k^q) \approx \int_0^1 u(\xi) d\xi \tag{1.3.25}$$

with equality for polynomials u of degree $2q - 1$. On $\widehat{K} = (0, 1)^2$, quadrature rules are obtained by tensor product construction:

$$GL_{q_1 q_2}(u) = \sum_{k=0}^{q_1} \sum_{\ell=0}^{q_2} w_k^{q_1}\, w_\ell^{q_2}\, u(x_1^{q_1}, x_2^{q_2}) . \tag{1.3.26}$$

We have always $x_0^q = 0$, $x_q^q = 1$ (unlike, for example, in Gauss-Legendre rules).

We approximate $\widehat{B}_K^{\alpha\beta rs}$ in (1.3.24) by

$$\widehat{B}_K^{\alpha\beta rs} \approx GL_{q_1 q_2}(\widehat{A}_{rs}^{\alpha\beta}\, \hat{\partial}_r\, \varphi_\alpha\, \hat{\partial}_s\, \psi_\beta) . \tag{1.3.27}$$

Crucial for computational efficiency in p-elements is **sum-factorization**. It requires that the basis of $S_p(\widehat{K})$ is of the form

$$\mathcal{E}^k = \{\chi^k_{1,i}(\xi_1)\,\chi^k_{2,j}(\xi_2)\,|\,1 \le i \le I_k,\ 1 \le j \le J_k\},\ k = 0,\dots,4 \quad (1.3.28a)$$

$$\mathcal{I} = \{\chi^{\mathcal{I}}_{1,i}(\xi_1)\,\chi^{\mathcal{I}}_{2,j}(\xi_2)\,|\,1 \le i \le I_{\mathcal{I}},\ 1 \le j \le J_{\mathcal{I}}\}, \quad (1.3.28b)$$

where $\mathcal{E}^0$ are the **vertex-shapefunctions** associated to $\hat{a}_1, \hat{a}_2, \hat{a}_3, \hat{a}_4$ in Figure 1.2, $\mathcal{E}^k, k = 1,\dots,4$ are the **side shapefunctions** associated with $\mathcal{E}^{k+4}$ and $\mathcal{I}$ are the **internal shapefunctions**.

The element stiffness matrix $\boldsymbol{A}_K$ has a 6×6 block structure, corresponding to the interaction of the 6 different types of shape functions introduced in (1.3.28). An algorithm that exploits the tensor product structure of the basis functions and is based on sum factorization is the following:

Algorithm 1.3.3.

```
initialize $\widehat{\boldsymbol{A}}_K^{\alpha\beta} = 0$
for r,s = 1:2
      for  $B_1 \in \{\mathcal{E}^0, \mathcal{E}^1, \mathcal{E}^2, \mathcal{E}^3, \mathcal{E}^4, \mathcal{I}\}$
           for $B_2 \in \{\mathcal{E}^0, \mathcal{E}^1, \mathcal{E}^2, \mathcal{E}^3, \mathcal{E}^4, \mathcal{I}\}$
           add sum_fact $\left(\frac{\partial}{\partial \xi_r} B_1, \frac{\partial}{\partial \xi_r} B_2, \widehat{\boldsymbol{A}}_{rs}^{\alpha\beta}\right)$ to the block $B_1 - B_2$ of $\widehat{\boldsymbol{A}}_K^{\alpha\beta}$
           end
      end
end
```

Here, the operators $\frac{\partial}{\partial \xi_r}, \frac{\partial}{\partial \xi_s}$ are understood to act elementwise on the sets B_1, B_2. Each call to sum_fact (see Algorithm (1.3.4) below) returns a $\dim B_1 \times \dim B_2$ matrix whose entries are just those calculated by the right hand side of (1.3.26) for all $\varphi \in B_1, \psi \in B_2$.

Algorithm 1.3.4. (sum factorization)

```
$A$  := sum_fact $(B_1, B_2, a)$
   %input:
   % $B_1 = \{\varphi^1_i(\xi_1)\varphi^2_j(\xi_2)\,|\,1 \le i \le I,\ 1 \le j \le J\}$
   % $B_2 = \{\varphi^1_i(\xi_1)\varphi^2_j(\xi_2)\,|\,1 \le i \le I',\ 1 \le j \le J'\}$
   % function $a : \widehat{K} \to \mathbb{R}$
   % output:  the matrix $A$ with entries
   % $A_{(i,j),(i',j')} = \sum_{k=0}^{q_1}\sum_{\ell=0}^{q_2} \varphi^1_i(x_{1,k})\,\varphi^2_j(x_{2,\ell})\psi^1_{i'}(x_{1,k})\psi^2_{j'}(x_{2,\ell})w_{1,k}w_{2,\ell}a(x_{1,k}, x_{2,\ell})$
 $S_1 := (1+q_1)(1+q_2)\,J \cdot J' + (1+q_1)\,I \cdot I' \cdot J \cdot J'$   %flop count for $\sum_k \sum_\ell$
 $S_2 := (1+q_1)(1+q_2)\,I \cdot I' + (1+q_2)\,I \cdot I' \cdot J \cdot J'$   %flop count for $\sum_\ell \sum_k$
 if $S_1 \le S_2$ then { % choose summation order to minimize flop count
   compute auxiliary array $H(k,j,j') := \sum_{\ell=0}^{q_2} \varphi^2_j(x_{2,\ell})\psi^2_{j'}(x_{2,\ell})w_{2,\ell}a(x_{1,k}, x_{2,\ell})$,
   $k \in \{0,\dots,q_1\}$, $j \in \{1,\dots,J\}$, $j' \in \{1,\dots,J'\}$
   compute entries $A_{(i,j),(i',j')} := \sum_{k=0}^{q_1} \varphi^1_i(x_{1,k})\psi^1_{i'}(x_{1,k})w_{1,k}H(k,j,j')$,
   $i \in \{1,\dots,I\}$, $j \in \{1,\dots,J\}$, $i' \in \{1,\dots,I'\}$, $j' \in \{1,\dots,J'\}$
```

```
}
else {
    compute auxiliary array H(ℓ,i,i') := Σ_{k=0}^{q1} φ_j^1(x_{1,k}) ψ_{j'}^1(x_{1,k}) w_{1,k} a(x_{1,k}, x_{2,ℓ}),
    ℓ ∈ {0,...,q2}, i ∈ {1,...,I}, i' ∈ {1,...,I'}
    compute entries A_{(i,j),(i',j')} := Σ_{ℓ=0}^{q2} φ_j^2(x_{2,ℓ}) ψ_{j'}^2(x_{2,ℓ}) w_{2,ℓ} H(ℓ,i,i'),
    i ∈ {1,...,I}, j ∈ {1,...,J}, i' ∈ {1,...,I'}, j' ∈ {1,...,J'}
}
return A
```

The Algorithm 1.3.4 can be used for any set of shapefunctions in $\widehat{K}$ and any family of quadrature formulas that is obtained from 1-d shapefunctions by tensor product construction, i.e. for which the sets $\mathcal{E}^k$ and $\mathcal{I}$ are as in (1.3.28). Note carefully that the quadrature (1.3.26) as well as the polynomial degrees in $\widehat{K}$ can be anisotropic (see (1.3.15) and Figure 1.2), which is particularly advantageous for "thin-solid" p-elements used in plate and shell moduling (see Section 2 below). Further, in (1.3.28) **any** family of 1-d shapefunctions, hierarchic or not, can be used, for example (1.3.16) or (1.3.17).

If, however, the nodal basis (1.3.17) is used, the property (1.3.18) can be exploited in the quadrature algorithm 1.3.4: by a suitable choice of the nodal set $\mathcal{N}$ in (1.3.17) (which must be unisolvent, but is otherwise at our disposal), many of the costly integrand evaluations in the application of (1.3.26) to (1.3.24) can be avoided: If, for example, $p_1 = p_2 = p$ and $S_p(\widehat{K}) = \mathcal{Q}_p(\widehat{K})$, and if $q_1 = q_2 = p$ in (1.3.26), the resulting scheme is just the **spectral** method for the evaluation of $\boldsymbol{A}_K$ (see, e.g. [35, 36]). The main points are:

Remark 1.3.5.

- For $bi - p$ elements, i.e. $S_p(\widehat{K}) = \mathcal{Q}_p(\widehat{K})$, and for (1.3.26) with $q_1 = q_2 = p$ and nodal shapefunctions (1.3.17), (1.3.18), the **mass-matrix becomes diagonal,**
- the spectral quadrature scheme *underintegrates* the element stiffness matrices,
- for scalar (antiplane) elasticity, this can be shown to nevertheless lead to SPD global stiffness matrices, and to preserve theoretical convergence properties of the hp-FEM (see Section 1.4 below),
- for vector-valued problems, such as 2 and 3-d elasticity and, more severely, plate and shell problems, the choice $q_1 = p_1$, $q_2 = p_2$ in GL_{q_1,q_2} as in spectral methods may lead to a loss of coercivity and to instability. This is particularly the case for mixed and hybrid FE-formulations (not discussed here).
- on triangles and tetrahedra, tensor product shapefunctions and sum-factorizations are also available [23].

To apply the spectral type quadrature in elasticity, it is desirable to decouple the quadrature orders q_1, q_2 in (1.3.26) from p, i.e. to allow for **overintegration**, while still exploiting the property (1.3.18). This is done by the following algorithm.

Algorithm 1.3.6. (Spectral Galerkin Sum-Factorization)

```
A   := spec_sum_fact (B1, B2, a)
    %input:
    % B1 = {φ^1_i(ξ1)φ^2_j(ξ2) | 1 ≤ i ≤ I, 1 ≤ j ≤ J}
    % B2 = {ψ^1_i(ξ1)ψ^2_j(ξ2) | 1 ≤ i ≤ I', 1 ≤ j ≤ J'}
    % function a : K̂ → ℝ
    %output:  the matrix A with entries
    % A_(i,j),(i',j') = Σ_{k=0}^{q1} Σ_{ℓ=0}^{q2} φ^1_i(x_{1,k})φ^2_j(x_{2,ℓ})ψ^1_{i'}(x_{1,k})ψ^2_{j'}(x_{2,ℓ})w_{1,k}w_{2,ℓ}a(x_{1,k}, x_{2,ℓ})

for each pair i,i' compute set K(i,i') := {k ∈ {0,...,q1} | φ^1_i(x_{1,k})ψ^1_{i'}(x_{1,k}) ≠ 0}
for each pair j,j' compute set L(j,j') := {ℓ ∈ {0,...,q2} | φ^2_j(x_{2,ℓ})ψ^2_{j'}(x_{2,ℓ}) ≠ 0}
S1 := (1+q1) Σ_{j,j'} |L(j,j')| + J·J'·Σ_{i,i'} |K(i,i')|   % flop count for Σ_k^{q1} Σ_ℓ^{q2}
S2 := (1+q2) Σ_{i,i'} |K(i,i')| + I·I'·Σ_{j,j'} |L(j,j')|   % flop count for Σ_ℓ^{q2} Σ_k^{q1}
if S1 ≤ S2 then { % choose summation order to minimize flop count

   compute auxiliary array H(k,j,j') := Σ_{ℓ∈L(j,j')} φ^2_j(x_{2,ℓ})ψ^2_{j'}(x_{2,ℓ})w_{2,ℓ}a(x_{1,k}, x_{2,ℓ}),
   compute entries A_(i,j),(i',j') := Σ_{k∈K(i,i')} φ^1_i(x_{1,k})ψ^1_{i'}(x_{1,k})w_{1,k}H(k,j,j'),
}
else {
   compute auxiliary array H(ℓ,i,i') := Σ_{k∈K(i,i')} φ^1_j(x_{1,k})ψ^1_{j'}(x_{1,k})w_{1,k}a(x_{1,k}, x_{2,ℓ}),
   compute entries A_(i,j),(i',j') := Σ_{ℓ∈L(j,j')} φ^2_j(x_{2,ℓ})ψ^2_{j'}(x_{2,ℓ})w_{2,ℓ}H(ℓ,i,i'),
}
return A
```

Here $|L(j,j')|$ and $|K(i,i')|$ stand for the number of elements of the sets $L(j,j')$, $K(i,i')$. We emphasize that Algorithm 1.3.6 is a generalization of Algorithm 1.3.4 as well as of the spectral method. If, for example, the basis functions (1.3.16) are used, Algorithm 1.3.6 reduces to the standard spectral Algorithm 1.3.4.

In choosing $q_1, q_2 > p$, the quadrature points in $GL_{q_1 q_2}$ change and the question arises **how to** choose the nodal set $\mathcal{N}$ in (1.3.17) so as to maximally exploit (1.3.18) in Algorithm 1.3.6. In [30], we selected $\mathcal{N}^p$ as subset of the Gauss-Lobatto points $\mathcal{GL}^q$ if $q > p$; many choices of $\mathcal{N}^p \subset \mathcal{GL}^q$ are of course possible - in [30], we gave for $0 < p, q \leq 20$ optimal choices of $\mathcal{N}^p$. Here "optimal" means that the condition number of the one-dimensional mass matrix corresponding to the basis $\ell_i^{(p)}(\xi; \mathcal{N})$ is minimal among all nodal sets $\mathcal{N}^p \subset \mathcal{GL}^q$. Tables of optimal nodal sets are given in [30].

The **work estimates** for Algorithms 1.3.4, 1.3.6 are in 2-d, for $S_p(\widehat{K}) = Q_p(\widehat{K})$, and for quadrature order $q_1 = q_2 = p + q$ in (1.3.26), given in Table 1.

algorithm	2D	3D
standard	$O(p^4(p+q)^2)$	$O(p^6(p+q)^3)$
sum fact.	$O(p^4(p+q)+p^2(p+q)^2)$	$O(p^6(p+1)+p^4(p+q)^2+p^2(p+q)^3)$
spect.-Gal.	$O(p^4(1+q)+p^2q^2)$	$O(qp^6+p^5+q^2p^4+q^3p^2)$

Table 1. Work estimates for stiffness matrix calculations

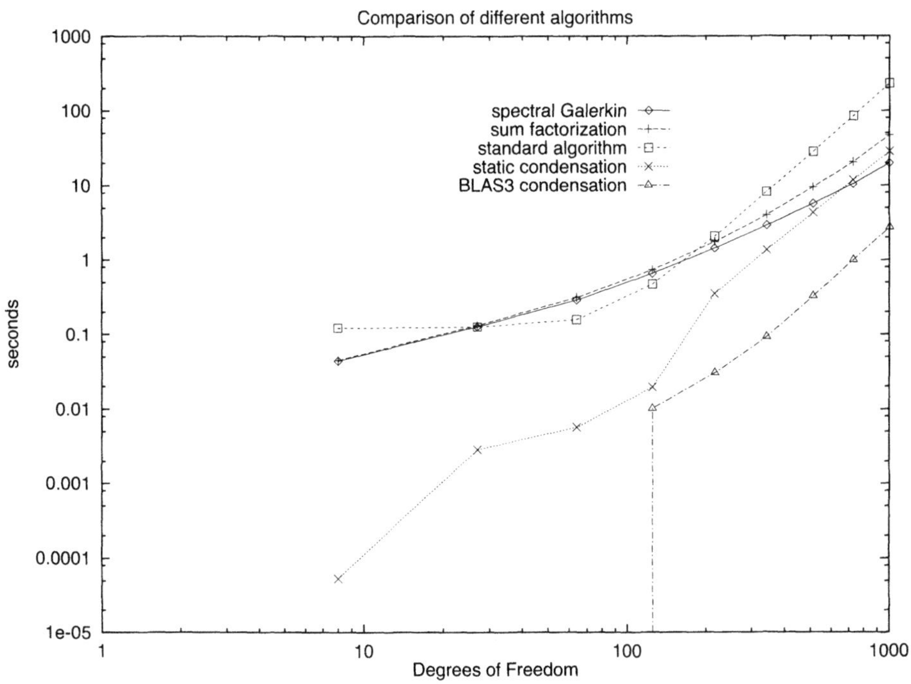

Figure 1.4. CPU-time comparisons for various algorithms

In Figure 1.4 we compare the CPU times of Algorithms 1.3.4 and 1.3.6 for different p, $1 \le p \le 9$. We see that

- in this range of p, static condensation of the (large number) of internal DOF is still a small part of the total CPU time per element and quadrature is by far the most expensive part of the element calculation.

- For $1 \leq p \leq 4$, the standard algorithm is, in fact, slightly cheaper than the spectral Algorithm 1.3.6.
- In the range $5 \leq p \leq 10$, the spectral algorithm 1.3.6 is best.
- Note that in Table 1, in 3-d the spectral Method ($q = 0$ in $\mathcal{GL}^{p+q}$ and $\mathcal{N}^p = \mathcal{GL}^p$) is special: the quadrature work is only $O(p^5)$ then.

Open problems in numerical integration for p-FEM are in particular the local and adaptive selection of the amount of overintegration q in $\mathcal{GL}^{p+q}$. A-posteriori quadrature error estimates are yet to be derived. Presently, e.g. in STRESSCHECK, one chooses the overintegration $q = 2$ generally and, in shells, $q = 3$ has proved successful for complex geometry.

1.3.5 Global solution algorithm

Once the element stiffness matrices $\boldsymbol{A}_K$ are computed, the global stiffness matrix $\boldsymbol{A}$ in (1.3.6) needs to be assembled and solved. This is sketched in

Algorithm 1.3.7. **(hp-spectral Galerkin Algorithm)**

```
loop over all mesh patches
   for each mesh patch, loop over all elements
      for each element, generate A_K, L_K,
                  condense internal DOF
      assemble patch stiffness-matrix
      (optional) condense DOF internal to mesh patch
   assemble patch stiffness matrix into global stiffness matrix
end loop over all mesh patches
solve global system
backsolve for inner patch DOF and inner element DOF
```

Remark 1.3.8.

- In Algorithm 1.3.7, **mesh patches** enter: they are groups of elements corresponding to patchwise structured meshes; the most important ones are **geometric corner**, **edge** and **vertex patches** and **boundary layer patches**. For unstructured meshes, the loop over all patches in Algorithm 1.3.7 becomes simply a loop over all elements.
- The condensation of internal DOF in steps 4 and 6 of 1.3.7 is known to **precondition** the resulting global stiffness matrices. In Step 9 of 1.3.7, therefore, also iterative solvers can be used. The condition number of the global stiffness matrix depends mainly on the geometry of the patches, but is rather insensitive to the meshes within the patches. Patches can be processed in parallel (see [33]).

1.3.6 Treatment of rigid body modes

We have seen in Section 1.2 that the solution $\underline{u}$ of the linearized elasticity problem (1.1.1)-(1.1.4) in Dimension d is only unique up to $d(d+1)/2$ rigid body motions $\mathcal{R}$, if pure traction boundary conditions are prescribed, i.e. if $\Gamma = \Gamma_T$. The same holds for the FE-solution $\underline{u}_S \in S$, provided

$$\mathcal{R} \subset S \tag{1.3.29}$$

i.e. if the FE-space can represent exactly rigid body nodes.

Remark 1.3.9. For affine or isoparametric hp-elements, (1.3.29) is satisfied, but for general, non-polynomial blending maps F_K in (1.3.22), (1.3.29) is not valid, in general. However, in the case of high p the rigid body motions are very well represented by FEM so that the usual method of constraining $\underline{u}_S$ in $\dim(\mathcal{R}) = d(d+1)/2$ points works also here.

1.4 Control of Discretization Errors

1.4.1 Hierarchic FE Spaces. Extensions

We are interested in quantitative control of Error C in Figure 0.1. Due to pessimistic bounds on constraints in a-priori estimates of the FE-error, sharp and computable error bounds must necessarily be **a-posteriori**, i.e. they are obtained from postprocessing (a sequence of) FE solutions. If the discretization error C is found too large, the FE-space is to be **increased** by adding degrees of freedom, i.e. by an **extension**.

Remark 1.4.1. 3 basic extensions are available

- h-**extension**: Discretization error control by **mesh-refinement**, resp. element subdivision, i.e. $h_{\max} \to 0$.
- p-**extension**: Discretization error control by **increase of the polynomial, degree**, i.e. $p_{\min} \to \infty$.
- hp-**extension**: Discretization error control by simultaneous h and p-extension.

A-posteriori error estimation and extension of the FE-space is based on a **hierarchic sequence** of **FE-spaces**

$$S_1 \subset S_2 \subset \cdots \subset S_\ell \subset \dots . \tag{1.4.1}$$

The **error measure of interest** is typically in a **functional of the solution**, i.e.

$$\Phi(\underline{u}) : \underline{u} \longrightarrow \mathbb{R} . \tag{1.4.2}$$

It could be any quantity of engineering interest, such as resultant force, bending moment, etc. but also pointwise stress, strains etc.

Remark 1.4.2. It is well-known that some functionals Φ are not bounded on all solutions of finite energy. For example:

i) **Pointwise maximum stress** is meaningless in linearized elasticity, if the domain Ω contains cracks or reentrant corners; near such points, the displacement has, in polar-coordinates (r, θ) centered at the singular point, the (formal) expansion

$$\underline{u} = \sum_{i=1}^{\infty} \frac{1}{2\mu} A_i^{(1)} r_r^{\lambda_i^{(1)}} \underline{\Psi}^{(1)}(\theta) + \sum_{i=1}^{\infty} \frac{1}{2\mu} A_i^{(2)} r^{\lambda_i^{(2)}} \underline{\Psi}^{(2)}(\theta) \tag{1.4.3}$$

where $\underline{\Psi}^{(1)}(\theta)$ and $\underline{\Psi}^{(2)}(\theta)$ are the mode I and mode II (generalized) singular fields. The **singularity exponents** $\lambda_i^{(1)}$, $\lambda_i^{(2)}$ are, in general, complex. If $Re\,\lambda_i^{(1)} < 1$, $Re\,\lambda_i^{(2)} < 1$, then the stresses $\sigma_{ij}[\underline{u}]$ at $r = 0$ are unbounded (e.g. in linear fracture, $\lambda_i^{(1)} = \lambda_i^{(2)} = 1/2$). In this case, meaningful description of $\boldsymbol{\sigma}$ in the vicinity of $r = 0$ is only possible through the **(generalized) stress-intensity factors** $A_i^{(1)}(\underline{u})$, $A_i^{(2)}(\underline{u})$ in (1.4.3), which are always finite for $\underline{u} \in E(\Omega)$, [48].

ii) **Pointwise displacements** $\underline{u}(x_0)$ are possibly unbounded, even if $\underline{u} \in E(\Omega)$ (one checks that $\underline{u}(x) = (\log|\log|x||, \log|\log|x||) \in H^1(\Omega)^2 \subset E(\Omega)$ but $\underline{u}(x)$ is unbounded for $|x| = 0$).

In Section 1.3.1, we discussed the **energy norm error**,

$$\|\underline{e}_S\|_{E(\Omega)} = \|\underline{u} - \underline{u}_S\|_{E(\Omega)} \,. \tag{1.4.4}$$

It can be expressed through the total potential energy $\Pi(\underline{u})$ in (1.2.20) as follows:

$$\|\underline{e}_S\|_{E(\Omega)} = (\Pi(\underline{u}_S) - \Pi(\underline{u}))^{\frac{1}{2}} \,, \tag{1.4.5}$$

i.e. if $\Phi(\underline{u}) = \Pi(\underline{u})$ were known, and since

$$\Pi(\underline{u}_S) = \frac{1}{2}\, B(\underline{u}_S, \underline{u}_S) - L(\underline{u}_S) = -\frac{1}{2}\, L(\underline{u}_S) \overset{(1.3.6)}{=} -\frac{1}{2}\, \underline{\alpha}^\top \underline{L} \tag{1.4.6}$$

can be computed cheaply, the error discretization $\|\underline{e}_S\|_E$ were known. Since $\Pi(\underline{u})$ is not known, an approximate value for $\Pi(\underline{u})$ must be found computationally, and in the following we sketch the p-extension and **extrapolation-technique**. It is based on a hierarchy (1.4.1) of FE-spaces that are obtained by

Algorithm 1.4.3. (hierarchic p-extension)

i) Fix a mesh $\mathcal{T}$ in Ω (taking into account singularities and boundary layers where applicable) and a vector of element mappings $\underline{F} = \{F_K : K \in \mathcal{T}\}$.

ii) Compute a sequence $\{\underline{u}^{(p)}\}_{p=1}^{p_{\max}}$ of FE-solutions in the FE-spaces $\{S^p(\Omega, \mathcal{T}, \underline{F})\}_{p=1}^{p_{\max}}$ of uniform polynomial degree p.

Remark 1.4.4.

- The p-**extension** is **optimal** in the sense that for practically all problems in linearized elasticity and all functionals Φ of engineering interest,

$$|\Phi(\underline{u}_S) - \Phi(\underline{u})| \to 0 \quad \text{as} \quad N = \dim S \to \infty \tag{1.4.7}$$

  at least as fast as in a h-extension, i.e. with uniform mesh refinement at fixed, low degree $p = 1, \ldots, p_{\max}$. Typically, $p_{\max} = 8$.

- Judicious combination of h- and p-extensions leads to **exponential convergence** in many problems of engineering interest. The h- or p-extensions alone lead to **algebraic rates of convergence** at best [8, 21, 42] for typical singular solutions like (1.4.3), as we will show next.

1.4.2 Solution categories

The proper mesh design for p -and hp-FEM depends on the **solution category** of $\underline{u}$:

Terminology 1.4.5. (Solution Categories)

- $\underline{u}$ in **category A**: $\underline{u}$ is analytic in the interior of the domain and on its boundaries; the domain does only exhibit smoothly curved boundaries and corners, edges and interfaces are absent.
- $\underline{u}$ in **category B**: $\underline{u}$ is analytic in most of Ω, but becomes singular at a set $\mathcal{G} \subset \overline{\Omega}$, the **generator**. It consists of lines and points, typically. Solutions (1.4.3) belong to category B.
- $\underline{u}$ in **category C**: $\underline{u}$ is nonanalytic in Ω or on "large" subsets thereof (e.g. fully developed turbulence etc.).

Most problems in engineering belong to category A and B. Some examples follow.

Example 1.4.6. (Problems with category B solutions)
We give several examples of category B solutions, and their generator sets $\mathcal{G}$. We assume that the loads are analytic here.

i) **Corner singularities**. Consider the curvilinear polygon Ω shown in Figure 1.5.

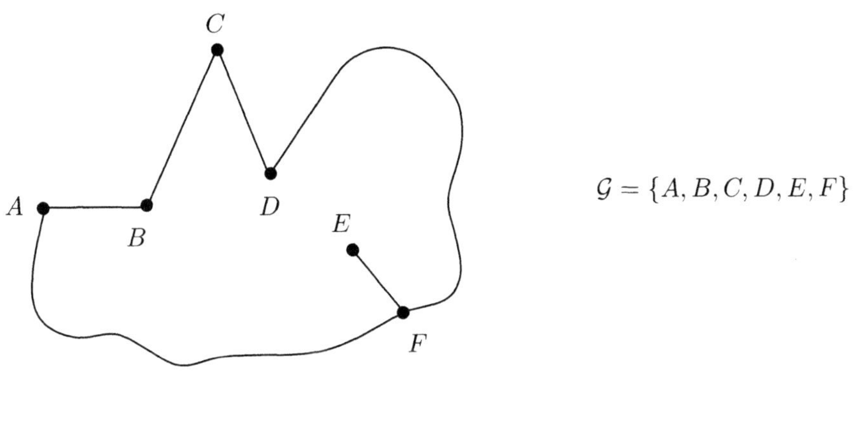

Figure 1.5. Corner singularities in curvilinear polygon $\Omega \subset \mathbb{R}^2$

Here, at each solid dot, the solution $\underline{u}$ is of the form (1.4.3), with different $\lambda_i^{(1)}$, $\lambda_i^{(2)}$, though. The generator $\mathcal{G}$ consists of **all** corners in the domain, not just the reentrant ones.

ii) **Thin solids** (plate and shell models with shear). Consider a thin plate/shell with midsurface $\mathcal{M} = F(\omega)$, where $\omega \subset \mathbb{R}^2$ is the reference domain and $F : \mathbb{R}^2 \to \mathbb{R}^3$ is the parametrization of $\mathcal{M}$. Then

$$\partial\omega \subseteq \mathcal{G}, \tag{1.4.8}$$

i.e. the solution behaves "singular" near the edge of the thin solid. The solution $\underline{u}$ then exhibits corner singularities and so-called **boundary layers**. Boundary layers are solution components with rapid variations near $\mathcal{G}$, of the generic form

$$\underline{u} = \underline{u}(s,\rho) = \underline{U}(s)\exp(-\rho(s)/\varepsilon) \tag{1.4.9}$$

with a smooth function $\underline{U}(s)$.

Here, (ρ, s) are curvilinear, orthogonal coordinates fitted to $\mathcal{G}$, see Figure 1.6.

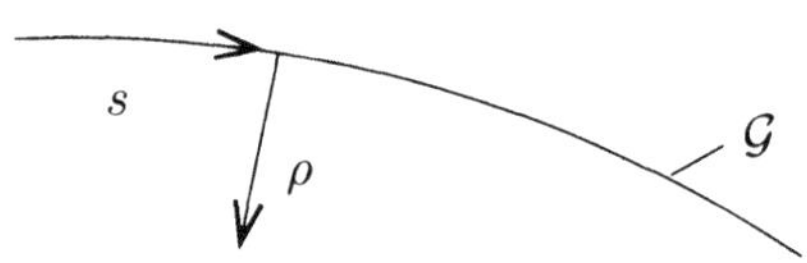

Figure 1.6. Line generator $\mathcal{G}$ and fitted coordinate system (ρ, s)

In (1.4.9), ε is the **length scale of the layer**. In beam and plate models with shear, we have the length scale [4, 37]

$$\varepsilon \sim t \tag{1.4.10}$$

where t is the plate-thickness. In shells with shear, the scales

$$\varepsilon \sim t \ \text{ and } \ \varepsilon \sim t^{\frac{1}{2}}, t^{\frac{1}{3}}, t^{\frac{1}{4}} \tag{1.4.11}$$

arise, $t^{\frac{1}{2}}$ due to the "simple edge effect" and the other length scales due to bending-membrane coupling in certain degenerate situations (see [37] for details).

iii) **3-d solids** Here, all edges and vertices are parts of $\mathcal{G}$.

1.4.3 Geometric mesh design for p- and hp-FEM

For solutions in category B, proper mesh design for p- and hp-FEM includes **geometric meshes**, or, more generally, **geometric mesh patches**.

Definition 1.4.7. *A mesh $\mathcal{T} = \{K\}$ is geometric with respect to the point x_0, if there is $c > 0$ such that*

$$\forall K \in \mathcal{T} :\ 0 < c \le \operatorname{dist}(x_0, K)/\operatorname{diam}(K) \le 1/c\,. \tag{1.4.12}$$

A mesh $\mathcal{T}$ is geometric with respect to a line, if the intersection of $\mathcal{T}$ with a normal plane to this line is geometric with respect to the intersection point x_0 of the line with this plane.

Example 1.4.8.

i) **1-d geometric mesh** with grading factor $1/2$ refined towards $\mathcal{G} = \{x_0\}$:

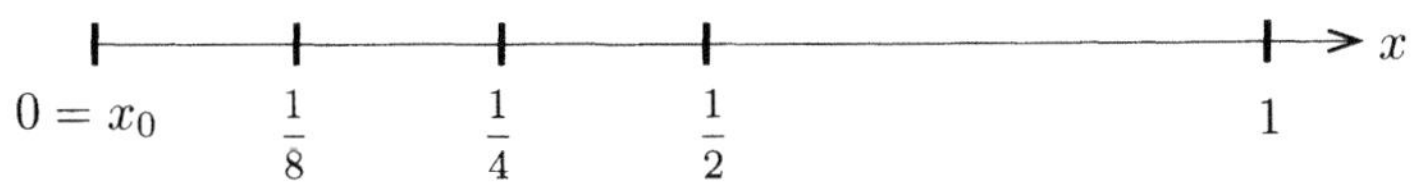

Figure 1.7. 1-d geometric mesh refinement towards $\mathcal{G} = \{x_0\}$, grading factor $1/2$

ii) **2-d geometric meshes**

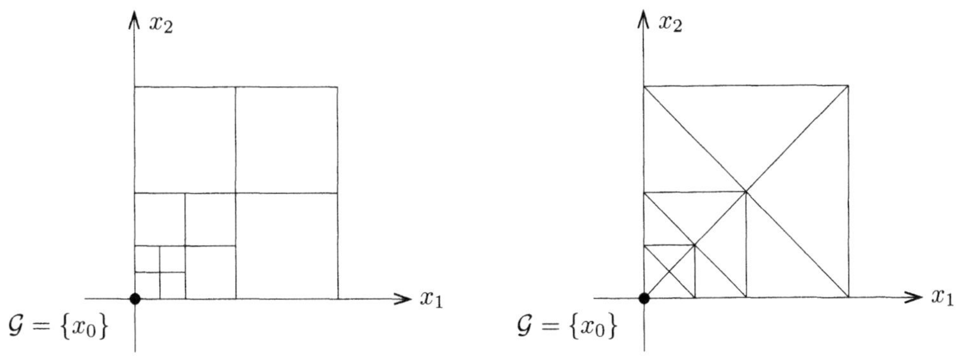

Figure 1.8. 2-d geometric meshes refined towards a point. Left: with hanging nodes, right: regular mesh without hanging nodes.

iii) **Geometric mesh with respect to lines meeting at a vertex**

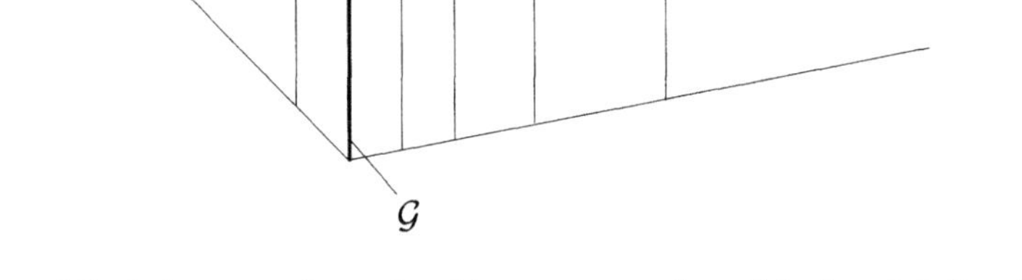

Figure 1.9. 3-d geometric mesh with respect to edges $\mathcal{G}$ in the vicinity of a vertex •

Remark 1.4.9.

- In a geometric mesh with respect to a point, the elements are shape-regular, whereas in meshes geometric with respect to a line, they are highly anisotropic.
- Geometric refinements with respect to a line are difficult to generate adaptively - anisotropic refinements and alignment to $\mathcal{G}$ is required.

In geometric meshes with respect to a point, elements of approximately equal size are combined to a **mesh layer**. (The meshes in Figure 1.8 have three layers). The following rules apply:

- To achieve exponential convergence in hp-extensions, geometric meshes with a number of mesh layers proportional to the degree p must be used.
- For p-extension on a fixed mesh, the mesh should be geometric with $p_{\max}$ layers; this will give exponential convergence in the practical range of $p \leq p_{\max}$.
- For plate and shell boundary layers of thickness ε, either a single needle element of thickness $p_{max}\varepsilon$ aligned with the edge $\mathcal{G}$ should be used, or a geometric mesh with $O(|\log(\varepsilon)|)$ layers aligned and anisotropically refined towards the edge must be used to assure exponential convergence (see [43, 44, 31]).

1.4.4 *hp*-interpolation estimates

For the extrapolation a-posteriori error estimation procedure below as well as for the correct scaling of residual and dual a-posteriori error estimates in the hp-context, certain sharp hp-a-priori estimates are required. Generically, for the FE subspace $S \subset E(\Omega)$ and for $\underline{\varphi} \in E(\Omega)$, an **interpolant** $\mathcal{I}_S \underline{\varphi} \in S$ must be constructed and the error $\|\underline{\varphi} - \mathcal{I}_S \underline{\varphi}\|_{E(\Omega)}$ estimated. An interpolant $\mathcal{I}_S \underline{u}$ of a statistically admissible displacement field $\underline{u} \in E_g(\Omega)$ is a finite element function, i.e. it belongs to S. However, its main feature is that its values in element $K \in \mathcal{T}$ depend only on those of $\underline{u}$ in a patch $\widetilde{K}$ of elements abutting at K. For h-version FEM, such interpolants are well-known [13]. Here, we present their hp-extensions (see also [8]).

By (1.2.13), $\|\underline{\varphi}\|_{E(\Omega)} \leq C \|\nabla \underline{\varphi}\|$ for any $\underline{\varphi} \in E(\Omega)$[1], and it is sufficient to estimate the H^1-norm of the interpolation error, $\|\underline{\varphi} - \mathcal{I}_S \underline{\varphi}\|_{H^1(\Omega)}$. There holds

Proposition 1.4.10. *Let $\Omega \subset \mathbb{R}^2$ and let $\mathcal{T}$ be a shape-regular mesh of triangles or quadrilaterals in Ω. Let further $u \in H^{k+1}(\Omega)$, $k \geq 1$ and $S = S^p(\Omega, \mathcal{T}, \underline{F})$ be the hp-FE space with polynomial degree $p \geq 1$. Then there exists an hp-interpolant $\mathcal{I}u \in S$ such that for any $K \in \mathcal{T}$*

$$\|u - \mathcal{I}u\|_{L^2(K)} + \frac{p}{h_K} \|\nabla(u - \mathcal{I}u)\|_{L^2(K)} \leq C_I(k)\Big(\frac{h_K}{p}\Big)^{\min(p,k)} \|u\|_{H^{k+1}(\widetilde{K})}$$

where $\widetilde{K} = \bigcup\{\overline{K}' : \overline{K}' \cap \overline{K} \neq \emptyset\}$ is the patch of elements in $\mathcal{T}$ that abut on K. Furthermore,

$$\|u - \mathcal{I}u\|_{L^2(\partial K)} \leq C_I(k) \left(\frac{h_K}{p}\right)^{\min(p,k)+1/2} \|u\|_{H^{k+1}(\widetilde{K})} \,.$$

[1] Note, however, that for plates and shells, $C(\Omega)$ may be very large at small thickness t.

For a proof, we refer to [8].

Proposition 1.4.10 gives algebraic convergence for $\mathcal{I}u$ if u has k bounded (precisely: square integrable) derivatives. If u is **analytic** in an element $\overline{K}$, exponential convergence is possible with p-extensions.

Proposition 1.4.11. *Let $\Omega \subset \mathbb{R}^2$ and $\mathcal{T}$ be as in Proposition 1.4.10. Assume that u is analytic in $\overline{K}$ and admits an analytic extension to a larger domain $\widetilde{K} \supset \overline{K}$ with*

$$\rho := \operatorname{dist}(K, \partial\widetilde{K}) \,/\, \operatorname{diam}(\overline{K}) > 0\,.$$

Assume that the element maps are analytic in $\overline{\widehat{K}}$. Then there exists an interpolant $\mathcal{I}u \in S^p(\Omega, \mathcal{T}, \underline{F})$ such that

$$\begin{aligned} \|u - \mathcal{I}u\|_{L^2(K)} + \frac{p}{h_K}\, \|\nabla(u - \mathcal{I}u)\|_{L^2(K)} &\leq C_I \exp(-bp) \\ \|u - \mathcal{I}u\|_{L^2(\partial K)} &\leq C_I \exp(-bp) \end{aligned}$$

where $b := \ln(1 + \rho) > 0$. Here C_I depends only on the maximum of u on $\overline{K}$ and on the shape-regularity of K.

We remark that in both Propositions 1.4.10, 1.4.11, the interpolant $\mathcal{I}u$ is local in the sense that it takes only function values of u from K or, at most, from a small neighborhood of K.

As we showed in (1.3.7), the rate of convergence of $\|\underline{e}_S\|_{E(\Omega)}$ is determined by the quality of an interpolation $\underline{v} = \mathcal{I}\underline{u} \in S$. From 1.4.10, 1.4.11, we get immediately

Proposition 1.4.12. *Let $\Omega \subset \mathbb{R}^2$, $\mathcal{T} = \mathcal{T}_h$ be quasiuniform with meshwidth $h > 0$ and $\underline{u} \in H^{k+1}(\Omega)$, $k \geq 1$, and $S = S^p(\Omega, \mathcal{T}_h, \underline{F})$. Then the local interpolants $\mathcal{I}$ in Propositions 1.4.10, 1.4.11 can be combined into a global interpolant $\mathcal{I}_S$ with*

$$\|\underline{e}_S\|_{E(\Omega)} \leq C_1 \, \|\underline{u} - \mathcal{I}_S \underline{u}\|_{H^1(\Omega)} \leq C_1 C(k) \Big(\frac{h}{p}\Big)^{\min(p,k)} \|\underline{u}\|_{H^{k+1}(\Omega)}\,.$$

1.4.5 Extrapolation a-posteriori error estimation in hierarchic p-extensions

The **Rate of Convergence of the FEM** measures, how **fast** the error C, the discretization error $|\Phi(\underline{u}_S) - \Phi(\underline{u})|$, tends to zero as $N = \dim(S) = \#$ DOF increases, i.e. the accuracy per work spent (admitting N as a crude work-measure). The rate of convergence is **algebraic**, if there are $C, \beta > 0$ with

$$|\Phi(\underline{u}_S) - \Phi(\underline{u})| \leq \frac{C}{N^\beta} \quad \text{for } N \to \infty\,. \tag{1.4.13}$$

The rate is **exponential** if, for $N \to \infty$, there are $C, \gamma, \theta > 0$ such that

$$|\Phi(\underline{u}_S) - \Phi(\underline{u})| \leq C \exp(-\gamma N^\theta)\,. \tag{1.4.14}$$

In a log-log plot of the error, we have qualitatively the situation in Figure 1.4. Exponential convergence is never achieveable by mesh refinement done. It requires a) least piecewise analytic solutions $\underline{u}$ and b) p- or hp-refinements, i.e. increasing polynomial approximation order elements that are not refined.

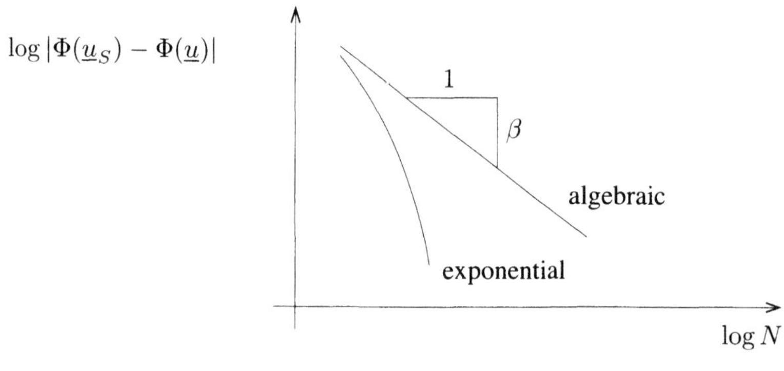

Figure 1.10. Algebraic and Exponential Convergence Rates

Remark 1.4.13.

Exponential convergence rates are achievable for problems of engineering interest only with **proper mesh design** and **hierarchic p-extensions** of the FE-spaces,

For problems in category B, (uniform or adaptive) h-extensions or p-extensions on a fixed mesh usually converge algebraically, since by Proposition 1.4.10

$$\mathcal{S} = \mathcal{S}^p(\Omega, \mathcal{T}, \underline{F})^d \Longrightarrow \|e_S\|_E \leq C/N_p^\beta, \quad p \to \infty . \tag{1.4.15}$$

For solutions in category B (cf. Terminology 1.4.5), $\beta > 0$ is typically small. Moreover, (1.4.15) is sharp in the sense that, as $N \to \infty$,

$$\|e_S\|_E - CN^{-\beta} \to 0 .$$

Using (1.4.5), it follows that

$$\Pi(\underline{u}_S) - \Pi(\underline{u}) \approx C^2 N^{-2\beta} \quad N \to \infty . \tag{1.4.16}$$

Denote the FE solutions $\{\underline{u}_S\}$ corresponding to $\{S^p\}_p = \{S^p(\Omega, \mathcal{T}, \underline{F})\}_{p=1}^{p_{\max}}$ (for fixed $\mathcal{T}$ and $\underline{F}$) by $\underline{u}_p$ and $\Pi(\underline{u}_p) = \Pi_p$, and $\Pi(\underline{u})$ by Π. Then we get from (1.4.16)

$$\frac{\Pi - \Pi_p}{\Pi - \Pi_{p-1}} \approx \left(\frac{\Pi - \Pi_{p-1}}{\Pi - \Pi_{p-2}}\right)^\alpha \tag{1.4.17}$$

where the exponent α is estimated by

$$\alpha = \log(N_{p-1}/N_p)/\log(N_{p-2}/N_{p-1})$$

and, as before,

$$N_p := \dim S^p(\Omega, \mathcal{T}, \underline{F}) .$$

The **error estimation procedure** consists now in

a) assuming that equality holds in (1.4.17), and b) given any 3 consecutive values of $\Pi_p, \Pi_{p-1}, \Pi_{p-2}$ from p-extension on a fixed mesh, solving for Π in (1.4.17).

Once an approximate value for Π has been obtained, the convergence rate β can be estimated from (1.4.16) with equality.

An error estimation procedure similar to (1.4.17) can be applied also for any bounded functional $\Phi(\underline{u})$ of the solution. Denoting once more the FE-solutions to a p-hierarchic sequence $\{S^p\}_{p=1}^{p_{\max}}$ of FE-spaces by $\{\underline{u}_p\}_{p=1}^{p_{\max}}$, $\Phi(\underline{u}_p)$ by Φ_p and $\Phi(\underline{u})$ simply by Φ, we have, upon hierarchic p-extension, at our disposal a sequence $\{\Phi_p\}_{p=1}^{p_{\max}}$ of approximate values for the functional. For category B problems, these converge once more algebraically as $N \to \infty$, i.e. (see Section 1.4.7 ahead).

$$|\Phi - \Phi_p| \sim CN^{-\beta} \quad N \to \infty \,. \tag{1.4.18}$$

Analogous to (1.4.17), the unknown limiting value Φ of the functional can be extrapolated by solving

$$\boxed{\frac{\Phi_{\text{est}} - \Phi_p}{\Phi_{\text{est}} - \Phi_{p-1}} = \left(\frac{\Phi_{\text{est}} - \Phi_{p-1}}{\Phi_{\text{est}} - \Phi_{p-2}}\right)^{\alpha}}\,, \tag{1.4.19}$$

where α is in (1.4.17), for any three successive values of $\{\Phi_p\}$, to obtain the value Φ_{est}.

A **discussion** of this error estimation procedure follows: first, we remark that the sequence $\{\Phi_p\}_{p=1}^{p_{\max}}$ need not be generated by p-extension - it can likewise be obtained by uniform h-refinement at fixed polynomial degree p. Crucial for the accuracy and the reliability of the estimated value Φ_{est} for Φ is the validity of the asymptotic convergence estimate (1.4.18), i.e. **that the error $\Phi - \Phi_p$ has reached the asymptotic range**. In solid mechanics problems which often depend on parameters and involve constraints (such as incompressibility, shear- and membrane constraints), low order elements and h-extensions are known to enter the asymptotic range either very late or not at all for realistic values of h. This phenomenon is known as **locking-effect** in the widest sense and has lead to numerous "unlocking" recipes (such as reduced integration etc.) for low order elements. As a rule, p-extensions on a fixed mesh are much less susceptible to locking effects and usually reach the asymptotic range where (1.4.18) holds already for moderate numbers of DOF and in practical range of p (see also Section 3.4.3 below).

Remark 1.4.14. Using (1.4.17), (1.4.19) for the extrapolation of the unknown exact values of $\Pi(\underline{u})$, $\Phi(\underline{u})$ requires monotonically convergent sequences $\{\Phi_p\}$, $\{\Pi_p\}$. For the potential energies $\{\Pi_p\}$, in the absence of quadrature errors, this monotonicity is assured, if a primal variational formulation and conforming FEM are chosen: since $\underline{u}_S$ is the minimizer of $\Pi(\underline{u})$,

$$\begin{gathered} S_1 \subset S_2 \subset \cdots \subset S_p \subset E(\Omega) \quad \Longrightarrow \\ \Pi_1 > \Pi_2 > \cdots > \Pi_p > \Pi = \Pi(\underline{u}) \end{gathered} \tag{1.4.20}$$

and, since $\|\underline{u}_S\|^2_{E(\Omega)} = B(\underline{u}_S, \underline{u}_S) = -2\Pi(\underline{u}_S)$,

$$\|\underline{u}_{S_1}\|_{E(\Omega)} < \|\underline{u}_{S_2}\|_{E(\Omega)} < \cdots < \|\underline{u}_{S_p}\|_{E(\Omega)} \leq \ldots \|\underline{u}\|_{E(\Omega)} \,. \tag{1.4.21}$$

For functionals other than the total potential energy or for mixed methods, such monotonicity can generally not be expected. Nevertheless, once the convergence has entered the asymptotic regime (usually, for properly designed meshes, this happens for $p \geq 4$), many functionals converge monotonically.

Remark 1.4.15. The use of (1.4.17), (1.4.19) for extrapolation is based on the assumption (1.4.13) of algebraic convergence. In practice, however, if meshes are used that are geometrically refined towards the generator $\mathcal{G}$, exponential convergence can realistically be achieved, however (cf. Remark 1.4.13 and Figure 1.4).

In this case, the extrapolation based on (1.4.17), (1.4.19) can still be used. It will indicate then **increasing rates β in the estimate (1.4.13) for each three successive p-levels.**

Summary 1.4.16.

- Hierarchic p-extensions on a suitably chosen mesh provide a systematic approach to reliability assessment of the computed information and for reduction of discretization errors (errors C). A commercial implementation of this procedure is available in the code STRESSCHECK.
- If only a single FE-solution is available, and no additional computation is done, accurate and quantitative error control is not possible (the cheap residual error estimators often grossly overestimate the discretization error).
- Alternatively to p-extension, accurate energy error estimates can be obtained by the **equilibrated residual method** or by duality techniques. In both cases, in principle the true discretization error can be computed, also for functionals (see [53, 47]), by solving local auxiliary problems exactly. However, in practice this must be once more done numerically, thereby introducing additional errors. This also can be controlled a-priori, however [47]. Nevertheless, the approaches also require additional calculations which could even be as expensive as the actual FE-solution.

1.4.6 Error Equation

Based on the weak formulation of (1.1.1) - (1.1.4),

$$\underline{u} \in E_g(\Omega): \; B(\underline{u}, \underline{v}) = L(\underline{v}) \quad \forall \underline{v} \in E_0(\Omega) \tag{1.4.22}$$

and the FE-approximation

$$\underline{u}_S \in S: \; B(\underline{u}_S, \underline{v}) = L(\underline{v}) \quad \forall \underline{v} \in S \tag{1.4.23}$$

we get by subtraction the **error equation**[1]

$$\underline{e} \in E_0(\Omega): \; B(\underline{e}_S, \underline{v}) = R(\underline{v}) \quad \forall v \in S \tag{1.4.24}$$

where the **weak residual** $R(\underline{v})$ is the loading defined by

$$R(\underline{v}) := B(\underline{u}, \underline{v}) - B(\underline{u}_S, \underline{v}) \overset{(1.2.1)}{=} L(\underline{v}) - B(\underline{u}_S, \underline{v})\,. \tag{1.4.25}$$

[1] We assume that the FE-solution $\underline{u}_S$ satisfies the kinematic constraint (1.1.3) exactly.

By the Galerkin orthogonality (1.3.4),

$$R(\underline{w}) = 0 \quad \forall \underline{w} \in S\,. \tag{1.4.26}$$

For later purposes, it is convenient to write $\mathcal{R}(\underline{v})$ in an element-by-element fashion. To this end, let $\mathcal{T}$ denote any triangulation of Ω. Then, for any $v \in E_0(\Omega)$

$$R(\underline{v}) = \sum_{K\in\mathcal{T}} R_K(\underline{v}) = \sum_{K\in\mathcal{T}} (\underline{f},\underline{v})_K - (\boldsymbol{\epsilon}[\underline{v}],\ \boldsymbol{C}\,\boldsymbol{\epsilon}[\underline{u}_S])_K$$

where we assume also that $\underline{t} = \underline{0}$ in (1.1.4), for simplicity. Then integrating by parts elementwise gives

$$R(\underline{v}) = \sum_{K\in\mathcal{T}} (\underline{f} + \operatorname{div}\boldsymbol{\sigma}(\underline{u}_S), \underline{v})_K + \frac{1}{2}\left([\boldsymbol{\sigma}(\underline{u}_S)\,\underline{n}_K],\ \underline{v}\right)_{\partial K}$$

where $[(\boldsymbol{\sigma}(\underline{u}_S)\underline{n}_K]$ denotes the jump in the normal traction of the FE-solution across ∂K if $\partial K \subset \Omega$ and twice the normal traction if $\partial K \subset \partial\Omega$. By (1.4.26), we get finally for any $\underline{w} \in S$, $\underline{v} \in E_0(\Omega)$:

$$R(\underline{v}) = \sum_{K\in\mathcal{T}} (\underline{f} + \operatorname{div}\boldsymbol{\sigma}(\underline{u}_S),\ \underline{v} - \underline{w})_K + \frac{1}{2}\left([\boldsymbol{\sigma}(\underline{u}_S)\,\underline{n}_K],\ \underline{v} - \underline{w}\right)_{\partial K}. \tag{1.4.27}$$

1.4.7 Duality based error estimation for functional output

So far, we considered only error control in the energy norm. Often, however, specific **functionals $\boldsymbol{\Phi}(\underline{u})$ of the solution** are of interest such as resultants, reactions etc. We assume for simplicity $\underline{g} = \underline{0}$ in (1.1.3) and that $\Phi(\cdot)$ is bounded

$$|\Phi(\underline{u})| \le C_\Phi\, \|\underline{u}\|_{E(\Omega)}\,. \tag{1.4.28}$$

Typically, C_Φ is not known explicitly.

Denote by $\underline{z}^\Phi$ the **dual solution** to Φ, defined by

$$\underline{z}^\Phi \in E_g(\Omega) :\ B(\underline{w}, \underline{z}^\Phi) = \Phi(\underline{w}) \quad \forall \underline{w} \in E_0(\Omega)\,. \tag{1.4.29}$$

By (1.4.28) and (1.2.13), $\underline{z}^\Phi$ exists and since

$$\|\underline{z}^\Phi\|^2_{E(\Omega)} = B(\underline{z}^\Phi, \underline{z}^\Phi) = |\Phi(\underline{z}^\Phi)| \le C_\Phi\, \|\underline{z}^\Phi\|_{E(\Omega)}$$

we get the a-priori estimate

$$\|\underline{z}^\Phi\|_{E(\Omega)} \le C_\Phi\,. \tag{1.4.30}$$

Let $S \subset E_0(\Omega)$ be any FE-subspace and $\underline{u}_S$ the FE-solution. Then the linearity of Φ gives

$$|\Phi(\underline{u}) - \Phi(\underline{u}_S)| = |\Phi(\underline{u} - \underline{u}_S)| = |B(\underline{e}_S, \underline{z}^\Phi)| = |B(\underline{e}_S, \underline{z}^\Phi - \underline{v})| \tag{1.4.31}$$

for any $\underline{v}$ in S, by (1.3.4). This relation is fundamental, and several statements will be deduced from it.

i) Picking in (1.4.21) $\underline{v} = \underline{0}$, we get

$$|\Phi(\underline{u}) - \Phi(\underline{u}_S)| = |\mathcal{B}(\underline{e}_S, \underline{z}^\Phi)| \le C_\Phi \, \|\underline{e}_S\|_{E(\Omega)} \,. \tag{1.4.32}$$

If we have a hierarchic sequence (1.4.1) of FE-spaces, the corresponding functional approximations $\{\Phi(\underline{u}_{S_\ell})\}_{\ell=1}^\infty$ converge at least as fast as the FE-error in energy norm. This justifies using the extrapolation (1.4.10) for monitoring the discretization error C in functional output.

ii) In fact, we usually have **superconvergence** of $\Phi(\underline{u}_S)$: since $\underline{v}$ in (1.4.16) was arbitrary,

$$|\Phi(\underline{u}) - \Phi(\underline{u}_S)| \le \|\underline{e}_S\|_{E(\Omega)} \inf_{\underline{v}\in S} \|\underline{z}^\Phi - \underline{v}\|_{E(\Omega)} \,. \tag{1.4.33}$$

Depending on the approximation properties of S, (1.4.33) indicates that the error of $\Phi(\underline{u}_S)$ decreases faster than the error in energy norm along a hierarchic sequence (1.4.1) of subspaces.

iii) More precise bounds are obtained generally by getting more information on $\underline{z}^\Phi$. Assuming that $\underline{z}^\Phi$ is known, we get from (1.4.29) and (1.4.24), (1.4.27)

$$\begin{aligned} |\Phi(\underline{u}) - \Phi(\underline{u}_S)| &= |\Phi(\underline{e}_S)| = |\mathcal{B}(\underline{e}_S, \underline{z}^\Phi)| = |\mathcal{R}(\underline{z}^\Phi - \underline{w})| \\ &= \Big| \sum_K (\underline{f} + \operatorname{div} \boldsymbol{\sigma}(\underline{u}_S), \underline{z}^\Phi - \underline{w})_K \\ &\qquad + \frac{1}{2} ([\boldsymbol{\sigma}(\underline{u}_S)\underline{n}_K], \underline{z}^\Phi - \underline{w})_{\partial K} \Big| \,. \end{aligned} \tag{1.4.34}$$

Here $\underline{w} \in S$ is arbitrary.

Estimates can be obtained from (1.4.34) by the Schwarz inequality:

$$|\Phi(\underline{u}) - \Phi(\underline{u}_S)| \le \sum_{K\in\mathcal{T}} \omega_K \, \rho_K + \omega_{\partial K} \, \rho_{\partial K} \tag{1.4.35}$$

where the element residuals are given by

$$\rho_K := \|\underline{f} + \operatorname{div} \boldsymbol{\sigma}(\underline{u}_S)\|_{L^2(K)}, \;\; \rho_{\partial K} := \|[\boldsymbol{\sigma}(\underline{u}_S)\underline{n}_K]\|_{L^2(\partial K)} \,, \tag{1.4.36}$$

and where the weights $\omega_K, \omega_{\partial K}$ are

$$\omega_K := \|\underline{z}^\Phi - \underline{w}\|_{L^2(K)}, \;\; \omega_{\partial K} := \|\underline{z}^\Phi - \underline{w}\|_{L^2(\partial K)} \,. \tag{1.4.37}$$

Here the dual solution $\underline{z}^\Phi$ is yet unknown and $\underline{w} \in S$ is arbitrary.

Now there are two choices, essentially: the first is based on a-priori interpolation error estimates like Propositions 1.4.10, 1.4.11. Assuming $S = S^p(\Omega, \mathcal{T})^d$ with uniform p and shape-regular mesh $\mathcal{T}$, we get from (1.4.35) - (1.4.37) with $\underline{w} = \mathcal{I}_S \, \underline{z}^\Phi$ that

$$\omega_K \le C_I \, \frac{h_K}{p_K} \, \|\underline{z}^\Phi\|_{H^1(\Omega)}, \;\; \omega_{\partial K} \le C_I \, \frac{h_K^{\frac{1}{2}}}{p_K^{\frac{1}{2}}} \, \|\underline{z}^\Phi\|_{H^1(\Omega)} \,.$$

Inserting in (1.4.35) gives the a-priori error estimate

$$\begin{aligned}
|\Phi(\underline{u}) - \Phi(\underline{u}_S)| &\leq C_I \|\underline{z}^\Phi\|_{E(\Omega)} \sum_{K\in\mathcal{T}} \left(\frac{h_K}{p_K}\,\omega_K + \frac{h_K^{\frac{1}{2}}}{p_K^{\frac{1}{2}}}\,\omega_{\partial K}\right) \\
&\leq C_I\, C_E\, \|\underline{z}^\Phi\|_{E(\Omega)} \sum_{K\in\mathcal{T}} \left(h_K\, p_K^{-1}\,\omega_K + h_K^{\frac{1}{2}}\, p_K^{-\frac{1}{2}}\,\omega_{\partial K}\right) \qquad (1.4.38) \\
&\overset{(1.4.30)}{\leq} C_I\, C_E\, C_\Phi \sum_{K\in\mathcal{T}} \left(h_K\, p_K^{-1}\,\omega_K + h_K^{\frac{1}{2}}\, p_K^{-\frac{1}{2}}\,\omega_{\partial K}\right).
\end{aligned}$$

Here, the constants are generally not explicitly known. The interpolation constant C_I depends strongly on the shape-regularity of the mesh and C_E on the conditioning of the problem (below, in Section 3.4.3, this is the locking factor).

Anisotropic meshes like in Figure 1.9 are not admitted. In addition, for generalized FE-spaces S_ℓ like the ones we design below for composites, analogs of Propositions 1.4.10, 1.4.11 are not available.

Sharper bounds and less dependence on the element shape is obtained by solving the dual problem (1.4.29) approximately. To this end, let $S \subset S^+ \subset E_0(\Omega)$ where S^+ is any "higher" space in the hierarchy (1.4.1). Then solving the FE-problem

$$\underline{z}^\Phi_{S^+} \in S^+ : \ B(\underline{w}, \underline{z}^\Phi_{S^+}) = \Phi(\underline{w}) \quad \forall \underline{w} \in S^+ , \qquad (1.4.39)$$

and also for S in place of S^+, we get in (1.4.32)

$$\omega_K = \|\underline{z}^\Phi_{S^+} - \underline{z}^\Phi_S\|_{L^2(K)}, \quad \omega_{\partial K} = \frac{1}{2}\, \|\underline{z}^\Phi_{S^+} - \underline{z}^\Phi_S\|_{L^2(\partial K)} . \qquad (1.4.40)$$

Note that due to the symmetry of $B(\cdot,\cdot)$ and for $S = S_p = S^p(\Omega, \mathcal{T}, \underline{F})$, i.e. for hierarchic p-extensions, one may take e.g. $S = S^p$, $S^+ = S^{p+2}$.[1] Solving (1.4.39) then involves, due to the hierarchic structure of the stiffness matrix (cf. Figure 1.3), merely several backsolves (always assuming that hierarchic shapefunctions are used). Alternatively, one may use the arbitrariness of $\underline{w}$ in (1.4.37) and insert a local projection of $\underline{z}^\Phi_{S^+}$ onto S, e.g. by a Gauss-Lobatto nodal interpolant. Generally then one takes efficiency versus robustness. We conclude that **in order to use the dual problem (1.4.29) for the functional error estimation, we must have at our disposal hierarchies (1.4.1) of FE-spaces**, obtained by any of the extensions in Remark 1.4.1.

2 Geometric and Material Nonlinearities

We discuss formulations of nonlinear problems suitable for the p-version FEM as they are implemented in STRESSCHECK. The exposition follows [49].

2.1 Geometrically Nonlinear Analysis

We present the formulation in two dimensions for simplicity; everything will also work in 3-d. We consider an elastic body in its reference state Ω_0 and denote the material point $P \in \Omega_0$ by (X, Y) (see Figure 2.1).

[1] It is generally not advisable to use $S^+ = S^{p+1}$, since in problems with symmetry we may actually have that $\underline{u} \in S^\perp_{p+1}$, i.e. the FE-solutions $\underline{u}_{S^p}$ and $\underline{u}_{S^{p+1}}$ may be identical.

Now $\underline{u} = \underline{u}(x, y)$ denotes the displacement with respect to $(x, y) \in \Omega$, the coordinates in the deformed configuration. The differential arc length in the undeformed and the deformed configurations are:

$$dS^2 = dX^2 + dY^2 \ \text{(undeformed)}, \quad ds^2 = dx^2 + dy^2 \ \text{(deformed)}. \tag{2.1.1}$$

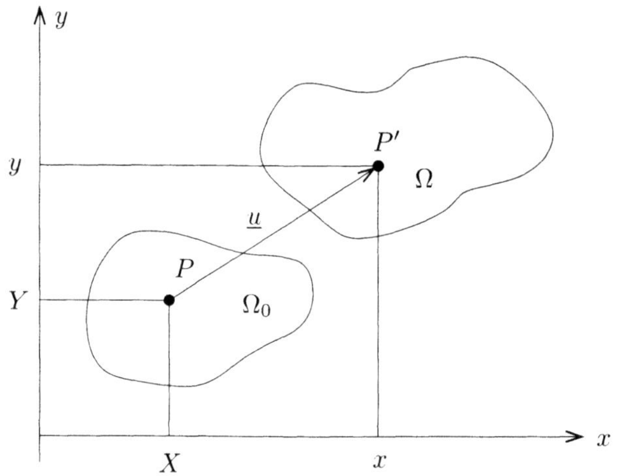

Figure 2.1. Kinematics for large deformation. Notation.

The deformation $\underline{\varphi} : \Omega_0 \to \Omega$ is assumed to be orientation preserving

$$\det|\mathbf{grad}\ \underline{\varphi}(P)| > 0 \qquad \forall P \in \Omega_0 . \tag{2.1.2}$$

The displacement vector $\underline{u}$ is defined by

$$\begin{pmatrix} x \\ y \end{pmatrix} = \underline{\varphi}\begin{pmatrix} X \\ Y \end{pmatrix} = \begin{pmatrix} X \\ Y \end{pmatrix} + \underline{u}(x, y), \tag{2.1.3}$$

see Figure 2.1. The **Almansi strain tensor** e measures the change in arc length induced by $\underline{\varphi}$ in terms of $(x, y) \in \Omega$:

$$ds^2 - dS^2 = 2\begin{pmatrix} dx \\ dy \end{pmatrix}^{\top}\begin{pmatrix} e_{xx} & e_{xy} \\ e_{yx} & e_{yy} \end{pmatrix}\begin{pmatrix} dx \\ dy \end{pmatrix}. \tag{2.1.4}$$

Turning to indicial notation, we find

$$ds^2 - dS^2 = (u_{i,j} + u_{j,i} - u_{k,i}\, u_{k,j})dx_i\, dx_j = 2e_{ij}\, dx_i dx_j \tag{2.1.5}$$

so that the Almansi strains are given by

$$e_{ij} = \frac{1}{2}\left(u_{i,j} + u_{j,i} - u_{k,i}\, u_{k,j}\right). \tag{2.1.6}$$

This expression is valid both in 2-d and 3-d. In dimension 2, the Almansi strains are

$$\begin{aligned} e_{11} &= \partial_1 u_1 - \frac{1}{2}\left((\partial_1 u_1)^2 + (\partial_1 u_2)^2\right) \\ e_{22} &= \partial_2 u_2 - \frac{1}{2}\left((\partial_2 u_1)^2 + (\partial_2 u_2)^2\right) \\ e_{12} &= e_{21} = \frac{1}{2}\left[(\partial_2 u_1 + \partial_1 u_2) - \partial_1 u_1\,\partial_2 u_1 - \partial_1 u_2\,\partial_2 u_2\right]. \end{aligned} \tag{2.1.7}$$

With the infinitesimal strain ϵ as in (1.1.5), we have

$$e = \epsilon + \text{higher order terms}. \tag{2.1.8}$$

The Almansi strain tensor is a strain measure in the deformed configuration Ω. Likewise, it is possible to express the strains also in the reference configuration Ω_0. This gives the **Green-Lagrange strain tensor E**: denoting by

$$\underline{U} = \underline{U}(X, Y) \tag{2.1.9}$$

the displacement in the reference coordinates (X, Y), we get in place of (2.1.4)

$$\begin{aligned} ds^2 - dS^2 &= dx_k dx_k - dX_k\, dX_k \\ &= (U_{i,j} + U_{j,i} + U_{k,j}\, U_{k,i})\, dX_i\, dX_j \end{aligned} \tag{2.1.10}$$

so that

$$E_{ij} := \frac{1}{2}\left(U_{i,j} + U_{j,i} + U_{k,i}\, U_{k,j}\right). \tag{2.1.11}$$

For the FE-formulation, we satisfy the principle of virtual work in the deformed configuration Ω: in indicial notation, the equilibrium equations are

$$\sigma_{ij,j} + f_i = 0 \text{ in } \Omega. \tag{2.1.12}$$

Multiplying by a test function v_i (a virtual displacement $\underline{v}$), integrating by parts in ω gives

$$\int_{\partial\Omega} (\sigma_{ij}\, n_j)\, v_i ds - \frac{1}{2}\int_{\Omega} \sigma_{ij}(v_{i,j} + v_{j,i}) dx + \int_{\Omega} f_i\, v_i\, dx = 0. \tag{2.1.13}$$

On $\partial\Omega$,

$$\sigma_{ij}\, n_j = t_i \tag{2.1.14}$$

are the surface loads (compare (1.1.4)), and $\underline{f}$ denotes the body force. Assuming the linear stress strain relationship (1.1.6), it must now be given in terms of the Almansi strains:

$$\sigma_{ij} := C_{ijkl}\, e_{k\ell} = \frac{1}{2}\, C_{ijk\ell}(u_{k,\ell} + u_{\ell,k} - u_{n,k}\, u_{n,\ell}). \tag{2.1.15}$$

Once more, we look for displacement fields of finite energy, the energy space $E(\Omega)$ is now such that $e_{ij} \in L^2(\Omega)$. The principle of virtual work reads then: Find $\underline{u} \in E(\Omega)$ such that

$$\int_{\Omega} \epsilon_{ij}[\underline{v}]\, C_{ijk\ell}\, e_{k\ell}(\underline{u}) dx = \int_{\Omega} f_i\, v_i\, dx + \int_{\partial\Omega} t_i\, v_i\, ds. \tag{2.1.16}$$

Notice that the left hand side of (2.1.16) is now nonlinear in $\underline{u}$, hence it is not a bilinear form any more and the solution must be obtained by iteration.

Algorithm 2.1.1.

i) Obtain a linear solution on the undeformed configuration Ω_0. Set $m = 0$ and denote the displacement by $\underline{u}^{(0)}$, and set $\Omega^{(-1)} := \Omega_0$.

ii) In iteration m, modify the domain according to

$$\Omega^{(m)} := \{\underline{x} \,|\, \underline{x} = \underline{\xi} + \underline{u}^{(m)}(\underline{\eta}),\ \ \underline{\xi} \in \Omega_0,\ \underline{\eta} \in \Omega^{(m-1)}\}\,. \qquad (2.1.17)$$

iii) Calculate $\underline{u}^{(m+1)}$ from the linear (!) problem

$$\int_{\Omega^{(i)}} \epsilon_{ij}[\underline{v}]\, C_{ijk\ell}\, \epsilon_{k\ell}[\underline{u}^{(m+1)}] dx = \int_{\Omega^{(i)}} f_i\, v_i\, dx + \int_{\partial\Omega^{(i)}} t_i\, v_i\, ds + \frac{1}{4} \int_{\Omega^{(i)}} C_{ijk\ell}\, u_{n,k}^{(m)}\, u_{n,\ell}^{(m)} (v_{i,j} + v_{j,i}) dx\,. \qquad (2.1.18)$$

iv) (Stopping criterion). Compute the relative difference between two successive displacement vectors: If it is less than a small tolerance τ, stop.

$$\|\underline{u}^{(m+1)} - \underline{u}^{(m)}\| \le \tau\, \|\underline{u}^{(m)}\|\,. \qquad (2.1.19)$$

Remark 2.1.2.

i) In each step (2.1.18) of the iteration, a new stiffness matrix must be evaluated.

ii) the geometry $\Omega^{(m)}$ is given in terms of the FE-solution from the previous step. To avoid numerical instabilities, choice of a FE-basis with good conditioning is essential.

2.2 Material Nonlinear Analysis

We address here the application of p and hp-FEM to elastic-plastic problems. A rather simple, nonlinear working model of plasticity, the deformation theory of plasticity, is used. Its main features are as follows:

- the von-Mises yield criterion is used
- the rather simple yield criterion provides estimates for the effects of nonlinearities at low computational cost
- memory requirements are considerably smaller than in the incremental theory
- no unloading events can in general be modelled

We briefly elaborate on the mathematical model. The main idea is to allow a dependence of the $C_{ijk\ell}$ in (2.1.18) on $\underline{u}$. To derive this dependence, consider first a typical uniaxial stress-strain curve as shown in Figure 2.2. The stress-strain law is given by

$$\epsilon = \epsilon^e + \epsilon^p = \sigma / E_s \tag{2.2.1}$$

where E_s is the **secant-modulus** and the indices "e" and "p" denote "elastic" and "plastic", respectively.

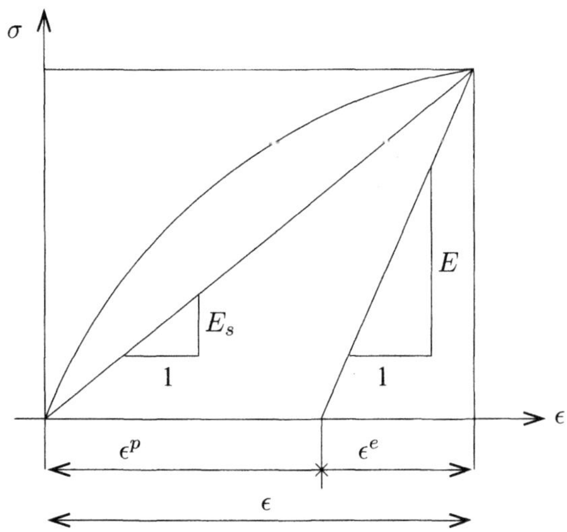

Figure 2.2. Uniaxial stress-strain curve

The elastic part of the strain is proportional to the stress, i.e (cf. Figure 2.2)

$$\epsilon^e = \sigma / E, \quad E > E_s \tag{2.2.2}$$

where E is the **Young modulus** of homogeneous, isotropic elasticity; from (2.2.1) and (2.2.2) it follows that

$$\epsilon^p = \left(\frac{1}{E_s} - \frac{1}{E} \right) \sigma \,. \tag{2.2.3}$$

This relation is now transferred to higher dimensions by assuming isotropy. We show this in two dimensions. The (infinitesimal) strain tensor is given by

$$\begin{aligned} \epsilon_{11} &= \partial_1 u_1 \,, \quad \epsilon_{22} = \partial_2 u_2 \,, \quad \epsilon_{33} = \partial_3 u_3 \,, \\ \epsilon_{ij} &= \frac{1}{2} \left(\partial_i u_j + \partial_j u_i \right), \quad \gamma_{ij} = 2\epsilon_{ij}, \ i \neq j \,. \end{aligned} \tag{2.2.4}$$

A crucial role will be played by the **deviatoric stress tensor**. Define

$$\hat{\sigma} := (\sigma_{11} + \sigma_{22} + \sigma_{33})/3 = \frac{1}{3} \operatorname{trace}(\boldsymbol{\sigma}) \,, \tag{2.2.5}$$

then the deviatoric stresses are given by

$$\widetilde{\boldsymbol{\sigma}} = \boldsymbol{\sigma} - \hat{\sigma}\,\mathbf{1}\,. \tag{2.2.6}$$

From the analogy to the uniaxial experiment,

$$\hat{\sigma} = \sigma/3 \tag{2.2.7}$$

it follows that

$$\widetilde{\sigma}_{ii} = \frac{2}{3}\,\sigma_{ii}\,. \tag{2.2.8}$$

Denoting the principal strains by ϵ_i, $i = 1, 2, 3$, the equivalent (to the uniaxial case) elastic and plastic strains are

$$\overline{\epsilon}^e = \frac{\sqrt{2}}{2(1+\nu)}\left((\epsilon_1^e - \epsilon_2^e)^2 + (\epsilon_2^e - \epsilon_3^e)^2 + (\epsilon_3^e - \epsilon_1^e)^2\right)^{\frac{1}{2}} \tag{2.2.9}$$

$$\overline{\epsilon}^p = \frac{\sqrt{2}}{3}\left((\epsilon_1^e - \epsilon_2^p)^2 + (\epsilon_2^p - \epsilon_3^p)^2 + (\epsilon_3^p - \epsilon_1^p)^2\right)^{\frac{1}{2}} \tag{2.2.10}$$

and we have, analogous to (2.2.1), the additive decomposition

$$\overline{\epsilon} = \overline{\epsilon}^e + \overline{\epsilon}^p\,. \tag{2.2.11}$$

Denoting the principal stress components by σ_i, the equivalent stress is given by

$$\overline{\sigma} = \frac{\sqrt{2}}{2}\left((\sigma_1 - \sigma_2)^2 + (\sigma_2 - \sigma_3)^2 + (\sigma_3 - \sigma_1)^2\right)^{\frac{1}{2}}. \tag{2.2.12}$$

the 3-d constitutive equations are then derived by the **assumptions** that

plastic strains are proportional to the deviatoric stress.

This yields in 2-d:

$$\boldsymbol{e}^p = \frac{2}{3}\left(\frac{1}{E_s} - \frac{1}{E}\right)\widetilde{\boldsymbol{\sigma}} \tag{2.2.13}$$

or, in components,

$$\begin{pmatrix} \epsilon_{11} \\ \epsilon_{12} \\ \gamma_{12} \end{pmatrix} = \underbrace{\left\{ \frac{1}{E}\begin{pmatrix} 1 & -\nu & 0 \\ -\nu & 1 & 0 \\ 0 & 0 & 2(1+\nu) \end{pmatrix} + \frac{E - E_s}{EE_s}\begin{pmatrix} 1 & -\frac{1}{2} & 0 \\ -\frac{1}{2} & 1 & 0 \\ 0 & 0 & 3 \end{pmatrix} \right\}}_{\text{elastic plastic compliance matrix}} \begin{pmatrix} \sigma_{11} \\ \sigma_{22} \\ \tau_{12} \end{pmatrix}. \tag{2.2.14}$$

For a displacement formulation, the elastic-plastic compliance matrix must be inverted.

We emphasize that (2.2.14) is nonlinear: E_s is obtained from the 1-d stress strain curve in Figure 2.2., once the equivalent elastic strain has been computed.

Algorithm 2.2.1.

i) Obtain a linearly elastic solution,

ii) Compute the equivalent elastic strain $\overline{\epsilon}^e$ by (2.2.9) at each integration point and set

$$\overline{\epsilon}^{(1)} := (\overline{\epsilon}^e)^{(1)} .$$

iii) For $k = 1, 2, 3, \ldots$ compute the secant modulus $E_s^{(k)}$ at each integration point from the uniaxial stress-strain curve.

iv) In each integration point where $\overline{\epsilon} > \epsilon_Y$, use the elastic-plastic matrix. Recompute the element stiffness matrices in all elements for which $\overline{\epsilon} > \epsilon_Y$ in at least one Gauss-point and solve for the new FE solution $\underline{u}_{\mathrm{FE}}^{(k-1)}$.

v) Given $E_s^{(k)}$ and $\underline{u}_{\mathrm{FE}}^{(k+1)}$, compute the stress tensor $\boldsymbol{\sigma}^{(k+1)}$ at each integration point, using the elastic-plastic compliance matrix and $\boldsymbol{\epsilon}^{(k+1)}$. Compute the elastic strains by subtracting the elastic strains from the total strains.

vi) Compute the new equivalent strains

$$(\overline{\epsilon}^e)^{(k+1)}, \ (\overline{\epsilon}^p)^{(k+1)}, \ (\overline{\epsilon})^{(k+1)} .$$

vii) Stopping criterion: If for every Gauss-point

$$|\overline{\epsilon}^{(k+1)} - \overline{\epsilon}^{(k)}| < \mathrm{tol}\, |\overline{\epsilon}^{(k+1)}| , \tag{2.2.15}$$

then stop, else compute $E_s^{(k+1)}$ using $\overline{\epsilon}^{(k+1)}$, increment $k \to k+1$ and return to iii).

Remark 2.2.2. A **typical elastic plastic analysis** in STRESSCHECK proceeds as follows:

- Obtain a converged sequence of linear FE-solutions by p-extension, verify smallness of discretization by extrapolation (1.4.17).
- Perform the elastic-plastic iteration 2.2.1.

Typical uniaxial stress-strain curves are given in the following:

- **elastic-perfectly plastic:** characterized by 3 parameters: E, ν and the yield stress σ_Y.
- **bilinear** (see Figure 2.3): linear-elastic, strain-hardening material

$$\epsilon = \frac{\sigma}{E} + \frac{3}{7}\frac{\sigma_{70E}}{E}\left(\frac{\sigma}{\sigma_{70E}}\right)^n . \tag{2.2.16}$$

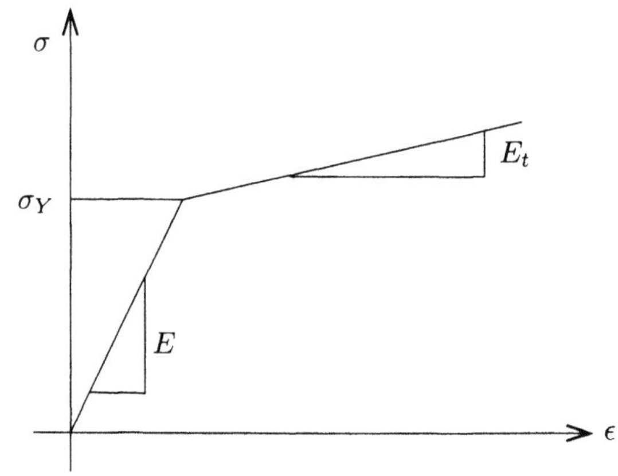

Figure 2.3. bilinear stress-strain curve

- **Ramberg-Osgood** (see Figure 2.4) $(E, \nu, \sigma_{70E}, n)$

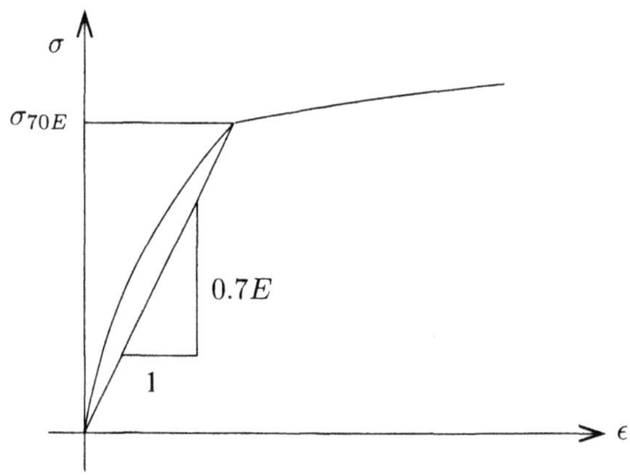

Figure 2.4. Ramberg-Osgood stress-strain curve

Here σ_{70E} is the stress corresponding to the intersection of the stress-strain curve with a line through the origin of slope $0.7E$.

- **5 parameter**: $(E, E_t, \nu, \sigma_Y$ and the starting point (ϵ_2, σ_2) of the linear strain-hardening curve).

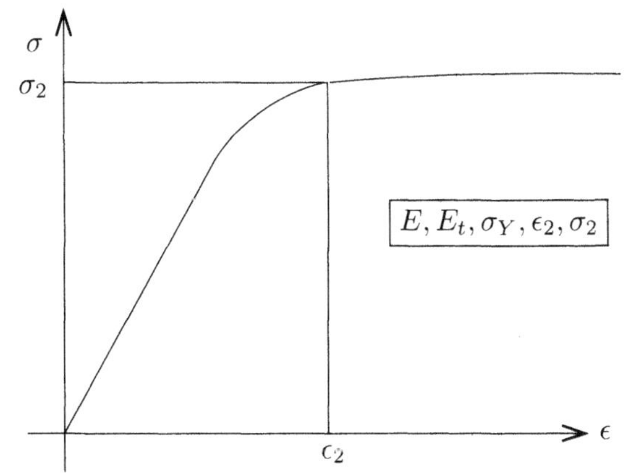

Figure 2.5. 5 parameter, uniaxial stress-strain curve

3 Thin Solids, in particular Plates and Shells

Plates and shells are three-dimensional solids Ω for which one dimension (the thickness) is much smaller than the others. **Plate and shell models** are dimensionally reduced working models, i.e. partial differential equations that are obtained from (1.1.1) - (1.1.4) by constraining the admissible displacements $\underline{u}$ and, possibly, by dropping certain terms in the $3-d$ strain energy, leading to working models (cf. Figure 0.1) that must be solved on the (two-dimensional) midsurface ω of Ω.

3.1 3-d Plate Problem

If $\omega \subset \mathbb{R}^2$ is the midsurface of the plate, the corresponding 3-d solid of thickness t is

$$\Omega_t = \omega \times \left(-\frac{t}{2}, \frac{t}{2} \right) = \{(x,y,z) : \ (x,y) \in \omega, \ |z| < t/2\} . \tag{3.1.1}$$

The 3-d plate problem is (1.1.1) - (1.1.4) posed on Ω_t. Laminated and graded materials are admitted; then $C_{ijk\ell}$ in (1.1.6) may depend discontinuously on the transverse coordinate z. More precisely, we assume that (1.1.6) takes the form

$$\boldsymbol{\sigma}[\underline{u}] = (\sigma_{11}, \sigma_{22}, \sigma_{33}, \sigma_{12}, \sigma_{13}, \sigma_{23})^\top, \ \ \boldsymbol{\epsilon}[\underline{u}] = (\epsilon_{11}, \epsilon_{12}, \epsilon_{33}, \epsilon_{12}, \epsilon_{13}, \sigma_{23})^\top \tag{3.1.2}$$

and

$$\boldsymbol{\sigma}[\underline{u}] = \begin{pmatrix} Q_{11} & Q_{12} & Q_{13} & Q_{14} & 0 & 0 \\ Q_{12} & Q_{22} & Q_{23} & Q_{24} & 0 & 0 \\ Q_{13} & Q_{23} & Q_{33} & Q_{34} & 0 & 0 \\ Q_{14} & Q_{24} & Q_{34} & Q_{44} & 0 & 0 \\ 0 & 0 & 0 & 0 & Q_{55} & Q_{56} \\ 0 & 0 & 0 & 0 & Q_{65} & Q_{66} \end{pmatrix} \boldsymbol{\epsilon}[\underline{u}] . \tag{3.1.3}$$

Here, the Q_{ij} are functions of the scaled thickness variable, i.e.

$$Q_{ij} = Q_{ij}(2z/t), \quad |z| < t/2 , \tag{3.1.4}$$

but are assumed to be independent of t. For homogeneous plates, the Q_{ij} are constants. For laminated plates, the $Q_{ij}(\zeta)$ are piecewise constant functions such that the matrix $\boldsymbol{Q}(\zeta)$ in (3.1.3) is symmetric, positive definite, uniformly for $|\zeta| \leq 1$. The Hooke's law (3.1.3) describes monoclinic materials; in the homogeneous isotropic case we have

$$\begin{aligned} Q_{11} &= Q_{22} = Q_{33} = \lambda + 2\mu , \\ Q_{12} &= Q_{13} = Q_{23} = Q_{21} = Q_{31} = Q_{32} = \lambda , \\ Q_{44} &= Q_{55} = Q_{66} = 2\mu , \\ Q_{14} &= Q_{24} = Q_{34} = Q_{56} = 0 . \end{aligned} \tag{3.1.5}$$

It is useful to separate **bending** and **membrane effects**. To this end, we write

$$\underline{u} = \underline{u}^I + \underline{u}^{II} \tag{3.1.6}$$

where the membrane part of the displacement is given by

$$u^I_\alpha(x,y,z) = u^I_\alpha(x,y,-z), \ \alpha = 1,2, \ u^I_3(x,y,z) = -u^I_3(x,y,-z) ,$$

and the bending part is

$$u^{II}_\alpha(x,y,z) = -u^{II}_\alpha(x,y,-z), \ \alpha = 1,2, \ u^{II}_3(x,y,z) = u^{II}_3(x,y,-z) .$$

Analogously, the energy space $E(\Omega_t)$ can be decomposed

$$E(\Omega_t) = E^I(\Omega_t) \oplus E^{II}(\Omega_t) \tag{3.1.7}$$

and this splitting is orthogonal in energy, i.e.

$$B(\underline{u},\underline{v}) = 0 \quad \forall \underline{u} \in E^I(\Omega_t), \quad \forall \underline{v} \in E^{II}(\Omega_t) . \tag{3.1.8}$$

(The constitutive law (3.1.3) is the most general one for which (3.1.8) holds; if, e.g. $Q_{15} \neq 0$, bending and membrane effects can not be decoupled by (3.1.6) anymore). Analogously, the loadings $L(\underline{v})$ in (1.2.2) can be decomposed:

$$L(\underline{v}) = L^I(\underline{v}) + L^{II}(\underline{v}) \tag{3.1.9}$$

such that $L^I(\underline{v}) = 0$ for all $\underline{v} \in E^{II}(\Omega_t)$ and vice versa. We also remark that such decoupling is in general not possible anymore for curved shells at positive thickness.

3.2 Kirchhoff and Reissner Mindlin Plates

The Kirchhoff-Love (KL) plate model is obtained by assuming an approximate 3-d displacement field of the form

$$\underline{u} = \underline{u}^{\mathrm{KL}} = \begin{pmatrix} u_0(x,z) - \partial_x w_0(x,y)\, z \\ v_0(x,y) - \partial_y w_0(x,y)\, z \\ w_0(x,y) \end{pmatrix}. \tag{3.2.1}$$

The kinematics in this case imply that

$$\epsilon_{33} = \sigma_{33} = 0, \quad \epsilon_{13} = \epsilon_{23} = 0\,. \tag{3.2.2}$$

The normal deflection $w_0(x,y)$ satisfies the biharmonic equation

$$D\Delta\Delta w_0 = t^{-3} g \tag{3.2.3}$$

where the **flexural rigidity** is given by

$$D = \frac{E}{12(1-\nu^2)} = \frac{8\mu(\lambda+\mu)}{3(\lambda+2\mu)}\,. \tag{3.2.4}$$

The in-plane displacements u_0, v_0 in (3.2.1) satisfy the so-called membrane equation, see [13] for more and also for a justification of (3.2.3) as scaled, asymptotic limit of $\underline{u}^{II}$ as $t \to 0^+$.

Kinematically, the assumption (3.2.1) means that fibers that are normal to the midsurface remain so while the plate undergoes bending, see Figure 3.1.

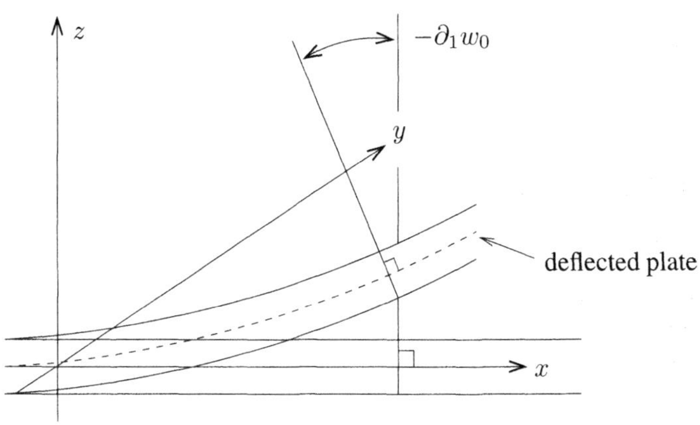

Figure 3.1. Illustration of the KL assumption (3.2.1)

In the Reissner-Mindlin (RM) plate model, the normality is relaxed by introducing additional unknowns u_1, v_1 as follows:

$$\underline{u}^{\mathrm{RM}}(x,y,z) = \begin{pmatrix} u_0(x,y) + u_1(x,y)\, z \\ v_0(x,y) + v_1(x,y)\, z \\ w_0(x,y) \end{pmatrix}. \tag{3.2.5}$$

Now the bending part of $\underline{u}^{\rm RM}$ satisfies a singularly perturbed elliptic system in ω:

$$\begin{aligned} -t^3 \operatorname{div} \boldsymbol{C}\, \boldsymbol{\epsilon}\,[u_1, v_1] - \lambda\kappa t\Big(\underline{\nabla} w_0 - \begin{pmatrix} u_1 \\ v_1 \end{pmatrix}\Big) &= \underline{f} \quad \text{in} \quad \omega\,, \\ -\lambda t \operatorname{div}\Big(\underline{\nabla} w_0 - \begin{pmatrix} u_1 \\ v_1 \end{pmatrix}\Big) &= g \quad \text{in} \quad \omega \end{aligned} \tag{3.2.6}$$

where $\epsilon_{\alpha\beta}[\varphi_1, \varphi_2] = \frac{1}{2}\,(\partial_\alpha \varphi_\beta + \partial_\beta \varphi_\alpha)$, $\alpha, \beta \in \{1, 2\}$, and where

$$\boldsymbol{C}\, \boldsymbol{\epsilon} := D[(1-\nu)\, \boldsymbol{\epsilon} + \nu \operatorname{trace}(\boldsymbol{\epsilon})\, \mathbf{1}]\,.$$

The (dimensionless) parameter κ in (3.2.6) is the so-called **shear correction factor**: some choices are $\kappa = \frac{5}{6}$ (Reissner), $\kappa = \frac{\pi^2}{12}$ (Mindlin), $\kappa = \frac{20(\lambda+\mu)}{17\lambda+24\mu} = \frac{10}{12-7\nu}$.

Remark 3.2.1.

- The system (3.2.6) can **not** be obtained by energy projection, i.e. by constraining the admissible displacements in the 3-d variational principle (1.2.1) to be of the form (3.2.5). The resulting Euler-Lagrange equations have the form (3.2.6), however with plane-strain in the constitutive equations.
- (3.2.6) has plane-stress in the constitutive equations.
- The RM and KL models are adequate for predicting **structural response** (deflections, shear forces, bending moments, etc.).
- The system (3.2.6) is, contrary to (3.2.3), singularly perturbed; the rotation variables (u_1, v_1) contain for small t boundary layers (see (1.4.9) and [4]). Their length scale is $\varepsilon = \sqrt{12\kappa}/t$.

3.3 Hierarchic Plate Models

3.3.1 Homogeneous plates

For homogeneous plates, the Q_{ij} in (3.1.3) are constant. Hierarchic plate models are obtained by **semidiscretization** of the 3-d elasticity in the transverse direction z of $\Omega_t = \omega \times (-\frac{t}{2}, \frac{t}{2})$. This means to constrain the admissible displacements $\underline{u}$ in the 3-d variational principle to be of the form [1, 2, 5, 41, 45, 51]

$$E^{\underline{n}}(\Omega_t) := \Big\{ \underline{u}(x, y, z) : \; u_i = \sum_{j=0}^{n_i} X_{ij}(x, y)\, L_j(2z/t),\; X_{ij} \in H^1(\omega) \Big\}\,. \tag{3.3.1}$$

Here $L_j(\zeta)$ are the Legendre polynomials and $\underline{n} = (n_1, n_2, n_3)$ are the polynomial degrees of semidiscrete approximations. If $n_1 = n_2 = n_3 = q$, we write also $E^q(\Omega_t)$ in (3.3.1). **Hierarchic plate models** are obtained as follows:

- select a sequence $\{\underline{n}(\ell)\}_{\ell=0}^{\infty}$ of increasing polynomial degrees,

- obtain approximate solutions $\underline{u}^{\underline{n}}$ to the problem of 3-d elasticity by minimizing the potential energy $\Pi(\underline{u})$ in (1.2.20) over $E^{\underline{n}}(\Omega_t)$:

$$\underline{u}^{\underline{n}} = \arg \min_{\underline{v} \in E^{\underline{n}}(\Omega_t)} \Pi(\underline{v}) . \tag{3.3.2}$$

By (1.2.21), the minimizer $\underline{u}^{\underline{n}} \in E^{\underline{n}}(\Omega_t)$ is obtained, once the unknown coefficient functions $X_{ij}(x, y) \in H^1(\omega)$ in (3.3.1) are determined from a two-dimensional, elliptic system in ω. We summarize the key points about this **dimension reduction procedure**.

Remark 3.3.1.

- The system in ω to compute the coefficients $X_{ij}(x, y)$, the so-called $\underline{n}$-model, is strongly elliptic, but in general singularly perturbed,
- if this system is solved by the p-version FEM, applied in ω, the resulting fully discrete problem corresponds to discretizing directly the 3-d problem by anisotropic brick elements with degree p in (x, y) and degrees $\underline{n}$, resp. q, in the transverse direction (see Figure 3.2 and Figure 1.2).

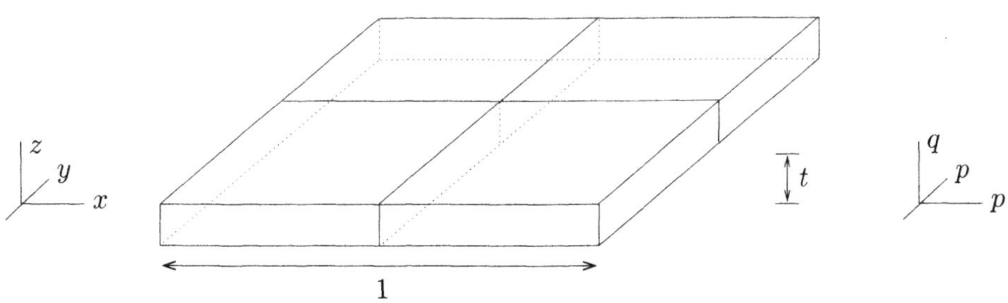

Figure 3.2. (p, q)-thin solid brick elements in $\omega \in (0, 1)^2$

- Let $n_1 = n_2 = n_3 = q = 0, 1, 2, \ldots$ and denote the FE-solutions by $\underline{u}^{p,q}$. Then, as $p \to \infty$, we have $\underline{u}^{p,q} \xrightarrow[E(\Omega_t)]{} \underline{u}^q$, i.e. hierarchic q-models (3.3.2) are **closures** in energy of (p, q) product FE-discretizations of the thin solids.
- Since the polynomials of degree q are dense in $L^2(-1, 1)$, the q-models converge, as $q \to \infty$, to the 3-d solution. Precisely: for any $t > 0$, we have **approximability**:

$$\lim_{q \to \infty} \frac{\|\underline{u}^q - \underline{u}_{3\mathrm{D}}\|_{E(\Omega_t)}}{\|\underline{u}_{3\mathrm{D}}\|_{E(\Omega_t)}} = 0 . \tag{3.3.3}$$

- More important for the performance of the hierarchic models is **asymptotic consistency**

$$\lim_{t\downarrow 0^+} \frac{\|\underline{u}^q - \underline{u}_{3\mathrm{D}}\|_{E(\Omega_t)}}{\|\underline{u}_{3\mathrm{D}}\|_{E(\Omega_t)}} = 0\,. \tag{3.3.4}$$

Proposition 3.3.2. *For $q = 0, 1$, the hierarchic models obtained from* (3.3.2) *are not asymptotically consistent; for $q \geq 2$ they are.*

Remark 3.3.3. For $q = 1$, $E^1(\Omega_t)$ has the kinematics (3.2.5) of the RM plate. However, (3.2.6) does **not** follow by (3.3.2) with the kinematics (3.2.5), (exercise!).

More important than the basic properties (3.3.3), (3.3.4) are the **convergence rates** as $t \downarrow 0^+$ resp. as $q \to \infty$ of $\underline{u}^q$. The inclusion of higher order terms (3.3.1) into the kinematics alone would give high order convergence as $t \downarrow 0^+$, **if the 3D-solution $\underline{u}_{3\mathrm{D}}$ were smooth uniformly in t.**

This is so in model cases ($\omega = (-\pi, \pi)^2$, periodic) but **generically the order of asymptotic consistency as $t \downarrow 0^+$ in** (3.3.4) **is limited to $O(t^{1/2})$, even for smooth domains ω and loadings**. The reason (and remedy) for this becomes apparent from recent asymptotic results on $\underline{u}_{3\mathrm{D}}$ [14]: $\underline{u}_{3\mathrm{D}}$ consists of two parts: the so-called **outer expansion** which is well-approximated by (3.3.1) and **boundary layers** (absent in the periodic setting) which have three-dimensional character and which are concentrated near the edge of the plate, $\partial\omega$, or near the singular support of the loadings. Thus, in order to generate plate models which are **asymptotically consistent of higher order**, i.e. for which

$$\frac{\|\underline{u}^q - \underline{u}_{3\mathrm{D}}\|_{E(\Omega_t)}}{\|\underline{u}_{3\mathrm{D}}\|_{E(\Omega_t)}} \leq C\, t^{\gamma(q)} \quad \text{as} \quad t \to 0^+ \tag{3.3.5}$$

holds with exponents $\gamma(q) > 1$, the **boundary layers of the 3-d solution $\underline{u}_{3\mathrm{D}}$ must be resolved.**

3.3.2 Boundary layer resolution [15, 40]

We describe the structure of the boundary layers. To this end, we introduce in a vicinity of the edge $\partial\omega$ boundary-fitted coordinates (r, s, z):

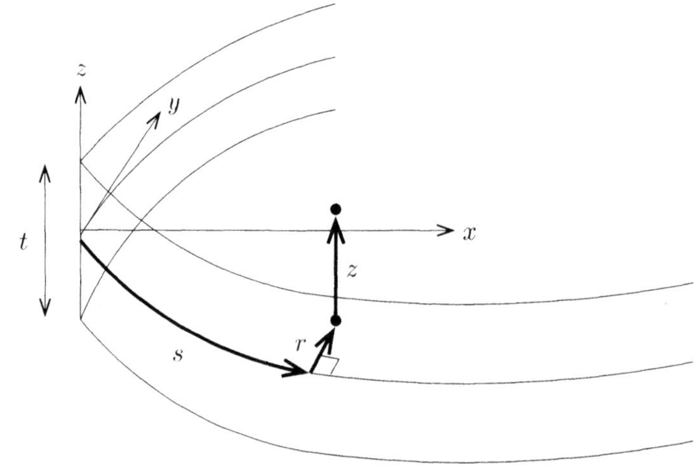

Figure 3.3. (r, s, z)-coordinates in a vicinity of $\partial\omega$.

In terms of the (r, s, z)-coordinates, the 3-d boundary layer has the form [14]

$$\underline{u}_{3D}^{BL}(s, r, z) \sim \sum_k t^k \, \underline{U}_k^{BL}\left(s, \frac{r}{t}, \frac{2z}{t}\right). \tag{3.3.6}$$

The *boundary layer profile* $\underline{U}_k^{BL}(s, \rho, \zeta)$ is independent of t, and depends analytically on s (if the data are analytic). For each s, $\underline{U}_k^{BL}(s, \rho, \zeta)$ solves a two-dimensional elasticity problem on the strip $\Pi^+ = \mathbb{R}_+ \times (-1, 1)$, with zero normal tractions, right hand sides. Up to some rigid body motions, $\underline{U}_k^{BL}(s, \rho, \zeta)$ decays exponentially in ρ:

$$|\underline{U}_k^{BL}(s, \rho, \zeta)| \le C e^{-b\rho}, \; \rho > 0, \; |\zeta| \le 1, \tag{3.3.7}$$

and $\underline{U}_k^{BL}$ has edge singularities near $(s, 0, \pm 1)$. Comparing (3.3.6), (3.3.7) we see that [15]

- boundary layer resolution in plates requires that the dimensional reduction kinematics in (3.3.1) is abandoned near $\partial\omega$,
- in an $O(t)$ neighborhood of $\partial\omega$, geometric and anisotropic mesh refinement towards the edges $\partial\omega \times \{\pm t/2\}$ has to be used (see Figure 3.4).
- If $n \ge 1$ denotes the number of mesh layers in Figure 3.4, for robust (in t) exponential converge, necessarily $n \ge |\log t|$. This is a weak requirement that cannot be dispensed with ([15, 40]).

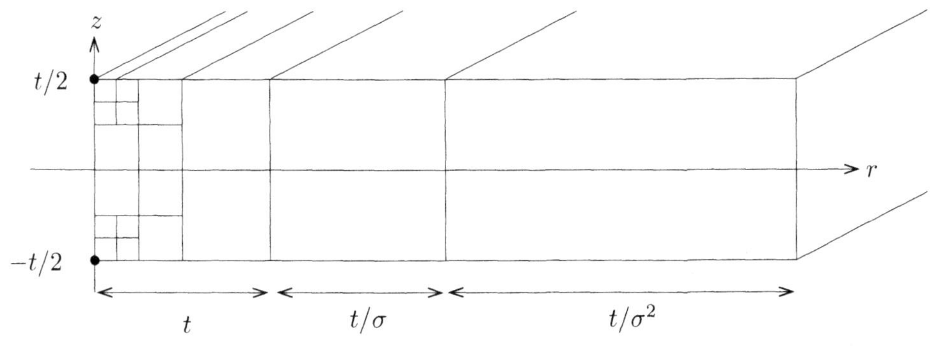

Figure 3.4. Boundary layer resolution by anisotropic, geometric mesh refinement towards $\partial\omega \times \{\pm t/2\}$. Cross-section through the plate thickness

We note that outside the $O(t)$ boundary layer zone, geometric mesh-refinement with grading factor $\sigma < 1$ has to be used in order to resolve the boundary layers of the $X_{ij}(x, z)$ in (3.3.1) (see also [31, 44]).

3.3.3 A-posteriori modeling error estimation

For homogeneous plates where (3.3.1) is used throughout ω, the modeling error $\|\underline{u}^q - \underline{u}_{3D}\|_{E(\Omega_t)}$, can be estimated in terms of the residual tractions on the faces of the plate alone (see [42] for more).

Theorem 3.3.4. *Assume that $n_1 = n_2 = n_3 = q$ in (3.3.1) and that the dimensionally reduced problem for the $X_{ij}(x, y)$ is solved exactly. Let the material constituting the plate be homogeneous and monoclinic, i.e. (3.1.3) holds. Then the modeling error $\underline{e}_q = \underline{e}_q^I + e_q^{II}$ can be decomposed into its bending and membrane parts as in (3.1.6) and the following a-posteriori error estimates hold:*

$$\|\underline{e}_q\|^2_{E(\Omega_t)} = \|\underline{e}_q^I\|_{E(\Omega_t)} + \|\underline{e}_q\|^2_{E(\Omega_t)} \tag{3.3.8}$$

and for $q \geq 1$ we have

$$\|\underline{e}_q^I\|^2_{E(\Omega_t)} \leq t \left\{ \frac{1}{2[\frac{q}{2}] + \frac{3}{2}} \left(\frac{\|r_1^I\|^2_{L^2(\omega)}}{Q_{55}} + \frac{\|r_2^I\|_{L^2(\omega)}}{Q_{66}} \right) + \frac{1}{2[\frac{q+1}{2}] + \frac{1}{2}} \frac{\|r_3^I]\|^2_{L^2(\omega)}}{2Q_{33}} \right\}$$

and, for $q \geq 2$,

$$\|\underline{e}_q^{II}\|^2_{E(\Omega_t)} \leq t \left\{ \frac{1}{2[\frac{q+1}{2}] + \frac{1}{2}} \left(\frac{\|r_1^{II}\|^2_{L^2(\omega)}}{Q_{55}} + \frac{\|r_2^{II}\|_{L^2(\omega)}}{Q_{66}} \right) + \frac{1}{2[\frac{q}{2}] + \frac{3}{2}} \frac{\|r_3^{II}]\|^2_{L^2(\omega)}}{2Q_{33}} \right\}.$$

Here $\underline{r}^j = \boldsymbol{\sigma}\left[\underline{u}^j\right] \underline{e}_3 \big|_{z=t/2}$, $j \in \{I, II\}$.

In [42], it was also shown that the estimates (3.3.8) are asymptotically exact, as $t \to 0^+$ and as $q \to \infty$.

3.3.4 Generalized hp-FEM for laminated plates

In laminated or sandwich materials, all coefficients Q_{ij} in (3.1.4) are piecewise constant and nonzero. It follows from the equilibrium equations (1.1.1) in Ω_t that, for smooth body forces $\underline{f}$, the stresses $\boldsymbol{\sigma}$ are smooth with respect to z (inside ω) whereas the strains $\boldsymbol{\epsilon}$ are discontinuous, in general, with respect to z. Correspondingly, through-thickness profiles of $\underline{u}$ are only piecewise smooth with respect to z, and generally have kinks at the interlamina boundaries where the Q_{ij} have jumps. It follows that even in the "interior" of ω (i.e. outside the boundary layer), the kinematics (3.3.1) with polynomial dependence in z is not asymptotically optimal in the sense that (3.3.5) is achievable. In a displacement-based approach to hierarchic modeling, the question arises **which displacement profiles $\underline{\psi}(\zeta)$ should be chosen** in the kinematics

$$E^{\underline{n}}(\Omega_t) = \left\{ \underline{u}(x, y, z) : u_i = \sum_{j=0}^{n_i} X_{ij}(x, y)\, \psi_{ij}(2z/t),\ X_{ij} \in H^1(\omega) \right\}. \tag{3.3.9}$$

In terms of mechanics, the issue is **the optimal selection of the director functions**. In terms of numerics, the issue is to find optimal (in the sense of (3.3.3), (3.3.4)) laminate-adapted, nonpolynomial subspaces in (3.3.9), which form once more a problem tailored-hierarchy.

The approach to finding these director functions is based on a **Fourier-analysis of the 3-d equilibrium equations (1.1.1)** which is explained in the following (we emphasize that a completely analogous procedure is also successful in homogenization problems, see [26, 27, 28]. Bearing in mind that in the boundary layers (3.3.6) the product ansatz (3.3.9) has to be abandoned anyway (see Section 3.3.2 and [40]), we consider the **infinite plate**

$$\Omega_t^\infty = \mathbb{R}^2 \times (-t/2, t/2)\,. \tag{3.3.10}$$

Since the Q_{ij} in (3.1.4) are independent of (x, y), the differential operator $Lu = -\operatorname{div}\boldsymbol{\sigma}[\underline{u}]$ in (1.1.1) is translation invariant in (x, y) and the solution $\underline{u}_{3D}^\infty(x, y, z)$ can be represented by superposition of Fourier-modes. Assuming that the volume forces are zero, and the surface tractions $\underline{t}^\pm$ on $z = \pm t/2$ are of the form

$$\underline{t}^\pm \exp(i(\xi_1 x + \xi_2 y)),\ \ \underline{t}^+ \in \mathbb{R}^3\,, \tag{3.3.11}$$

we find that

$$\underline{u}_{3D}^\infty(x, y, z) = \exp(i\xi_1 x_1 + \xi_2 x_2)\{\boldsymbol{G}^+(t\underline{\xi}, 2z/t)\,\underline{t}^+ + \boldsymbol{G}^-(t\underline{\xi}, 2z/t)\,\underline{t}^-\} \tag{3.3.12}$$

where the kernels $\boldsymbol{G}^\pm(\underline{\eta}, \zeta)$ do not depend on t and solve (in the weak sense) the two-parameter family of two point boundary-value problems

$$e^{-i(x\eta_1 + y\eta_2)} \operatorname{div}\boldsymbol{\sigma}\left[\boldsymbol{G}^\pm(\underline{\eta}, \zeta)\, e^{i(x\eta_1 + y\eta_2)}\right] = \boldsymbol{0} \ \text{ in } \ (-1, 1) \tag{3.3.13}$$

with the boundary conditions

$$\pm e^{-i(x\eta_1 + y\eta_2)} \operatorname{div}\underline{e}_z \cdot \boldsymbol{\sigma}\left[\boldsymbol{G}^+(\underline{\eta}, \zeta)\, e^{i(x\eta_1 + y\eta_2)}\right]\Big|_{\zeta = \pm 1} = \begin{cases} \boldsymbol{1} & \zeta = 1 \\ \boldsymbol{0} & \zeta = -1\,, \end{cases} \tag{3.3.14}$$

and vice versa for $\boldsymbol{G}^-$. The kernels $\boldsymbol{G}^\pm$ for homogeneous and isotropic material were first explicitly computed by Ufl'yand and Lur'e. For the general, laminated materials (3.1.4) with $Q_{ij}(\zeta) \in L^\infty(-1, 1)$, closed form solution of (3.3.13), (3.3.14) appears not to be possible. (3.3.13), (3.3.14) can be considered the "**unit-cell problem for laminated plates**". Optimal, non-polynomial director functions can be obtained from it as follows: from (3.3.12), we see that **asymptotic optimality as $t \downarrow 0^+$ is closely related to the behavior of $\boldsymbol{G}^\pm(\underline{\eta}, \zeta)$ at $\underline{\eta} = \underline{0}$**. Shape functions $\psi_{ij}(\zeta)$ in (3.3.9) can be obtained by sampling the kernels $\boldsymbol{G}^\pm(\eta, \zeta)$ near $\eta = 0$. Actis in [1] proceeds as follows: select $\eta_1 = \beta\eta_2$ and calculate asymptotic expansions of $\boldsymbol{G}^\pm(\beta\eta_2, \eta_2; \zeta)$ about $\eta_2 = 0$. The unit cell problem (3.3.13), (3.3.14) is then solved by symbolic manipulation and yields with $\beta = 1$ the following $\psi_{ij}(\zeta)$ in (3.3.9) (see also [2]):

$$\psi_{10}(\zeta) = \psi_{20}(\zeta) = \psi_{30}(\zeta) = 1,\ \psi_{11}(\zeta) = \psi_{21}(\zeta) = \zeta\,, \tag{3.3.15a}$$

i.e. the RM-kinematics (3.2.5). The next six transverse shapefunctions in (3.3.9) are:

$$\begin{aligned} &\psi_{12}(\zeta) = \frac{1}{2}\,(\varphi_2(\zeta) - \varphi_2(0)),\ \psi_{22}(\zeta) = \frac{1}{2}\,(\psi_2(\zeta) - \psi_2(0)), \\ &\psi_{31}(\zeta) = \frac{1}{2}\,(\rho_1(\zeta) - \rho_1(0)),\ \psi_{13}(\zeta) = \frac{1}{4}\,\varphi_3(\zeta), \\ &\psi_{23}(\zeta) = \frac{1}{4}\,\varphi_3(\zeta),\ \psi_{32}(\zeta) = \frac{1}{4}\,\rho_2(\zeta) \end{aligned} \tag{3.3.15b}$$

where we define, for $|w| = 1$,

$$
\begin{aligned}
\varphi_2(w) &= \int_{-1}^{w} \frac{Q_{44} - Q_{45}}{Q_{44}Q_{55} - Q_{45}^2}\, d\zeta, \quad \psi_2(w) = \int_{-1}^{w} \frac{Q_{55} - Q_{45}}{Q_{44}Q_{55} - Q_{45}^2}\, d\zeta, \quad \rho_1(w) = \int_{-1}^{w} \frac{1}{Q_{33}}\, d\zeta \\
\varphi_3(w) &= \int_{0}^{w} \left(\rho_1(w) + \frac{Q_{44}\gamma_a(w)}{Q_{44}Q_{55} - Q_{45}^2} - \frac{Q_{45}\gamma_b(w)}{Q_{44}Q_{55} - Q_{45}^2} \right) d\zeta \\
\psi_3(w) &= \int_{-1}^{w} \left(\rho_1(w) + \frac{Q_{55}\gamma_b(w)}{Q_{44}Q_{55} - Q_{45}^2} - \frac{Q_{45}\gamma_a(w)}{Q_{44}Q_{55} - Q_{45}^2} \right) d\zeta \\
\rho_2(w) &= \int_{0}^{w} \left(\frac{2w}{Q_{33}} + \frac{(Q_{13} + Q_{36})\,\varphi_2(w)}{Q_{33}} - \frac{(Q_{23} + Q_{36})\,\psi_2(w)}{Q_{33}} \right) d\zeta \\
\gamma_a(w) &= \int_{0}^{w} \frac{Q_{13} + Q_{36}}{Q_{33}}\, d\zeta, \quad \gamma_b(w) = \int_{0}^{w} \frac{Q_{23} + Q_{36}}{Q_{33}}\, d\zeta\,.
\end{aligned}
$$

One readily verifies that for homogeneous materials, i.e. constant Q_{ij}, the ψ_{ij} are polynomials in ζ, i.e. the kinematics (3.3.9) reduce to (3.3.1) then. This justifies the notion of generalized hp-FEM for (3.3.9)-solution adapted shapefunctions are used in place of polynomials. The choice $\beta = 1$ in $\eta_1 = \beta\eta_2$ may seem arbitrary at first, but it was checked in [1] that the span of $\psi_I(\zeta)$ is indeed very weakly depending on the choice of β.

In the case of **vibrating plates**, some eigenfunctions must, in addition, be included into the kinematics (3.3.9), in order to prevent error pollution phenomena at high wave-number. The number of eigenfunctions depends on the dimensionless wave-number kt (see [18] for vibrating membranes).

3.4 Hierarchic Shell Models

We briefly discuss the extension of the hierarchic modeling to shells. Everything that was said for plates holds also here - new effects arise, however, due to

- bending and membrane coupling due to curvature effects and
- the dimensionally reduced problems have boundary layer length scales $O(t)$ due to shear deformation, as in plates, **and** possibly $O(t^{\frac{1}{n}})$, $n \in \{2, 3, 4\}$.

Robust a-posteriori error estimation with performance independent of t for FE-discretizations of general shells is, at present, open. Nevertheless, by hierarchic p-extensions on properly designed meshes and by extrapolation error estimates, locking can be detected and the discretization error be controlled a-posteriori.

3.4.1 Dimensional reduction for shells

The midsurface $\mathcal{S}$ of the shell is an orientable surface piece in $\mathbb{R}^3$ that is an image of a parameter domain $\omega \subset \mathbb{R}^2$: $\mathcal{S} = \underline{\varphi}(\omega)$, $\underline{\varphi} : \mathbb{R}^2 \to \mathbb{R}^3$. We assume that $\underline{\varphi}$ is piecewise smooth nondegenerate and invertible in $\overline{\omega}$. By (r, s) we denote coordinates in $\mathcal{S}$ which are induced by $\underline{\varphi}$ (see Figure 3.5). Since $\underline{\varphi}$ is nondegenerate, the unit tangent vectors $\underline{e}_r$, $\underline{e}_s$ to $\mathcal{S}$ are linearly independent. We assume that $\mathcal{S}$ is orientable and denote by $\underline{e}_w := \underline{e}_r \times \underline{e}_s$ the unit normal to $\mathcal{S}$.

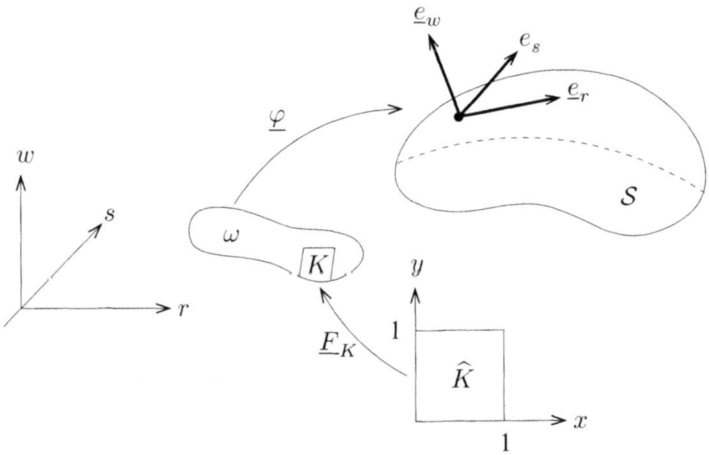

Figure 3.5. Curvilinear coordinates in the shell and the notation

Remark 3.4.1. The shell mapping $\underline{\varphi}$ is in practice not explicitly available. Often it is hidden in CAD systems and FE programs can only interrogate the CAD system for function values, and possibly derivatives, of $\underline{\varphi}$. However, the assumption of piecewise smoothness of $\underline{\varphi}$ is satisfied by the internal representations of $\underline{\varphi}$ by B-splines or as NURBS-surfaces. See [2] and the references there for more.

In dimensional reduction methods, the displacements $\underline{u}$ of the shell are written as products, analogous to (3.3.1): the shell kinematics are approximated by

$$E^{\underline{n}}(\Omega_t) = \left\{ \underline{u}(r,s,w) | u_i = \sum_{j=0}^{n_i} X_{ij}(r,s)\, \psi_{ij}(2w/t),\ i \in \{r,s,w\} \right\}. \tag{3.4.1}$$

Note that the semidiscretization is performed in the lamina basis, i.e. $\underline{u} = u_r\, \underline{e}_r + u_s\, \underline{e}_s + u_w\, \underline{e}_w$, and **not** in the Cartesian one. For homogeneous shell, the Cartesian displacements could also be used, but for multilayer sandwich shells, use of the lamina basis is essential. This advantage comes at the expense of having the rotation matrix $\boldsymbol{R}$

$$\{\underline{u}\}_{(xyz)} = \boldsymbol{R}\, \{\underline{u}\}_{(rsw)} \tag{3.4.2}$$

in the formulation. The bilinear form (1.2.3) can then be written the in curvilinear coordinates (r, s, w) (see Figure 3.5) as follows:

$$B(\underline{u}, \underline{v}) = \int_{\mathcal{S}} \int_{-t/2}^{t/2} (\boldsymbol{D}\,\boldsymbol{R}\,\underline{v}_{(rsw)})^{\top}\, \boldsymbol{Q}\Big(\frac{\partial w}{t}\Big)(\boldsymbol{D}\,\boldsymbol{R}\,\underline{u}_{(rsw)})\, dw dS \qquad (3.4.3)$$

$$L(\underline{v}) = \int_{\mathcal{S}} (\boldsymbol{R}\,\underline{v}_{(rsw)})^{\top}\, \underline{t}\, \big|_{w=\pm t/2} dS \qquad (3.4.4)$$

where dS is the surface element of $\mathcal{S}$ and $\underline{t}$ the surface tractions acting on the shell Ω_t. Here

$$\boldsymbol{D} = \begin{pmatrix} \partial_x & 0 & 0 & 0 & \partial_z & \partial_y \\ 0 & \partial_y & 0 & \partial_z & 0 & \partial_x \\ 0 & 0 & \partial_z & \partial_y & \partial_x & 0 \end{pmatrix}^{\top} . \qquad (3.4.5)$$

The full FE-discretization of the problem requires once more the use of FEM in the mid-surface $\mathcal{S}$. This is done by using p-FEM in the parametric domain ω (see Figure 3.5). We shall not discuss the shell models which arise as $p \to \infty$ and $\underline{n}$ is finite in (3.4.1).

For more on the FE-formulation, see [2]. In particular, the shape functions ψ_{ij} in (3.4.1) are as in (3.3.15).

3.4.2 Mesh design for p-FE shell analysis

Even more important than in the case of plates, for good performance of hierarchic p-extensions in shell problems, it is essential that layers in shells are resolved (see [38, 22, 19] and the references there for more). We explain this briefly:

- In shell models with shear (Naghdi-type shells) as well as in hierarchical shell models, there is always an $O(t)$ length scale near $\partial\mathcal{S}$, analogous to what we observed above in plates [4]. Therefore, in a $O(t)$ vicinity of $\partial\mathcal{S}$ the kinematics (3.4.1) should be abandoned and a mesh as in Figure 3.4 adopted, if the shell is homogeneous. In laminated (sandwich) plates and shells, additional geometric refinements towards the interfaces has to be performed for accurate stress calculation near the edge (see Figure 3.6). See [15, 40] for analytical results.

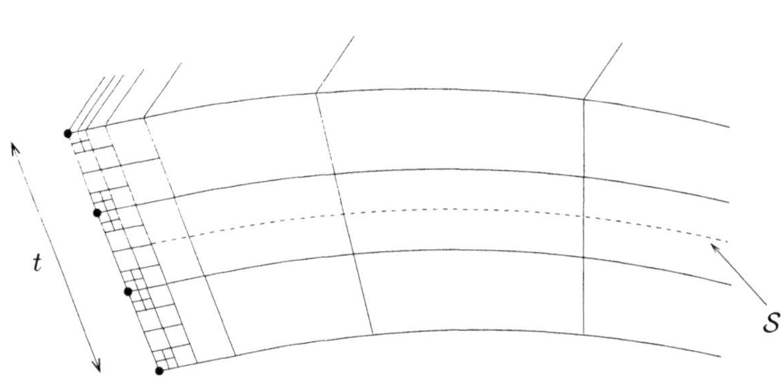

Figure 3.6. Boundary layer resolution for laminated shells by anistropic hp-refinement towards the edges. Cut through shell shown.

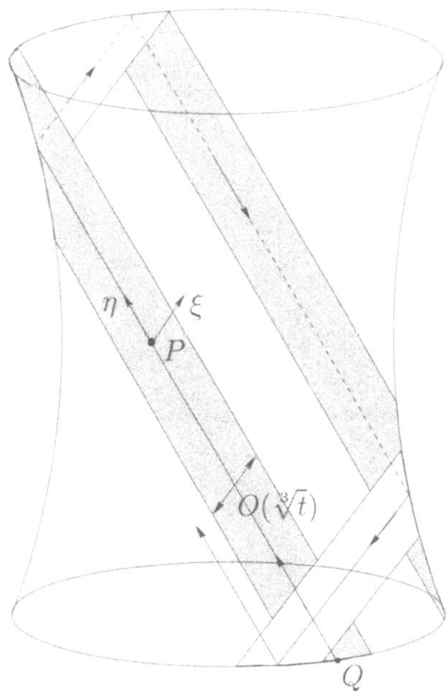

Figure 3.7. $O(\sqrt[3]{t})$ layer in hyperbolic shell (cooling tower). $O(\sqrt{t})$-layers at top and bottom edge are not displayed. Propagation of point singularities at $Q \in \partial\mathcal{G}$ along characteristic line of principal curvature.

- In addition, there is a bending-membrane layer ("simple edge effect") of scale $O(\sqrt{t})$, present in all shell models with bending energy. Its generator $\mathcal{G}$ (see Terminology 1.4.5 and Example 1.4.6) is all of ∂S and points of non-analyticity of the data.

- Finally, there are for parabolic and hyperbolic shells further degenerate length scale of $O(\sqrt[3]{t})$ and $O(\sqrt[4]{t})$. The latter penetrate deeply into the shell and carry a nonnegligible part of the strain energy as $t \to 0$. They are, however, usually resolved without special mesh refinement by hp-FEM (see Figures 3.7 and 3.8).

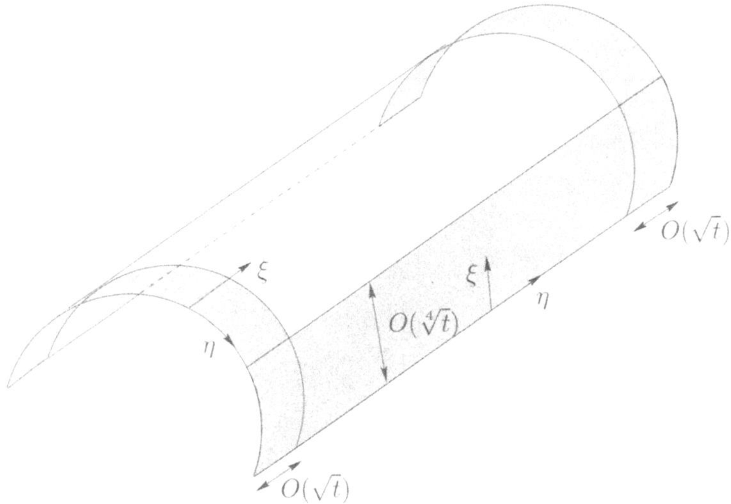

Figure 3.8. $O(\sqrt[4]{t})$ layer in parabolic shell (barrel vault)

3.4.3 Locking and high order FEM

It was asserted previously that high order FEM are advantageous for the solution of plate and shell problems, since the various locking effects in low order discretizations tend to disappear with high order without resorting to special devices (such as reduced integration / mixed formulations / strain-enhancement etc.). We will elaborate on this assertion in more details. As stated in Section 1.3, the FEM is optimal in energy: denoting by $\underline{u}^t$ the exact solution of the shell model and by $\underline{u}^t_S$ its FE-approximation, we have

$$\|\underline{u}^t - \underline{u}^t_S\|_{E(\omega)} \leq \|\underline{u}^t - \underline{v}\|_{E(\omega)} \quad \forall \underline{v} \in S\,. \tag{3.4.6}$$

However, now the norm depends strongly on t and may, e.g. in membrane-dominated cases, become very weak as $t \to 0^+$. Since, for $t > 0$, the energy norm is equivalent to the $H^1(\omega)$-norm, we may by appropriate scaling assume that $\|\underline{u}^t\|_{E(\omega)} \leq \|\underline{u}^t\|_{H^1(\omega)}$,

$$\|\underline{u}^t - \underline{u}^t_S\|_{E(\omega)} \leq \|\underline{u}^t - \underline{v}\|_{H^1(\omega)} \quad \forall \underline{v} \in S \tag{3.4.7}$$

holds. Note, however, that $\|\underline{u}^t\|_{H^1(\omega)} \leq \mathcal{L}(t)\|\underline{u}^t\|_{E(\omega)}$ with $\mathcal{L}(t) \uparrow \infty$ as $t \downarrow 0$ (in bending dominated situations as fast as t^{-1}). Following [37, 38, 19], we have for the relative error (independent of scalings) in energy for every $\underline{v} \in S \subset E(\omega)$

$$\frac{\|\underline{u}^t - \underline{u}^t_N\|_{E(\omega)}}{\|\underline{u}^t\|_{E(\omega)}} \leq \frac{\|\underline{u}^t\|_{H^1(\omega)}}{\|\underline{u}^t\|_{E(\omega)}} \; \frac{\|\underline{u}^t - \underline{v}\|_{E(\omega)}}{\|\underline{u}^t\|_{H^1(\omega)}} \leq \mathcal{L}(t) \; \frac{\|\underline{u}^t - \underline{v}\|_{H^1(\omega)}}{\|\underline{u}^t\|_{H^1(\omega)}}\,. \tag{3.4.8}$$

The constant $\mathcal{L}(t)$ in (3.4.8) measures the t-sensitivity of the problem and may be termed **locking-factor**. We claim that in plate and shell elements, high order FEM are less susceptible of locking than low order elements. This is due to the following, heuristic argument of Juhani Pitkäranta.

To see this, let $S = S^p(\omega, \mathcal{T}_h)$ with constant p and quasiuniform mesh $\mathcal{T}_h$ in the plate/shell midsurface. We assume that the exact solution $\underline{u}^t$ is relatively smooth in ω, $\underline{u}^t \in H^k(\omega)$ for some $k > 1$, and varies on a **length scale** L with

$$0 < L \leq \mathrm{diam}(\omega) = O(1)\,, \tag{3.4.9}$$

e.g. $\underline{u}^t \sim \exp(-\rho/\varepsilon) \Longrightarrow L \sim \varepsilon$. Then, for any $\ell \geq 1$, higher derivatives of the solution $\underline{u}^t$ scale in terms of the length scale L as follows:

$$\|\underline{u}^t\|_{H^{\ell+1}(\omega)} \leq C_\ell\, L^{-\ell} \;\; \|\underline{u}^t\|_{H^1(\omega)}\,. \tag{3.4.10}$$

From Proposition 1.4.10 and (3.4.8) we get then

$$\frac{\|\underline{u}^t - \underline{u}^t_S\|_{E(\omega)}}{\|\underline{u}^t\|_{E(\omega)}} \leq \mathcal{L}(t) \left(\frac{h}{p}\right)^{\min(p,k)} L^{-k} \leq \frac{C}{t} \left(\frac{h}{p}\right)^{\min(p,k)} L^{-k}\,. \tag{3.4.11}$$

Let us assume first that $\underline{u}^t$ is smooth, i.e. that $L = 1$. Then (3.4.11) gives the bound

$$\frac{\|\underline{u}^t - \underline{u}^t_S\|_{E(\omega)}}{\|\underline{u}^t\|_{E(\omega)}} \leq C\, \min\{1, (h/p)^{\min(p,k)} t^{-1}\}\,. \tag{3.4.12}$$

Here, we also used (3.4.8) with $\underline{v} \equiv \underline{0}$.

Let us now consider the convergence behaviour of a low order FEM as the mesh is refined. We consider $p = 1$ and $h \to 0$ in (3.4.12). Then we obtain a qualitative behaviour as shown in Figure 3.8, provided the solution $\underline{u}^t$ is smooth and belongs to $H^2(\omega)$.

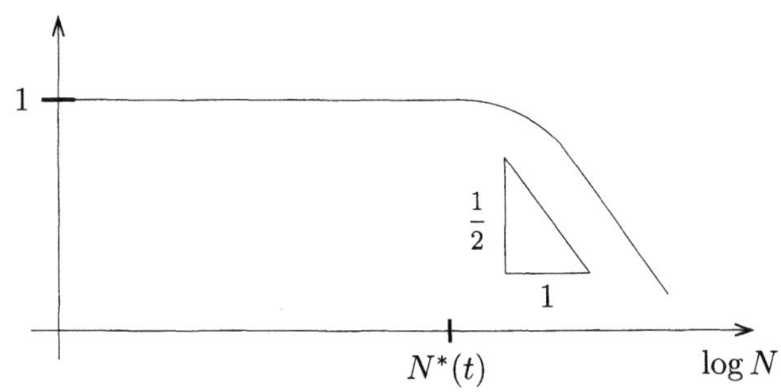

Figure 3.9. Schematic log-log graph of asymptotic convergence according to the upper bound (3.4.12) with $p = 1$, $t \ll 1$ as $h \to 0$. Relative error in energy norm or number N of DOF.

We note that for very small t, there exists $N^*(t)$ such that convergence is stalled until the number of DOF is so large that the consistency error matches the locking factor t^{-1} in the bound.

Note that in particular, the FEM still converges asymptotically - but this convergence occurs for small values of h possibly outside the computational range. Since the interpolation error decreases only as $O(h) = O(N^{-\frac{1}{2}})$, asymptotic convergence occurs when $(N^*(t))^{-\frac{1}{2}} \sim t$ or when

$$N \geq N^*(t) \sim t^{-2} . \tag{3.4.13}$$

For higher order elements and at least for smooth solution (in the sense that (3.4.10) holds with $L = 1$), locking effects are reduced since then (3.4.12) give the requirement

$$h^p t^{-1} \overset{!}{<} 1 \Longrightarrow h < t^{\frac{1}{p}} \Longrightarrow N^*(t) \sim t^{-\frac{2}{p}} . \tag{3.4.14}$$

An analogous argument using the exponential convergence rates (1.4.14) even gives only $N^*(t) \sim |\log t|$ (see [19] for cylindrical shells).

4 Eigenvalue Buckling Analysis

4.1 General Considerations

We describe a working model for buckling (a very similar formulation can also be used for pre-stress model analysis to investigate the impact of pre-stressing on the stiffness of the solid, see [50] for details). In conventional FE Analysis, buckling analyses are generally performed for thin structures, such as beams, plates and shells. If there are described by dimensionally reduced models without shear deformation (such as Bernoulli beam, Kirchhoff plate and the Koiter shell), the buckling load is given as smallest eigenvalue of a generalized eigenvalue problem of the type

$$\mathcal{B}u = \lambda \mathcal{M}u . \tag{4.1.1}$$

The function u is the buckling mode, λ the buckling load factor and $\mathcal{B}$ a fourth order elliptic operator on a closed subset $H \subset H^2(\omega)$ and $\mathcal{M}$ a second order differential operator continuous from $H^2(\omega) \to L^2(\omega)$. In weak form,

$$u \in H,\ \lambda \in \mathbb{C} : B(u,v) = \lambda G(u,v) \quad \forall v \in H . \tag{4.1.2}$$

In these cases, the bilinear form $B(\cdot,\cdot)$ is continuous on H and $G(\cdot,\cdot)$ is compact; (4.1.2) has thus a discrete spectrum and standard theory ensures convergence of FE-approximations (u_N, λ_N) to eigenpairs (u, λ) of (4.1.1).

In the hierarchic models in Section 3, the domain is not necessarily thin anymore, and thus a buckling formulation for thick domains cannot be based on the absence of shear deformation. It must rather accommodate fully three-dimensional kinematics. Then, again an eigenvalue problem of the form (4.1.1), (4.1.2) results. However, now $\mathcal{B}$ and $\mathcal{M}$ are both second order operators and (4.1.1), (4.1.2) have, in general, a partly continuous spectrum for which eigenvectors (i.e. buckling states) may or may not exist. However, as long as the domain Ω is slender, the parts of the spectrum that are critical for buckling, appear to be point spectra. We present in the following the buckling formulation in [50] which goes back to Trefftz [52].

4.2 Assumptions and Derivation of the Buckling Model

We assume that the three-dimensional body occupying the reference configuration $\Omega \subset \mathbb{R}^3$ is subjected to an initial stress field $\boldsymbol{\sigma}^0$. The boundary $\Gamma = \partial\Omega$ is partitioned into Γ_D, Γ_N and Γ_S where, as before, on Γ_D essential (displacement) boundary conditions are prescribed and on $\Gamma_N \cup \Gamma_S$, the normal stress acts on Ω either through the tractions (on Γ_N) or via a distributed spring.

It is assumed that $\boldsymbol{\sigma}^0$ is a pre-existing stress state, independent of $\underline{u}$ and of $\boldsymbol{C}$. In particular, $\boldsymbol{\sigma}^0$ satisfies the equilibrium equations of linearized elasticity, i.e. equilibrium

$$-\operatorname{div} \boldsymbol{\sigma}^0 = \underline{F}^0 \ \text{ in } \ \Omega\,, \tag{4.2.1}$$

and the traction boundary conditions

$$\boldsymbol{\sigma}^0\, \underline{n} = \underline{T}^0 \ \text{ on } \ \Gamma_N \cup \Gamma_S\,. \tag{4.2.2}$$

In particular, therefore, $\boldsymbol{\sigma}^0$ satisfies the variational equations

$$(\boldsymbol{\sigma}^0, \boldsymbol{\epsilon}\,[\underline{v}])_\Omega = (\underline{F}^0, \underline{v})_\Omega + (\underline{T}^0, \underline{v})_{\Gamma_N \cup \Gamma_S} \quad \forall \underline{v} \in H^1(\Omega)^3\,. \tag{4.2.3}$$

We assume now that the reference configuration Ω is deformed by a small change $\overline{\underline{F}}$ in the body force, or a change $\overline{\underline{u}}^*$ in the boundary displacements on Γ_D, or by a change $\overline{\underline{T}}$ in the surface tractions on Γ_N, which are causing (small) kinematically admissible displacement $\overline{\underline{u}}$ of Ω. Therefore

- the analysis assumes that Ω is in a small neighborhood of the elastic equilibrium of $\boldsymbol{\sigma}^0$,
- the perturbations are "dead" loads, i.e. they are independent of $\overline{\underline{u}}$.

The stress-strain relationship of the deformed body is of the form

$$\boldsymbol{\sigma} = \boldsymbol{\sigma}^0 + \boldsymbol{C}\,\overline{\boldsymbol{\epsilon}}\,[\underline{u}]\,. \tag{4.2.4}$$

Consider the principle of virtual work in Ω. The deformations $\overline{\underline{u}}$ give rise to the **Green-Lagrange strains** (since the derivation proceeds in reference configuration Ω)

$$E_{ij} = \epsilon_{ij}\,[\overline{\underline{u}}] + \frac{1}{2}\,\partial_i\,\overline{u}_\alpha\,\partial_j\,\overline{u}_\alpha\,. \tag{4.2.5}$$

In terms of the strain energy, $\overline{\underline{u}}$ should minimize the potential energy

$$\Pi(\overline{\underline{u}}) = \mathcal{U}(\overline{\underline{u}}) - (\overline{\underline{F}}, \overline{\underline{u}})_\Omega - (\overline{\underline{T}}, \overline{\underline{u}})_{\Gamma_N} - (\boldsymbol{\kappa}\,\overline{\underline{\delta}}, \overline{\underline{u}})_{\Gamma_S}\,, \tag{4.2.6}$$

where the strain energy $\mathcal{U}(\overline{\underline{u}})$ is, using (4.2.3), given by

$$\mathcal{U}(\overline{\underline{u}}) = \frac{1}{2}\,(\overline{\epsilon}_{ij},\, C_{ijk\ell}\,\overline{\epsilon}_{k\ell})_\Omega + \frac{1}{2}\,(\kappa_{ij}\,\overline{u}_i, \overline{u}_j)_{\Gamma_S} + \frac{1}{2}\,(\sigma^0_{ij}, \partial_i\,\overline{u}_\alpha\,\partial_j\,\overline{u}_\alpha)_\Omega \tag{4.2.7}$$

(summation over repeated indices). The last integral in (4.2.7) stems from $\boldsymbol{\sigma}\colon \boldsymbol{E}$ with $\boldsymbol{\sigma}$ in (4.2.4) and $\boldsymbol{E}$ as in (4.2.5) (note that $\boldsymbol{\sigma}\colon \overline{\boldsymbol{\epsilon}}$ cancels due to (4.2.3)). Making $\Pi(\underline{\overline{u}})$ stationary yields the principle of virtual work:

$$\begin{aligned}&\text{Find } \underline{\overline{u}} \in E(\Omega) \text{ such that}\\ &(\boldsymbol{\epsilon}[\underline{v}],\ \boldsymbol{C}\,\boldsymbol{\epsilon}[\underline{\overline{u}}])_\Omega + (\boldsymbol{\kappa}\,\underline{\overline{u}}, \underline{v})_{\Gamma_S} + (\sigma^0_{ij}\,\partial_j\,\overline{u}_\alpha, \partial_i\, v_\alpha)_\Omega\\ &= (\underline{\overline{F}}, \underline{v})_\Omega + (\underline{\overline{T}}, \underline{v})_{\Gamma_N} + (\boldsymbol{\kappa}\,\underline{\overline{\delta}}, \underline{v})_{\Gamma_S} \quad \forall \underline{v} \in E_0(\Omega)\,.\end{aligned} \tag{4.2.8}$$

This linear problem has a unique solution if and only if the left hand side bilinear form satisfies an inf-sup condition. To obtain a buckling criterion from (4.2.8), we assume that

$$\boldsymbol{\sigma}^0 = \lambda\,\boldsymbol{\sigma}^* \tag{4.2.9}$$

where $\boldsymbol{\sigma}^*$ is the **pre-buckling stress state** and λ the **buckling load factor**. Inserting (4.2.9) into (4.2.8), we obtain the variational form of the buckling problem:

$$\begin{aligned} B_\lambda(\underline{\overline{u}}, \underline{v}) &:= B(\underline{\overline{u}}, \underline{v}) - G_\lambda(\underline{\overline{u}}, \underline{v}), \quad \text{where}\\ B(\underline{\overline{u}}, \underline{v}) &:= (\boldsymbol{\epsilon}[\underline{v}],\ \boldsymbol{C}\,\boldsymbol{\epsilon}[\underline{\overline{u}}])_\Omega + (\boldsymbol{\kappa}\,\underline{\overline{u}}, \underline{v})_{\Gamma_S}\,,\\ G_\lambda(\underline{\overline{u}}, \underline{v}) &:= -\lambda\,(\sigma^*_{ij}\,\partial_j\,\overline{u}_\alpha, \partial_i\, v_\alpha)_\Omega\,.\end{aligned} \tag{4.2.10}$$

(4.2.8) is well-posed if and only if there are $c_1(\lambda), c_2(\lambda) > 0$ with

- $B_\lambda(\cdot,\cdot)$ is continuous, i.e. if $$|B_\lambda(\underline{v}, \underline{w})| \le c_1(\lambda)\|\underline{v}\|_{E(\Omega)}\,\|\underline{w}\|_{E(\Omega)} \quad \forall \underline{v}, \underline{w} \in E(\Omega),$$
- $$\inf_{\underline{v}\in E_0(\Omega)}\ \sup_{\underline{w}\in E_0(\Omega)} \frac{B_\lambda(\underline{v}, \underline{w})}{\|\underline{v}\|_{E(\Omega)}\,\|\underline{w}\|_{E(\Omega)}} \ge c_2(\lambda) > 0\,. \tag{4.2.11}$$

Loss of stability corresponds to a failure of (4.2.11). We are thus lead to a generalized eigenvalue problem for the buckling load factor λ in (4.2.9):

Find $\lambda \in \mathbb{R}$ and $\underline{0} \neq \underline{\overline{u}} \in E_0(\Omega)$ such that

$$B(\underline{\overline{u}}, \underline{v}) = \lambda\, G(\underline{\overline{u}}, \underline{v}) \quad \forall \underline{v} \in E_0(\Omega)\,. \tag{4.2.12}$$

Since neither $B(\cdot,\cdot)$ nor $G(\cdot,\cdot)$ are compact on $E(\Omega)$, (4.2.12) has continuous spectrum, in general, and the existence of a buckling mode $\underline{\overline{u}}$ is not assured. Nevertheless, in practice one observes that **for slender structures**, the smallest λ are discrete eigenvalues with corresponding eigenvectors, the buckling modes [1, 49, 50].

References

[1] R. Actis: Doctoral Dissertation, Dept. of Mech. Engg., Washington University, St. Louis, MO, USA, (1991).

[2] R. Actis, B. Szabó, C. Schwab: Hierarchic models for laminated plates and shells. *Comput. Meths. Appl. Mech. Engg.*, **172** (1999), 79-107.

[3] B. Andersson, U. Falk, I. Babuška and T. von Petersdorff: Reliable stress and fracture mechanics analysis of complex aircraft components using a hp-version FEM. *Int. J. Num. Meth. Engg.*, **38** (1995), 2135-2163.

[4] D.N. Arnold and R.S. Falk: Asymptotic Analysis of the boundary layer of the Reissner-Mindlin plate model. *SIAM J. Math. Anal.*, **27** (1996), 486-514.

[5] I. Babuška, B.A. Szabo and R. Actis: Hierarchic Models of laminated composites. *Int. J. Num. Meth. Engg.*, **33**, (1992), 503-536.

[6] I. Babuška, B. Anderson, P. J. Smith, K. Levin: Damage Analysis of Fiber Composites. Part I: Statistical Analysis on Fiber Scale. Preprint No. 15, The Aeronautical Research Institute of Sweden, 1998.

[7] I. Babuška, A. Craig, J. Mandel, and J. Pitkäranta: Efficient preconditioning for the p version finite element method in two dimensions. *SIAM J. Numer. Anal.*, **28** (1991), 624-661.

[8] I. Babuška and M. Suri: The p and hp Versions of the Finite Element Methods, Basic Principles and Properties. *SIAM review*, **36** (1994), 578-632.

[9] C. Bernardi and Y. Maday: Approximations spectrales de problèmes aux limites elliptiques. Springer, 1992.

[10] E. Bertoti and B.A. Szabo: Adaptive Selection of polynomial degrees on a Finite Element Mesh. *Int. J. Num. Math. Engg.*, **42**, (1998), 561-578.

[11] F. Brezzi and M. Fortin: Mixed and hybrid Finite Element Methods. Springer Series in Comp. Mathematics vol. 15 , Springer Verlag New York, 1991.

[12] C. Canuto, M. Y. Hussaini, A. Quarteroni and T. A. Zhang: Spectral Methods in Fluid Dynamics, Springer Verlag, 1986.

[13] P. G. Ciarlet: The Finite Element Method for elliptic problems. North-Holland, 1978.

[14] M. Dauge and I. Gruais: Asymptotics of arbitrary order for a thin elastic damped plate, parts I and II. *Asymptotic Anal.*, **13** (1996), 167-197 and **16** (1998), 99-124.

[15] M. Dauge and C. Schwab: Boundary layer resolution in plates by hp-Finite Elements (in preparation).

[16] L. Demkowicz, J.T. Oden, W. Rachowicz and O. Hardy: Toward a Universal hp Adaptive Finite Element Strategy. Part 1: Constrained Approximation and Data Structure. *Computer Methods in Applied Mechanics and Engineering*, **77** (1989), 79-112.

[17] L. Demkowicz, K. Gerdes, C. Schwab, A. Bajer and T. Walsh: HP90: A general and flexible Fortran 90 hp-FE code. *Computing and Visualization in Science*, **1** (1998), 145-163.

[18] K. Gerdes and C. Schwab: Hierarchic Models of Helmholtz problems on thin domains. *Math. Mod. Meth. Appl. Sci.*, **8** (1998), 1-35.

[19] K. Gerdes, A.M. Matache and C. Schwab: Analysis of membrane locking in hp-FEM for a cylindrical shell. ZAMM **78** (1998), 663-686.

[20] W.J. Gordon and Ch.A. Hall: Transfinite Element Methods: Blending Function Interpolation over Arbitrary Curved Element Domains. *Numer. Math.*, **21** (1973), 109-129

[21] B. Guo and I. Babuška: The $hp-$version FEM I: The Basic Approximation Results; and Part II: General Results and Applications. *Comp. Mech.*, **1** (1986), 21-41 and 203-226.

[22] H. Hakula, Y. Leino and J. Pitkäranta: Scale Resolution, Locking and high-order FE Modeling of shells. *Comp. Meth. Appl. Mech. Engg.*, **133** (1996), 157-182.

[23] G.E. Karniadakis and S.J. Sherwin: Spectral/hp Element Methods for CFD. Oxford University Press, 1999

[24] Y. Maday and A.T. Patera: Spectral Element Method for Navier-Stokes Equations. In: *State of the Art Surveys in Computational Mechanics*, (Ed. Noor), 71-143.

[25] Y. Maday and E.M. Rønquist: Optimal error analysis of spectral method with emphasis on non-constant coefficients and deformed geometries. *Comput. Meths. Appl. Mech. Engrg.*, **80** (1990), 91-115.

[26] A. M. Matache: Spectral- and p-Finite Elements for problems with microstructure. Doctoral Dissertation, ETH Zürich (in preparation).

[27] A. M. Matache, I. Babuška and C. Schwab: Generalized p-FEM in homogenization. To appear in Numer. Math.

[28] A. M. Matache and C. Schwab: Homogenization via p-FEM for problems with microstructure. Report 99-09, Seminar für Angewandte Mathematik, ETH Zürich, *Applied Numerical Math.*, **33** (2000), 32-59.

[29] J.M. Melenk: On n-widths for elliptic problems. Tech. Report 98-02, Seminar für Angewandte Mathematik, ETH Zürich (in press in J. Math. Anal. Applic.).

[30] M. Melenk, K. Gerdes and C. Schwab: Fully discrete hp-FEM I: Fast Quadrature. (in press in *Comp. Meth. Appl. Mech. Engg.*).

[31] J.M. Melenk and C. Schwab: hp-FEM for Reaction-Diffusion Equations. I: Robust Exponential Convergence. *SIAM J. Numer. Anal.*, **35** (1998), 1520-1557.

[32] J.T. Oden, L. Demkowicz, W. Rachowicz and T. A. Westermann: Toward a Universal hp Adaptive Finite Element Strategy. Part 2: A Posteriori Error Estimation. *Computer Methods in Applied Mechanics and Engineering*, **77** (1989), 113-180.

[33] J. T. Oden, A. Patra and Y. Feng: Parallel Domain Decomposition Solver for Adaptive hp Finite Element Methods. *SIAM J. Numer. Anal.*, **34** (1997), 2090-2118.

[34] O. A. Oleinik, A.S. Shamaev, G. A. Yosifian: Mathematical Problems in Elasticity and Homogenization. North-Holland, 1992.

[35] S.A. Orszag: Spectral methods for problems in complex geometries. *J. Comput. Phys.*, **37** (1980), 70-92.

[36] A.T. Patera: Spectral Element Method for Fluid Dynamics: Laminar Flow in a Channel Expansion. *J. Comput. Phys.*, **54** (1984), 568-488.

[37] J. Pitkäranta, A.-M. Matache and C. Schwab: Fourier Modal Analysis of layers in shallow shell deformations. Report 99-18, Seminar for Applied Mathematics, ETH Zürich http://www.sam.math.ethz.ch/Reports/prepr99.html. (to appear in Comp. Meth. Appl. Mech. Engg.).

[38] J. Pitkäranta, Y. Leino, O. Ovaskainen and J. Piila: Shell deformation states and the FEM. Benchmark study of cylindrical shells. *Comp. Meth. Appl. Mech. Engg.*, **128** (1995), 81-121.

[39] R. Rannacher: Lectures on Error Estimation and Adaptation Techniques in FEM, these lecture notes.

[40] C. Schwab: Boundary layer resolution in hierarchical models of laminated composites. *RAIRO Anal. Numérique*, **28** (1994), 517-537.

[41] C. Schwab: A-posteriori modeling error estimation for hierarchic plate models. *Num. Math.*, **74** (1996), 221-259.

[42] C. Schwab: p and hp FEM. Oxford University Press, 1998.

[43] C. Schwab and M. Suri. The p and hp Versions of the Finite Element Method for Problems with Boundary Layers. *Math. Comp.*, **65** (1996) 1403-1429.

[44] C. Schwab, M. Suri and C.A. Xenophontos: The hp-FEM for problems in mechanics with boundary layers. *Comp. Meth. Appl. Mech. Engg.*, **157** (1998), 311-333.

[45] User Manual to STRIPE. Aeronautic Institute, Sweden.

[46] Stresscheck. ESRD Inc., www.esrd.com.

[47] T. Strouboulis, I. Babuška and S.K. Gangaraj: Guaranteed computable bounds for the exact error in the finite element solution, II: Bounds for the energy norm of the error in two dimensions. Int. J. Num. Meth. Engg. (1999), to appear.

[48] B. Szabó and I. Babuška: Finite Element Analysis. Wiley, 1991.

[49] B. Szabó: Large Strain and Deformation, Lecture Notes, Washington University, Dept. of Mech. Engg. (1998).

[50] B.A. Szabó and G. Kiralyfalvi: Linear Models of buckling and stress-stiffening. *Comput. Methods Appl. Mech. Engg.*, **171** (1999), 43-59.

[51] B.A. Szabó and G.J. Sahrmann: Hierarchic Plate and shell models based on p-Extension. *Int. J. Num. Meth. Engg.*, **26** (1998), 1855-1881.

[52] E. Trefftz: Zur Theorie der Stabilität des Elastischen Gleichgewichts. ZAMM **13** (1933), 160-165.

[53] R. Verfürth: A Review of a-posteriori error estimation techniques for elasticity problem. Report 223/1997, Dept. of Mathematics, Ruhr-Universität Bochum, Deutschland.

Constitutive Relation Error Estimators and Adaptivity in Structural Engineering

Pierre Ladevèze

LMT-Cachan (ENS Cachan / CNRS / Univ. Paris 6)
61, avenue du Président Wilson
94235 Cachan Cedex - France
E-Mail : ladeveze@lmt.ens-cachan.fr

Abstract: The basic aspects of error estimators based on residuals on the constitutive relations which have been developed for the past 20 years in particular at Cachan are given. The crucial point is detailed; it is the construction of equilibrated stress fields from the computed finite element solution and the data. Adaptive computational strategies are also described and applied to 2 and 3D engineering examples.

Key words: Error estimator, Adaptivity, Finite element, Elasticity, Plasticity, Damage, Dynamic, Modelling error, Model updating.

1. Introduction

Quite a lot of error estimators are available for mastering finite element calculation; the state of the art can be found in (Ainsworth and Oden 1993) (Babuska et al. 1994) (Ladevèze and Oden 1998). We deal herein with a particular family introduced in Ladevèze (1975) (1977) which has been developed, in particular at Cachan, for more than 20 years (see also (Ladevèze 1995)). It is named Constitutive Relation Errors (CRE) because these a posteriori error estimators are based on residual errors on constitutive relations. The principle is simple; to the finite element displacement-stress solution, one associates a new one which verifies exactly the most reliable equations (kinematic constraints, equilibrium equations). Therefore, its quality, i.e. the quality of the f.e. analysis is defined by the residual error on the constitutive relation. Their worth lies first in their easy extension to time-dependent non linear problems : (visco)-plasticity, damage, large displacements, dynamics. The CRE also have a strong physical interpretation. Another important property is the fact that the global effectivity index in elasticity is greater than 1. The CRE are then deemed conservative. For an enhanced CRE estimator, this property also holds locally for elasticity problems (Ladevèze and Rougeot 1997).

A central and crucial point is the construction of equilibrated stresses for which we have developed explicit or quasi-explicit general methods. The recovered stress is determined in terms of the computed finite element solution and of the data. The key is what we call "prolongation conditions" which are the link between the f.e. solution and the stress to recover.

Comparison between CRE estimators and other error estimators are also given in addition to the work done in (Strouboulis and Haque 1992) (Babuska et al. 1994, 1997). The estimator ERpB3 developed and used by Babuska, Strouboulis et al. coincides with our standard and practical CRE estimator for linear problems.

Adaptive computational strategies are also described and applied to 2D and 3D engineering examples. For time-dependent nonlinear problems, specific error indicators (time discretization, space discretization, iteration stopping) are introduced to adapt all the computational parameters for a given accuracy.

Summary

2 Constitutive Relation Error Estimators in Elasticity

CRE have been introduced for linear problems in Ladevèze (1975) (1977) and developed for thermal problems in (Ladevèze and Leguillon 1983) and for elasticity problems in (Ladevèze, Coffignal and Pelle 1986) and (Ladevèze, Pelle and Rougeot 1992). Nearly incompressible elastic materials have been studied in (Gastine et al. 1992). A further study introducing an enhanced CRE estimator can be found in (Ladevèze and Rougeot 1997).

2.1 The Reference Problem to be Solved - Notations

The studied structure initially occupies a domain Ω with a boundary $\partial\Omega$. We will consider small displacements, quasi-static loadings and isothermal conditions. The time interval of interest is denoted by [0,T]. The structure, for all t belonging to [0,T], is placed in an "environment" characterized by a displacement $\underline{U}_d$ on a part $\partial_1\Omega$ of the boundary of Ω, a surface force density $\underline{F}_d$ on $\partial_2\Omega$, the part of $\partial\Omega$ complementary to $\partial_1\Omega$, and a volume force density $\underline{f}_d$ on the domain Ω.

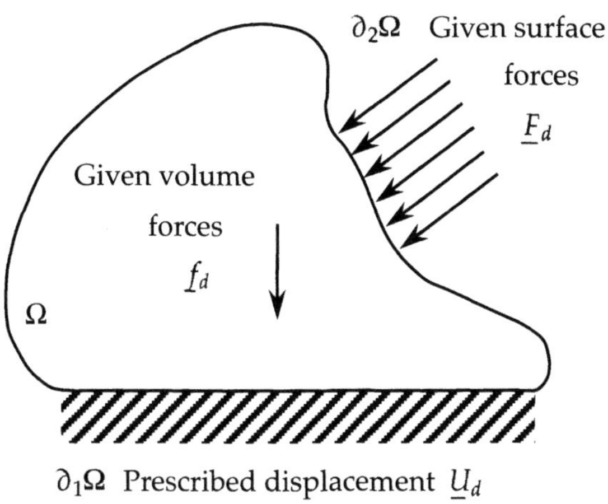

Figure 1. Schematic diagram of the environment (i.e. the imposed conditions)

The problem which describes the evolution of the structure on [0,T] is:

Find the displacement $\underline{U}(\underline{M}, t)$ and the stress $\sigma(\underline{M}, t)$ with $(t \in [0, T])$ and $(\underline{M} \in \Omega)$, which satisfy:

- kinematic constraints:

$$\underline{U} \in \mathcal{U}^{[0,T]}; \ \underline{U}|_{\partial_1\Omega} = \underline{U}_d \text{ on }]0,T[\tag{1}$$

- equilibrium equations (principle of virtual work):

$$\sigma \in \mathcal{S}^{[0,T]};\ \forall t \in]0,T[\ \forall \underline{U}^* \in \mathcal{U}_{ad,0}$$
$$\int_\Omega \mathrm{Tr}\left[\sigma\, \varepsilon\left(\underline{U}^*\right)\right] d\Omega = \int_\Omega \underline{f}_d \circ \underline{U}^* d\Omega + \int_{\partial_2\Omega} \underline{F}_d \circ \underline{U}^* dS \qquad (2)$$

- constitutive relation:

$$\forall t \in [0,T]\ \forall \underline{M} \in \Omega \quad \sigma|_t = \mathbf{A}\left(\varepsilon\left(\dot{\underline{U}}_{|\tau}\right); \tau \le t\right) \qquad (3)$$

$\varepsilon(\underline{U})$ denotes the strain associated with the displacement $\left(\varepsilon(\underline{U})_{ij} = \frac{1}{2}\left(U_{i,j} + U_{j,i}\right)\right)$. $\mathcal{U}^{[0,T]}$ is the space containing the displacement field $\underline{U}$ defined on $\Omega \times]0,T[$, and $\mathcal{S}^{[0,T]}$ is the space containing the stresses, likewise defined on $\Omega \times]0,T[$. Finally, $\mathcal{U}_{ad,0}$ is the vector space of the specified virtual velocities. The operator **A**, which is given and generally single-valued, characterizes the mechanical behavior of the material.

Let us denote by $\mathcal{U}_{ad}^{[0,T]}$ the space of displacements verifying the kinematic constraints (1). $\mathcal{S}_{ad}^{[0,T]}$ is the space of stresses which are solutions to the equilibrium equations (2).

Let us start with the simplest family of problems, the elasticity problems. One looks for the final state of the structure at $t = T$; then, the problem to be solved does not depend on the time. Moreover, the constitutive relation (3) becomes:

$$\sigma = \mathbf{K}\, \varepsilon\left(\underline{U}\right) \qquad (4)$$

where **K** denotes the Hooke's tensor which is symmetric and positive definite. The spaces of admissible displacement and stress are $\mathcal{U}_{ad}$ and $\mathcal{S}_{ad}$.

The densities $\underline{U}_d, \underline{f}_d, \underline{F}_d$ are known at $t = T$.

2.2 Principle of Constitutive Relation Errors

Let us consider that the finite element solution has been computed using a displacement approach. The finite element displacement-stress pair $\left(\underline{U}_h, \sigma_h\right)$ is then known, the stress σ_h being f.e. equilibrated.

The principle behind our approach is to associate a new displacement-stress pair denoted by $\left(\hat{\underline{U}}_h, \hat{\sigma}_h\right)$, which is admissible i.e. belonging to $\mathcal{U}_{ad} \times \mathcal{S}_{ad}$, to the data and the finite element displacement-stress pair. This pair verifies the most reliable equations of the reference problem to be solved ; its quality, and thus the quality of the f.e. solution is measured through the residue related to the verification of the constitutive relation. It verifies also what we call the "prolongation conditions" which are relations with the finite element solution.

Definition of CRE. The displacement $\hat{\underline{U}}_h$ is generally taken as $\underline{U}_h$. The crucial point is to build a stress $\hat{\sigma}_h$ associated to both σ_h and the data, which verifies the equilibrium equations. We will return later to this problem ; let $\left(\hat{\underline{U}}_h, \hat{\sigma}_h\right)$ be a displacement-stress pair belonging to $\mathcal{U}_{ad} \times \mathcal{S}_{ad}$. The relative constitutive relation error is:

$$\varepsilon^2 = \frac{\left\|\hat{\sigma}_h - \sigma\left(\hat{\underline{U}}_h\right)\right\|^2}{\frac{1}{2}\left(\left\|\hat{\sigma}_h\right\|^2 + \left\|\sigma\left(\hat{\underline{U}}_h\right)\right\|^2\right)} \tag{5}$$

where:

$$\|\sigma\|^2 = \int_\Omega \mathrm{Tr}\left[\sigma\, \mathbf{K}^{-1}\, \sigma\right] d\Omega \text{ and } \sigma\left(\hat{\underline{U}}_h\right) = \mathrm{K}\, \varepsilon\left(\hat{\underline{U}}_h\right)$$

Introducing the contribution to the error of the element $\mathrm{E} \in \mathbf{E}$, one gets:

$$\varepsilon^2 = \sum_{\mathrm{E} \in \mathbf{E}} \varepsilon_{\mathrm{E}}^2 \tag{6}$$

It is also possible to introduce a local error :

$$\left[\varepsilon_{loc}\right]^2 = \sup_{\mathrm{E} \in \mathbf{E}} \left[\varepsilon_{\mathrm{E}}^2 \frac{\Omega}{\Omega_{\mathrm{E}}}\right]$$

Stress prolongation conditions for a f.e. displacement approach. In the standard recovery method, the prolongation condition is the strong one defined by :

$$\begin{array}{ll} \forall\, \mathrm{E} \in \mathbf{E} & \\ \forall\, \mathrm{i} \in \mathbf{I} & \displaystyle\int_{\mathrm{E}} \left(\hat{\sigma}_h - \sigma_h\right) \underline{\mathrm{grad}}\, \omega_i \, d\Omega = 0 \text{ (strong P.C.)} \end{array}$$

where ω_i $\mathrm{i} \in \mathbf{I}$ are the f.e. shape functions.

In a last version, the prolongation condition is weaker :

$$\begin{array}{ll} \forall\, \mathrm{E} \in \mathbf{E} & \\ \forall\, \mathrm{i} \in \bar{\mathbf{I}} \subset \mathbf{I} & \displaystyle\int_{\mathrm{E}} \left(\hat{\sigma}_h - \sigma_h\right) \underline{\mathrm{grad}}\, \omega_i \, d\Omega = 0 \text{ (weak P.C.)} \end{array}$$

$\bar{\mathbf{I}}$ is associated with non-vertex nodes.

Link to the classical error. The CRE is related to the classical error by the Prager-Synge theorem :

$$\frac{\left\| \sigma_{ex} - \sigma_h^* \right\|^2}{\frac{1}{2}\left(\left\| \hat{\sigma}_h \right\|^2 + \left\| \sigma\left(\hat{\underline{U}}_h\right) \right\|^2 \right)} = \frac{\varepsilon^2}{4}$$

where $\sigma_h^* = \frac{1}{2}\left[\hat{\sigma}_h + \sigma\left(\hat{\underline{U}}_h\right) \right]$ and σ_{ex} is the exact stress.

2.3 Quality of Constitutive Relation Errors

Let us first recall the global effectivity index : $\dfrac{\left\| \hat{\sigma}_h - \sigma_h \right\|}{\left\| \sigma_{ex} - \sigma_h \right\|}$

A first comparison for thermal problems. The results have been derived in (Babuska et al. 1994). They concern a 2D thermal problem solved with a periodic mesh, the shape ratio being b/a (Figure 2)

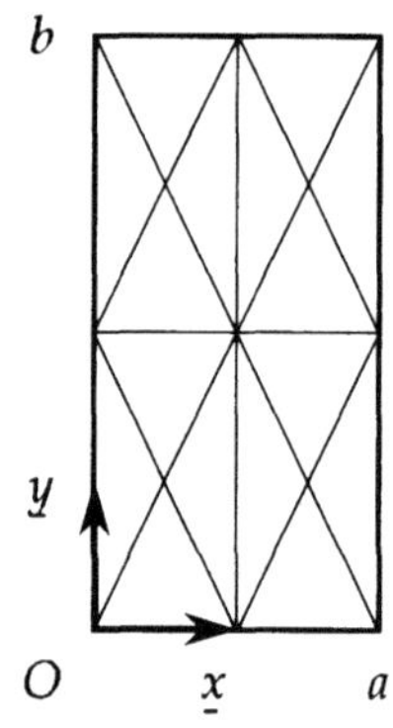

Figure 2. Cell mesh

Figure 3 shows that the constitutive relation has an excellent behavior for thermal problems.

Test 1: f.e. calculations with anisotropic mesh. For the sake of simplicity, the Hooke's tensor is the identity. Two CRE estimators have been compared among themselves and with the ZZ2 error estimator. It is the original ZZ2 introduced in (Zienkiewicz and Zhu 1992) which is the one being implemented.

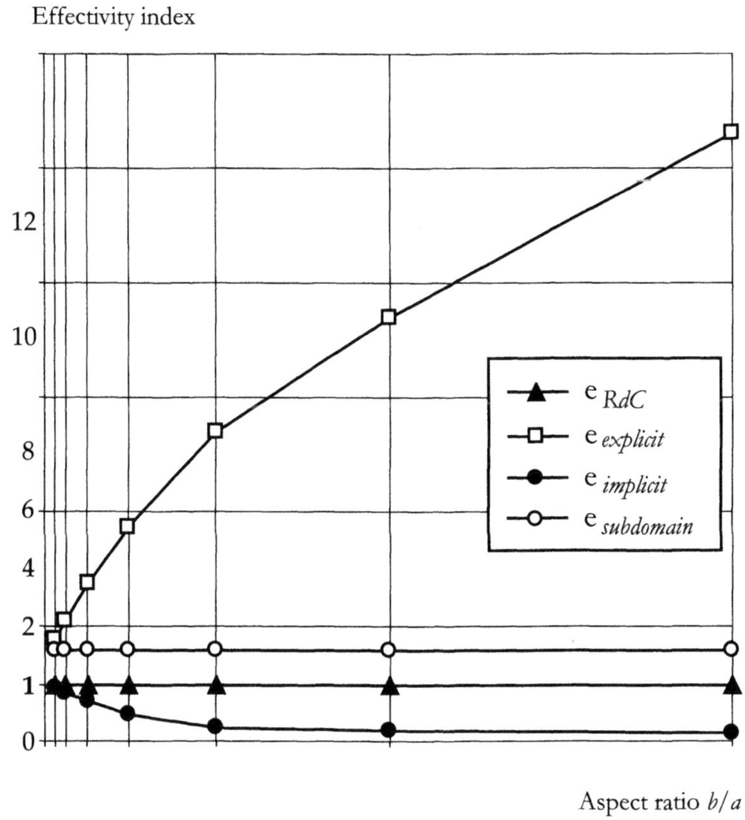

Figure 3. Effectivity indexes for different error estimators (thermal problem)

The initial standard CRE estimator is associated to the strong prolongation condition and a completely explicit recovering technique. The "new" CRE recovering estimator uses a weak prolongation condition and a semi-explicit recovering technique which will be described later.

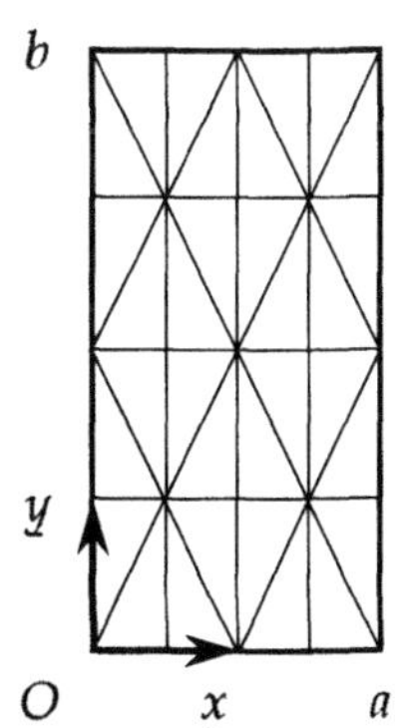

Figure 4. Anisotropic mesh

Figure 4 shows the studied structure and its mesh; the shape ratio 1/H is a parameter. Four loadings are considered; they are defined by 2D prestresses which are polynomials of degree 2 and satisfy the interior equilibrium and compatibility equations.

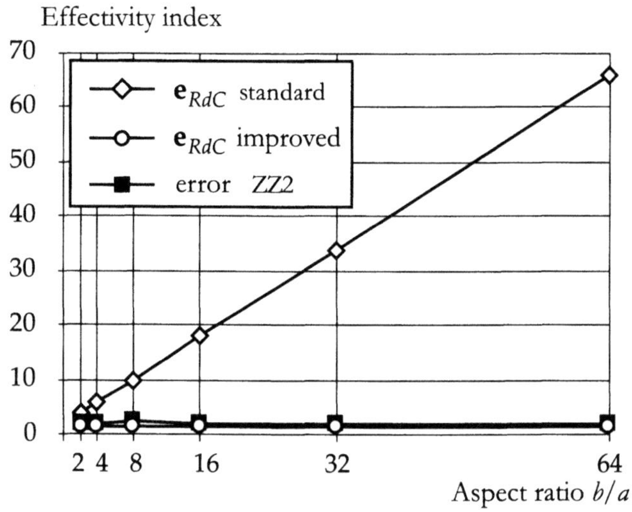

Prestress : $\begin{bmatrix} -2\,xy & y^2-x^2 \\ y^2-x^2 & 2\,xy \end{bmatrix}$

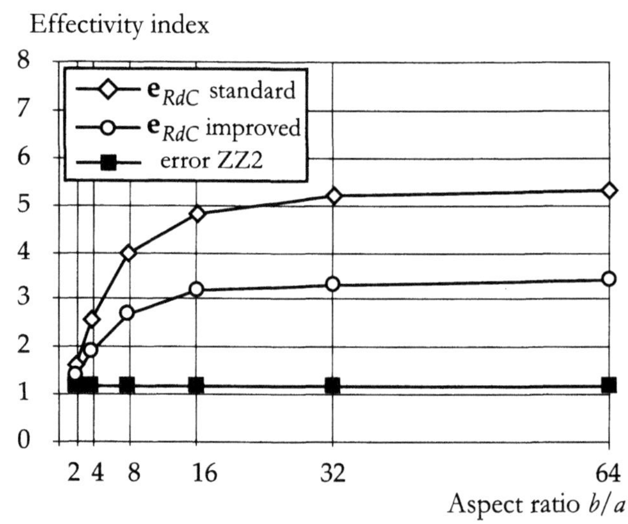

Prestress : $\begin{bmatrix} y^2-x^2 & 2\,xy \\ 2\,xy & x^2-y^2 \end{bmatrix}$

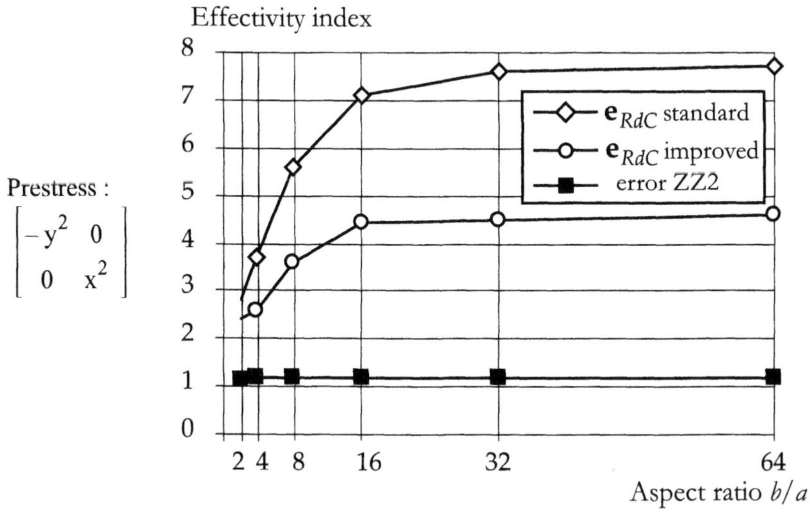

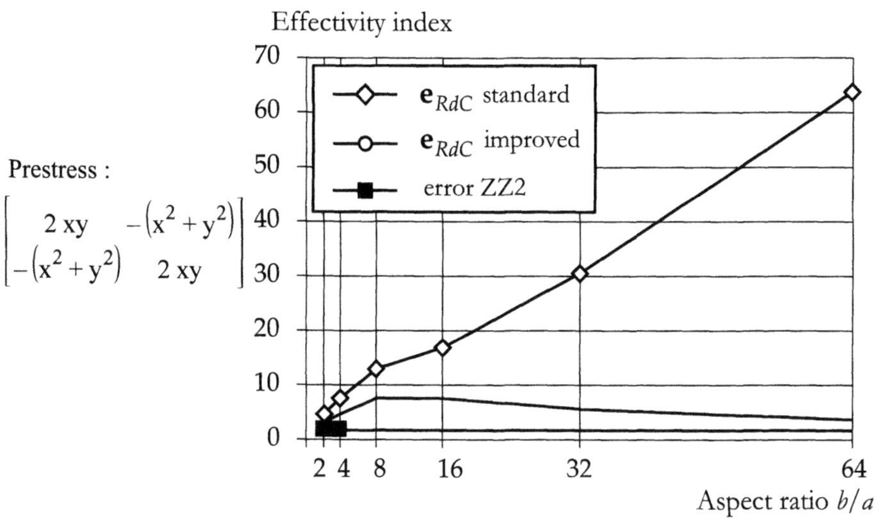

Figure 5. Efficitivity index versus shape ratio for different error estimators (P2 element)

It appears that the last error estimator is much better than the standard one for large values of the shape ratio. Its behavior is satisfactory. When the ZZ2 error estimator is included in the comparison, the best error estimator becomes the ZZ2 for the three last cases, and they are equivalent for the first one.

A deeper study is given in [Ladevèze and Pelle 2000]. It is proved that with a slight modification, the standard CRE estimator becomes satisfactory for large values of the shape ratio. This modified standard CRE estimator follows an idea of Babuska and Strouboulis which is to solve numerically the stress recovery problem that one gets on each element. We will come back on this point later.

Test 2: 2D elasticity problems with stiff stress gradient.

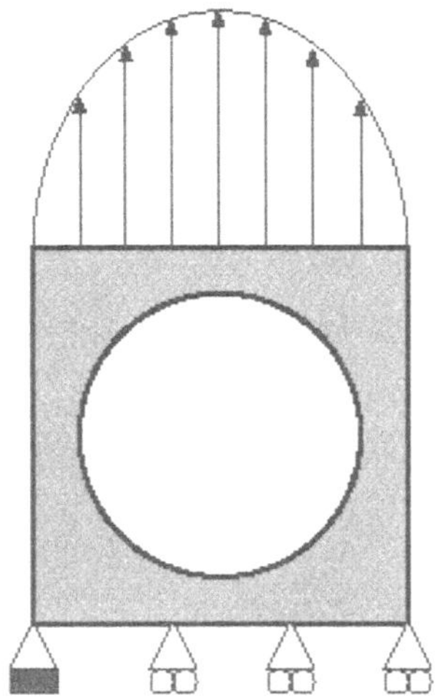

Figure 6. Test 2

This example is a better representation of the commonly-encountered f.e. problems. Figure 6 displays the reference problem to be solved. Figure 7 gives the local effectivity index maps for the three error estimators. The global effectivities are:

constitutive relation error (standard version)	: 2.6
constitutive relation error (new version)	: 2.0
ZZ2	: 1.6

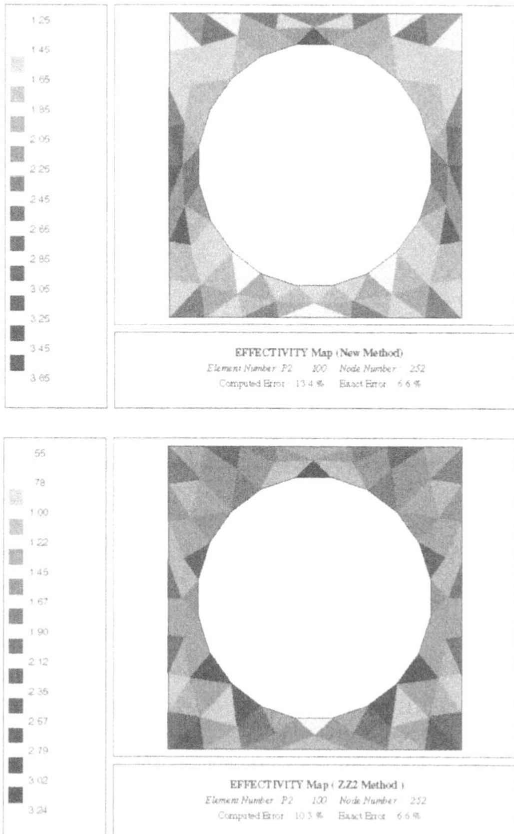

Figure 7. Local effectivity map for different estimators

Figure 8 presents the number of elements versus local element effectivity for two error estimators: ZZ2 and the last version of the error on constitutive relation. It appears, and this is an important point, that the local effectivity indexes are greater than 1. This property has been verified in many other tests; this error concerning stresses is then to be deemed "conservative" locally. Globally, this property is a mathematical result, consequence of the Prager-Synge theorem.

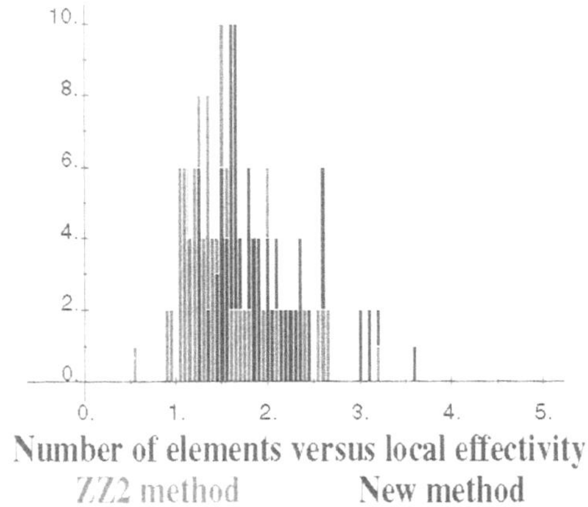

Figure 8. Number of elements versus local element effectivity for both ZZ2 error estimator and the new one

Remarks.

- The error estimator ERpB+3 developed by Babuska and Strouboulis coincides with our modified standard CRE estimator for which the stress recovery problem on each element is solved numerically. They have done numerous studies concerning its quality.
- An iterative technique has been introduced recently by (Strouboulis et al. 2000) in order to get guaranteed upper and lower bounds of the error.
- To derive global error estimators is a first necessary step. A further one is to derive error estimators for engineering local quantities such as stresses, displacements, intensity factors... Several methods have recently been proposed; the idea is to approximate numerically an extractor operator which is a kind of Green operator. A general method based on CRE estimators has been proposed in (Ladevèze et al. 1999). For the particular case of stresses, our new CRE estimator seems to be sufficient for getting upper bound of the local error; further studies are in progress. Other works concerning local error estimators can be found in (Rannacher and Suttmeier 1998) (Peraire and Patera 1998) (Oden and Prudhomme 1999) (Ciral and Ramm 1998) (Strouboulis et al. 2000);

3 A General Method for Recovering Equilibrated Stress from the f.e. Stress and the Data

Other research works have been carried out in order to recover more or less equilibrated stress: (Ainsworth and Oden 1997) (Bank and Weiser 1985), (Babuska, Strouboulis et al. 1994), (Kelly 1984), (Stein et al. 1998), (Wiberg and Abdulwahab 1992)...

Our recovery method has been introduced in Ladevèze (1975) (1977) and further developed in several papers: Ladevèze and Leguillon (1983), Ladevèze et al. (1986) (1991), Ladevèze and Maunder (1996) (1997). The key is what has been termed: the "prolongation condition".

3.1 Prolongation Conditions

In the standard recovery method, the strong prolongation condition is:

$$\begin{matrix} \forall E \in \mathbf{E} \\ \forall i \in \mathbf{I} \end{matrix} \quad \int_E \left(\hat{\sigma}_h - \sigma_h\right) \underline{\text{grad}}\, \omega_i \, d\Omega = 0 \quad \text{(strong P.C)} \tag{7}$$

$\hat{\sigma}_h$ is the S.A. stress we want to build. E denotes an arbitrary element and $\left(\omega_i \, ; i \in \mathbf{I}\right)$ the f.e. scalar shape functions. I is the set of the f.e. nodes.

In the last version, the prolongation condition is weaker:

$$\begin{matrix} \forall E \in \mathbf{E} \\ \forall i \in \bar{\mathbf{I}} \subset \mathbf{I} \end{matrix} \quad \int_E \left(\hat{\sigma}_h - \sigma_h\right) \underline{\text{grad}}\, \omega_i \, d\Omega = 0 \quad \text{(weak P.C)} \tag{8}$$

$\bar{\mathbf{I}}$ is associated with non-vertex nodes.

3.2 Main features of the Standard Recovery Method

Main property. Let us recall the strong prolongation condition:

$$\begin{matrix} \forall E \in \mathbf{E} \\ \forall i \in \mathbf{I} \end{matrix} \quad \int_E \left(\hat{\sigma}_h - \sigma_h\right) \underline{\text{grad}}\, \omega_i \, d\Omega = 0$$

which can be written using the Stokes formula and $\hat{\sigma}_h \in \mathcal{S}_{ad}$:

$$\forall E \in \mathbf{E} \quad \forall i \in \mathbf{I}$$

$$\int_{\partial E} \hat{\sigma}_h \underline{n}\, \omega_i \, dS = -\int_E \underline{f}_d\, \omega_i \, d\Omega + \int_E \sigma_h \, \underline{\text{grad}}\, \omega_i \, d\Omega \tag{9}$$

This last relation introduces the element tractions on the boundaries of the elements, representing in fact their projections on the shape functions. The element traction due to the environment of element E is denoted by $\eta_E \hat{\underline{F}}_h$, where $\hat{\underline{F}}_h$ is defined on the element's boundaries. η_E is a scalar function also defined on the element's edges. The edge's set will be denoted Γ.

The η_E function depends on the element E; it has on each side a constant value: +1, -1 or 0. Moreover, for two adjacent elements, we prescribe that: $\eta_E + \eta_{E'} = 1$ on the common edge $\Gamma_{EE'}$ (Figure 9).

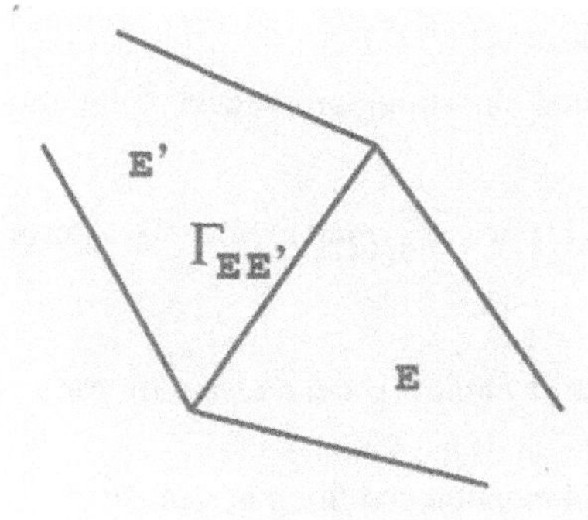

Figure 9. Definition of η_E

Taking:

$$\eta_E \underline{\hat{F}}_h = \left[\hat{\sigma}_h|_E \underline{n}\right]_{\partial E} \tag{10}$$

the strong prolongation condition then becomes:

$$\int_{\partial E} \eta_E \underline{\hat{F}}_h \omega_i \, dS = -\int_E \underline{f}_d \, \omega_i \, d\Omega + \int_E \sigma_h \, \underline{\text{grad}} \, \omega_i \, d\Omega \equiv Q_E(i) \tag{11}$$

Let $\mathbf{F}_{ad}$ be the equilibrated element's traction space; these tractions verify the prescribed external loading on $\partial_2 \Omega$ and the global equilibrium conditions on each element. Let us now present the main property which demonstrates that the prolongation condition is in fact the key.

Property. The element tractions $\underline{\hat{F}}_h$ are equilibrated if the following conditions hold:

$\forall E \in \mathbf{E} \;\; \forall i \in \mathbf{I}$

$$\int_{\partial E} \eta_E \underline{\hat{F}}_h \omega_i \, dS = \int_E \sigma_h \, \underline{\text{grad}} \, \omega_i \, d\Omega - \int_E \underline{f}_d \, \omega_i \, d\Omega$$

where:

- $\eta_E \underline{\hat{F}}_h$ verifies the prescribed traction on $\partial_2 \Omega$, and

- the rigid body displacements for each element can be expressed in terms of the shape - functions. The proof can be found in (Ladevèze and Leguillon 1983). Let us note that the third condition is practically always satisfied.

Equilibrated traction recovery. Several cases should be distinguished:

➢ *Non-vertex node interior to Ω:* Let $\Gamma_{EE'}$ be a common boundary to adjacent elements E and E'. ω_i is the scalar shape function associated with an internal node to $\Gamma_{EE'}$. The f.e. equilibrium leads to:

$$\int_E \sigma_h \,\underline{\underline{grad}}\, \omega_i \, d\Omega + \int_{E'} \sigma_h \,\underline{\underline{grad}}\, \omega_i \, d\Omega = \int_E \underline{f}_d \, \omega_i \, d\Omega + \int_{E'} \underline{f}_d \, \omega_i \, d\Omega \tag{12}$$

and then to:

$$\underline{Q}_E(i) + \underline{Q}_{E'}(i) = 0 \tag{13}$$

Using the prolongation condition, one obtains:

$$\int_{\Gamma_{EE'}} \eta_E \, \underline{\hat{F}}_h \, \omega_i \, dS = \underline{Q}_E(i)$$

$$\int_{\Gamma_{EE'}} \eta_{E'} \, \underline{\hat{F}}_h \, \omega_i \, dS = \underline{Q}_{E'}(i)$$

It follows that:

$$\int_{\Gamma_{EE'}} \underline{\hat{F}}_h \, \omega_i \, dS = \eta_E \, \underline{Q}_E(i) = \eta_{E'} \, \underline{Q}_{E'}(i) \tag{14}$$

➢ *Vertex interior to Ω:* Figure 10 specifies the notations. Let:

$$\underline{b}_j = \int_{\Gamma_j} \eta_{E_j} \, \underline{\hat{F}}_h \, \omega_i \, dS \tag{15}$$

The prolongation condition, written for the elements connected to vertex i, leads to the reduced linear system:

$$\begin{cases} \underline{b}_1 - \underline{b}_2 = \underline{Q}_{E_1}(i) \\ \underline{b}_2 - \underline{b}_3 = \underline{Q}_{E_2}(i) \\ \quad \cdots \\ \underline{b}_n - \underline{b}_1 = \underline{Q}_{E_n}(i) \end{cases} \tag{16}$$

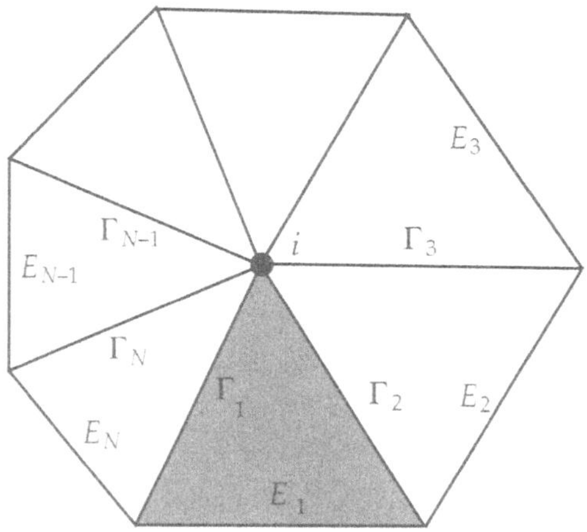

Figure 10. Projection on a vertex

This system admits a solution if the sum of the second members is equal to zero:

$$\sum_{i=1}^{n} \underline{Q}_{E_j}(i) = 0 \tag{17}$$

σ_h verifies the f.e. equilibrium; it follows that condition (17) holds. However, the linear system (16) admits an infinite number of solutions. Among them, we choose the one which minimizes:

$$\left(\underline{b}_1, \ldots, \underline{b}_n\right) \rightarrow \sum_{j=1}^{n} \frac{\left(\underline{b}_j - \underline{b}_j^m\right)^2}{p_j^2} \tag{18}$$

where:

- $\underline{b}_j^m$ is an average computed with the f.e. stress vectors for two adjacent elements E_j, E_{j-1} having a common edge Γ_j.
- p_j is a weight $\left(p_j^2 = L_j^2 \,;\, L_j : \text{length of } \Gamma_j\right)$.

➢ *Other cases*: The treatment of all other cases is very similar to that for the previous ones.

Remarks:

- The cost function (18) does not have a strong mechanical interpretation.
- Several variants concerning traction recovery can be easily introduced (see (Ladevèze and Maunder 1996, 1997)).
-

To recover element tractions from their projections on the shape functions, the process is straightforward.

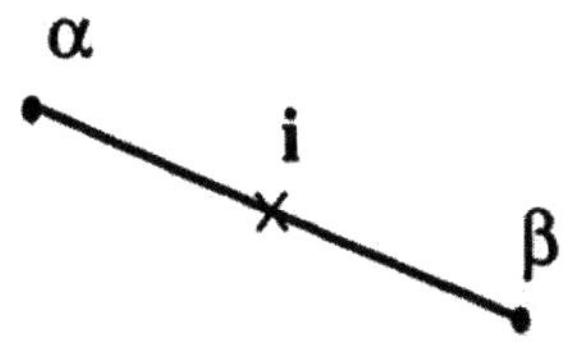

Figure 11. Element boundary Γ

One supposes that the element traction on Γ is a linear combination of the restrictions on Γ of the shape functions (figure 11).

$$\underline{\hat{F}}_h\big|_\Gamma = \underline{A}_\alpha\,\omega_\alpha\big|_\Gamma + \underline{A}_\beta\,\omega_\beta\big|_\Gamma + \sum_{i\in\bar{I}} \underline{A}_i\,\omega_i\big|_\Gamma \tag{19}$$

Equilibrated stress recovery. This work is to be performed element-by-element. For the element E, Figure 12 specifies the problem to be solved.

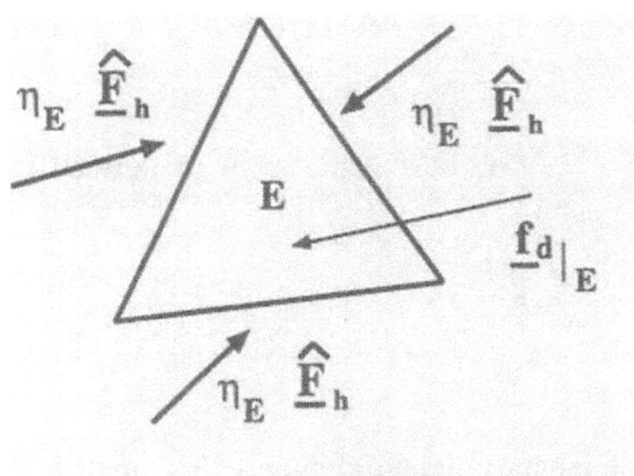

Figure 12. Equilibrated element stress recovery

Several techniques have been developed in order to derive $\hat{\sigma}_h\big|_E$ explicitly. At the present time, we prefer to use numerical techniques involving high order elements; generally we use one element of degree p+3, p being the element degree of the displacement approach. Such a modified standard error estimator coincides with ERpB+3 estimator of Babuska and Strouboulis.

3.3 Main Features of the New Version

Let us recall the weak prolongation condition:

$$\begin{matrix} \forall E \in \mathbf{E} \\ \forall i \in \bar{\mathbf{I}} \subset \mathbf{I} \end{matrix} \int_E \left(\hat{\sigma}_h - \sigma_h\right) \underline{\underline{\text{grad}}}\, \omega_i \, d\Omega = \mathbf{0} \tag{20}$$

$\bar{\mathbf{I}}$ does not contain any vertices.

Equilibrated traction recovery. The element traction projections are completely defined for the shape functions ω_i with $\omega_i\ i \in \bar{\mathbf{I}}$. Following the method used previously, it results that the element traction on Γ (see Figure 11) can be written:

$$\hat{\underline{F}}_h\big|_\Gamma = \underline{A}_\alpha\, \omega_\alpha\big|_\Gamma + \underline{A}_\beta\, \omega_\beta\big|_\Gamma + \sum_{i \in \bar{\mathbf{I}}} \underline{A}_i\, \omega_i\big|_\Gamma$$

where:

- the high-degree part $\Sigma_{i \in \bar{\mathbf{I}}} \underline{A}_i\, \omega_i\big|_\Gamma$ is completely defined. If necessary, additional prolongation conditions are introduced.
- $\underline{A}_\alpha$, $\underline{A}_\beta$ are parameters.

Let $\underline{A}$ be the column formed with all $\underline{A}_\alpha$, $\underline{A}_\beta$; the corresponding space is $\mathbf{A}$. $\hat{\underline{F}}_h$ has to be admissible $\left(\hat{\underline{F}}_h \in \mathbf{F}_{ad}\right)$; then, $\underline{A}$ has to belong to a certain subspace of $\mathbf{A}$ denoted by $\mathbf{A}_{ad}$.

Equilibrated stress recovery. For fixed $\underline{A} \in \mathbf{A}_{ad}$, $\hat{\sigma}_h$ is built as in the previous fashion element-by-element.

Optimal value for $\underline{A}$. The best value of $\underline{A}$ is defined by the following minimization problem:

$$\begin{matrix} \underline{A} \\ \mathbf{A}_{ad} \end{matrix} \rightarrow J_2\left(\underline{A}\right) = \frac{1}{2} \underline{A}^t\, \mathbb{H}\, \underline{A} - \underline{Z}^t\, \underline{A}$$

$J_2\left(\underline{A}\right)$ is the complementary potential energy. The matrix $\mathbb{H}$ and the second member $\underline{Z}$ can be built up element-by-element. More details can be found in Ladevèze and Rougeot (1997).

Practical method. Let us recall that $\underline{A}$ has been defined on the set of element boundaries, denoted by Γ. For "smooth" zones of $\bar{\Omega}$, we take the values given by the standard recovery method for the components of $\underline{A}$ associated with the boundaries. "Non-smooth" zones are those zones located around corner points, deformed elements,... For such zones, better values for the corresponding components of $\underline{A}$ are computed. In particular, $\underline{A}$ can be split into:

$$\begin{bmatrix} \underline{\tilde{A}} \\ \underline{\approx}\\ \underline{A}_0 \end{bmatrix}$$

where $\left[\underline{\tilde{A}}\right]$ (resp. $\underline{\overset{\approx}{A}}_0$) is related to the non-smooth zones (resp. smooth zones).

Using the conjugate gradient technique, and with some iterations, one then minimizes:

$$\begin{array}{ccc} \begin{bmatrix} \underline{\tilde{A}} \\ \approx \\ \underline{A}_0 \end{bmatrix} & \rightarrow & J_2\left(\begin{bmatrix} \underline{\tilde{A}} \\ \approx \\ \underline{A}_0 \end{bmatrix}\right) \\ \mathbf{A}_{ad} & \rightarrow & R \end{array}$$

The initial value for $\underline{\tilde{A}}$ is the one given by the standard recovery method. This practical method is driven by the following quality indicator:

$$\underline{r} = \mathbb{H}\,\underline{A}_0 - \underline{Z} \tag{21}$$

where $\underline{A}_0$ is the value given by the standard recovery method.

4. Adaptivity and Optimal Mesh for Steep Gradient Elasticity Problems

4.1 Influence of the Regions with Steep Gradients

To give a mesh with a criterion of prescribed error, one must know the convergence rate q of the error ε as a function of the element size h:

$$\varepsilon = O\left(h^q\right)$$

Theoretical results show that for smooth problems the convergence rate q depends only on the type of element used:

$$q = p$$

with, for example:

- $p = 1$ for 3-node elements
- $p = 2$ for 6-node elements.

Figure 13 shows the results obtained on a cracked plate, meshed with 6-node elements. For the optimization, we impose on each element q = 2. We notice that the prescribed error (2%) is not achieved; therefore, the mesh is not optimal.

For this example, the error does not converge with the degree of the shape functions. In fact, when steep gradient regions appear in the structure (singularities, for instance), the convergence rate q is determined by the following formula (Ciarlet 1978) (Azziz and Babuska 1972):

$$q = \min(\alpha, p)$$

where α is the strength of the singularity (local solution such as $u = r^{\alpha} f(\theta)$) and p the degree of the shape functions. If we impose $q = \alpha = 1/2$ (strength of the singularity in crack tip: $1/2$), we obtain a mesh with 53.907 elements. Without making the finite element analysis, which is too expensive, it is evident that this mesh is not optimal.

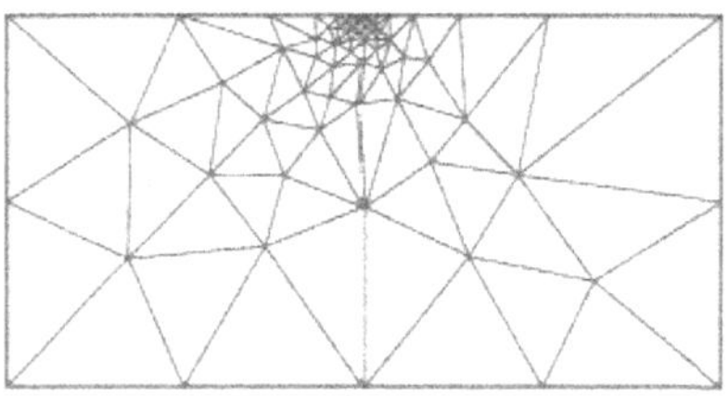

Initial mesh
124 6 node elements
273 nodes ε = *11,34 %*

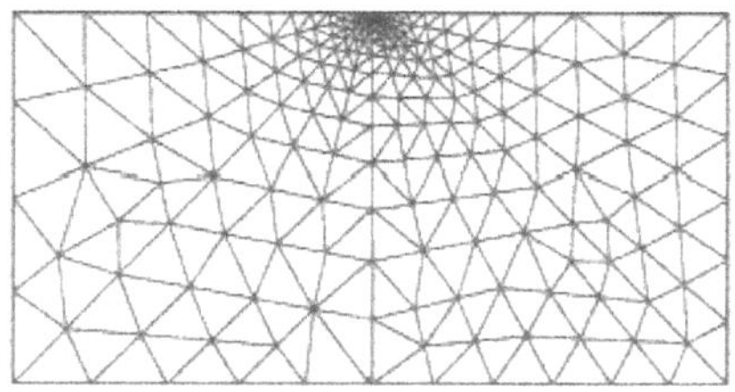

Optimized mesh
448 6 node elements
953 nodes ε = *4,2 %*

Figure 13. Optimization without taking into account steep gradients (prescribed error : 2%)

The explanation is that theoretical results have been demonstrated only for h tending to 0, so this behavior is asymptotic. We consider a cracked plate of width 4b and of height b loaded in mode I with a parabolic effort (Figure 14). When we change the length a of the crack, one notices a different evolution of the error as a function of the number of elements (Figure 15). For a = 0, the problem of a plate without crack loaded in traction, the convergence rate is about 1.8, a value close to 2 which is the theoretical rate of the 6-node triangular element. On the other hand, for a = b/12, the convergence rate varies from 1.33 to 0.79 and for a = b/2 from 0.85 to 0.52, a value near α which is only obtained asymptotically when h is very small.

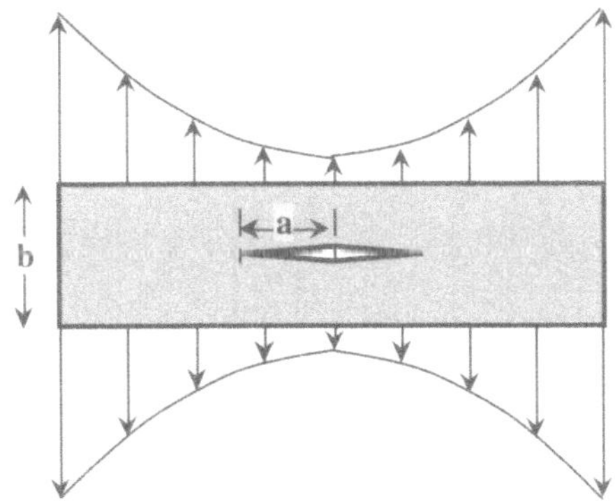

Figure 14. Cracked plate with a parabolic given traction

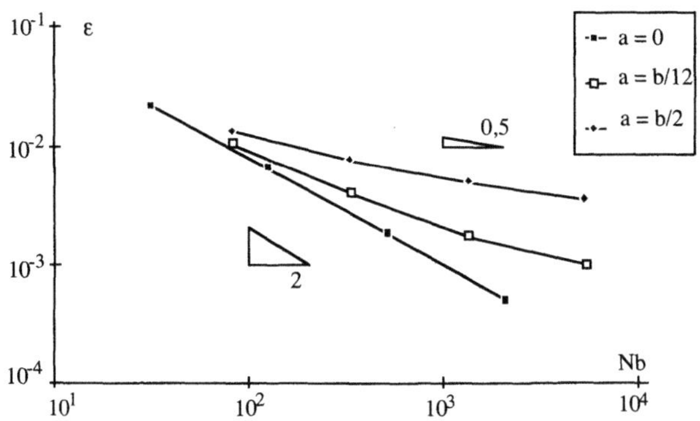

Figure 15. Error versus number of elements for different lengths of crack

4.2 Calculation of the Convergence Coefficient Density

The definition and the calculation method are given in (Coorevits et al. 1992). To illustrate it, let us consider the case of the cracked plate.

The computation of the convergence coefficients requires two steps:

- detection of the singularities and, more generally, of the steep gradient regions ;
- computation of the convergence coefficients p_E in these regions.

The computation of the local error indicator $\bar{\varepsilon}_E$ defined by:

$$\bar{\varepsilon}_E^2 = \frac{|\Omega|}{|E|}\varepsilon_E^2 \tag{22}$$

(where ε_E denotes the contribution to the relative error, $|E|$ the area of the element, $|\Omega|$ the area of the structure) allows the detection of the steep gradient regions. Indeed, we notice that the local errors are larger in these areas (Figure 16) than in the interior. Therefore, a comparison of this quantity with an *average error* identifies regions where the coefficient p_E must be computed.

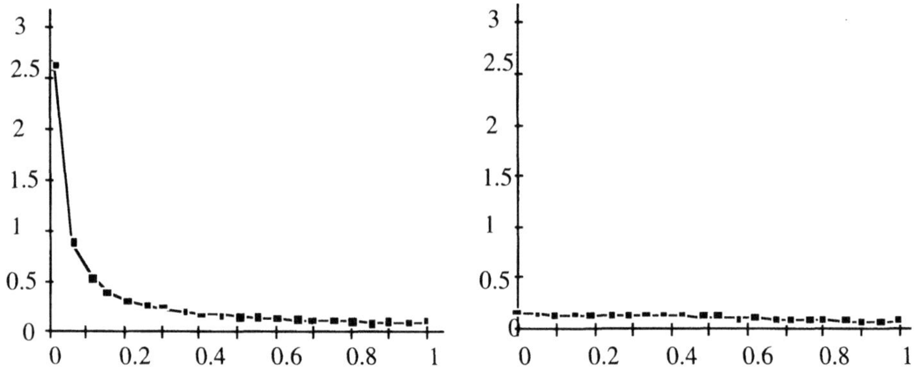

Figure 16. Local error at the crack tip and at an interior node in the X_1 -direction

To illustrate the method, let us consider the case of a cracked plate. To compute p_E, we use the energy density $\bar{e}_h$ defined at every node by:

$$\bar{e}_h(r) = \frac{1}{|A|} \int_A \mathrm{Tr}\left[\sigma_h \mathbf{K}^{-1} \sigma_h\right] \frac{dA}{2}$$

where $|A|$ denotes the measure of the area of the radius r (figure 17).

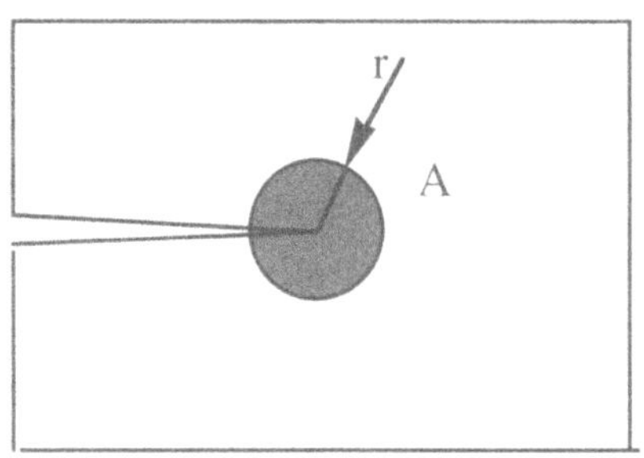

Figure 17. Area A of radius r at the crack tip

Figure 18 shows this energy density $\bar{e}_h$ as a function of the distance r from the crack tip. The theoretical energy density e can be written as:

$$e(r) = k\, r^{2(\alpha - 1)} + c \tag{23}$$

where k and c denote constants depending on the mechanical problem, and α the strength of the singularity, since the singular part of the displacement solution is $u = r^{\alpha} f(\theta)$. The identification of the theoretical expression of e with the computed energy density $\bar{e}_h$ by the least square method gives a coefficient $\bar{\alpha}$ close to the theoretical value α. In the case of the crack, we obtain $\bar{\alpha} = 0.52$ ($\alpha = 0.5$).

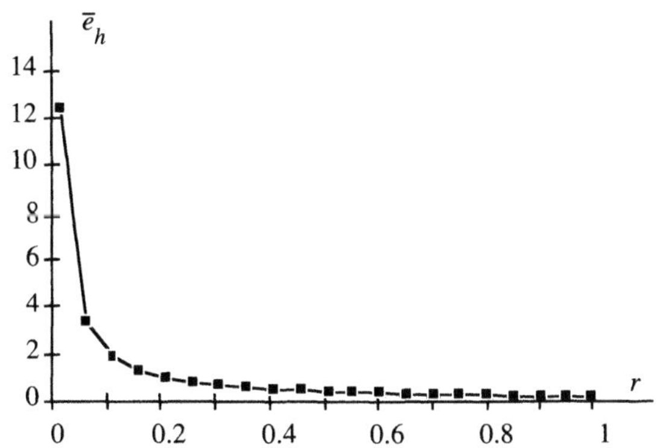

Figure 18. Energy density e_h as a function of the distance r from the crack tip

For the elements in the singularity region, we impose $p_E = \bar{\alpha}$; for the other elements of the structure $p_E = p$.

In practice, this method of identification allows us to account for not only the singularities but also the steep gradients which are not singularities. In addition, we find that the steep gradients are at the edge of the structure. The computation is made only for the nodes close to the boundary and consequently needs very little CPU time.

4.3 Optimal Mesh: Definition and Calculation

Let us consider that a first finite element analysis has lead to the global error ε and the element contributions ε_E .

Definition. A mesh T^* is named optimal if the corresponding global error ε^* is equal to the error ε_0 prescribed by the user and if its number of elements N^* is minimum.

This definition introduced by Pelle can be found in (Ladevèze, Coffignal and Pelle 1986). It improves the previous one developed in (Ladevèze and Leguillon 1981, 1983) which is still classically used and for which the local contributions ε_E^* are prescribed E^* uniform.

Calculation. Since the topology of the optimal mesh T^* is better described by a map of nodal sizes, the problem of optimization will be written as a function of those. In addition, the meshes are isotropic (for example, the ideal element is the equilateral triangle), but the elements may have rapidly varying size. The mesher, therefore, respects the prescribed surface at every node of an element E of the initial mesh T, and generates a regular distribution of the new elements of the mesh T^* inside the element E. The size for an isotropic mesh depends on the square root of the area, so we assume a linear distribution of the square of the sizes. The number of elements N_E^* can be estimated by:

$$N_E^* = \frac{1}{|E|} \int_E \frac{h_E^2}{\sum_{i=1}^{3} h_i^{*2} \lambda_i} \, dE \tag{24}$$

and the contribution ε_E^* of an element to the global error by:

$$\varepsilon_E^* = \frac{1}{|E|} \int_E \frac{\left(\sum_{i=1}^{3} h_i^{*2} \lambda_i\right)^{p_E}}{h_E^{2p}} \varepsilon_E^2 \, dE \tag{25}$$

where $|E|$ denotes the surface of the element, h_E its size, p_E its computed convergence coefficient, h_i^* the desired sizes at the three nodes of the element E and λ_i the barycentric coordinates. Consequently we have to solve the following problem:

$$\text{Minimize } N^* = \sum_E N_E^* \text{ with } \sum_E \varepsilon_E^{*2} = \varepsilon_0^2$$

This problem of optimization is solved using Lagrange multipliers; the conditions of optimality are numerically solved with an iterative process.

The control of optimality is crucial in judging the efficiency of the method. In some cases, the prescribed accuracy can be obtained without a minimum number of elements. The control is obtained by the following method: on the optimized mesh if the computed error ε^* is near the prescribed error ε_0, we recompute the size map for an imposed error equal to ε^*. If the mesh T^* is perfectly optimal, the size modification coefficient r_E, defined by:

$$r_E = \frac{\frac{1}{3} \sum_{i=1}^{3} h_i^*}{h_E} \tag{26}$$

must be equal to 1. The map of these coefficients allows us to appreciate the quality of the obtained meshes.

Remark : If the first mesh is too coarse, several intermediate meshes could be needed.

4.4 2D Examples

Quite a lot of examples have been done in (Ladevèze and Leguillon 1981, 1983) (Ladevèze, Coffignal and Pelle 1986) (Ladevèze, Pelle and Rougeot 1991) (Coorevits et al. 1992) (Coorevits, Dumeau and Pelle 1997). Two of them are shown here: the cracked plate and the gear tooth.

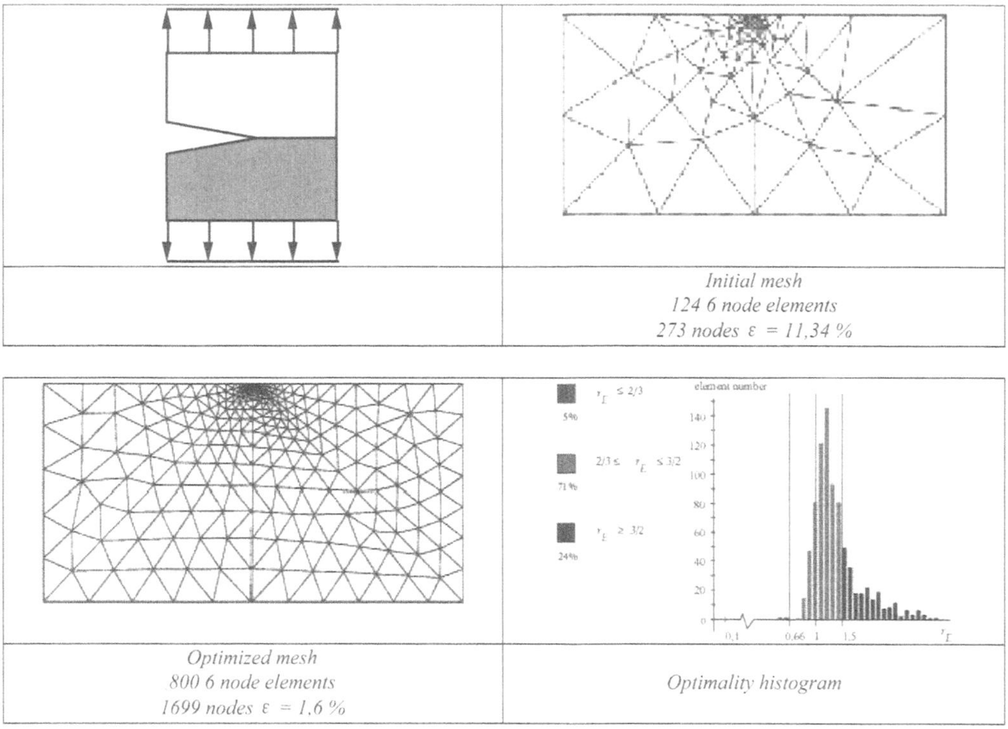

Initial mesh
124 6 node elements
273 nodes ε = 11,34 %

Optimized mesh
800 6 node elements
1699 nodes ε = 1,6 %

Optimality histogram

Figure 19. Cracked plate - prescribed error $\varepsilon_0 = 2\%$

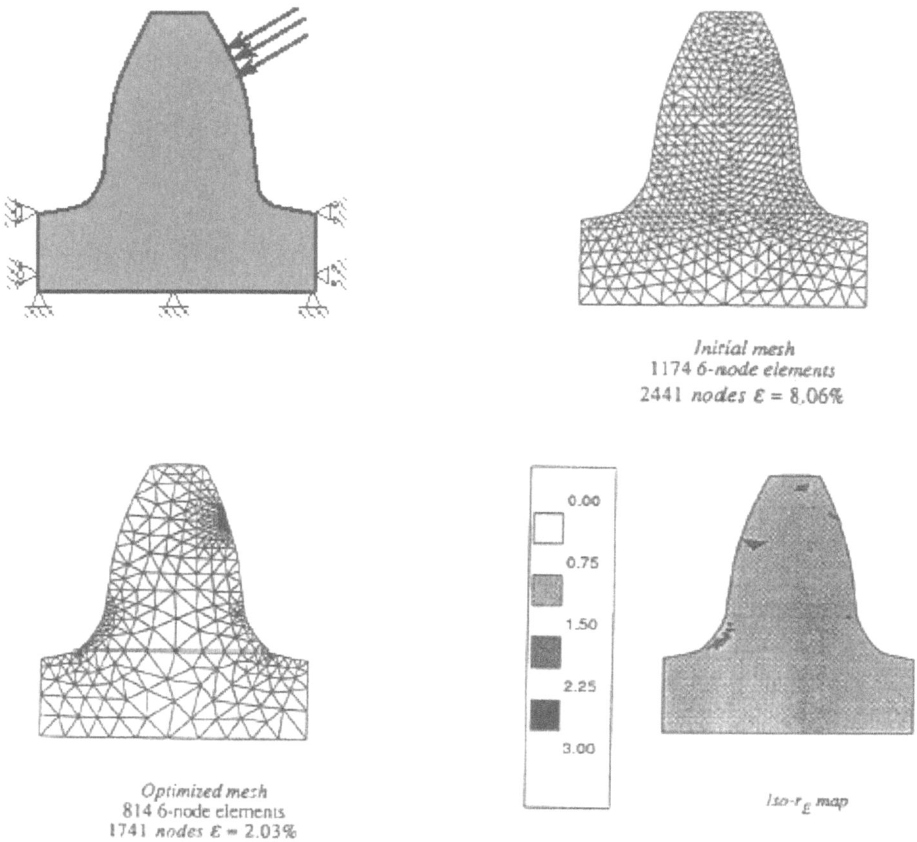

Figure 20. Gear tooth - prescribed error $\varepsilon_0 = 2\%$

4.5 3D Examples

Several 3D mesh adaptations are done in (Coorevits, Dumeau and Pelle 1996, 1997). Here is shown a structure meshed with 10-nodes tetrahedra. The prescribed error is 7%.

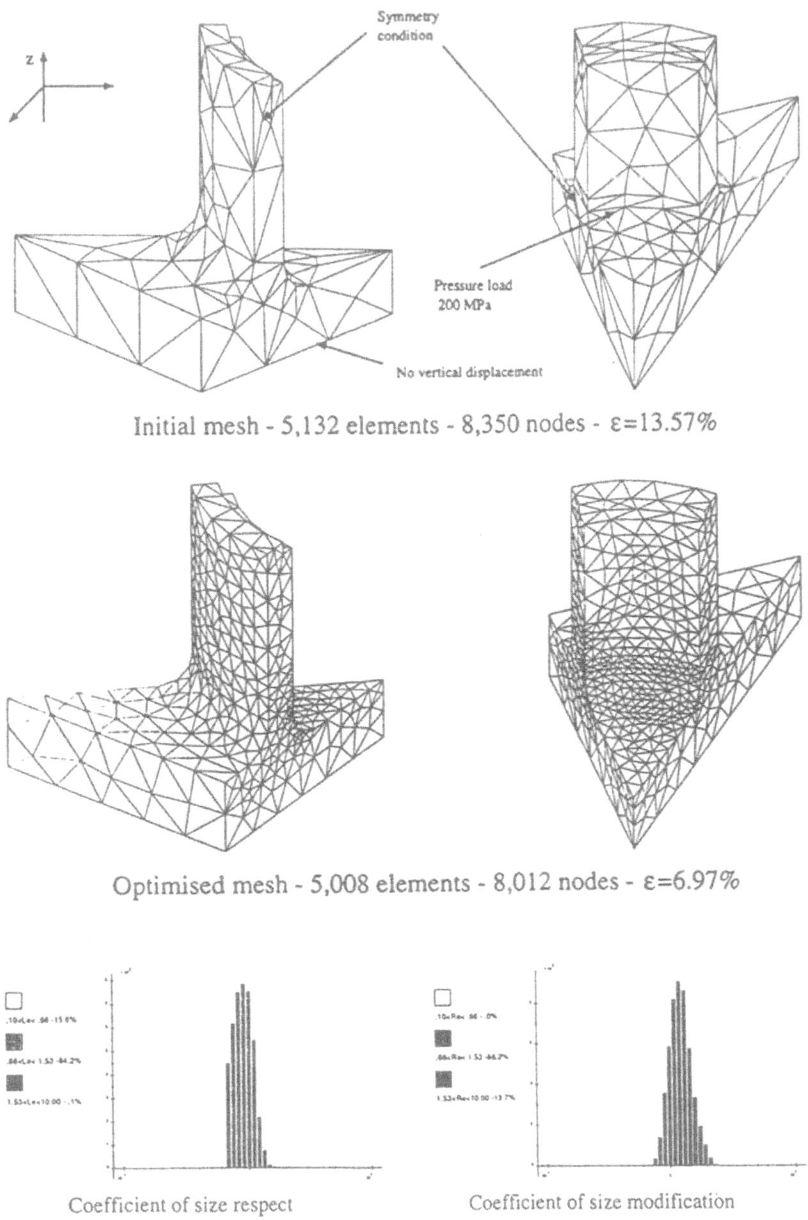

Figure 21. 3D mesh adaptation - prescribed error : 7%

4.6 Remarks

Other examples can be found in (Ladevèze and Pelle 2000). Extension of CRE estimators to plate elements has been done in (Ladevèze 1975) (Boisse and Coffignal 1998) (Boisse et al. 1999).

5. Constitutive Relation Errors for f.e. Analysis Considering (Visco)Plasticity and Damage - Adaptivity

A posteriori error estimators based on constitutive relation residuals are developed to measure the quality of finite element computations of structures which exhibit plastic / viscoplastic behavior with softening. All the classical error sources involved in the computation over the studied time interval are taken into account: the space discretization (the mesh), the time discretization and the iterative technique used to solve the nonlinear discrete problem.

Concerning nonlinear problems, it is important to distinguish between time-dependent and time-independent problems. For the late type, let us mention Babuska and Rheinboldt (1982) for the design of estimates for nonlinear elasticity problems of rods, Johnson and Hansbo (1991) for Hencky-type plasticity problems, and Verfurth (1996) for the proof of bounds on the error. For nonlinear time-dependent problems, plasticity and viscoplasticity problems were treated in (Bass and Oden 1987) (Rannacher and Suttmeier 1998) (Boroomand and Zienkiewicz 1998), strain localization in (Wunderlich et al. 1998) (Aubry et al. 1992) (Peric et al. 1994) (Pijaudier-Cabot et al. 1995) and large strains in (Bussy et al. 1985) (Fourment and Chenot 1995) (Brink and Stein 1998). Most of the proposed error estimators for time-dependent nonlinear problems are simply error estimators devised for usual linear or nonlinear problems and applied at each time step. Thus, the error due to the time discretization can not be taken into account. At the present time, only the constitutive relation error estimators are able to take into account all sources of error.

For plasticity and viscoplasticity problems without softening, two kinds of error estimators - named "Drucker error estimator" and "dissipation error estimator" - have been respectively introduced in (Ladevèze, Coffignal and Pelle 1986) (Gallimard, Ladevèze and Pelle 1996) and in (Ladevèze and Moes 1998, 1999). It is important to note that for the dissipation constitutive relation error estimator, a link with the "solution error " similar to the Prager-Synge theorem has been proved (Ladevèze 1999).

An extended damage model coupled with plasticity and viscoplasticity is investigated for the dissipation constitutive relation error estimator. A first attempt is given in (Ladevèze, Moes and Douchin 1999). New error estimators are introduced; they are based on the convexity properties of the state equations and of the state evolution laws. Numerical examples are shown.

5.1 The Drucker CRE Estimator

Let us come back to the time-dependent problem defined by the equations (1), (2) (3).

The densities $\underline{U}_d$, $\underline{f}_d$ and $\underline{F}_d$ are known on $[0, T]$. We also consider that the finite element solution has been computed using a displacement approach. The finite element displacement-stress pair $(\underline{U}_h, \sigma_h)$ is then known at the time value t_m with $m \in [0, 1, \cdots, n]$. Supposing that the data are piecewise linear on $[0, T]$, it is easy to extend the f.e. solution on $[0, T]$. Let us also denote it by $(\underline{U}_h, \sigma_h)$. Several f.e. stresses are available ; here, the stress σ_h is f.e. equilibrated on $[0, T]$.

The principle behind our approach is to associate a new displacement-stress pair denoted by $(\hat{\underline{U}}_h, \hat{\sigma}_h)$, which is admissible i.c. belonging to $\mathcal{U}_{ad}^{[0,T]} \times \mathcal{S}_{ad}^{[0,T]}$, to the data and the finite element displacement-stress pair. This pair verifies the most reliable equations of the reference problem to be solved ; its quality, and thus the quality of the f.e. solution, is measured through the residue related to the verification of the constitutive relation on $[0, T]$.

The displacement $\hat{\underline{U}}_h$ is generally taken as $\underline{U}_h$ on $[0, T]$. The crucial point is always to build a stress $\hat{\sigma}_h$ which verifies the equilibrium equations on $[0, T]$. The recovery method described previously is still used for any time value t_m with $m \in [0, 1, \cdots, n]$. $\hat{\sigma}_h$ is taken piecewise linear on $[0, T]$. Let $(\hat{\underline{U}}_h, \hat{\sigma}_h)$ be a displacement-stress pair belonging to $\mathcal{U}_{ad}^{[0,T]} \times \mathcal{S}_{ad}^{[0,T]}$.

For materials satisfying Drucker's stability inequality, a constitutive relation error has been introduced in (Ladevèze 1985) and associated error estimators in (Ladevèze et al. 1986) and (Gallimard et al.1996). A broad class of plasticity and viscoplasticity models enter into this category. Let us assume that this inequality is satisfied; denoting $(\hat{\underline{U}}_h, \hat{\sigma}_h)$ by $(\underline{U}_{KA}, \sigma_{SA})$, we can associate to it, by using the constitutive relation, a stress σ_{KA} and a strain ε_{KA}, defined on $\Omega \times [0, T]$. Then, the following is derived:

$$\eta(\underline{M}, t) = \int_0^t dt \, \mathrm{Tr}\left[(\sigma_{SA} - \sigma_{KA})(\dot{\varepsilon}_{SA} - \dot{\varepsilon}_{KA})\right] \tag{27}$$

The main results are:

- η is positive or null on $\Omega \times [0, T]$,
- $(\underline{U}_{KA}, \sigma_{KA})$ is the exact solution on $\Omega \times [0, T]$ if and only if η is equal to zero on $\Omega \times [0, T]$.

The global error on constitutive relation at T is then:

$$\varepsilon_T = \left[\sup_{t \in [0,T]} \int_\Omega \eta(\underline{M}, t) \, d\Omega \right]^{1/2} \Big/ D \tag{28}$$

$$D^2 = \frac{1}{2} \int_0^T \int_\Omega dt \, d\Omega \left[\mathrm{Tr}\left[\sigma_{KA} \dot{\varepsilon}_{KA}\right] + \mathrm{Tr}\left[\sigma_{SA} \dot{\varepsilon}_{SA}\right] \right]$$

This includes all sources of error: space discretization, time discretization, etc. A local error can also be introduced:

$$\varepsilon_{T,loc} = \sup_{\substack{t \in [0,T] \\ \underline{M} \in \Omega}} \left[\eta\left(\underline{M}, t\right) \text{mes}\Omega\right]^{1/2} \Big/ D \tag{29}$$

Remark : Most applications which have been done, have used the standard recovery method; this method does not depend on the constitutive relation. Some works in progress use the weak prolongation condition; a minimization procedure is then introduced. For example, one minimizes at any time t_m, $m \in [1, 2 \cdots, n]$ the free energy written in term of stresses (Gallimard et al. 2000).

5.2 The Dissipation CRE Estimator

This kind of error has been introduced in (Ladevèze 1989) and detailed in (Ladevèze 1999). Associated error estimators are described in (Ladevèze and Moës 1998, 1999).

The reference problem to be solved is reformulated introducing the internal variables framework. Then, the state of the material at the point $\underline{M}$ and at the time t is defined by the inelastic strain ε_p and additional internal variables denoted by $\mathbf{X}$ (hardening variables, damage variables, ...). The force associated to $\mathbf{X}$ is denoted by $\mathbf{Y}$. Therefore, one is looking for $s \equiv \left(\underline{U}, \dot{\varepsilon}_p, \dot{\mathbf{X}}, \sigma, \mathbf{Y}\right)$ defined on the time-space domain $[0, T] \times \Omega$, the corresponding space being $\mathbf{S}^{[0,T]}$.

Moreover, equations are divided into two groups :

- The first group - related to the free energy - defines the *admissibility* of a solution. It combines the equilibrium equations, the kinematic constraints and the state equations. The corresponding subspace is $\mathbf{S}_{ad}^{[0,T]}$.

$$\begin{aligned} &\bullet\ \underline{U} \in \mathcal{U}_{ad}^{[0,T]} && \sigma \in \mathcal{S}_{ad}^{[0,T]} \\ &\bullet\ \sigma = \mathbf{K}\left(\varepsilon\left(\underline{U}\right) - \varepsilon_p\right) && \mathbf{Y} = \mathbf{\Lambda}\,\mathbf{X}\ \ \left(\text{on } \Omega \times [0, T]\right) \end{aligned} \tag{30}$$

- The second group - related to the dissipation - only includes the evolution laws. The quality of an admissible solution will be estimated through the quality of satisfaction of these laws.

$$\begin{vmatrix} \dot{\varepsilon}_p \\ -\dot{\mathbf{X}} \end{vmatrix} = \mathbf{B}\left(\begin{vmatrix} \sigma \\ \mathbf{Y} \end{vmatrix}\right) \qquad X|_{t=0} = 0 \tag{31}$$

The operators $\mathbf{\Lambda}$ and $\mathbf{B}$ characterize the material. It is not restrictive to suppose that the material model is normal i.e. the operator $\mathbf{\Lambda}$ is linear (Ladevèze 1999).

Following the recovery technique used in the case of the Drucker error, one associates to the finite element solution and the data an admissible solution $\hat{s}_h$ $\left(\hat{s}_h \in \mathbf{S}_{ad}^{[0,T]}\right)$. The only equations which are not a priori satisfied by $\hat{s}_h$ are the evolution laws. The practical definition of such an error depends on the form of the evolution laws. Most materials can be described by standard models; in this case, the dissipation error for any admissible solution $\hat{s}$ is:

$$\eta\left(\hat{s}\right)_{|t,\underline{M}} \equiv \varphi\left(\hat{\dot{\varepsilon}}_p, -\hat{\dot{\mathbf{X}}}\right) + \varphi^*\left(\hat{\sigma}, \hat{\mathbf{Y}}\right) - \mathrm{Tr}\left[\hat{\sigma}\, \hat{\dot{\varepsilon}}_p\right] + \hat{\mathbf{Y}}^t \hat{\dot{\mathbf{X}}} \tag{32}$$

where φ, φ^* are a couple of convex dual functions, named potentials of the evolution laws.

The main results are:

- η is positive or null on $\Omega \times [0, T]$
- $\hat{s}$ is the exact solution on $\Omega \times [0, T]$ if and only if η is equal to zero on $\Omega \times [0, T]$.

The global dissipation error at time T is then:

$$\varepsilon_T \equiv \frac{\int_0^T \int_\Omega dt\, d\Omega\, \eta\left(\hat{s}\right)}{D^2} \tag{33}$$

with:

$$D^2 \equiv 4 \sup_{t \in [0, T]} \left[\frac{1}{2} \int_0^t \int_\Omega dt\, d\Omega \left[\varphi\left(\hat{\dot{\varepsilon}}_p, -\hat{\dot{\mathbf{X}}}\right) + \varphi^*\left(\hat{\sigma}, \hat{\mathbf{Y}}\right)\right] + \frac{1}{2} \int_\Omega \left(\mathrm{Tr}\left[\mathbf{K}^{-1}\, \hat{\sigma}\, \hat{\sigma}\right] + \left(\Lambda^{-1}\, \hat{\mathbf{Y}}\right)^t \mathrm{o}\, \hat{\mathbf{Y}}\right)_{|t} d\Omega \right.$$

The implementation of the errors for the f.e. method is easy. Let us consider the classical incremental finite element method. The problem to be solved on $[0, T]$ is divided into a succession of resolutions over $[t_n, t_{n+1}]$ $(n = 0, \ldots, N\,; t_0 = 0\,, t_{N+1} = T)$. For the sake of simplicity, displacements, stresses prescribed forces and displacements are supposed to be time-continuous and piecewise linear within the subintervals. It follows that the recovery of an equilibrated stress over $[0, T]$ can be done easily supposing also it is piecewise linear: the recovery method proposed in elasticity is applied at the time-step t_n.

Applied to the recovered $\hat{s}_h$, one obtains the value ε_{hT} of the dissipation error. A remarkable property of such an error is its link with the error between the exact and the admissible solutions which has been established in (Ladevèze 1999). This property is an extension of the famous Prager-Synge theorem valid in elasticity.

Several examples are detailed in (Ladevèze and Moës 1998, 1999).

Specific error indicators have been also introduced in (Ladevèze and Moës [1998, 1999]; they characterize:

- the space discretization: i_{spce},
- the time discretization: i_{time},
- the iteration stopping: i_{ite}.

Remark.

Extensions to non standard material can be easily done using the concept of bipotential introduced by de Saxcé (1992). CRE estimators using this formalism have been developed by Hjiaj (1999).

5.3 Extension of the Dissipation CRE Estimator to (Visco-)Plasticity and Damage

For the sake of simplicity, let us consider first elasticity coupled to isotropic damage.

Elasticity coupled to isotropic damage. One has :

$$\begin{aligned} &\sigma = (1-d)\,\mathbf{K}_0\,\varepsilon \qquad 0 \leq d \leq 1 \\ &Y = \frac{1}{2}\,\mathrm{Tr}\left[\mathbf{K}_0\,\varepsilon\,\varepsilon\right] \\ &\dot{d} = h'\left(\langle Y - Y_0 - \omega(d)\rangle_+\right) \end{aligned} \tag{34}$$

where $h'(0) = 0$, $\omega(0) = 0$; $h'(x) \geq 0$, $\omega'(x) \geq 0$, $\forall\, x \geq 0$.

The scalar d is the damage variable whose value must stay between 0 (non-damaged material) and 1 (completely damaged material). $\mathbf{K}_0$ is the elastic stiffness of the non-damaged material. It is a positive definite symmetric tensor. The evolution of d is driven by the strain energy Y through the positive material function h'. The damage may only grow (irreversible phenomena). The material parameter Y_0 is the threshold under which d may not grow. As the damage proceeds, this threshold rises through the scalar material function $\omega(.)$ which introduces a "hardening" effect. Characterizing the "hardening " effect by the additional internal variable α and denoting its associate variable by β, we may rewrite an equivalent form in which we distinguish the state laws:

$$\begin{aligned} &\sigma = (1-d)\,\mathbf{K}_0\,\varepsilon \qquad 0 \leq d \leq 1 \\ &Y = \frac{1}{2}\,\mathrm{Tr}\left[\mathbf{K}_0\,\varepsilon\,\varepsilon\right] \\ &\beta = \omega(\alpha) \end{aligned} \tag{35}$$

from the evolution laws:

$$\begin{aligned} &\dot{d} = h'\left(\langle Y - Y_0 - \beta\rangle_+\right) \\ &\dot{\alpha} = -\,h\left(\langle Y - Y_0 - \beta\rangle_+\right) \end{aligned}$$

and the initial conditions : $d = \alpha = 0$ at $t = 0$.

Two bilinear forms are then introduced. First, a work bilinear form is introduced for describing the state laws:

$$\left([\varepsilon,\, Y,\, \beta],\, [\sigma,\, d,\, \alpha]\right) \rightarrow \mathrm{Tr}\left[\sigma\,\varepsilon\right] + Yd + \beta\alpha \tag{36}$$

We have proved in (Ladevèze, Moes and Douchin 1999) that the state laws can be defined through two dual potentials, $\Psi(\sigma,\, d,\, \alpha)$ and $\Psi^*(\varepsilon,\, Y,\, \beta)$, which are convex functions. This result is not trivial. If the state laws are not verified, the constitutive relation residual is:

$$\eta_\Psi(\mathbb{\varepsilon}, Y, \beta; \sigma, d, \alpha) = \Psi(\sigma, d, \alpha) + \Psi^*(\mathbb{\varepsilon}, Y, \beta) - \mathrm{Tr}[\sigma\,\mathbb{\varepsilon}] - Yd - \beta\alpha \geq 0 \tag{37}$$

For the evolution laws, the dissipation bilinear form is used :

$$([\dot{d}, -\dot{\alpha}]; [Y, \beta]) \rightarrow Y\dot{d} - \beta\dot{\alpha}$$

Here, too, the evolution laws can be defined by two convex dual potentials, $\varphi^*(Y, \beta)$ and $\varphi(\dot{d}, -\dot{\alpha})$; the corresponding constitutive relation residual is:

$$\eta_\varphi(\dot{d}, -\dot{\alpha}, Y, \beta) = \varphi(\dot{d}, -\dot{\alpha}) + \varphi^*(Y, \beta) - Y\dot{d} + \beta\dot{\alpha} \geq 0 \tag{38}$$

A solution $s \equiv (\mathbb{\varepsilon}, Y, \beta, \sigma, d, \alpha)$ is called admissible on $\Omega \times [0, T]$ if it verifies the equilibrium equations and the kinematic constraints. To be exact, it has to verify the state laws and the evolution laws too ; the value of the following constitutive relation residual must be zero:

$$e_t^2(s) \quad t \in [0, T] \tag{39}$$

with :

- $e_t = \sup_{0 \leq \tau \leq t} e_\tau$

- $e_t(s) = \int_\Omega \eta_\Psi(\mathbb{\varepsilon}, Y, \beta; \sigma, d, \alpha)|_t \, d\Omega + \int_0^t dt \int_\Omega \eta_\varphi(\dot{d}, -\dot{\alpha}; Y, \beta)\, d\Omega$

The relative global error is:

$$\varepsilon_T^2(s) = \frac{e^2_{[0,T]}(s)}{\sup_{t \in (0,T)} d_t^2} \tag{40}$$

$$d_t^2 \equiv \int_\Omega [\varphi(\mathbb{\varepsilon}, Y, \beta) + \varphi^*(\sigma, d, \alpha)]|_t \, d\Omega + \int_0^t dt \int_\Omega \varphi(\dot{d}, -\dot{\alpha}) + \varphi^*(Y, \beta)\, d\Omega$$

Remark:

In the case of anisotropic damage, the damage needs to be expressed in a more complex manner than with a scalar d, for instance by a n-vector $\underline{d}$. Two dual convex potentials can be still introduced for defining the sate laws if :

- $\underline{d} \rightarrow \mathbf{K}(\underline{d})$ is a linear function
- $\mathbf{K}(0)$ is a positive definite operator.

The proof is given in Ladevèze et al. (1999).

Extension to (visco)-plasticity coupled to damage

(Visco)-plastic state laws and evolution laws have to be introduced. A posteriori error estimators are constructed following the previous approach which supposes two types of convexity. Let us note that bipotentials can be also used for describing both the state laws and evolution laws (De Saxcé 1992).

5.4 Error Estimation and Adaptivity for (Visco)-Plastic Structures without Softening

Several experiments concerning both the Drucker error and the dissipation error have been done. Only, one example is reported here: the frame problem (Ladevèze and Moes 1998).

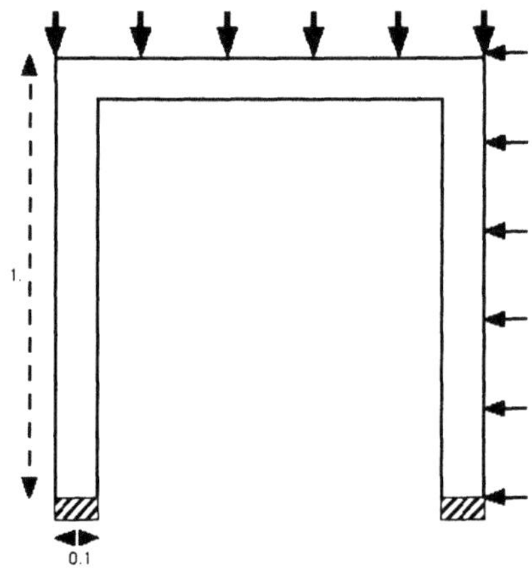

Figure 22. The geometry and applied loading on the frame

The frame, figure 22, is submitted to a growing then decreasing load on its right side (thin line in figure 23). Then, an increasing pressure is applied to its upper part (thick line in figure 23). The Prandtl-Reuss plastic model with linear hardening is considered. The optimized mesh in elasticity for an error of 10% is given figure 23. It is composed of 194 six-node triangular elements. Three time steps are used and the tolerance for the iterative technique is set to $\delta_{tol} = 10^{-2}$. The errors obtained are 28.15% for the dissipation error and 26.90% for the Drucker error.

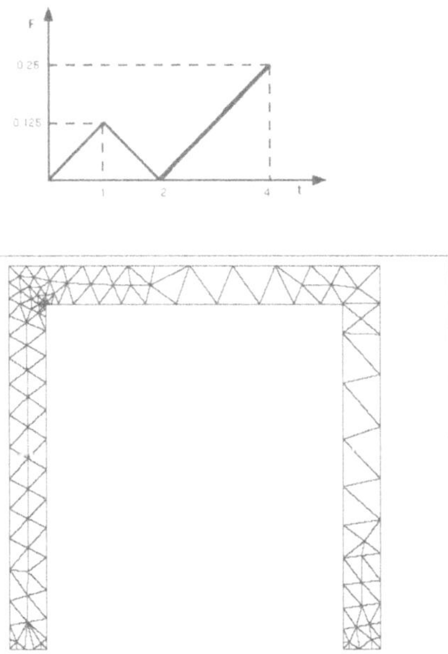

Figure 23. Evolution of the loading and initial mesh optimized in elasticity, 194 six-nodes triangles

Adaptive procedures have been implemented for optimizing both the time discretization and the mesh discretization. For an asked accuracy of 5.10^{-2}, figures 24 and 25 give the final meshes. Figure 26 shows the evolutions of the two errors for the successive adaptive steps.

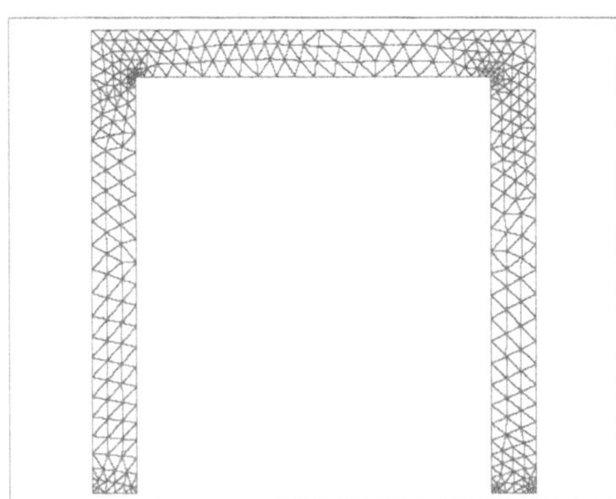

Figure 24. Final mesh optimized with the dissipation error, 661 six-node triangles, 11 time steps

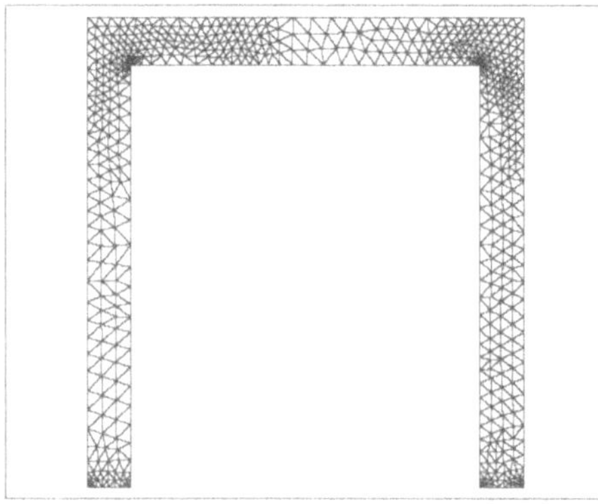

Figure 25. Final mesh optimized with Drucker error, 1286 six-node triangles, 12 time steps

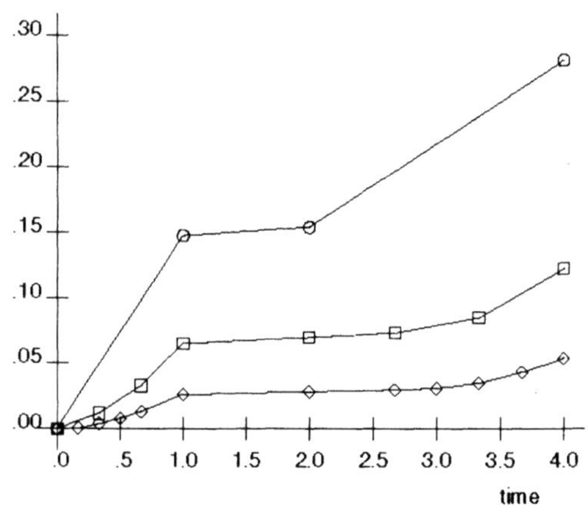

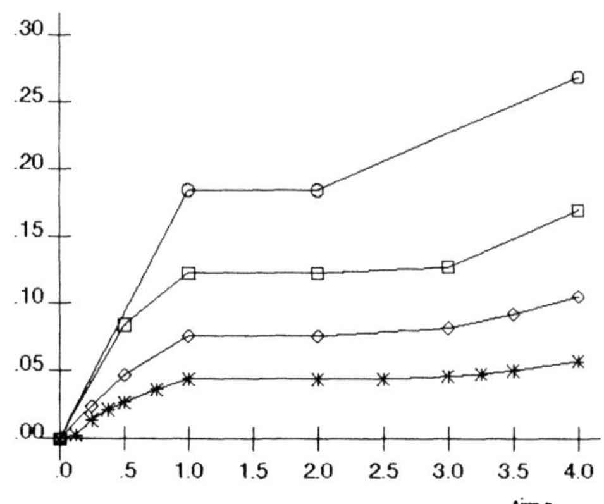

Figure 26. Time evolution of the error for the successive adaptive steps - the dissipation error on the top and the Drucker error on the bottom

5.5 Enhanced Drucker Error Estimator for (Visco)-Plasticity without Softening

The weak prolongation condition is introduced. The corresponding effectivity index is much better for quasi-perfect plasticity and for deformed element. The example developed in (Gallimard et al. 2000) is used ; the mesh can be seen figure 4 where 1/H is the shape ratio. The loading is increasing; the studied prestress is the first one. Figures 27 and 28 give the effectivity indexes for standard error estimators and the enhanced one as a function of the shape ratio.

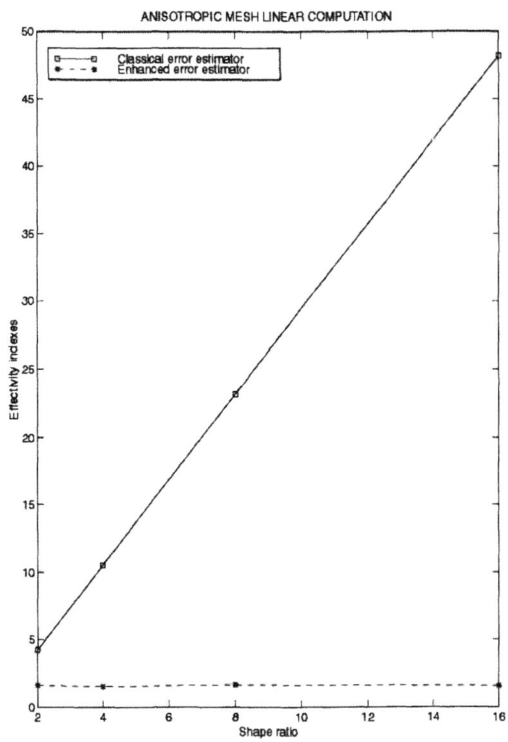

Figure 27. Effectivity index in elasticity versus shape ratio, for the classical and enhanced error estimators

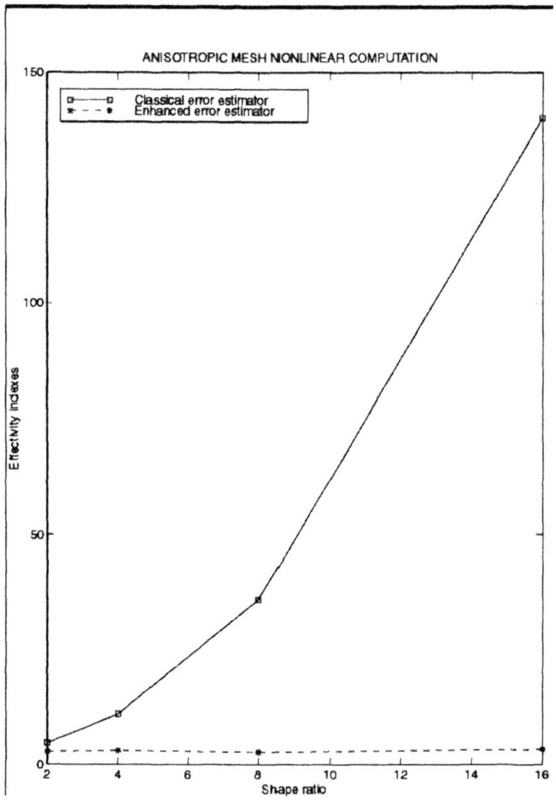

Figure 28. Effectivity index in elastoplasticity versus shape ratio, for the classical and enhanced error estimator

5.6 First Numerical Experiments for (Visco)-Plastic Structures with Softening

We consider a beam with a flaw shown figure 29, fixed on its left end and submitted to a growing displacement on its right end described figure 30. The length of the beam is 1 (dimensionless treatment of the problem). The behavior of the beam material is elastoplastic coupled to isotropic damage and follows equations (34). In fact, the initial yield stress R_0 is chosen high enough to avoid plasticity. The state laws and the evolution laws derive from the potentials given in (37)-(38). The material parameters are:

$$E_0 = 6350, \quad Y_0 = 0, \quad Y_c = 0.13, \quad k = 7000, \quad a = 1, \quad R_0 = 1000, \quad k_y = 22582, \quad m = 1$$

where E_0 is the Young's modulus of the non damaged material.

Figure 33 gives the evolution of the damage for a computation carried out with 100 time steps and a uniform mesh of 50 elements. A strong localization of the damage in the flaw region occurs as soon as d reaches 0.5.

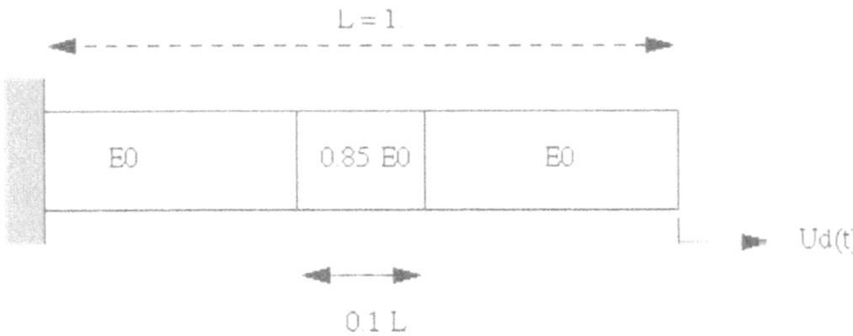

Figure 29. Geometry and loading for the beam problem

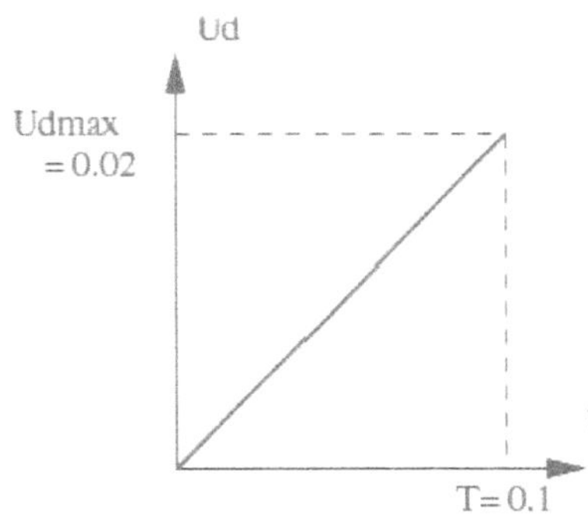

Figure 30. Evolution of the loading on the beam as a function of time

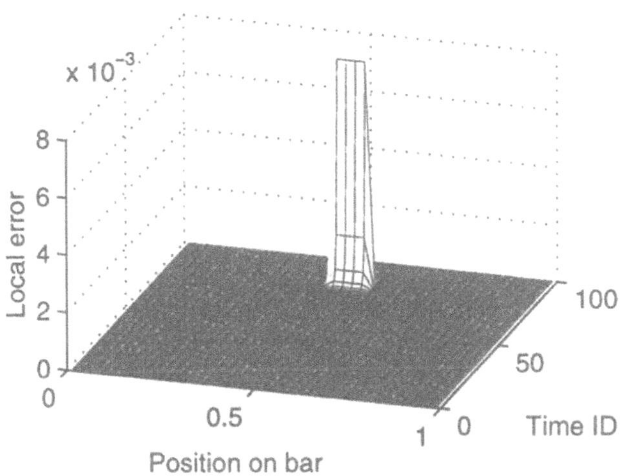

Figure 31. Local contributions to the error

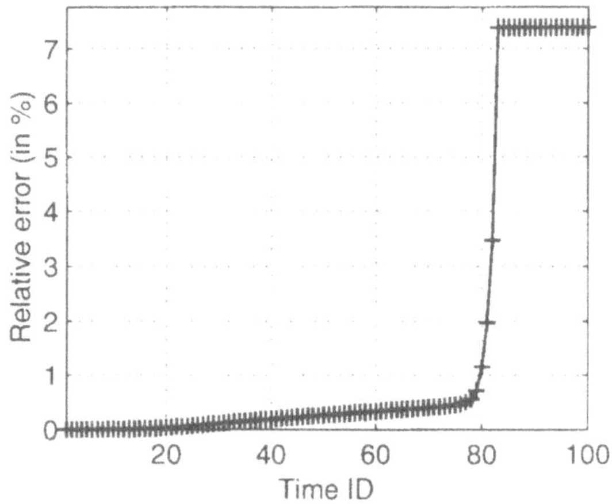

Figure 32. Contribution to the relative error through time

Figure 31 shows the space-time map of the local contributions to the error defined in (40). We note than the error is localized in the vicinity of the rupture. The evolution of the contribution to the relative error through time is given figure 32.

Table 1 shows the influence of the numerical parameters on the error.

Table 1. Influence of the numerical parameters on the error

nb steps	50	100	200	400	800	1600	3200	6400
error (%)	8,37	7,53	5,64	3,26	1,76	1,00	0,49	0,25

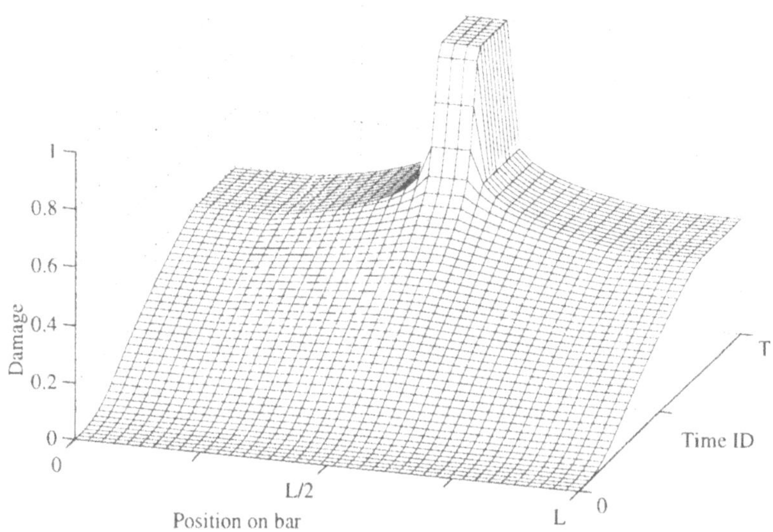

Figure 33. Damage scalar variable d as a function of time and space location on the bar

6 Constitutive Relation Error for f.e. Dynamic Analysis

Chapters 6 and 7 deal with a central and crucial question for structural dynamic models which, in fact, concerns the whole Mechanics. Three kinds of models can be distinguished: the continuum mechanics model, the numerical model and experimental simulations. Two categories of quality control can be differentiated.

First, one can study the situation where the reference is the continuum mechanics model, the approximated model being the numerical one. This situation is the usual one for which quite a lot of a posteriori error estimators have been derived. For dynamic problems, the main difficulty lies in building error estimators which maintain a satisfactory effectivity for problems encountered in practice. This difficulty has been previously highlighted in the literature, in particular in (Bouillard and Ihlenburg 1999). The error estimator described herein, is based on a recovering technique which gives an equilibrated displacement, stress and acceleration triplet; it fully takes advantage of the properties of the either explicit or implicit integration scheme used (Combe, Ladevèze and Pelle 1999).

The second situation is apparently only quite different, the approximated and reference models being respectively the continuum mechanics model and the experimental data. Error estimators can be built as extension to a posteriori error estimators quantifying the quality of a f.e. model with respect to the continua one. This is the line that we have been following for nearly 15 years leading to specific constitutive relation error estimators. More common are other works which enter into two categories. The first consists of the "direct" method (Berman and Nagy 1983) and (Kaouk and Zimmerman 1994). The model is corrected without really taking into account either the geometrical or mechanical characteristics of the structure. The second category encompasses "indirect" methods, which are also called sensitivity methods (Farhat and Hemez 1993). The model is corrected by acting on the structural parameters, such as the Young's modulus, density, section etc. Most of the previous methods are particular cases of regularization techniques evoked in (Natke 1991, 1994).

For both situations the common stone introduced in this chapter is the concept of CRE in dynamics.

6.1 The Reference Problem to Be Solved

Let us consider again the problem defined by equations (1) (2) (3) with some modifications. At the initial time t = 0, the initial position is $\underline{U}_0$ and the initial velocity $\underline{V}_0$. The quantity $\rho \frac{d^2 \underline{U}}{dt^2}$ where ρ is the density is introduced as a full unknown, denoted by $\underline{\Gamma}$. The problem can be written as follows:

Find $\underline{U}(\underline{M}, t), \sigma(\underline{M}, t), \underline{\Gamma}(\underline{M}, t)$ defined over $\Omega \times [0, T]$ such that they verify:

- Kinematics constraints and initial conditions :

$$\underline{U} \in \mathcal{U}^{[0,T]}, \underline{U}\Big|_{\partial_1\Omega} = \underline{U}_d \text{ on }]0, T[\tag{41}$$

$$\underline{U}\Big|_{t=0} = \underline{U}_0, \quad \frac{d\underline{U}}{dt}\Big|_{t=0} = \underline{V}_0$$

- Equilibrium equations :

$$\sigma \in \mathcal{S}^{[0,T]} \quad \underline{\Gamma}_d \in \mathcal{T}^{[0,T]}$$

$$\forall \underline{U}^* \in \mathcal{U}_{ad,0}$$
$$\forall t \in]0,T[\int_\Omega \underline{\Gamma} \circ \underline{U}^* \, d\Omega = -\int_\Omega \mathrm{Tr}\left[\sigma \, \varepsilon(\underline{U}^*)\right] d\Omega + \int_\Omega \underline{f}_d \circ \underline{U}^* \, d\Omega + \int_\Omega \underline{F}_d \circ \underline{U}^* \circ ds \tag{42}$$

- Constitutive relations :

$$\forall t \in]0,T[\quad \forall t\underline{M} \in \Omega$$
$$\sigma\Big|_t = \mathbf{A}\left(\varepsilon\left(\dot{\underline{U}}\Big|_\tau\right), \tau \le t\right) \tag{43}$$

$$\Gamma\Big|_t = \rho \, \ddot{\underline{U}}\Big|_t \tag{44}$$

The use of $\underline{\Gamma} = \rho \, \ddot{\underline{U}}$ as a constitutive relation, makes it possible to consider the acceleration quantity $\underline{\Gamma}$ as being independent of the displacement field $\underline{U}$, and the link is recreated by this constitutive relation. The advantage of such an approach is to clearly separate the kinematic quantity $\underline{U}$, which satisfies the kinematic constraints and the initial conditions, from the dynamic fields $(\sigma, \underline{\Gamma})$ which satisfy the equilibrium equation.

6.2 Constitutive Relation Error Estimators

Admissible quantities. The notion of constitutive relation error is based on a partition of the reference problem's equations into two groups:

- kinematic constraints, initial conditions and equilibrium equation
- constitutive relations

Definition: A triplet $s_{Ad} = (\underline{U}_{KA}, \sigma_{DA}, \underline{\Gamma}_{DA})$ is admissible if:

- $\underline{U}_{KA}$ satisfies the kinematic constraints and the initial conditions (41), and
- $(\sigma_{DA}, \underline{\Gamma}_{DA})$ satisfy the equilibrium equation (42).

The quality of s_{Ad} as an approximate solution to the reference problem can be assessed by the way in which the constitutive relations (43) and (44) are satisfied.

The Drucker CRE estimator. Given an admissible solution $s_{Ad} = \left(\underline{U}_{KA}, \sigma_{DA}, \underline{\Gamma}_{DA}\right)$, to the "kinematically-admissible" field $\underline{U}_{KA}$, the constitutive relations enable associating a stress field σ_{KA} such that:

$$\sigma_{KA}\Big|_t - \mathbf{A}\left(\varepsilon\left(\dot{\underline{U}}_{KA}\right)\Big|_\tau, \tau \in [0, t]\right) \quad \forall t \in [0, T] \tag{45}$$

along with an acceleration quantity field $\underline{\Gamma}_{KA}$ such that:

$$\underline{\Gamma}_{KA} = \rho\, \ddot{\underline{U}}_{KA}$$

With respect to the "dynamically-admissible" fields $\left(\sigma_{DA}, \underline{\Gamma}_{DA}\right)$, the inverse constitutive relations also allow associating a strain field $\dot{\varepsilon}_{DA}$ such that:

$$\sigma_{DA}(t) = \mathbf{A}\left(\dot{\varepsilon}_{DA}\Big|_\tau, \tau \in [0, t]\right) \quad \forall t \in [0, T] \tag{46}$$

and a displacement field $\underline{U}_{DA}$ such that:

$$\ddot{\underline{U}}_{DA} = \frac{1}{\rho}\underline{\Gamma}_{DA} \quad \underline{U}_{DA}\Big|_{t=0} = \underline{U}_0 \quad \dot{\underline{U}}_{DA}\Big|_{t=0} = \underline{V}_0$$

The extended Drucker residue is:

$$\eta\left(t, s_{Ad}\right) = \gamma \int_0^t \left(\underline{\Gamma}_{DA} - \underline{\Gamma}_{KA}\right)\left(\dot{\underline{U}}_{DA} - \dot{\underline{U}}_{KA}\right) d\Omega\, d\tau + (1-\gamma) \int_0^t \mathrm{Tr}\left[\left(\sigma_{DA} - \sigma_{KA}\right)\left(\dot{\varepsilon}_{DA} - \dot{\varepsilon}\left(\underline{U}_{KA}\right)\right)\right] d\Omega\, d\tau \tag{47}$$

where γ is a parameter ($0 \le \gamma \le 1$) which enables balancing the two terms of η. A typical value is: $\gamma = 0.5$.

It is supposed that the material model verifies the Drucker's stability conditions. Then, we have :

$$\begin{aligned} &\bullet\ \forall\, t \in [0, T] \quad \eta\left(t, s_{Ad}\right) \ge 0 \\ &\quad \forall\, \underline{M} \in \Omega \\ &\bullet\ \left\{\eta\left(t, s_{Ad}\right) = 0 \text{ on } \Omega \times [0, T]\right\} \Leftrightarrow \left\{s_{Ad} : \text{exact solution}\right\} \end{aligned} \tag{48}$$

The global Drucker CRE estimator is defined by:

$$\varepsilon = \left| \sup_{t\in[0,T]} \int_\Omega \eta\left(t, s_{Ad}\right) d\Omega \right|^{1/2} \Big/ D \tag{49}$$

with:

$$D^2 = \sup_{t\in[0,T]} \int_0^t \int_\Omega dt\, d\Omega \left[\gamma\left(\underline{\Gamma}_{DA} + \underline{\Gamma}_{KA}\right) \circ \left(\dot{\underline{U}}_{DA} + \dot{\underline{U}}_{KA}\right) + (1-\gamma)\, \mathrm{Tr}\left[\left(\sigma_{DA} + \sigma_{KA}\right)\left(\dot{\varepsilon}_{DA} + \dot{\varepsilon}_{KA}\right)\right]\right]$$

For elastic materials, one gets:

$$\eta\left(t, s_{Ad}\right) = \frac{\gamma}{2}\, \rho \left(\dot{\underline{U}}_{DA} - \dot{\underline{U}}_{KA}\right)^2 \Big|_t + \frac{(1-\gamma)}{2}\, \mathrm{Tr}\left[\left(\varepsilon_{DA} - \varepsilon\left(\underline{U}_{KA}\right)\right) \mathbf{K} \left(\varepsilon_{DA} - \varepsilon\left(\underline{U}_{KA}\right)\right)\right]\Big|_t \tag{50}$$

In this case, η is interpreted as the distance between the KA and DA fields in terms of both the strain energy and the kinematic energy.

Remarks:

- If the admissible fields have also to satisfy $\underline{\Gamma}_{DA} = \underline{\Gamma}_{CA} = \rho\, \ddot{\underline{U}}_{CA}$, the error in the constitutive relation used in (Coorevits et al. 1992) is then found again.
- Let the exact solution be $s_{ex} = \left(\underline{U}_{Ex}, \sigma_{Ex}, \underline{\Gamma}_{Ex}\right)$. The effectivity index is calculated considering that the "exact" error is:

$$\left| \sup_{t\in[0,T]} \int_\Omega \bar{\eta}\left(s_{Ex}, s_{Ad}\right) d\Omega \right|^{1/2} \tag{51}$$

with $\bar{\eta}\left(s_{Ex}, s_{Ad}\right) = \frac{1}{2}\left[\eta\left(t, \underline{U}_{Ex}, \sigma_{DA}, \underline{\Gamma}_{DA}\right) + \eta\left(t, \underline{U}_{KA}, \sigma_{Ex}, \underline{\Gamma}_{Ex}\right)\right]$, η being defined by (47).

Recovery of an admissible $\hat{s}_h$ from the data and the f.e. solution. Generally speaking, practical approximations apply a finite element choice as well as a time integration scheme: the explicit central differences method, the implicit Newmark scheme, …

The data are supposed to be piecewise linear over [0, T]; therefore it is easy to recover $\hat{s}_h$ over [0, T] from its values at the time t_n $n = 1, 2, \ldots, N$ supposing the same property. Let s_{hn} $n = 1, 2, \ldots, N$ be the finite element solution.

- Recovery of kinematically-admissible quantities

It is the displacement $\hat{\underline{U}}_{KA}$. Let us consider that the central difference method is used. One prescribes :

$$\frac{d}{dt}\hat{\underline{U}}_{KA}\Big|_t = \underline{V}_{ht} \quad t = t_0, t_{1/2}, t_1, \ldots, t_N$$

The values $\underline{V}_{ht}$ for $t = t_0, t_{1/2}, t_1, \ldots, t_N$ are derived from the f.e. solution $\underline{V}_{ht}$ for $t = t_{1/2}, \ldots, t_{N-1/2}$ and the data $\underline{V}_0$.

One defines :

$$\underline{V}_{htn} = \frac{1}{2}\left(\underline{V}_{htn-1/2} + \underline{V}_{htn+1/2}\right) n \in \{1, 2, \ldots, N\}$$

Supposing $\hat{\underline{U}}_h$ piecewise linear, on the subinterval $\left[t_n, t_{n+1/2}\right]\left[t_{n-1/2}, t_n\right] n \in 1, 2, \ldots, N$, $\hat{\underline{U}}_{CA}$ and then $\hat{\underline{U}}_{KA}$ are defined over [0, T].

- Recovery of dynamically-admissible quantities

They are the stress $\hat{\sigma}_{DA}$ and the acceleration quantity $\hat{\Gamma}_{DA}$. The corresponding quantities computed from the f.e. solution are σ_{hn} and $\underline{\Gamma}_{hn}$ at the time t_n $n \in 1, 2, \ldots, N$. $\hat{\sigma}_{hn}$ and $\hat{\Gamma}_{hn}$ are related to the f.e. solution by the strong prolongation condition which becomes in dynamics:

$$\int_E \left(\hat{\sigma}_{DAn} - \sigma_{hn}\right) \underline{\underline{grad}}\, \omega_i \, dE + \int_E \left(\hat{\underline{\Gamma}}_{DAn} - \underline{\Gamma}_{hn}\right) \omega_i \, dE = 0 \tag{52}$$

The tractions $\hat{\underline{F}}_{DAn}$ on the element edge are still linear combination of the restrictions of the shape functions ω_i on the edges. For $\hat{\underline{\Gamma}}_{DAn}$, several possibilities occur: it is a piecewise polynomial of degree k ($0 \le k \le p$). k being a choice, there are several variants. Let $\hat{\mathcal{F}}_E^k$ be the corresponding subspace of admissible $\hat{\underline{\Gamma}}_{DA}$, i.e. verifying (52); an additional prolongation condition is :

$$\begin{aligned} &\forall E \in \mathbf{E} \quad \forall n \in 1, 2, \ldots, N \\ &\forall \hat{\underline{\Gamma}}^* \in \hat{\mathcal{F}}_E^k \quad \int_E \frac{1}{\rho}\left(\hat{\underline{\Gamma}}_{DAn} - \underline{\Gamma}_{hn}\right) \circ \left(\hat{\underline{\Gamma}}^* - \hat{\underline{\Gamma}}_{DAn}\right) d\Omega = 0 \end{aligned} \tag{53}$$

This recovery technique is very similar to the one previously described in statics.

6.3 Test: One Dimensional f.e. Dynamic Analysis

We consider the problem described in figure 35.

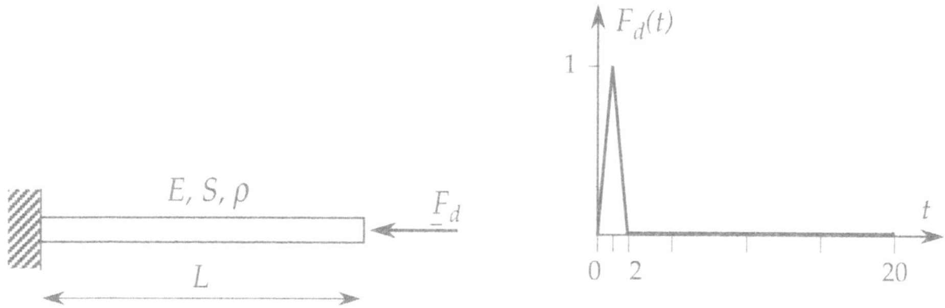

Figure 35. One dimensional dynamic problem

The finite element solution is computed using an explicit (0; 0.5) Newmark integration scheme. Since the analytical solution to such a problem is well-known, we compared the values of the estimator with the value of the exact error for various uniform meshes of the beam, as well as for different time step increments.

The time step size is evaluated in terms of its deviation from the courant condition : β_{CFL} . Then, the stability condition is written:

$$\Delta t = \beta_{CFL} \times \Delta t_{crit.} \leq \Delta t_{crit.} \tag{54}$$

Figure 36 shows the effectivity index in terms of the number of elements and of the deviation to the Courant condition. Over the zone where the time step has been chosen close to the critical time step $\left(0.8 \leq \beta_{CFL} \leq 1\right)$, the proposed estimator is very efficient.

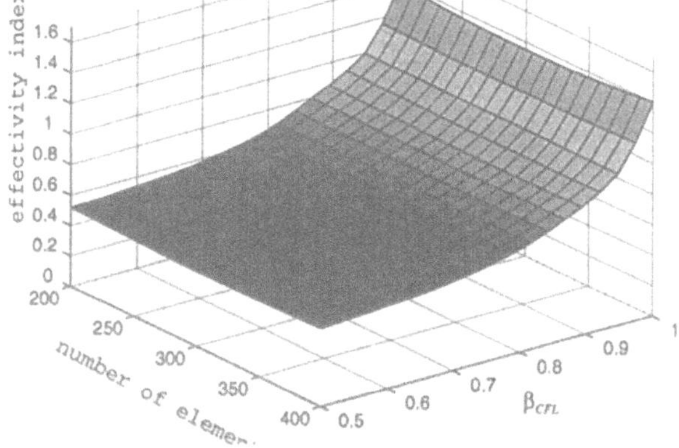

Figure 36 . Effectivity index versus the number of elements and the deviation to the Courant condition

6.4 Comparison with Other Indicators

The previous test is simplified; one takes one element. It is a single degree of freedom problem where space does not play any role.

We compare the constitutive relation error estimator with other error indicators:

- the (Zienkiewicz et al. 1984) (Zienkiewicz and Xie 1991) indicator denoted here ZWX
- the (Li et al. 1993) indicator denoted LZW
- the (Wiberg and Li 1993) indicator denoted WL.

More details can be found in (Combe et al. 1999). Figure 37 shows the time evolution of the different estimated errors. It appears that the effectivity index of the CRE estimator is rather good.

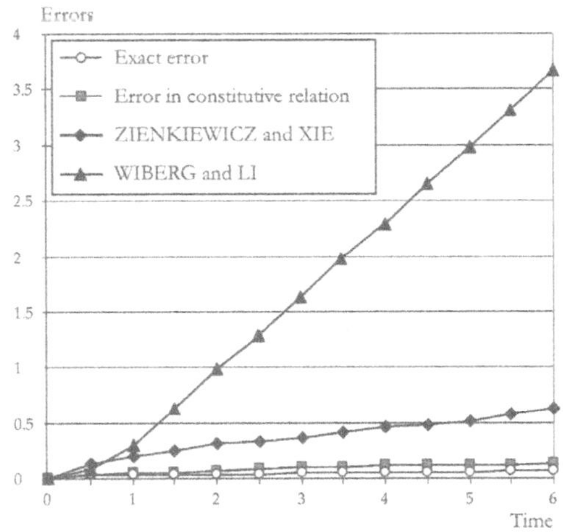

Figure 37. Time evolution of various estimated errors

7 Constitutive Relation Modelling Error in Dynamics and Model Updating

The principles of the constitutive relation modelling error are only recalled for the forced vibration problem. For further details, see (Ladevèze 1998).

The reference is made up of reliable quantities, such as reliable equations (e.g. the equilibrium equation) and accurate experimental data (e.g. frequencies and sensor locations). The constitutive relations and the less accurate experimental data (eigenmodes components and frequency response functions magnitudes) are treated separately. Therefore, a modified error on the constitutive relation is defined, in order to quantify the model quality with respect to our reference. In contrast with a lot of methods, the error defined herein has a strong mechanical sense. Hence, the most erroneous regions are those providing the most significant contribution to the error. Nearly space uniform erroneous structural parameters can also be observed.

The parameter updating is performed by minimizing the modified constitutive relation error. Such an inverse problem without addition is ill-posed. Our updating process can be thought of as a regularization because it leads to a unique solution of the problem. It is iterative, and each iteration includes two steps. The first consists in localizing the most erroneous regions and then the most erroneous structural parameters, and the second is the correction of the parameters belonging to these regions. The updating process is stopped when a threshold error is reached. Several updating examples are shown.

The basic concept is the modelling CRE which has been introduced in (Ladevèze 1983, 1991) and developed in (Ladevèze, Reynier and Maia 1994) (Chouaki, Ladevèze and Proslier 1998) (Ladevèze and Chouaki 1999).

7.1 The forced vibration problem

The prescribed forces and displacements (complex numbers) are harmonic:

- $\underline{f}_{d,\omega} \exp(i\,\omega\, t)$ $\underline{F}_{d,\omega} \exp(i\,\omega\, t)$
- $\underline{U}_{d,\omega} \exp(i\,\omega\, t)$

where ω is the angular frequency. The forced vibration problem is:

Find the displacement, stress, acceleration force amplitudes and phases $\left(\underline{U}(\underline{M}), \underline{\sigma}(\underline{M}), \underline{\Gamma}(\underline{M})\right)_\omega$ $\underline{M} \in \Omega$ such that they satisfy:

- the kinematic constraints: $\underline{U} \in \mathcal{U}$

$$\underline{U}\Big|_{\partial_1\Omega} = \underline{U}_{d,\omega} \tag{55}$$

- the equilibrium equation: $\left(\underline{\Gamma}, \underline{\sigma}\right) \in \mathcal{S}$

$$-\int_\Omega \operatorname{tr}\left|\sigma\, \varepsilon\left(\underline{U}^*\right)\right| d\Omega + \int_\Omega \underline{f}_{d,\omega} \circ \underline{U}^*\, d\Omega + \int_{\partial_1\Omega} \underline{F}_{d,\omega} \circ \underline{U}^*\, dS = \int_\Omega \underline{\Gamma} \circ \underline{U}^*\, d\Omega \tag{56}$$

$$\forall\, \underline{U}^* \in \mathcal{U}_{ad,0} = \left\{\underline{U}^* / \underline{U}^*_{|\partial_1\Omega} = 0,\ \underline{U}^* \in \mathcal{U}\right\}$$

- the constitutive relations:

$$\sigma = \mathbf{K}\, \varepsilon\left(\underline{U}\right) + \mathbf{B}\, \varepsilon\left(\underline{U}\right) i\,\omega = \overline{K}\, \varepsilon\left(\underline{U}\right) \tag{57a}$$

$$\underline{\Gamma} = -\rho\, \omega^2\, \underline{U} + \mathbb{a}\left(\underline{U}\right) i\,\omega = \overline{\mathbb{a}}\, \underline{U} \tag{57b}$$

$\mathcal{U}$ denotes the space of the displacement amplitudes defined over Ω which possess a finite energy; $\mathbf{K}$ is the Hooke's tensor which is positive definite and symmetric. $\mathbf{B}$ and $\mathbb{a}$ are real, linear, symmetric and positive definite operators satisfying the Drucker's stability conditions; ρ is the mass density and ε is the strain operator.

The Drucker stability conditions which are assumed, ensure the uniqueness of the solution, and they are satisfied by a large class of materials.

7.2 Classical CRE Estimator

Let us come back to the CRE definition in dynamics §6.2. For the given frequency ω, we consider an amplitude triplet $s = \left(\underline{U}_{KA}, \sigma_{DA}, \underline{\Gamma}_{DA}\right)$ which is admissible. It follows that s verifies the kinematic constraints (55) and the equilibrium equations (56). Let $\mathbf{S}_{ad,\,\omega}$ be the corresponding space. However, s does not verify the constitutive relations (57a) and (57b). Its quality is then defined by the following CRE:

$$\eta_{\omega}^{2}(s) = \frac{\gamma}{2}\int_{\Omega} \mathrm{Tr}\left[\left(\mathbf{K} + T\,\omega^{2}\,\mathbf{B}\right)\left(\varepsilon\left(\underline{U}_{KA}\right) - \varepsilon_{DA}\right)\left(\varepsilon\left(\underline{U}_{KA} - \varepsilon_{DA}\right)\right)\right] d\Omega$$
$$+ \frac{1-\gamma}{2}\int_{\Omega} \omega^{2}\left(\rho\,\mathbb{1} + T\,\mathbb{a}\right)\left(\underline{U}_{KA} - \underline{U}_{DA}\right)\circ\left(\underline{U}_{KA} - \underline{U}_{DA}\right)^{*} d\Omega \qquad (58)$$

(*) signifies here conjugate. T is a parameter which is a characteristic time determining the dissipation part in the error. Moreover, one has:

$$\varepsilon_{DA} = \overline{K}^{-1}\sigma_{DA}$$
$$\underline{U}_{DA} = \overline{\mathbb{a}}^{-1}\underline{\Gamma}_{DA} \qquad (59)$$

γ is a parameter belonging to [0, 1] that depends on the relative reliability of the constitutive relations (57a) and (57b). Its current value is 0.5. Then, the initial problem can be rewritten as follows:

$$\text{Find } s_{\omega} \in \mathbf{S}_{ad,\,\omega}$$
$$\text{minimizing } \eta_{\omega}^{2}(s') \text{ where } s' \in \mathbf{S}_{ad,\,\omega} \qquad (60)$$

s admissible $\left(s_{\omega} \in \mathbf{S}_{ad,\,\omega}\right)$ means that the triplet satisfies exactly the reliable equations of the problem, the kinematical constraints and the equilibrium equation.

7.3 Reliable and Less Reliable Experimental Data

For the sake of simplicity, let us consider the problem where only one single excitation (or displacement) is prescribed at one point $\underline{P}$; its direction is $\underline{n}$. At a given frequency ω, the components on $\underline{n}$ of both the displacement and the force at $\underline{P}$ are denoted u_ω and f_ω, respectively, the measured quantities being $\tilde{u}_\omega$ and $\tilde{f}_\omega$ ($\tilde{()}$ means an experimental quantity).

Hence, we must distinguish between the reliable data and the less reliable data. It is considered here that the reliable ones are:

- the location of the excitation point;
- the excitation direction.

The less reliable experimental quantities are:

- $\tilde{u}_\omega$, $\tilde{f}_\omega$ or the transfer function $\tilde{\alpha}$ defined as follows:

$$\tilde{u}_\omega = \tilde{\alpha}_\omega \tilde{f}_\omega \tag{61}$$

- the other measured response information: the displacements denoted $\Pi \underline{\tilde{U}}_\omega$ and the forces denoted $\Pi' \underline{\tilde{\Gamma}}_\omega$. Π and Π' are projection operators that specify the parts of the measured displacements $\underline{\tilde{U}}_\omega$ and forces $\underline{\tilde{\Gamma}}_\omega$, respectively.

To go further, a slight modification of the constitutive relation error must be carried out in order to include the fact that the data $\left(\underline{U}_{d,\,\omega}, \underline{F}_{d,\,\omega}, \underline{f}_{d,\,\omega}\right)$ are not entirely reliable quantities. First, the solution s of the problem is modified introducing two new variables u_ω and f_ω:

$$s = \left(\underline{U}, \sigma, \underline{\Gamma}, u_\omega, f_\omega\right)$$

We still denote $\mathbf{S}_{ad,\,\omega}$ the corresponding space of admissible s; it contains only reliable data. Then, we add the constitutive relation error related to the model and to the quantities u_ω, f_ω :

$$\bar{\eta}^2_\omega(s) = \eta^2_\omega(s) + \left|\alpha_\omega f_\omega - u_\omega\right|^2 / \left|\alpha_\omega\right|$$

α_ω is the model's transfer function between f_ω and u_ω .

Finally the problem (60) remains as follows:

Find $s_\omega \in \mathbf{S}_{ad,\,\omega}$

minimizing $\bar{\eta}^2_\omega(s')$ where $s' \in \mathbf{S}_{ad,\,\omega}$

7.4 Modelling Constitutive Relation Error

- Modified error on the constitutive relation at a given frequency ω

Now, in order to take into account the less reliable experimental information, a modified error on the constitutive relation is introduced. At a given frequency ω, this error is written as follows:

$$\mathbf{e}_{\omega}^{2}(s)=\bar{\eta}_{\omega}^{2}(s)+\frac{r}{1-r}\left\{\gamma\left\|\Pi\,\underline{U}_{c}-\Pi\,\underline{\tilde{U}}_{\omega}\right\|^{2}+(1-\gamma)\left\|\Pi'\,\underline{\Gamma}_{s}-\Pi'\,\underline{\tilde{\Gamma}}_{\omega}\right\|^{2}\right.$$
$$\left.\times\frac{1}{2}\left[\left|f-\tilde{f}\right|^{2}\left|\alpha_{\omega}\right|+\frac{\left|u-\tilde{u}\right|^{2}}{\left|\alpha_{\omega}\right|}\right]\right\} \tag{62}$$

Let us introduce the problem:

$$\begin{aligned}&\text{Find } s_{\omega}\in\mathbf{S}_{\text{ad},\,\omega}\\&\text{minimizing } e_{\omega}^{2}(s')\text{ where } s'\in\mathbf{S}_{\text{ad},\,\omega}\end{aligned} \tag{63}$$

All the accurate quantities and equations have been introduced in $\mathbf{S}_{\text{ad},\,\omega}$ and are then verified exactly by the solution to problem (63). The less reliable equations and measured quantities are verified in a mean sense through the minimization of e_{ω}^{2}.

The firsts term of $e_{\omega}^{2}(s)$ contains the constitutive relations of the first analytical model. The second term is the square of a distance to the less reliable measured quantities. $r/(1-r)$ depends on the relative quality. $r = 0.5$ is the current value. $\|\ \|$ and $\left\|\ \right\|$ are energetic norms obtained after a standard condensation technique whose choice is not so critical.

- Modelling constitutive relation error on the studied frequency bandwidth

The error ε defines the quality of the model with respect to experimental data :

$$\varepsilon^{2}=\int_{\omega_{\min}}^{\omega_{\max}} z(\omega)\,\bar{\eta}_{\omega}^{2}(s_{\omega})\,/\,D_{\omega}^{2}(s_{\omega})\,d\omega \tag{64}$$

The function $z(\omega)$ is a weighting factor introduced in order to focus on some selected frequencies. This function is scaled such that:

$$\int_{\omega_{\min}}^{\omega_{\max}} z(\omega)\,d\omega=1 \qquad z(\omega)\geq 0 \tag{65}$$

Weighting factor examples can be given as follows:

$$\bullet\ z(\omega) = \frac{1}{\omega_{max} - \omega_{min}} \qquad \text{over} \left[\omega_{min}, \omega_{max}\right]$$
$$\bullet\ z(\omega) = \frac{1}{m} \sum_{i=1}^{m} \delta_{\omega_i}(\omega) \tag{66}$$

where ω_i are the eigenfrequencies belonging to $\left[\omega_{min}, \omega_{max}\right]$, and δ_{ω_i} is the Dirac distribution associated with the value ω_i.

$D_\omega(s_\omega)$ is used to introduce the relative error $\bar{\eta}_\omega(s_\omega)/D_\omega(s_\omega)$ (without dimension). One uses a quantity similar to D in (49).

Then, if the structure is divided into substructures $E \in \mathbf{E}$, it is possible to define $\varepsilon_{E,\omega}$, the contribution of the substructure E to the error such that:

$$\varepsilon_\omega^2(s_\omega) = \sum_{E \in \mathbf{E}} \varepsilon_{E,\omega}^2 \tag{67}$$

Remarks:

- The extension to multiple excitations does not involve additional difficulties.
- The transposition to the finite element framework is very straightforward; greater detail is provided in (Chouaki et al. 1998).
- The problem formulated above can be described using inverse problem terminology as follows:
  - the objective cost function: ε^2
  - the state variables: $(\underline{U}, \sigma, \underline{\Gamma}, u_\omega, f_\omega)$ over $\omega \in \left[\omega_{min}, \omega_{max}\right]$
  - the state equation: equilibrium equation
  - the constraints: kinematic constraints (1)
  - the model's parameters: Young's modulus, mass density, section, etc.

7.5 Organization of the Updating Process

The mathematical model depends on numerous structural parameters. Recalling that inverse problems are ill-posed, it is obvious that the correction of all the parameters will be very critical. Thus, a reasonable approach consists in improving only a few structural parameters at a time. This leads to an iterative process which is stopped when a threshold error is reached.

At the nth iteration, our method is staggered as follow.

- *The localization step.* The most erroneous regions are localized by solving the problem:

$$\text{find the substructures } E \in \mathbf{E} \text{ such that:} \qquad \varepsilon_E \geq 0.8 \max_{E \in \mathbf{E}} \varepsilon_E \tag{68}$$

- *The correction step.* Once the most erroneous regions have been localized, some structural parameters belonging to these regions can then be selected. The corresponding structural parameter subset is $[\mathcal{K}]_n$. Hence, denoting by $\underline{k}$ the structural parameter vector, the correction step is expressed as :

$$\left| \begin{array}{l} \text{find } \underline{k} \in [\mathcal{K}]_n \\ \text{minimizing the total error } e^2(\underline{k}') \quad \underline{k}' \in [\mathcal{K}]_n \end{array} \right. \tag{69}$$

where:

$$e^2(\underline{k}') = \int_{\omega_{max}}^{\omega_{min}} d\omega\, z(\omega)\, e_\omega^2(s_\omega, \underline{k}') / D_\omega^2(s_\omega, \underline{k}')$$

- *The iteration stopping error*

The iteration process is stopped when:

$$e(\underline{k}_n) \leq e_0$$

where e_0 is a threshold which depends essentially on the experiment quality.

Remark : There is no regularization term in the modified error on the constitutive relation. Hence, the minimization of this error remains an ill-posed problem. Then, the updating process can be thought of as a regularization because it leads to an unique solution.

7.6 Test 1: Influence of the Amount of Experimental Information

The model is the finite element one. Discretization errors are not considered. The "experimental" data have been simulated from the finite element model in which some structural parameters have been modified.

Test 1 concerns the clamped-free beam in figure 38. The structure is discretized using 15 beam elements (see table 2). The experimental data consist of frequency response functions, collected over the frequency bandwidth [0,20] Hz, comprising the first five eigenfrequencies. The "measured" frequency response functions have been simulated using the finite-element model, in which the Young's modulus of the eighth element has been increased by 100%. Figure 39 shows the results of the localization step with respect to the amount of experimental information. The results obtained with 15% measured degrees of freedom differ greatly from those obtained with 100% measured degrees of freedom.

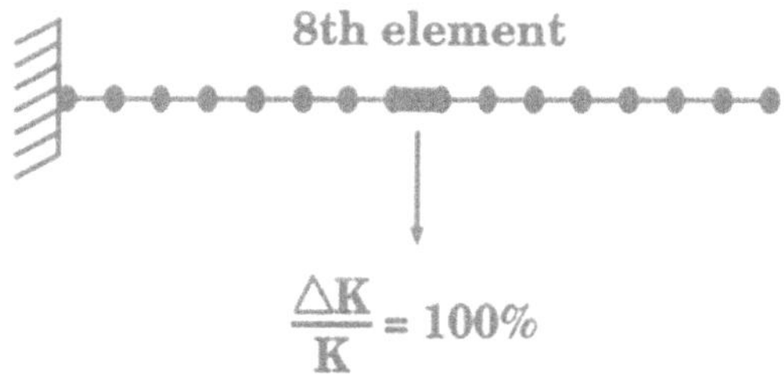

Figure 38. Clamped-free beam

Total length	0.8 m
Section	10^{-4} m^2
Young's modulus	0.31 x 10^8 N m^2
Density	7800 kg m^3
Damping	hysteretic
Damping loss factor	1%

Table 2. Characteristics of the structure

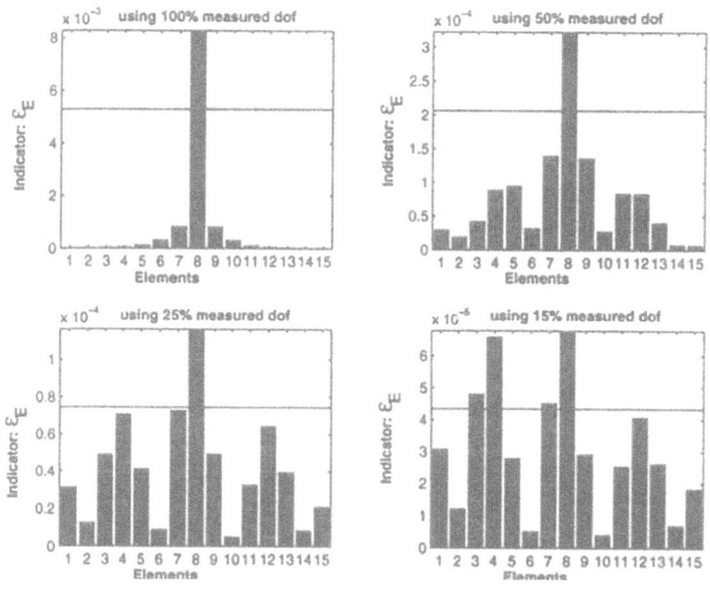

Figure 39. The localization capabilities with respect to the amount of experimental information

Now, in order to illustrate that the error on the constitutive relation is also able to detect uniform defects, the Young's moduli of the 15 elements have been increased by 100%. Then, figure 40 shows that the contributions of the various elements to the error are very close.

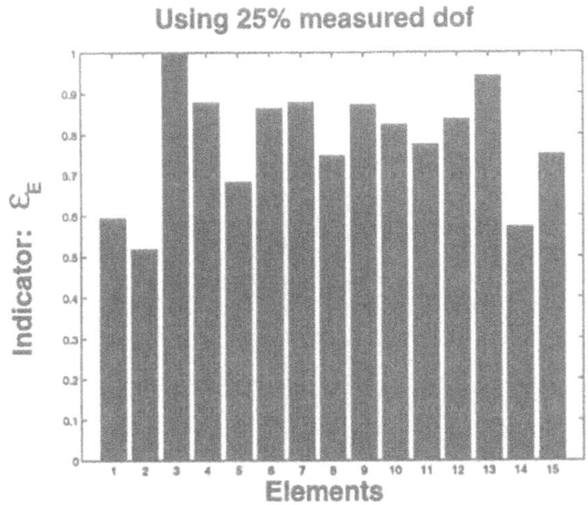

Figure 40. Uniform stiffness defect

7.7 Test 2: Eigh-Bay Truss

Industrial applications concern satellites, launchers, planes, etc. Here, we study the "Eight-Bay Truss" whose real configuration is described in Figure 41. The "experimental results" are computed with such a configuration. Tests conducted and experimental data collected are shown in Figure 42.

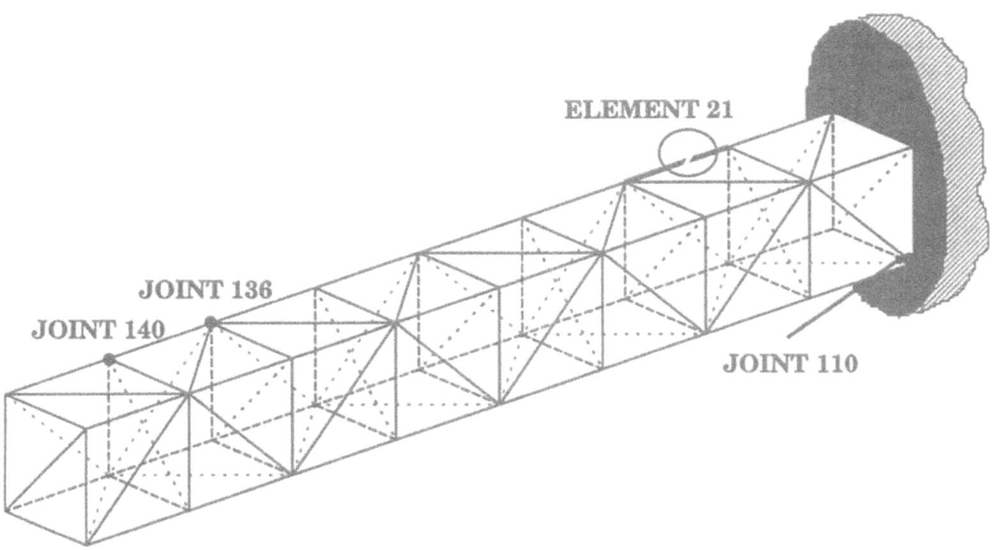

Figure 41. The real Eight-Bay Truss

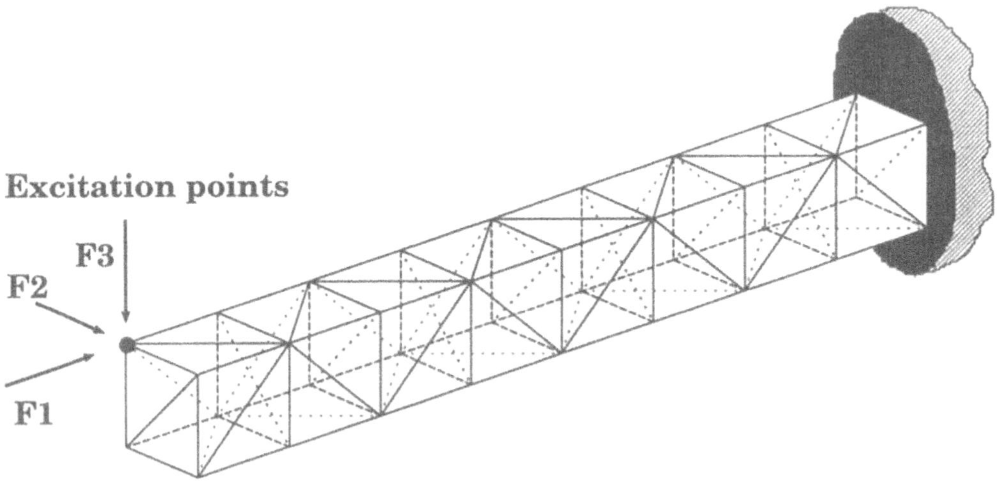

Figure 42. Tests conducted and experimental data collected

Elements and initial values of the structural parameters	First step	Second step	Exact values
Bar 21			
- Stiffness $p_k = 1$	$p_k := 0.0002$	corrected	$p_k := 0.0000$
- Mass : $p_m = 1$	$p_m = 1.0467$		$p_m = 1.0000$
Joint 110			
- Stiffness : $p_k = 1$	Not localized	$p_k = 0.5025$	$p_k = 0.5000$
- Mass : $p_m = 1$		$p_m = 1.0007$	$p_m = 1.0000$
Error	1.109	0.147	-

Figure 43. First and second updating steps

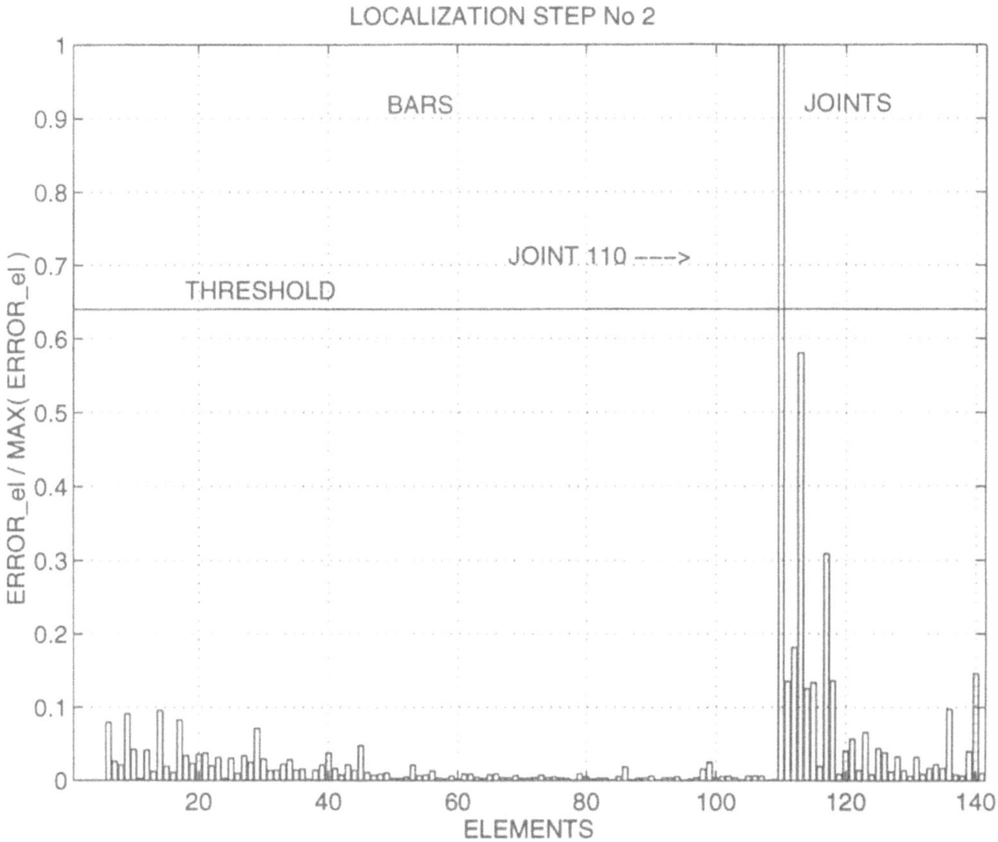

Figure 44. Error contributions of the various elements - second updating step

Figure 43 presents the structural parametric corrections and the global error value for the first two updating steps. The chart of error contributions to the global error of the various elements is given in Figure 44 for the second updating step; joint 110 appears clearly as an incorrectly modelled zone. The updating procedure normally has to be stopped after the second step because the error value (0.053) is small in comparison with the quality of the experimental results which have been contaminated by noise (level: ± 3%). During this step, the modelling errors related to joints 136 and 140 have not yet been detected. If one were to proceed further, it would be possible to detect and then correct these errors. Figure 45 shows, for the third updating step, the chart of error contribution to the global error of the various elements; modelling errors for joints 136 and 140 are clearly indicated.

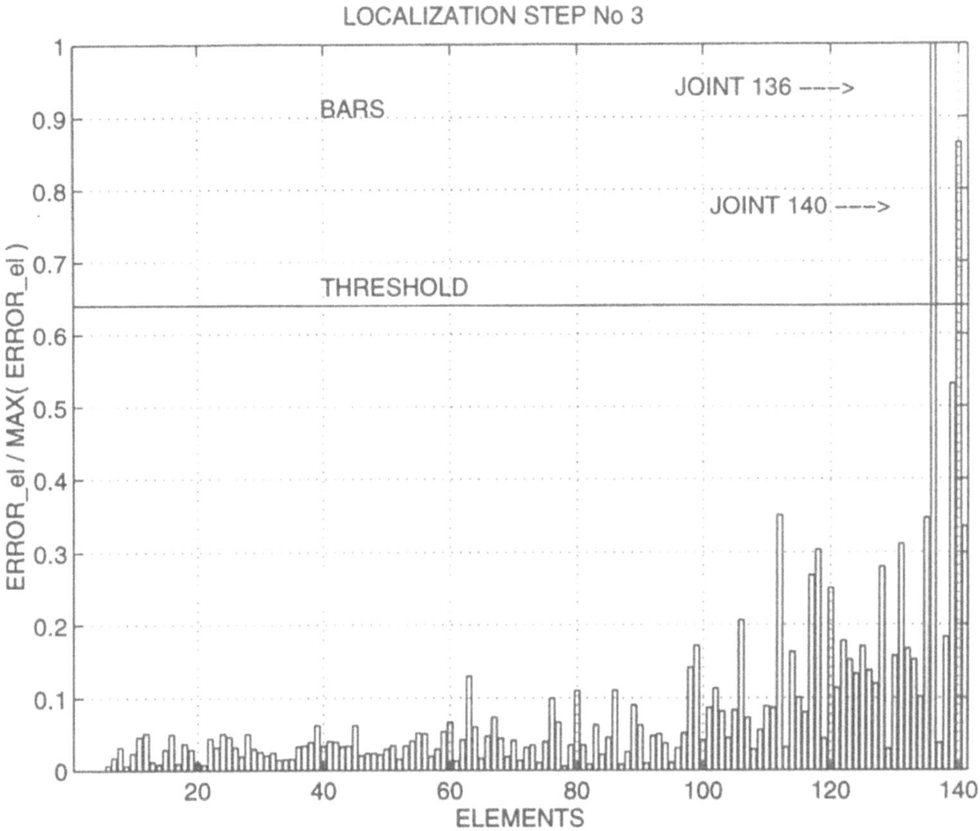

Figure 45. Error contributions of the various elements - third eventual updating step

References

Ainsworth, M., and Oden, J.T. (1997). A Posteriori Error Estimation in Finite Element Analysis, *Computational Methods Applied, Mechanical Engineering*, 142, 1-88

Aubry, D., and Tie, B. (1992). Error Estimates, h-Adaptive Strategy and Hierarchical Concept for Non Linear Finite Element Method, *Proceedings of the First European Conference on Numerical Methods in Engineering*, Hirsch, Zienkiewicz and Oñate Editors, Bruxelles (Belgique), 17-24

Azziz, K., and Babuska, I. (1972). *The Mathematical Foundations of the Finite Element Method with Applications to Partial Differential Equations*, Academic Press, New-York

Babuska, I., and Rheinboldt, W.C. (1982). Computational Error Estimates and Adaptive Processes for Some Nonlinear Structural Problems, *Computational Methods Applied Mechanical Engineering*, 34, 895-937

Babuska, I., Strouboulis, T., and Gangaraj, S.K. (1997). A Model Study of the Quality of a Posteriori Error Estimators for Finite Element Solutions of Linear Elliptic Problems with Particular Reference to the Behavior near the Boundary, *International Journal Numerical Method Engineering,* 40, 2521-2577

Babuska, I., Strouboulis, T., Upadhyay, C.S., and Gangaraj, S.K. (1994). A Model Study of the Quality of a Posteriori Estimators for Linear Elliptic Problems: Error Estimation in the Interior of Patchwise Uniform Grids of Triangles, *Computational Methods Applied Mechanical Engineering*, 114, 307-378

Babuska, I., Strouboulis, T., Upadhyay, C.S., Gangaraj, S.K., and Copps, K. (1994). Validation of a Posteriori Error Estimators by Numerical Approach, *International Journal Numerical Method Engineering*, 37, 1073-1123

Bank, R.E., and Weiser, A. (1985). Some a Posteriori Error Estimators for Elliptic Partial Differential Equations, *Mathematics of Computations*, 44, 283-301

Bass, J.M., and Oden, J.T. (1987). Adaptive Finite Element Methods for a Class of Evolution Problems in Viscoplasticity, *International Journal of Engineering. Sciences*, 25 (6), 623-653

Berman, A., and E.J., Nagy (1983). Improvement of a Large Analytical Model Using Test Data, *AIAA Journal*, 21 1168-1173

Boisse, P., and Coffignal, G. (1998). Error through the Constitutive Relation for Beams or C° Plate Finite Element: Static and Vibration Analyses, *In Advances in Adaptive Computational Methods in Mechanics*, 493-512, Ladeveze and Oden Editors, Elsevier

Boisse, P., Perrin, S., Coffignal, G., and Hadjeb, K. (1999). Error Estimation through the Constitutive Relation for Ressner-Mindlin Plate Bending Element, *Computers and Structures*, 73, 615-627

Boroomand, B. and Zienkiewicz, O.C. (1998). Recovery Procedures in Error Estimation and Adaptivity: Adaptivity in Nonlinear Problems of Elasto-Plasticity Behaviour, *In Advances in Adaptive Computational Methods in Mechanics*, 383-410, Ladeveze and Oden Editors, Elsevier

Bouillard, P., and F., Ihlenburg (1998). Error Estimation and Adaptivity for the Finite Element Method in Acoustics, In: *Advances in Adaptive Computational Methods in Mechanics*; P. Ladevèze and J.T. Oden (Eds.), Elsevier 477-492

Bouillard, P., and Ihlenburg, F. (1999). Error Estimation and Adaptivity for the Finite Element Method in Acoustic: 2d and 3d Applications, *Computational Methods Applied Mechanical Engineering,* 176, 147-164

Brink, U., and E., Stein (1998). A posteriori error estimation in large-strain elasticity using equilibrated local Neumann problems, *Computational Methods in Applied Mechanics*, 161, 77-101

Bussy, P., G., Coffignal, and P., Ladevèze (1985). Reliability and optimization of nonlinear finite element analyses, *in Proceedings of NUMETA '85*, Swansea

Chouaki, A., P., Ladevèze, and L., Proslier (1998). Updating Structural Dynamic Models with Emphasis on the Damping Properties, *AIAA Journal*, 36, 1094-1099

Ciarlet, P.G. (1978). *The Finite Element Method for Elliptic Problems*, North-Holland

Cirak, F., and Ramm, E. (1998). A Posteriori Error Estimation and Adaptivity for Linear Elasticity Using the Reciprocal Theorem, *Computational Methods Applied Mechanical Engineering*, 156, 351-362

Combe, J.P., Ladeveze, P., and Pelle, J.P. (1998). Constitutive Relation Error Estimator for Transient Finite Element Analysis, CD-ROM of Proceedings of *the Fourth World Congress on Computational Mechanics*, Buenos Aires (Argentine)

Combe, J.P., Ladeveze, P., and Pelle, J.P. (1999). Constitutive Relation Error Estimator for Transient Finite Element Analysis, *Computational Methods Applied Mechanical Engineering,* 176 (1-4), 165-185

Combe, J.P., P., Ladevèze, and J.P. Pelle (1999). Constitutive Relation Error Estimator for Transient Finite Element Analysis, *Proceeding ECCM 99*, Munich

Coorevits, P., Dumeau, J.P., and Pelle, J.P. (1995). Maillages 3D Adaptatifs: Quelques Procédures pour Respecter une Carte de Taille, *Proceedings of Strucome 95*, Paris, 117-127

Coorevits, P., Dumeau, J.P., and Pelle, J.P. (1996). Analyses Éléments Finis Adaptatives pour les Structures Tridimensionnelles en Élasticité, *Revue Européenne des Eléments Finis*, 5, n°3, 341-374

Coorevits, P., Dumeau, J.P., and Pelle, J.P. (1997). Error Estimator and Adaptivity for three-dimensional Finite Element Analyses, In *Advances in Adaptive Computational Methods in Mechanics*, 443-458, Ladeveze and Oden Editors, Elsevier

Coorevits, P., Ladeveze, P., and Pelle, J.P. (1992). Anisotropic Adaptive Mesh: Application to Transient Problems, *Proceedings of First European Conference on Numerical Methods in Engineering*, Bruxelles (Belgique)

Coorevits, P., P., Ladevèze, and J.P., Pelle (1992). Anisotropic Adaptive Mesh: Application to Transient Problems, *Proceedings of the 1st European Conference on Numerical Methods in Engineering*, Brussels, 209-215

Coorevits, P., Ladevèze, P., Pelle, J.P., and Rougeot, P. (1992). Some New Applications of a Method for the Control and Optimization of Finite Element Computations, *In New Advances in Computational Structural Mechanics*, Ladeveze and Zienkiewicz Editors, Elsevier, 205-217

De Saxce, G. (1992). Une Généralisation de l'Inégalité de Fenchel et ses Applications aux Lois Constitutives, *Comptes-Rendus Académie des Sciences*, Paris, 314, Ii, 125-129

Farhat, C., and F.M., Hemez (1993). Updating Finite-Element Dynamic Models using an Element-by-Element Sensitivity Methodology, *AIAA Journal*, 31, 1702-1711

Fourment, L., and Chenot, J.L. (1995). Error estimators for viscoplastic materials: application to forming processes, *International Journal for Numerical Methods in Engineering*, 12(5), 469-490

Gallimard, L., Ladeveze, P., and Pelle, J.P. (1996). Error Estimation and Adaptivity in Elastoplasticity, *International Journal for Numerical Method in Engineering*, 39, 189-217

Gallimard, L., Ladeveze, P., and Pelle, J.P. (2000). An Enhanced Error Estimator on the Constitutive Relation for Plasticity Problems, *Computer and Structures*, à paraître

Hjiaj, M. (1999). Sur la Classe de Matériaux Standards Implicites: Concept, Aspect Discrétisés et Estimation de l'Erreur a Posteriori, *Thèse de la Faculté Polytechnique*, Mons, Belgique

Johnson, C., and Hansbo, P. (1991). Adaptive Finite Element Methods for Small Strain Elasto-Plasticity, *Proceedings of Iutam Conference on Finite Inelastic Deformations Theory and Applications*, Hannover (Allemagne)

Kaouk, M., and D.C., Zimmerman (1994) Structural Damage Assessment Using a Generalized Minimum Rank Perturbation, *AIAA Journal*, 32, 836-842

Kelly, D.W. (1984). The Self-Equilibration of Residuals and Complementary a Posteriori Error Estimates in the Finite Element Method, *International Journal Numerical Method Engineering*, 20, 1491-1506

Ladevèze, P. (1983). Recalage de Modélisations des Structures Complexes, *Technical Report 33.11.01.4*, Aérospatiale, Les Mureaux

Ladeveze, P. (1985). Sur une Famille d'Algorithmes en Mécanique des Structures, *Comptes-Rendus Académie des Sciences*, Paris, 300, Ii, 41-44

Ladeveze, P. (1989). La Méthode à Grand Incrément de Temps pour l'Analyse des Structures à Comportement Non Linéaire décrit par Variables Internes, *Comptes-Rendus Académie des Sciences*, Paris, 309, II, 1095-1099

Ladevèze, P. (1993). Constitutive Relation Error in Dynamics Theory and Application to Model Updating, *Internal Report 150* (in french)

Ladevèze, P. (1998). A Modelling Error Estimator for Dynamic Structural Model Updating, In: *Advances in Adaptive Computational Methods in Mechanics*, P. Ladevèze and J.T. Oden (Eds.), Elsevier 135-51

Ladeveze, P. (1999). *Nonlinear Computation Structural Mechanics, New Approaches and non Incremental Methods of Calculation*, Springer

Ladeveze, P., and Chouaki, A. (1999). Application of a Posteriori Error Estimation for Structural Model Updating, *Inverse Problems*, 15, 49-00

Ladevèze, P., Coffignal, G., and Pelle, J.P. (1986). Accuracy of Elastoplastic and Dynamic Analysis, In *Accuracy Estimates and Adaptative Refinements in Finite Element Computations*, Chap 11, 181-203, Babuska, Gago, Oliveira and Zienkiewicz Editors, J. Wiley

Ladevèze, P., and Leguillon, D. (1981). Error Computation in Finite Element Method and Application, *Proceedings Gamni*, Dunod, Paris (France)

Ladevèze, P., and Leguillon, D. (1983). Error Estimate Procedure in Finite Element Method and Application, *Siam Journal Num. Anal*, 20 (3), 485-509

Ladeveze, P., and Maunder, E.A.W. (1996). A General Method for Recovering Equilibrating Element Tractions, *Computational Methods Applied Mechanical Engineering*, 137, 111-151

Ladeveze, P., and Maunder, E.A.W. (1997). A General Methodology for Recovering Equilibrating Finite Element Tractions and Stress Fields for Plates and Solid Elements, *Computational Ass. Mechanical and Engineering Sciences*, 4, 533-548

Ladeveze, P., and Moes, N. (1998). A new a Posteriori Error Estimation for Nonlinear Time-Dependent Finite Element Analysis, *Computational Methods Applied Mechanical Engineering*, 157, 45-68

Ladeveze, P., Moes, N., and Douchin, B. (1999). Constitutive Relation Error Estimations for (Visco) Plasticity Finite Element Analysis with Softening, *Computational Methods Applied Mechanical Engineering*, 176, 247-264

Ladeveze, P., and Oden, J.T. (1998). *Advances in Adaptive Computational Methods in Mechanics*, Elsevier

Ladevèze, P., Pelle, J.P., and Rougeot, P. (1991). Error Estimation and Mesh Optimization for Classical Finite Elements, *Engineering Computations*, 8, 69-80

Ladevèze, P., M., Reynier, and N.M.M., Maia (1994). *Error on the Constitutive Relation in Dynamics Inverse Problems in Engineering*, 1st edition, ed. H.D. Bui et al. (Rotterdam: Balkema) 251-256

Ladevèze, P., and Rougeot, P. (1997). New Advances on a Posteriori Error on Constitutive Relation in Finite Element Analysis, *Computational Methods Applied Mechanical Engineering*, 150, 239-249

Ladevèze, P., Rougeot, P., Blanchard, P., and Moreau, J.P. (1999). Local Error Estimators for Finite Element Linear Analysis, *Computational Methods Applied Mechanical Engineering*, 176, 231-246

Li, X.D., L.F., Zeng, and N.E., Wiberg (1993). A Simple Local Error Estimator and an Adaptive Time-Stepping Procedure for Direct Integration Method in Dynamic Analysis, *Com. in Numerical Method in Engineering*, 9 273-292

Natke, H.G. (1991). Error Localization within a Spatially Finite-Dimensional Mathematical Model, *Computational Mechanics*, 8, 153-160

Natke, H.G. (1994). Model Verification and Validation in Engineering, *Bulletin of Polish Academy Sciences Tech. Sciences,* 42 505-519

Nobari, A.S., D.A., Robb, and D.J., Ewins (1993). Model Updating and Joint Identification Methods: Applications, Restrictions and Overlap, *Journal Modal. Anal.* 8, 93

Oden, J.T., and Prudhomme, S. (1999). Goal-Oriented Error Estimation and Adaptivity for the Finite Element Method, *TICAM Report*, 99-15, Austin (U.S.A.)

Peraire, J., and Patera, A.T. (1998). Bounds for Linear-Functional Outputs of Coercive Partial Differential Equations; Local Indicators and Adaptive Refinements, *In Advances in Adaptive Computational Methods in Mechanics*, 199-216, Ladeveze and Oden Editors, Elsevier

Peric, D., Yu, J., and Owen, D.R.J. (1994). On Error Estimates and Adaptivity in Elastoplastic Solids: Application to the Numerical Simulation of Strain Localization in Classical and Cosserat Continua, *International Journal Numerical Method Engineering*, 37, 1351-1379

Pijaudier-Cabot, G., Bode, L., and Huerta, A. (1995). Arbitrary Lagrangian Eulerian Finite Element Analysis of Strain Localization in Transient Problems, *International Journal Numerical Method Engineering*, 38 (24), 4171-4191

Rannacher, R., and Suttmeier, F.T. (1998). A Posteriori Error Control and Mesh Adaptation for FE Models in Elasticity and Elasto-Plasticity, In *Advances in Adaptive Computational Methods in Mechanics*, 275-292, Ladeveze and Oden Editors, Elsevier

Ren, Y., and C.F. ,Beards (1995). Identification of Joint Properties of a Structure using F.R.F. Data, *Journal Sound Vibration*, 186, 567-587

Stein, E., F.J., Bathold, and M., Schmitdt (1997). Error Estimation Mesh Adaptivity for Elastoplastic Deformations, in D.R.J. Owen, E. Onate and E. Hinton (Eds.*), Proceedings of the Fifth International Conference on Computational Plasticity*, Barcelona 597-602

Stein, E., Barthold, F.J., Ohnimus, S., and Schmidt, M. (1998). Adaptive Finite Element in Elastoplasticity with Mechanical Error Indicators and Neumann-Type Indicators, In *Advances in Adaptive Computational Methods in Mechanics*, 81-100, Ladeveze and Oden Editors, Elsevier

Strouboulis, T., Babuska, I., Datta, D., Copps, K., and Gangaraj, S. (2000). A Posteriori Estimation and Adaptive Control of the Error in the Quantity of Interest, Part I: A Posteriori Estimation of the Error in the Von Mises Stress and the Stress Intensity Factors, *Computational Methods Applied Mechanical Engineering*, 181, 261-294

Strouboulis, T., Babuska, I., and Gangaraj, S.K. (2000). Guaranteed Computable Error Bounds for the Exact Error in the Finite Element Solution, Part II: Bounds for the Energy Norm of the Error in two Dimension, *International Journal Numerical Method Engineering*,

Wiberg, N.E., and Abdulwahab, F. (1992). An Efficient Post-Processing Technique for Stress Problems Based on Superconvergent Derivatives and Equilibrium, *In Numerical Methods in Engineering*, Hirsch, Periaux and Oñate Editors, Elsevier, 25-32

Wiberg, N.B., and Li, X.D. (1993). A Post-Processing Technique and a Posteriori Error Estimate for the NEWMARK Method in Dynamic Analysis, *Earthquake Engineering. Structural Dynamic*, 22, 465-489

Wunderlich, W., Cramer, H., and Steinl, G. (1998). An Adaptive Finite Element Approach in Associated and Non-Associated Plasticity Considering Localization Phenomena, In *Advances in Adaptive Computational Methods in Mechanics*, 293-308, Ladeveze and Oden Editors, Elsevier

Zienkiewicz, O.C., Wood, W., Hine, N., and Taylor, R. (1984). A Unified Set of Single Step Algorithms. Part I: General Formulation and Applications, *International Journal Numerical Method Engineering*, 20, 1529-1552

Zienkiewicz, O.C., and Xie, Y. (1991). A Simple Error Estimator and Adaptive Time Stepping Procedure for Dynamic Analysis, *Earthquake Engineering Structural Dynamic*, 20, 871-887

Zienkiewicz, O.C., and Zhu, J.Z. (1992). The Superconvergent Patch Recovery and a Posteriori Error Estimates. Part I: The Recovery Technique, Part II: Error Estimates and Adaptivity, *International Journal Numerical Method Engineering*, 33, 1331-1382

Adaptive Methods for Contact Problems

Peter Wriggers
Institut für Baumechanik und Numerische Mechanik, Universität Hannover, Germany

Abstract This contribution is concerned with the adaptive finite-element-formulation of contact problems. For this, first the theoretical background of continuum mechanics and contact kinematics is given. Next the basic finite element formulation for large deformation contact processes is presented for two-dimensional problems. The development of the discretization of contact contributions follows. Here standard approaches are discussed for the case of frictional and frictionless contact. Finally different adaptive strategies involving as well residual based error estimators as error indicators based on superconvergence patch recovery are derived and formulated to include the contributions due to contact. Also error estimators based on dual principles are discussed. Examples show the performance of the different formulations and adaptive strategies.

1 Introduction

Boundary value problems involving contact are of great importance in industrial applications in mechanical and civil engineering. The range of application starts with relatively simple problems like foundations or bearings and includes metal forming processes, rubber sealings, drilling problems, crash analysis of cars, rolling contact between car tires and the road. Other applications are related to biomechanics where human joints, implantats or teeth are of consideration. Due to this variety contact problems are today combined either with large elastic or inelastic deformations including time dependent responses. Thermal coupling might have to be considered, see cooling of electronic devices, heat removal within nuclear power plant vessels or thermal insulation of astronautic vehicles. For most industrial applications numerical methods have to be applied since the contacting bodies have complex geometries or undergo large deformations. Due to that the appropriate meshing for a problem can often not be found a priori to the computation. Hence there is a need to develop adaptive methods for contact problems.
The main idea of adaptive mesh refinement is depicted in Fig. 1. It shows the basic ingredients of an adaptive method for contact problems. Part 1 is the determination of the contact, part 2 a solution of the problem with a given mesh and finally the adaptive process in part 3 leads to an automatic mesh refinement of the discretization.
Adaptive methods rely on a posteriori error estimators which estimate e.g. the difference between the exact solution $\mathbf{u}$ and the approximate solution $\mathbf{u}_h$ of the displacement field. For an elliptic variational problem of order m with complete interpolation polynomials of order k in a norm containing derivatives of degree s, see Strang and Fix (1973), one

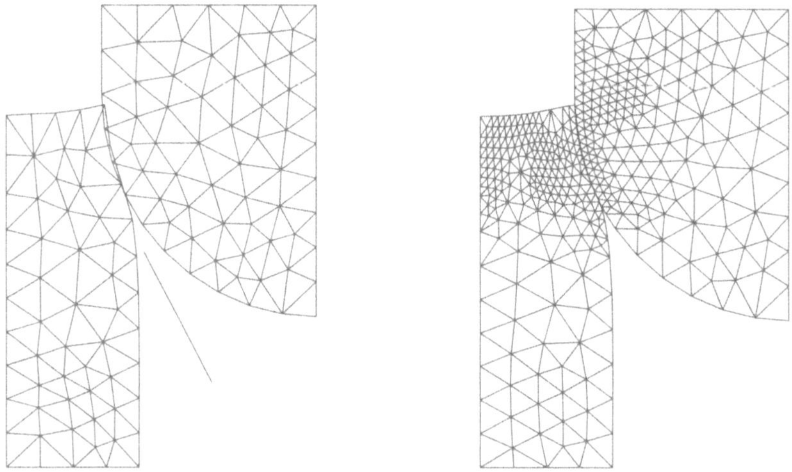

Figure 1. Adaptive process for contact problems.

has the result

$$\| \mathbf{u} - \mathbf{u}_h \|_s = O(h^{k-s} + h^{2(k-m)}) \tag{1.1}$$

Error estimates have to be derived knowing the underlying mathematical structure of the problem, for details see e.g. Strang and Fix (1973) or Verfürth (1996). In case that mathematics cannot provide such estimators (e.g. in complex nonlinear problems like finite elasticity or plasticity etc.) one can apply error measures which are constructed based on extensions of the error estimators for the linear problem to the nonlinear range or on purely heuristic observations. Such error measures provide a tool for refining the mesh at the right spots, however they do not estimate the error and thus do not provide a sharp stopping criterion for a given error tolerance.

The adaptive finite element method will be formulated here first for frictionless contact problems in linear elasticity. For this class of problems it is still possible to develop mathematically sound error estimators. Since standard contact applications include often friction, we will also state an error indicator for frictional contact and discuss its extension in case of large deformations.

Basically one has several different possibilities to derive error estimators which can be applied within adaptive methods to refine the finite element mesh. These are the residual based error estimators, projection or defect correction methods which rely on super convergence properties or error estimators which control local quantities. Most techniques will be discussed first for geometrically linear problems and then in the presence of large deformations.

Let $\mathbf{u}$ denote the exact solution of and let $\mathbf{u}_h$ be the discrete finite element solution. Then we can define the error in the displacement field by

$$\mathbf{e}_u = \mathbf{u} - \mathbf{u}_h \, . \tag{1.2}$$

In the same way the error in the stress field can be defined as

$$\mathbf{e}_\sigma = \boldsymbol{\sigma} - \boldsymbol{\sigma}_h \,. \tag{1.3}$$

Adaptive techniques rely on indicators and/or estimators which are able to predict the error given in (1.2) or (1.3). These quantities display the error distribution of the finite element solution, see e.g. Johnson (1987) and references therein. Based on the error distribution a new refined mesh can then be constructed which yields a better approximate solution.

The methods rely on error estimators which have been developed so far in different versions. Estimators which are most frequently used for elastic problems in solid mechanics are residual based error estimators, see e.g. Babuska and Rheinboldt (1978) or Johnson and Hansbo (1992), or error estimators which use superconvergence properties, see e.g. Zienkiewicz and Zhu (1987).

Becker and Rannacher (1996) or Rannacher and Suttmeier (1997) have developed error estimators for solid mechanics problems which are derived from dual principles. These estimates can be used to control local as well as global quantities. Applications of the associated estimators to practical engineering problems can be found in Ramm and Cirak (1997) and for contact problems in Wriggers et al. (2000).

For frictionless contact problems *a priori* error estimators have been derived for linear elastic bodies, see e.g. Kikuchi and Oden (1988) or Hlavacek et al. (1988). An adaptive method for problems with unilateral constraints has been developed by Lee et al. (1991) who treated as an example a free surface flow problem. In Wriggers et al. (1994) a residual based error estimator has been developed following an approach pursued by Johnson and Hansbo (1992) for unilateral membrane problems, see also Carstensen et al. (1999). However, as shown in Wriggers and Scherf (1998), also the Z^2 error indicators, due to Zienkiewicz and Zhu (1987), can be applied to contact problems. The advantages and disadvantages of residual based error estimators and Z^2 indicators have been discussed in Wriggers and Scherf (1998) for frictionless contact.

In this contribution we formulate first the contact kinematics and associated discretizations then error measures for contact problems based on the residual method, the projections schemes and the dual method. These error indicators and estimator will then be compared by means of two– and three–dimensional examples.

2 Contact Formulation

Many technical contact problems involve large deformations. Thus we will formulate all contact relations for finite deformations and will look at problems where two or more bodies B^α approach each other during a finite deformation process which come into contact on parts of their boundaries denotes by Γ_c, see Fig. 2. We observe that two points, $\mathbf{X}^1$ and $\mathbf{X}^2$, in the initial configuration of the bodies which are distinct, can occupy the same position in the current configuration, $\varphi(\mathbf{X}^2, t) = \varphi(\mathbf{X}^1, t)$, within the deformation process.

Hence in this chapter we will summarize the basic kinematical relations, the weak forms and discretization techniques needed to formulate large deformation contact.

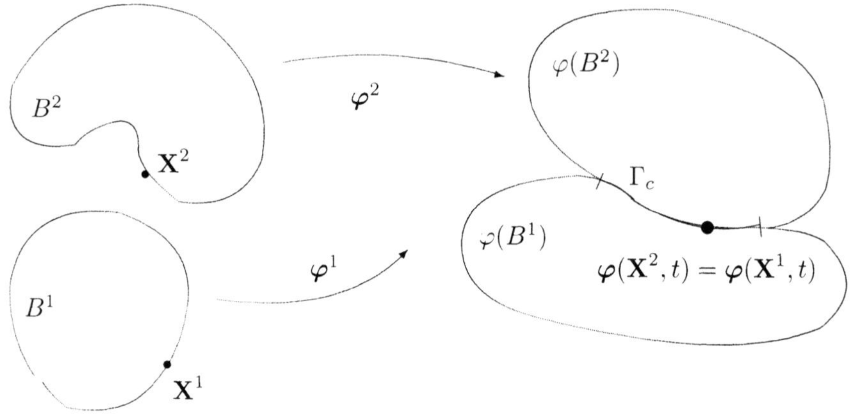

Figure 2. : Finite deformation contact.

2.1 Kinematics

In general two steps have to be performed to detect whether contact takes place or not. These are the global search for contact and the set–up of local kinematical relations which are needed to formulate the contact constraints. Here we will focus on the latter. In a large deformation, continuum based formulation of contact kinematics the distance between the bodies being in contact is measured by minimization of the distance between a point on the boundary of one body and another point on the boundary of the other body. This yields a non–penetration condition see e.g. Curnier and Alart (1988). This non–penetration function plays also a significant role for the definition of the tangential velocity in the contact interface which is needed to formulate frictional problems, see e.g. Simo and Laursen (1992), Wriggers and Miehe (1992), Laursen and Simo (1993).

Normal contact Assume that two bodies come into contact. In that case the non–penetration condition can be developed according to Fig. 3 by a minimum distance problem between point $\mathbf{x}^2$ and $\bar{\mathbf{x}}^1$. $\mathbf{x}^\alpha$ denote the coordinates of the current configuration $\varphi(B^\alpha)$ of body B^α: $\mathbf{x}^\alpha = \mathbf{X}^\alpha + \mathbf{u}^\alpha$ where $\mathbf{X}^\alpha$ is related to the initial configuration B^α and $\mathbf{u}^\alpha$ is the displacement field. The normal vector $\mathbf{n}^1$ is associated with body B^1. By assuming that the contact boundary describes, at least locally, a convex region, we can relate to every point $\mathbf{x}^2$ on Γ^2 a point $\bar{\mathbf{x}}^1 = \mathbf{x}^1(\bar{\boldsymbol{\xi}})$ on Γ^1 via the minimal distance problem

$$\|\mathbf{x}^2 - \bar{\mathbf{x}}^1\| = \min_{\mathbf{x}^1 \subseteq \Gamma^1} \|\mathbf{x}^2 - \mathbf{x}^1(\boldsymbol{\xi})\| , \tag{2.1}$$

see Fig 3 b which illustrates the two–dimensional case. $\boldsymbol{\xi} = (\xi^1, \xi^2)$ denotes the parametrization of the boundary Γ^1 via convective coordinates, see e.g. Wriggers and Miehe (1994). The parametrized surface Γ^1 is also called master surface in the following. (2.1) yields

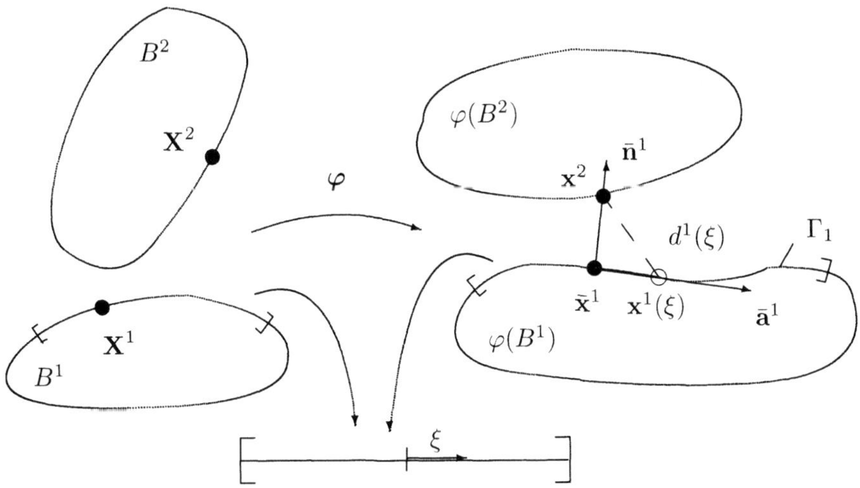

Figure 3. : a) Deformed configuration of bodies $\mathcal{B}^\alpha$, b) Minimum distance

the condition

$$\frac{d}{d\,\xi^\alpha}\,\|\,\mathbf{x}^2-\mathbf{x}^1(\xi^1,\xi^2)\,\| = \frac{\mathbf{x}^2-\mathbf{x}^1(\xi^1,\xi^2)}{\|\,\mathbf{x}^2-\mathbf{x}^1(\xi^1,\xi^2)\,\|}\cdot\mathbf{x}^1_{,\alpha}(\xi^1,\xi^2)=0 \tag{2.2}$$

$\mathbf{x}^1_{,\alpha}=\mathbf{a}^1_\alpha$ is a vector tangent to the master surface. It has to be, at the solution point $\bar{\mathbf{x}}^1$ of (2.1), perpendicular to $\mathbf{x}^2-\mathbf{x}^1(\xi^1,\xi^2)$ which thus is normal to the master surface, see Fig. 3 b.
Once the point $\bar{\mathbf{x}}^1$ is known, we can define the inequality constraint of the non–penetration condition by introducing the gap function between the two materials $g_N=[\,\mathbf{x}^2-\mathbf{x}^1(\bar{\boldsymbol{\xi}})\,]\cdot\mathbf{n}^1(\bar{\boldsymbol{\xi}})$. This function then describes the state at the interface which is expressed by an inequality constraint

$$g_N=(\mathbf{x}^2-\bar{\mathbf{x}}^1)\cdot\bar{\mathbf{n}}^1\geq 0 \tag{2.3}$$

with $g_N>0$ means gap opening, $g_N=0$ denotes perfect contact and $g_N<0$ depicts a penetration.
For some formulations which describe the contact it is useful to define a penetration function

$$g_N^-=\begin{cases}(\mathbf{x}^2-\bar{\mathbf{x}}^1)\cdot\bar{\mathbf{n}}^1 & \text{if }\ (\mathbf{x}^2-\bar{\mathbf{x}}^1)\cdot\bar{\mathbf{n}}^1<0\\ 0 & \text{otherwise}\,.\end{cases} \tag{2.4}$$

It defines the magnitude of penetration of one body into the other. Note that (2.4) can be used to determine the contact area $\Gamma^\alpha_c\subseteq\Gamma^\alpha$.

Tangential kinematical contact relations The tangential kinematical relations for contact have to represent as well stick as slip conditions. The stick condition means that no relative displacement is possible in tangential direction. Hence, by defining the relative

tangential displacement we can formulate this condition in form of a constraint. In case of slip a tangential movement is possible, which is defined by the relative tangential velocity.
The stick condition is given by the constraint

$$\mathbf{g}_T = \mathbf{0}\,. \tag{2.5}$$

To obtain the amount of sliding we have to compute the relative tangential slip and furthermore need a constitutive equation for the frictional behaviour in the contact interface.
The tangential relative displacement between two bodies is related to the change of the solution point $\bar{\boldsymbol{\xi}} = (\bar{\xi}^1, \bar{\xi}^2)$ which has been obtained via the minimal distance problem (2.1) at time t. We define as the tangential relative velocity on the current surface

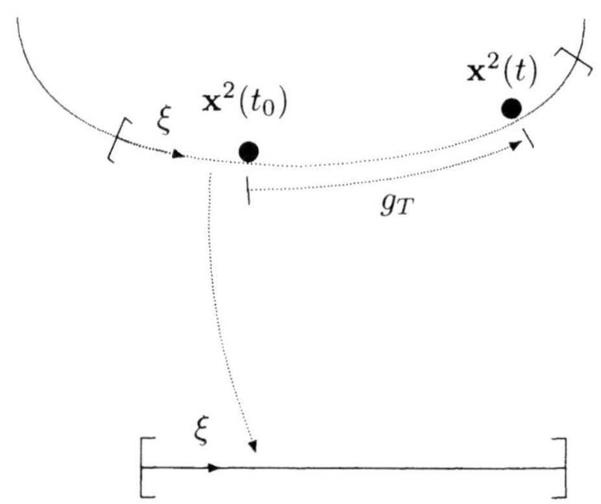

Figure 4. Relative tangential movement of point $\mathbf{x}^2$

$\varphi^1(\Gamma^1_c)$ by

$$\dot{\mathbf{g}}_T \;:=\; \dot{\bar{\xi}}^{\alpha}\,\bar{\mathbf{a}}_\alpha \tag{2.6}$$

(2.6) determines the evolution of the tangential slip $\mathbf{g}_T$ which may enter as a local kinematical the constitutive function for the contact tangential stress. With respect to Fig. 4 we obtain for the total slip between point $\mathbf{x}^2$ and the deformed master surface

$$g_T = \int_{t_0}^{t} \| \dot{\bar{\xi}}^{\alpha} \bar{\mathbf{a}}^1_\alpha \| \, dt \tag{2.7}$$

where t is the time, here used to parameterize the path of point $\mathbf{x}^2$. For the evaluation of this integral we have to compute the time derivative of ξ^α at the projection point $\bar{\mathbf{x}}^1$ of point x^2. This can be done by using (2.2). Thus we can compute the time derivative of ξ^α from the relation

$$[\mathbf{x}^2 - \bar{\mathbf{x}}^1(\xi^2, \xi^2)] \cdot \bar{\mathbf{a}}^1_\alpha = 0 \tag{2.8}$$

which is valid at the contact point since the difference $\mathbf{x}^2 - \bar{\mathbf{x}}^1$ is normal to the contact surface and $\bar{\mathbf{a}}^1_\alpha$ denotes the tangent vector to the surface Γ^1 at the minimal distance point, see Fig. 3 b. This yields the following result

$$\frac{d}{dt}\left[\mathbf{x}^2 - \bar{\mathbf{x}}^1(\bar{\xi^1}, \bar{\xi^2})\right] \cdot \bar{\mathbf{a}}^1_\alpha = \left[\mathbf{v}^2 - \bar{\mathbf{v}}^1 - \bar{\mathbf{a}}_\beta \dot{\bar{\xi^\beta}}\right] \cdot \bar{\mathbf{a}}^1_\alpha + \left[\mathbf{x}^2 - \bar{\mathbf{x}}^1\right] \cdot \dot{\bar{\mathbf{a}}}^1_\alpha = 0\,. \tag{2.9}$$

With $\dot{\bar{\mathbf{a}}}^1_\alpha = \bar{\mathbf{v}}^1_{,\alpha} + \hat{\mathbf{x}}^1_{,\alpha\beta}\, \dot{\bar{\xi^\beta}}$ we obtain $\dot{\bar{\xi^\beta}}$ from (2.9) which leads to the following system of equations

$$\bar{H}_{\alpha\beta}\, \dot{\bar{\xi^\beta}} = \bar{R}_\alpha \tag{2.10}$$

with

$$\begin{aligned} \bar{H}_{\alpha\beta} &= \left[\bar{a}_{\alpha\beta} + g_N\, \bar{\mathbf{x}}^1_{,\alpha\beta} \cdot \bar{\mathbf{n}}^1\right] \\ \bar{R}_\alpha &= \left[\mathbf{v}^2 - \bar{\mathbf{v}}^1\right] \cdot \bar{\mathbf{a}}^1_\alpha + g_N\, \bar{\mathbf{n}}^1 \cdot \bar{\mathbf{v}}^1_{,\alpha} \end{aligned} \tag{2.11}$$

A well known result from differential geometry introduces for $\bar{\mathbf{x}}^1_{,\alpha\beta}(\bar{\xi^1}, \bar{\xi^2}) \cdot \bar{\mathbf{n}}^1$ the curvature tensor $\bar{b}_{\alpha\beta}$. Thus we can rewrite $\bar{H}_{\alpha\beta} = \left[\bar{a}_{\alpha\beta} + g_N\, \bar{b}_{\alpha\beta}\right]$. For a penetration $g_N < 0$ we have to take into account the terms depending on g_N in (2.11).

REMARK

1. Note that the second terms on the right hand side of (2.10) depends on the penetration g_N. Thus in the case of a the strong enforcement of the non–penetration condition ($g_N = 0$) with Lagrangian multipliers this term vanishes. This yields

$$\dot{\bar{\xi^\beta}} = \left[\mathbf{v}^2 - \bar{\mathbf{v}}^1\right] \cdot \bar{\mathbf{a}}^{1\,\beta} \tag{2.12}$$

   Then the evolution relative tangential velocity in (2.6) is given by the projection of the spatial velocities $\mathbf{v}^2$ and $\hat{\mathbf{v}}^1(\bar{\boldsymbol{\xi}})$ evaluated at the contact points onto the tangential direction of the contact surface.

$$\dot{\mathbf{g}}_T = \left(\bar{\mathbf{a}}^1_\alpha \otimes \bar{\mathbf{a}}^{1\,\alpha}\right)\left[\mathbf{v}^2 - \bar{\mathbf{v}}^1\right] \tag{2.13}$$

2. If we have a flat contact surface the curvature tensor $\bar{b}_{\alpha\beta}$ is zero.

3 Weak forms and solution algorithms for solids in contact

For the formulation of the weak form we will discuss only the additional terms due to contact in detail.

For a numerical solution of the nonlinear boundary value problem we will use the finite element method. Thus we need the weak form of the local field equations. Due to the fact that the constraint condition (2.3) is represented by an inequality we obtain in general a variational inequality for the mechanical part, see e. g. Wriggers (2001). The general form can be written as

$$\sum_{\gamma=1}^{2} \int_{\Omega^\gamma} \boldsymbol{\tau}^\gamma \cdot \nabla^S (\boldsymbol{\eta}^\gamma - \boldsymbol{\varphi}^\gamma)\, dV \geq \sum_{\gamma=1}^{2} \int_{\Omega^\gamma} \rho_0\, \bar{\mathbf{b}}^\gamma \cdot (\boldsymbol{\eta}^\gamma - \boldsymbol{\varphi}^\gamma)\, dV - \int_{\Gamma_\sigma{}^\gamma} \bar{\mathbf{t}}^\gamma \cdot (\boldsymbol{\eta}^\gamma - \boldsymbol{\varphi}^\gamma)\, dA \tag{3.1}$$

where the integration is performed with respect to the domain Ω^γ occupied by the body $\mathcal{B}^\gamma$ in the reference configuration. The KIRCHHOFF stress tensor $\boldsymbol{\tau}$ and the gradient operator "∇" are evaluated with respect to the current coordinates, for details see also next chapter.
We now have to find the deformation $(\boldsymbol{\varphi}^1 . \boldsymbol{\varphi}^2) \in \mathbf{K}$ such that (2.3) is fulfilled for all $(\boldsymbol{\eta}^1 . \boldsymbol{\eta}^2) \in \mathbf{K}$ with

$$\mathbf{K} = \{ (\boldsymbol{\eta}^1 . \boldsymbol{\eta}^2) \in \mathbf{V} \,|\, [\boldsymbol{\eta}^2 - \hat{\boldsymbol{\eta}}^1(\bar{\xi}^1 . \bar{\xi}^2)] \cdot \bar{\mathbf{n}}_1 \geq 0 \}, \tag{3.2}$$

In case of finite elasticity the existence of the solution of (3.2) can be proved, see e.g. Ciarlet (1988) or Curnier et al. (1992).

REMARK

1. In the geometrically linear case these equations can be recast in a weak or variational formulation as follows

$$a(\mathbf{u}. \mathbf{v} - \mathbf{u}) \geq f(\mathbf{v} - \mathbf{u}). \tag{3.3}$$

with

$$a(\mathbf{u}. \mathbf{w}) = \int_\Omega \boldsymbol{\varepsilon}(\mathbf{u}) : \mathbb{C}_0 : \boldsymbol{\varepsilon}(\mathbf{w}) \, d\Omega, \tag{3.4}$$

$$f(\mathbf{w}) = \int_\Omega \hat{\mathbf{b}} \cdot \mathbf{w} \, d\Omega + \int_{\Gamma_\sigma} \hat{\mathbf{t}} \cdot \mathbf{w} \, d\Gamma \tag{3.5}$$

and $\Omega = \cup_\gamma \mathcal{B}^\gamma$. $\mathbb{C}_0$ is the elasticity tensor for the classical linear elastic constitutive equation. The linear strain tensor is defined by $\boldsymbol{\varepsilon}(\mathbf{u}) = \frac{1}{2}(\nabla \mathbf{u} + \nabla^T \mathbf{u})$.
The problem is now, as in the nonlinear case, to find $\mathbf{u} \in \mathbf{K}$ such that (3.3) is fulfilled for all $\mathbf{v} \in \mathbf{K}$ with

$$\mathbf{K} = \{ \mathbf{v} \in \mathbf{V} \,|\, (\mathbf{v}^2 - \bar{\mathbf{v}}^1) \cdot \bar{\mathbf{n}}^1 + g_0 \geq 0 \text{ on } \Gamma_c \} \tag{3.6}$$

and

$$\mathbf{V} = \{ \mathbf{v} \in [H^{1/2}(\Omega)]^{dim} \,|\, \mathbf{v} = \mathbf{0} \text{ on } \Gamma_u \}. \tag{3.7}$$

The mathematical structure of the variational inequality (3.3) is discussed in detail in e. g. Duvaut and Lions (1976) or Kikuchi and Oden (1988). Due to the inequality constraint on the deformation field the contact problem is nonlinear even in the case of linear elasticity.

Algorithms for solving variational inequalities are given by mathematical programming, active set strategies or sequential quadratic programming methods, to name only a few. Each of these methods is well known from optimization theory, see e. g. Luenberger (1984) or Bertsekas (1984). Mathematical programming was applied to solve contact problems in e. g. Klarbring (1986). The sequential quadratic programming approach has been considered by e.g. Barthold and Bischoff (1988) and lately with application to large strain elasticity by Björkman et al. (1995).

Here we will concentrate on the active set strategies which are basis of the contact formulation in many existing finite element codes. Within this approach the contact constraints are not changed during the iteration for equilibrium and thus can be treated like equality constraints. These can be introduced via Lagrangian multipliers or penalty terms.

Within an active set strategy we can write the weak form as an equality since we know the active set within an incremental solution step. Then equation (3.1) can be recast in the following form

$$\sum_{\gamma=1}^{2} \{ \int_{\Omega^\gamma} \boldsymbol{\tau}^\gamma \cdot \nabla^S \boldsymbol{\eta}^\gamma \, dV - \int_{\Omega^\gamma} \bar{\mathbf{f}}^\gamma \cdot \boldsymbol{\eta}^\gamma \, dV - \int_{\Gamma_\sigma{}^\gamma} \bar{\mathbf{t}}^\gamma \cdot \boldsymbol{\eta}^\gamma \, dA \} + C_c = 0 \tag{3.8}$$

where the contact contributions C_c are associated with the active constraint set Γ_c^{act}. $\boldsymbol{\eta}^\gamma \in V$ is the so called test function or virtual displacement which is zero at the boundary Γ_φ^γ where the deformations are prescribed.

For two bodies being in contact we obtain the weak form of the interface contributions by assuming that contact is active at the surface Γ_c^{act}.

1. **Lagrangian multiplier method:**

$$C_c = \int_{\Gamma_c^{act}} (\lambda_N \, \delta g_N + \boldsymbol{\lambda}_T \cdot \delta \mathbf{g}_T) \, dA \tag{3.9}$$

   Here λ_N denotes the Lagrangian multiplier which can be identified as the contact pressure. δg_N is the variation of the normal gap. The term $\boldsymbol{\lambda}_T \cdot \delta \mathbf{g}_T$ is associated with the tangential stick or slip motion and needs further discussion. In case of pure stick the relative tangential slip $\mathbf{g}_T$ is zero which yields a constraint equation from which $\boldsymbol{\lambda}_T$ follows as a reaction. In case of sliding the tangential stress vector $\mathbf{t}_T$ is determined by the constitutive law for frictional slip and thus we should write instead of $\boldsymbol{\lambda}_T \cdot \delta \mathbf{g}_T \longrightarrow \mathbf{t}_T \cdot \delta \mathbf{g}_T$.
2. **Penalty method:** In this formulation a penalty term due to the constraint condition is added to the weak form (3.8). This means that once the constraint equation for $g_N{}^-$ is violated

$$C_c = \int_{\Gamma_c^{act}} \epsilon_N \, g_N{}^- \, \delta g_N{}^- \, dA \,, \quad \epsilon_N > 0 \tag{3.10}$$

   has to be considered for normal contact. It can be shown, see e.g. Luenberger (1984), that the solution of the Lagrangian multiplier method can be recovered from this formulation for $\epsilon_N \to \infty$, however high values of ϵ_N will lead to an ill–conditioned numerical problem. As in the Lagrangian multiplier method we have to distinguish between pure stick in the contact interface which yields a penalty term also for the tangential direction

$$C_c = \int_{\Gamma_c^{act}} (\epsilon_N \, g_N{}^- \, \delta g_N{}^- + \epsilon_T \, \mathbf{g}_T \cdot \delta \mathbf{g}_T) \, dA \,, \quad \epsilon_N > 0 \,, \epsilon_T > 0 \tag{3.11}$$

and the slip condition which leads to

$$C_c = \int_{\Gamma_c^{act}} (\epsilon_N \, g_N^{\,-} \, \delta g_N^{\,-} + \mathbf{t}_T \cdot \delta \mathbf{g}_T \,) \, dA \,, \qquad \epsilon_N > 0 \tag{3.12}$$

In the latter equation a frictional laws has to be applied.
In equations (3.8) to (3.12) the variation of the normal gap function g_N is needed which yields

$$\delta g_N = [\,\boldsymbol{\eta}^2 - \boldsymbol{\eta}^1(\bar{\xi}_1 . \bar{\xi}_2)] \cdot \bar{\mathbf{n}}_1 \,. \tag{3.13}$$

Furthermore the variation of the tangential slip can be stated as

$$\delta \mathbf{g}_T = \delta \bar{\xi}^\alpha \, \bar{\mathbf{a}}_\alpha^1 \,. \tag{3.14}$$

The latter relation follows simply from (2.6) by replacing the velocities $\mathbf{v}$ by the test function $\boldsymbol{\eta}$ in (2.11).

REMARK: A major problem associated with the numerical treatment of the penalty method is the ill conditioning which arises when the penalty parameter ϵ_N assumes large values. A standard method to overcome the problem of ill–conditioning is based on the augmented Lagrangian technique. well known in optimization theory. This technique has been considered extensively within the context of incompressibility constraints Glowinski and Le Tallec (1984) and was also applied to contact problems, see Wriggers et al. (1985) or Kikuchi and Oden (1988) for frictionless contact. Recently this approach has been extended successfully also to large displacement contact problems including friction, see Alart and Curnier (1991) or Laursen and Simo (1993). A formulation which accounts for micromechanical interface laws can be found in Wriggers and Zavarise (1993a).

3.1 Local integration of constitutive equations

For the normal contact a mere function evaluation –like for finite elasticity– can be used to obtain within the penalty method the contact pressure $p_N = \epsilon_N \, g_N^{\,-}$.
For the treatment of the frictional contact one has to specify a constitutive relation at the contact interface. This can be done analogous to plasticity. By using the kinematical split for the relative tangential velocities into a stick (elastic) and a slip (plastic part

$$\dot{\mathbf{g}}_T = \dot{\mathbf{g}}_T^{st} + \dot{\mathbf{g}}_T^{sl} \tag{3.15}$$

one can specify an "elastic" constitutive relation by

$$\mathbf{t}_T = \epsilon_T \, (\, \dot{\mathbf{g}}_T - \dot{\mathbf{g}}_T^{sl} \,) \tag{3.16}$$

and evolution equations for the plastic slip, see e. g. Wriggers and Miehe (1994),

$$\dot{\mathbf{g}}_T^{sl} = \dot{\lambda} \, \frac{\partial \hat{g}_s(\mathbf{t}_T)}{\partial \mathbf{t}_T} = \dot{\lambda} \, \mathbf{n}_T \qquad \text{with} \quad \mathbf{n}_T = \frac{\mathbf{t}_T}{\|\mathbf{t}_T\|} \tag{3.17}$$

$$\dot{g}_v = \dot{\lambda} \tag{3.18}$$

which denote the normality rule for fixed contact pressure and the evolution of total slip. The function $\hat{g}_s = \| \mathbf{t}_T \|$ in (3.17) defines the so-called plastic potential which is different from the slip rule

$$\hat{f}_s(\mathbf{t}_T\,, t_N) = \| \mathbf{t}_T \| - \mu\, t_N \leq 0 \tag{3.19}$$

to account for non-associativity. The parameter μ is the frictional coefficient. Additionally we have the loading–unloading conditions in Kuhn–Tucker form

$$\dot{\lambda} \geq 0\,, \quad \hat{f}_s(\mathbf{t}_T) \leq 0\,, \quad \dot{\lambda}\, \hat{f}_s(\mathbf{t}_T) = 0\,. \tag{3.20}$$

which can be used to determine the plastic parameter $\dot{\lambda}$.
The algorithmic computation of the tangential stress $\mathbf{t}_{T\,n+1}$ due to friction is performed by the return algorithm based on an objective (backward Euler) integration of the evolution equation for the plastic slip, see e.g. Wriggers (1987), Ju and Taylor (1988), Giannokopoulos (1989), Wriggers et al. (1990).
The results can be summarized as follows: Integration of (2.6) gives the increment of the total slip within the time step Δt_{n+1}

$$\Delta \mathbf{g}_{T\,n+1} = (\,\bar{\xi}^{\alpha}_{n+1} - \bar{\xi}^{\alpha}_{n}\,)\bar{\mathbf{a}}_{\alpha\,n+1}\,. \tag{3.21}$$

The total slip has is decomposed into a stick and slip part, see , $\mathbf{g}^e$, and a slip, $\mathbf{g}^s$, part at time t_{n+1}. Then we can compute an elastic trial state from the elastic or stick part using (3.16)

$$\mathbf{t}^{tr}_{T\,n+1} = \epsilon_T\, \mathbf{g}^{st}_{n+1} \tag{3.22}$$

and evaluate the COULOMB slip criterion $f_s = \| \mathbf{t}_T \| - \mu\, p_N \leq 0$ at time t_{n+1}

$$\begin{aligned} \mathbf{t}^{tr}_{T\,n+1} &:= \epsilon_T\,(\,\mathbf{g}_{T\,n+1} - \mathbf{g}^{sl}_{T\,n}\,) = \mathbf{t}_{T\,n} + \epsilon_T\, \Delta \mathbf{g}_{T\,n+1}\,, \\ f^{tr}_{s\,n+1} &:= \|\mathbf{t}^{tr}_{T\,n+1}\| - \mu\, p_{N\,n+1}\,. \end{aligned}$$

If this state is elastic ($f^{tr}_{s\,n+1} \leq 0$) then no friction takes place and we have to use the elastic relation (3.22). In case that $f^{tr}_{s\,n+1} > 0$ we have to perform the return mapping. Using the implicit Euler scheme, the integration of the slip rule (3.17) yields

$$\mathbf{g}^{s}_{T\,n+1} = \mathbf{g}^{sl}_{T\,n} + \lambda\, \mathbf{n}_{T\,n+1}\,, \qquad g_{v\,n+1} = g_{v\,n} + \lambda \tag{3.23}$$

Using (3.16) and (3.17) the tangential stresses are given by

$$\mathbf{t}_{T\,n+1} = \mathbf{t}^{tr}_{t\,n+1} - \lambda\, \epsilon_T\, \mathbf{n}_{T\,n+1}\,, \qquad \mathbf{n}_{T\,n+1} = \mathbf{n}^{tr}_{T\,n+1}\,, \tag{3.24}$$

The slip parameterλ can be computed from the condition

$$f_s(\lambda) = \| \mathbf{t}^{tr}_{T\,n+1}\| - \mu\, p_{N\,n+1} - \epsilon_T\, \lambda = 0 \tag{3.25}$$

which can be directly solved for λ in case of Coulomb's law

$$\lambda = \frac{1}{\epsilon_T}\,(\,\| \mathbf{t}^{tr}_{t\,n+1} \| - \mu\, p_{N\,n+1}\,) \tag{3.26}$$

Knowing λ the stress update follows from (3.24) and the frictional slip from (3.23). Again we state the explicit results for Coulomb's model

$$\begin{aligned} \mathbf{t}_{T\,n+1} &= \mu\, p_{N\,n+1}\, \mathbf{n}^{tr}_{T\,n+1}\,, \\ \mathbf{g}^{sl}_{T\,n+1} &= \mathbf{g}^{sl}_{T\,n} + \frac{1}{\epsilon_T}\,(\,\|\,\mathbf{t}^{tr}_{t\,n+1}\,\| - \mu\, p_{N\,n+1}\,)\, \mathbf{n}^{tr}_{T\,n+1}\,. \end{aligned}$$

which completes the algorithm for the frictional interface law.

4 Discretization of contact contributions

The discretization of the domain contributions of the bodies being in contact is not the objective of this work. For a detailed treatment with respect to the finite element implementations of boundary-value-problems regarding large deformations, see e.g. Crisfield (1991), Zienkiewicz and Taylor (1991), or Bathe (1986) and references therein. However, in the next chapter the discretization of continua undergoing large strains using isoparametric elements is discussed briefly for completeness.

Within the finite element method we have different approximations. These are geometrical approximations of the domain B on which the boundary value problem is defined. Furthermore the associated fields, deformations or stresses, have to be approximated. Also the integrals are not evaluated exactly, since such as they are evaluated for the weak form, they have to be computed via numerical integration procedures. Collectively, these approximations are sources for errors inherent in the finite element method. Estimation of these errors is contents of chapter 3.

Here we focus now on the discretization of the contact constraints. For reasons of simplicity we will restrict ourselves here to two dimensional formulations. Three dimensional contact discretizations can be found in e.g. Hallquist et al. (1985), Laursen and Simo (1993) or Heegaard and Curnier (1993).

In the first applications of finite elements to contact problems of two deformable bodies only small changes in the geometry were assumed so that the geometrically linear theory could be applied. Then it is possible to incorporate the contact constraints on a purely nodal basis, see e.g. Francavilla and Zienkiewicz (1975). Later also contact elements were developed which resulted from a degenerated solid element, see e.g. Stadter and Weiss (1979) or the textbook of Kikuchi and Oden (1988). A mathematical study of these classes of elements which also accounts for the correct integration rules can be found in Oden (1981). All of the above mentioned elements need a discretization in which the element nodes match each other in the contact interface. For the general case of nodes being arbitrary distributed along the possible contact interface between two bodies, which can occur when automatic meshing is used for two different bodies, Simo et al. (1985) developed a segment approach to discretize the contact interface.

For the general case of contact including large deformations finite element nodes on the contacting boundaries do not assume the same position, see Fig. 5 Then we define the active part of the contact interface Γ_c where $g_N \leq 0$ for the solution process. Right now most algorithms rely on discretizations which are based on nodes on the boundary of the bodies. Thus the algorithm has to find the active contact constraints denoted by $\mathcal{J}_A \in$

$\mathcal{J}_C$, where $\mathcal{J}_C$ are all possible contact nodes. The most frequently used discretization is the so called node–to–segment approach. Here arbitrary sliding of a node over the entire contact area is allowed. Early implementations can be found in Hughes et al. (1977) which have been developed for more and more general cases, Hallquist et al. (1985), Bathe and Chaudhary (1985) and Wriggers et al. (1990). Now some finite element codes include also self–contact, see e. g. Hallquist et al. (1992). Also the idea of contact segments proposed by Simo et al. (1985) has been followed up and applied to problems involving large deformations, see Papadopoulos and Taylor (1992). The basic difference

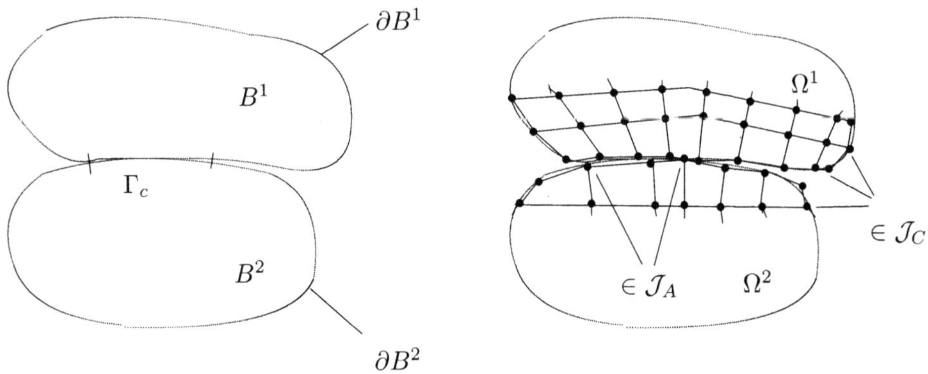

Figure 5. Discretization for large deformation contact.

between the Lagrangian multiplier method (3.9) and the penalty approach (3.10) lies in the fact that the Lagrangian multiplier formulation is a mixed method which means that both variables λ_N and δg_N have to be discretized

$$\int_{\Gamma_c} \lambda_N \, \delta g_N \, d\Gamma \longrightarrow \int_{\Gamma_c^h} \lambda_N^h \, \delta g_N^h \, d\Gamma \tag{4.1}$$

with interpolations for λ_N^h and δg_N^h

$$\lambda_N^h = \sum_K M_K(\xi) \, \lambda_{NK} \quad \text{and} \quad \delta g_N^h = \sum_I N_I(\xi) \, \delta g_{N\,I} \tag{4.2}$$

Note that the interpolations have to be chosen in such a way that they fulfill the Babuska-Brezzi (BB) condition since the Lagrangian multiplier method is a mixed formulation, see e.g. Kikuchi and Oden (1988).
Contrary, the penalty method needs only the discretization of the displacement variables

$$\int_{\Gamma_c} \epsilon_N \, g_N^- \, \delta g_N^- \, d\Gamma \longrightarrow \int_{\Gamma_c^h} \epsilon_N \, g_N^h \, \delta g_N^h \, d\Gamma \tag{4.3}$$

where usually the same interpolation is introduced for the gap function and its variation

$$g_N^h = \sum_I N_I(\xi) \, g_{N\,I} \quad \text{and} \quad \delta g_N^h = \sum_I N_I(\xi) \, \delta g_{N\,I} \tag{4.4}$$

In the following we will only discuss discretizations related to the penalty method and to the formulation using constitutive equations in the contact interface. But also here one has to be careful when choosing the interpolation for the continuous contact. This follows from the fact that the penalty method is equivalent to a mixed method and hence the BB condition plays again a role, see e.g. Kikuchi and Oden (1988), for a detailed discussion of this matter.

4.1 Small deformation contact

In small deformation contact we do not need search algorithms since no large slip can occur on the contact surface. Thus the mesh topology does not change and it is possible to develop contact interface elements.

In the following we will discuss the case of two deformable bodies being in contact. The simplest possibility of defining the contact constraints is via a node-to-node contact element. This can only be applied to geometrically linear problems since a large relative tangential movement of the nodes is not allowed in the contact area. The geometrical

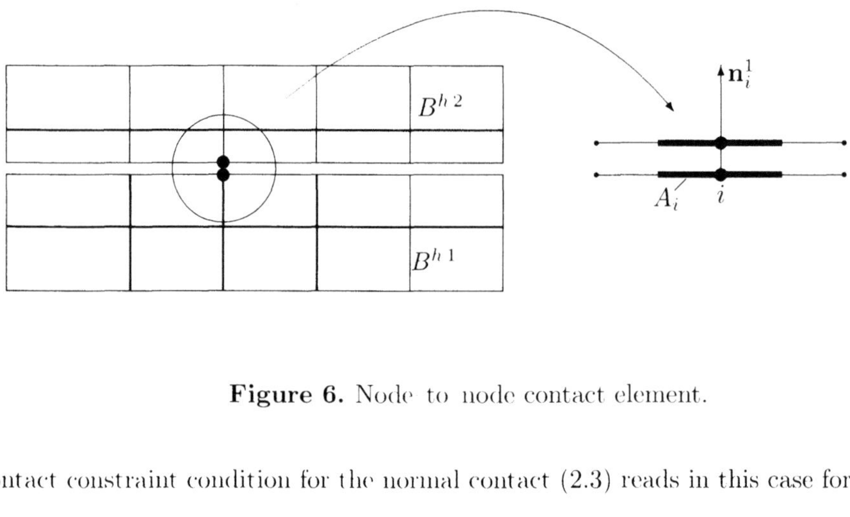

Figure 6. Node to node contact element.

contact constraint condition for the normal contact (2.3) reads in this case for one node i

$$g_{N\,i} = (\mathbf{u}_i^2 - \mathbf{u}_i^1) \cdot \mathbf{n}_i^1 + G_0 \geq 0 \tag{4.5}$$

In the case of small strain the change in the normal is neglected within the computations. Thus the variation of (4.5) is

$$\delta g_{N\,i} = (\boldsymbol{\eta}_i^2 - \boldsymbol{\eta}_i^1) \cdot \mathbf{n}_i^1 \tag{4.6}$$

Due to its simplicity and assuming that the contact constraint is active for n_c nodes we can express the integral (4.3) for the contact contributions in the weak for by a sum over all active contact nodes

$$\int_{\Gamma_c} \epsilon_N \, g_N \, \delta g_N \, d\Gamma \longrightarrow \sum_{i=1}^{n_c} \epsilon_N \, g_{N\,i} \, \delta g_{N\,i} \, A_i = \sum_{i=1}^{n_c} \epsilon_N \, g_{N\,i} \, (\boldsymbol{\eta}_i^2 - \boldsymbol{\eta}_i^1) \cdot \mathbf{n}_i^1 \, A_i \tag{4.7}$$

where n_c are the contact nodes in Γ_c^h. The test function $\boldsymbol{\eta}_i^\alpha$ and the normal vector $\bar{\mathbf{n}}_i^1$ is defined for the node i, see e.g. Wriggers and Zavarise (1993b). Often the area A_i which is associated with the contact point i, see Fig. 6 is neglected (or "hidden" in the penalty parameter ϵ_N) in the node–to–node contact formulation. This means that the contact stress p_N becomes a contact (nodal) force $f_{N\,i} = \epsilon_N \, g_{N\,i}$. The associated matrix formulation leads in the geometrically linear case for the contact element i to the definition of the contact residual $G_i^c = \boldsymbol{\eta}^T \mathbf{G}_i^c$ and its associated tangent matrix $\mathbf{K}_i^c$ with

$$\mathbf{G}_i^c = \epsilon_N \, g_{N\,i} \, \mathbf{N}_i \,, \quad \mathbf{K}_i^c = \epsilon_N \, \mathbf{N}_i \, \mathbf{N}_i^T \,, \quad \text{with} \quad \mathbf{N}_i = \begin{Bmatrix} \mathbf{n}_i^1 \\ -\mathbf{n}_i^1 \end{Bmatrix} \tag{4.8}$$

Let us remark that it is not always obvious how to define the normal vector for a given interpolation of the contact boundary. This is especially true when a linear interpolation, as shown in Fig. 7, is used. Here it can be seen that there is a jump in the normal at point i. Thus the definition of the normal at point i is not unique. A way out is given by using a normal $\mathbf{n}_i^1$ which is obtained from a smooth interpolation by e. g. BEZIER or spline functions as depicted in Fig. 7.

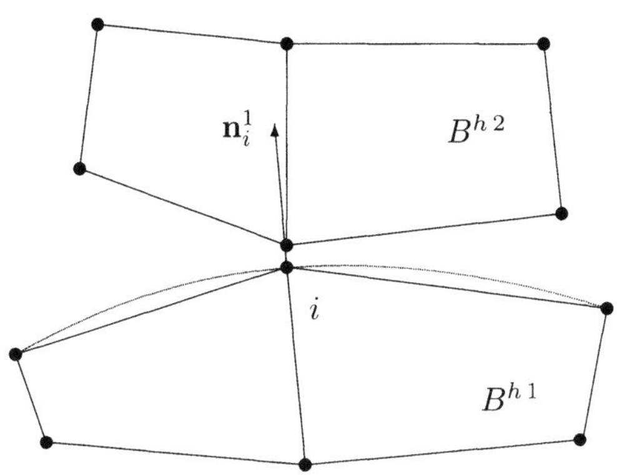

Figure 7. Definition of the contact normal.

The problem of defining the normal at nodal points can also be circumvented by introducing an isoparametric interpolation for the contact surface.

4.2 Large deformation contact

For the case of nonlinear deformations we will discuss the most simple element which is widely used in standard finite element codes. This discretization is named node–to–segment contact element and is widely used in nonlinear finite element simulations of contact problems.

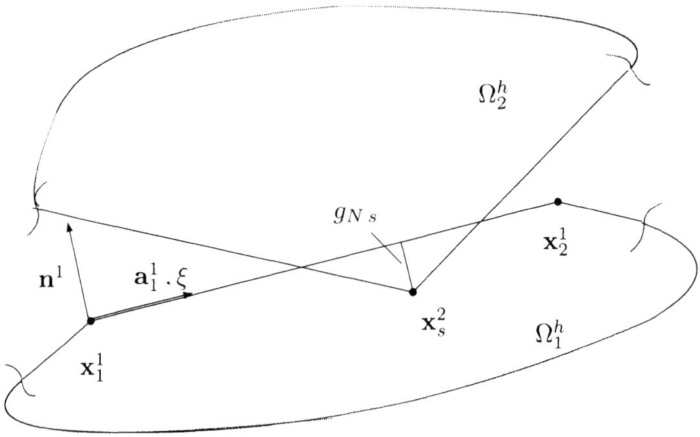

Figure 8. Node–to–segment contact element.

Due to its importance we like to consider this contact element in more detail. Assume that the discrete slave point (s) of body B^2 with coordinate $\mathbf{x}_s^2$ comes into contact with the master segment (1)–(2) of body B^1 defined by the nodal coordinates $\mathbf{x}_1^1$ and $\mathbf{x}_2^1$, see Fig. 8. Then the kinematical relations can be directly computed using the equations stated in section 2. With linear interpolation for the master segment based on the introduction of the surface coordinate ξ along the master surface

$$\hat{\mathbf{x}}^1(\xi) = \mathbf{x}_1^1 + (\mathbf{x}_2^1 - \mathbf{x}_1^1)\,\xi \tag{4.9}$$

we derive that each segment has a constant tangent vector

$$\bar{\mathbf{a}}_1^1 = \hat{\mathbf{x}}^1(\xi)_{,1} = (\mathbf{x}_2^1 - \mathbf{x}_1^1) \tag{4.10}$$

It is connected to a base vector $\mathbf{a}_1^1$ by $\mathbf{a}_1^1 = \bar{\mathbf{a}}_1^1 \,/\, l$ with $l = \parallel \mathbf{x}_2^1 - \mathbf{x}_1^1 \parallel$ being the current length of the master segment. With the unit tangent vector $\mathbf{a}_1^1$ the unit normal to the segment (1)–(2) can be defined as $\mathbf{n}^1 = \mathbf{e}_3 \times \mathbf{a}_1^1$.
$\bar{\xi}$ and $g_{N\,s}$ are given by the solution of the minimal distance problem, i.e. by the projection of the slave node $\mathbf{x}_s$ in (s) onto the master segment (1)–(2)

$$\bar{\xi} = \frac{1}{l}(\mathbf{x}_s^2 - \mathbf{x}_1^1) \cdot \mathbf{a}_1^1 \qquad \text{and} \qquad g_{N\,s} = \left[\,\mathbf{x}_s^2 - (1 - \bar{\xi})\,\mathbf{x}_1^1 - \bar{\xi}\,\mathbf{x}_2^1\,\right] \cdot \mathbf{n}^1 \,. \tag{4.11}$$

From these equations and the local continuous formulation (3.13) we compute directly the variation of the gap function δg_N on the straight master segment (1)–(2).

$$\delta g_{N\,s} = \left[\,\boldsymbol{\eta}_s^2 - (1 - \bar{\xi})\,\boldsymbol{\eta}_1^1 - \bar{\xi}\,\boldsymbol{\eta}_2^1\,\right] \cdot \mathbf{n}^1 \,. \tag{4.12}$$

The local equation (3.14) yields the expression for $\delta\bar{\xi}$ which however can also be computed directly from (4.11). The total relative displacement is given by (2.7) which yields for the straight master segment

$$g_T = \int_{\xi_0}^{\xi} \| \bar{a}_{11} \| \, d\xi = (\, \xi - \xi_0 \,) \, l \, . \tag{4.13}$$

This leads to the variation

$$\delta g_T = \delta \xi \, l + (\, \xi - \xi_0 \,) \, \delta l \, . \tag{4.14}$$

With the interpolation for the variation $\hat{\boldsymbol{\eta}}^1(\xi) = \boldsymbol{\eta}_1^1 + \xi \, (\boldsymbol{\eta}_2^1 - \boldsymbol{\eta}_1^1)$ on the straight master segment (1)–(2) we specialize this relation

$$\delta g_T = [\, \boldsymbol{\eta}_s^2 - (1 - \bar{\xi}) \, \boldsymbol{\eta}_1^1 - \bar{\xi} \, \boldsymbol{\eta}_2^1 \,] \cdot \mathbf{a}_1^1 + \frac{g_{N\,s}}{l} \, [\, \boldsymbol{\eta}_2^1 - \boldsymbol{\eta}_1^1 \,] \cdot \mathbf{n}^1 + \frac{g_{T\,s}}{l} \, [\, \boldsymbol{\eta}_2^1 - \boldsymbol{\eta}_1^1 \,] \cdot \mathbf{a}^1 \, . \tag{4.15}$$

Equations (4.12) and (4.15) characterize the main kinematical relations of the contact element in Figure 10.

In what follows we compute the contribution of the node–to–segment element to the weak form (3.10). The basic formulation for this discretization is analogous to the node–to–node element. Thus we assume that we know the normal force $P_{Ns} = p_{Ns} \, A_s$ and the tangential force $T_{Ts} = t_{Ts} \, A_s$ at the discrete contact point (s) of the contact element under consideration where A_s denotes the area of the contact element. Both forces, $P_{N\,s}$ and $T_{T\,s}$, can be obtained from the constitutive relations discussed above. This leads to

$$\int_{\Gamma_c} (\, p_N \, \delta g_N + t_T \, \delta g_T \,) \, d\Gamma \longrightarrow \sum_{s=1}^{n_c} (\, P_{N\,s} \, \delta g_{N\,s} + T_{T\,s} \, \delta g_{T\,s} \,) \tag{4.16}$$

In practice we compute the normal force $P_{N\,s}$ from the penalty update $P_{N\,s} = \epsilon \, g_{N\,s}$ multiplied by the area of the contact element. For the tangential force $T_{T\,s}$ we have to perform an algorithmic update, see last section. Thus the contributions of one contact element takes the form

$$\delta g_{N\,s} \, P_{N\,s} \; + \; \delta g_{T\,s} \, T_{T\,s} \tag{4.17}$$

for the discrete contact point (s) with the mechanical relative variations analogous to (4.12) and (4.15). This equations can now be cast into a matrix formulation. For the normal part $(4.17)_1$ we set for the variation (4.12) of the penetration

$$\delta g_{N\,s} = \boldsymbol{\eta}^T \mathbf{N}_s \, . \tag{4.18}$$

With the same notation we can express the variation (4.15) of the tangential gap

$$\delta g_{T\,s} = \boldsymbol{\eta}^T \left(\mathbf{T}_s + \frac{g_{N\,s}}{l} \, \mathbf{N}_{0\,s} + \frac{g_{T\,s}}{l} \, \mathbf{T}_{0\,s} \right) . \tag{4.19}$$

In these equations the following vectors have been used

$$\boldsymbol{\eta} = (\boldsymbol{\eta}_s^2 \quad \boldsymbol{\eta}_1^1 \quad \boldsymbol{\eta}_2^1)^T \, , \tag{4.20}$$

$$\mathbf{N}_s = \left\{ \begin{matrix} \mathbf{n}^1 \\ -(1 - \bar{\xi}) \, \mathbf{n}^1 \\ -\bar{\xi} \, \mathbf{n}^1 \end{matrix} \right\}_s , \qquad \mathbf{N}_{0\,s} = \left\{ \begin{matrix} \mathbf{0} \\ -\mathbf{n}^1 \\ \mathbf{n}^1 \end{matrix} \right\}_s , \tag{4.21}$$

and

$$\mathbf{T}_s = \begin{Bmatrix} \mathbf{a}_1^1 \\ -(1-\bar{\xi})\,\mathbf{a}_1^1 \\ -\bar{\xi}\,\mathbf{a}_1^1 \end{Bmatrix}_s , \qquad \mathbf{T}_{0\,s} = \begin{Bmatrix} \mathbf{0} \\ -\mathbf{a}_1^1 \\ \mathbf{a}_1^1 \end{Bmatrix}_s . \tag{4.22}$$

Thus the virtual mechanical work of the contact element can be written in the matrix formulation $\boldsymbol{\eta}^T \mathbf{G}_s^c$ with the contact element residual

$$\mathbf{G}_s^c = P_{N\,s}\,\mathbf{N}_s + T_{T\,s}\,(\mathbf{T}_s + \frac{g_{N\,s}}{l}\mathbf{N}_{0\,s} + \frac{g_{T\,s}}{l}\,\mathbf{T}_{0\,s}\,) . \tag{4.23}$$

Due to this approach a pure displacement formulation of the contact problem is possible by expressing $P_{N\,s}$ through the penalty relation $P_{N\,s} = \epsilon_N\,g_{N\,s}$. This is in contrast to the Lagrangian multiplier technique, where $P_{N\,s} = \lambda_{N\,s}$. But we observe that this discretization can be applied to both methods.
Often a Newton Raphson iteration is used to solve the global set of equations. Then the linearization of (4.23) is needed to achieve quadratic convergence near the solution point. The associated derivation is a little bit cumbersome and thus only the final results will be summarized for this discretization. Details of the frictionless case can be found in Wriggers and Simo (1985) and for contact including friction in Wriggers et al. (1990). The tangent matrix for the normal contact is derived from the term $\delta g_{N\,s}\,P_{N\,s}$. Note that in (4.12) the change in $\bar{\xi}$ has to be considered as well as the change of the normal $\mathbf{n}^1$. For the penalty approach with $P_{N\,s} = \epsilon_N\,g_{N\,s}$ we obtain the tangent matrix

$$\mathbf{K}_{N\,s}^c = \epsilon_N \left[\mathbf{N}_s\,\mathbf{N}_s^T - \frac{g_{N\,s}}{l}\left(\mathbf{N}_{0\,s}\,\mathbf{T}_s^T + \mathbf{T}_s\,\mathbf{N}_{0\,s}^T + \frac{g_{N\,s}}{l}\,\mathbf{N}_{0\,s}\,\mathbf{N}_{0\,s}^T \right) \right] \tag{4.24}$$

The used matrices have been defined in (4.21) and (4.22). Note that in a geometrically linear case all terms vanish which are multiplied by $g_{N\,s}$. This gives the simple matrix $\mathbf{K}_{N\,s}^{L\,c} = \epsilon_N\,\mathbf{N}_s\,\mathbf{N}_s^T$.
For the tangential contributions in the contact area we have to linearize the term $\delta g_{T\,s}\,T_{T\,s}$ which yields for the pure stick

$$\begin{aligned} \mathbf{K}_{T\,s}^c \;=\; & c_T \Big\{ (\mathbf{T}_s + \frac{g_{N\,s}}{l}\,\mathbf{N}_{0\,s} + \frac{g_{T\,s}}{l}\,\mathbf{T}_{0\,s})\,(\mathbf{T}_s + \frac{g_{N\,s}}{l}\,\mathbf{N}_{0\,s} + \frac{g_{T\,s}}{l}\,\mathbf{T}_{0\,s})^T \\ & + \frac{g_{N\,s}}{l}\,\Big[\mathbf{N}_{0\,s}\,\mathbf{N}_s^T + \mathbf{N}_s\,\mathbf{N}_{0\,s}^T - \mathbf{T}_{0\,s}\,\mathbf{T}_s^T - \mathbf{T}_s\,\mathbf{T}_{0\,s}^T \\ & - 2\,\frac{g_{N\,s}}{l}\,(\mathbf{N}_{0\,s}\,\mathbf{T}_{0\,s}^T + \mathbf{T}_{0\,s}\,\mathbf{N}_{0\,s}^T\,)\Big]\Big\} \end{aligned} \tag{4.25}$$

Also in this case all terms containing $g_{N\,s}$ disappear in a geometrically linear situation which yields $\mathbf{K}_{T\,s}^{L\,c} = c_T\,\mathbf{T}_s\,\mathbf{T}_s^T$. The case of frictional slip leads to an additional contribution in (4.25).
This expression is obtained together with the local update algorithm for the frictional contact, see (3.23). In the two dimensional case of the node–to–segment contact element as discussed above, the explicit matrix form results from the term $\delta g_{T\,s\,n+1}\,T_{T\,s\,n+1}$ and can be stated for a contacting node (s), with $\mathbf{K}_{T\,s}^c$ as

$$\mathbf{K}_{T\,s}^{S\,c} = \mathbf{K}_{T\,s}^c + \mu\,\epsilon_N \left(\mathbf{T}_s + \frac{g_{N\,s}}{l}\,\mathbf{N}_{0\,s} \right) \mathbf{N}_s^T \tag{4.26}$$

Note that this matrix is unsymmetric which corresponds to the non–associativity of Coulomb's frictional law.
All derivations have been obtained for the two–dimensional large deformation contact. The associated tangent matrices for the frictionless three–dimensional case of a node–to–surface discretizations is developed in Parisch (1989). The special case of the contact of a body with a rigid obstacle is treated in Hansson and Klarbring (1990), Wriggers and Imhof (1993) or Heegaard and Curnier (1993). The consistent linearization for a continuum based approach to contact problems has been derived in Laursen and Simo (1993).

5 Adaptive methods for contact problems

Since numerical methods like the finite element method yield approximate solutions it is necessary to control the errors inherited in these methods. This is also true for contact problems, especially when large deformations can occur. Due to this demand, research activities have been focused during the last years on adaptive techniques in order to provide numerical models which yield accurate and reliable solutions. The objective of adaptive techniques is to obtain a mesh which is optimal in the sense that the computational costs involved are minimal under the constraint that the error in the finite element solution is beyond a certain limit.
In this chapter we formulate error measures for contact problems based on the residual method, the projections schemes and the dual method. These error indicators and estimator will then be compared by means of two– and three–dimensional examples.

5.1 Boundary value problem

Let us recall the boundary value problem of solid mechanics which is represented by the variational inequality

$$\sum_{\alpha=1}^{2} \int_{\Omega^\alpha} \boldsymbol{\tau}^\alpha \cdot \nabla_x^S (\boldsymbol{\eta}^\alpha - \boldsymbol{\varphi}^\alpha)\, dV \geq \sum_{\alpha=1}^{2} \int_{\Omega^\alpha} \bar{\boldsymbol{f}}^\alpha \cdot (\boldsymbol{\eta}^\alpha - \boldsymbol{\varphi}^\alpha)\, dV + \int_{\Gamma^\alpha} \bar{\boldsymbol{t}}^\alpha \cdot (\boldsymbol{\eta}^\alpha - \boldsymbol{\varphi}^\alpha)\, dA\,, \quad (5.1)$$

see section 2.2 in which also all quantities appearing in (5.1) are defined.
There are several methods to solve (5.1). One of the mostly used approaches is an active set strategy combined with the penalty method, see e.g. Oden (1981) or Kikuchi and Oden (1988), which also will be employed in this paper. This leads instead of (5.1) to the equation

$$R\,(\boldsymbol{\varphi}\,,\boldsymbol{\eta}) = G\,(\boldsymbol{\varphi}\,,\boldsymbol{\eta}) - \lambda\, f\,(\boldsymbol{\eta}) = 0\,. \quad (5.2)$$

$\boldsymbol{\varphi}$ and $\boldsymbol{\eta}$ denote the deformation fields of the two bodies and their variations, respectively. λ is the load parameter . Now we can split the functional G into a part associated with the bodies G_i^α and a part which describes the contact interface G_c:

$$G\,(\boldsymbol{\varphi}\,,\boldsymbol{\eta}) = \sum_{\alpha=1}^{2} G_i^\alpha\,(\boldsymbol{\varphi}^\alpha\,,\boldsymbol{\eta}^\alpha) + G_c\,(\boldsymbol{\varphi}^1\,,\boldsymbol{\varphi}^2\,,\boldsymbol{\eta}^1\,,\boldsymbol{\eta}^2) \quad (5.3)$$

with the explicit expressions for (5.2) and (5.3)

$$
\begin{aligned}
G_i^\alpha(\boldsymbol{\varphi}^\alpha,\boldsymbol{\eta}^\alpha) &= \int_{\Omega^\alpha} \boldsymbol{\tau}^\alpha \cdot \nabla_x^S \boldsymbol{\eta}^\alpha \, dV \,, \\
G_c(\boldsymbol{\varphi}^1,\boldsymbol{\varphi}^2,\boldsymbol{\eta}^1,\boldsymbol{\eta}^2) &= \int_{\Gamma_c^1} (\epsilon_N \, g_N^- \,(\boldsymbol{\eta}^2 - \bar{\boldsymbol{\eta}}^1) \cdot \bar{\mathbf{n}}^1 + t_{T\beta}\, \delta\bar{\xi}^\beta)\, dA \,, \\
f(\boldsymbol{\eta}) &= \sum_{\alpha=1}^{2} \int_{\Omega^\alpha} \bar{\boldsymbol{f}}^\alpha \cdot \boldsymbol{\eta}^\alpha \, dV + \int_{\Gamma_\sigma{}^\alpha} \bar{\mathbf{t}}^\alpha \cdot \boldsymbol{\eta}^\alpha \, dA \,,
\end{aligned} \tag{5.4}
$$

where $\epsilon_N > 0$ is the penalty parameter, $\delta\xi^\beta$ the variation of the tangential gap, see (3.14) and $t_{T\beta}$ are the components of the frictional force.
Linearization of the weak form (5.2) will be performed around a given deformation state $\bar{\boldsymbol{\varphi}}$

$$
R(\bar{\boldsymbol{\varphi}} + \mathbf{u}, \boldsymbol{\eta}) = R(\bar{\boldsymbol{\varphi}}, \boldsymbol{\eta}) + DG(\bar{\boldsymbol{\varphi}}, \boldsymbol{\eta}) \cdot \mathbf{u} + {}'' \, higher \; order \; terms'' = 0 \tag{5.5}
$$

with $G(\bar{\boldsymbol{\varphi}}, \boldsymbol{\eta})$ being defined in (5.4). Note that applied loads which depend on the deformations are not considered. The explicit form of the linearization $DG(\bar{\boldsymbol{\varphi}}, \boldsymbol{\eta}) \cdot \mathbf{u}$ follows by employing the directional derivative to $G(\bar{\boldsymbol{\varphi}}, \boldsymbol{\eta})$. This yields with (5.3)

$$
DG(\bar{\boldsymbol{\varphi}}, \boldsymbol{\eta}) \cdot \mathbf{u} = \sum_{\alpha=1}^{2} DG_i^\alpha(\bar{\boldsymbol{\varphi}}^\alpha, \boldsymbol{\eta}^\alpha) \cdot \mathbf{u}^\alpha + DG_c(\bar{\boldsymbol{\varphi}}, \boldsymbol{\eta}) \cdot \mathbf{u} \,, \tag{5.6}
$$

where the explicit form of the linearization for the solid is as follows

$$
DG_i^\alpha(\bar{\boldsymbol{\varphi}}^\alpha, \boldsymbol{\eta}^\alpha) \cdot \mathbf{u}^\alpha = \int_{\Omega^\alpha} [\nabla_{\bar{x}}^S \boldsymbol{\eta}^\alpha \cdot \mathbf{C}_{\bar{x}}^\alpha [\nabla_{\bar{x}}^S \mathbf{u}^\alpha] + \nabla_{\bar{x}}^S \boldsymbol{\eta}^\alpha \cdot \bar{\boldsymbol{\tau}}^\alpha \nabla_{\bar{x}}^S \mathbf{u}^\alpha \,]\, dV \,, \tag{5.7}
$$

In (5.7) all derivatives have to be performed with respect to the state $\bar{\boldsymbol{\varphi}}^\alpha$ denoted by $\bar{x}$. For the contact contributions see section 2.3.
Let us recall, for later use, the incremental boundary value problem associated with (5.6). By applying to (5.7) the Gauss theorem we obtain the incremental operator at state $\bar{\boldsymbol{\varphi}}$

$$
\mathbf{L}_{\bar{x}}(\mathbf{u}^\alpha) = \mathrm{div}_{\bar{x}} \left[\bar{\mathbf{C}}_{\bar{x}}^\alpha \, \nabla_{\bar{x}}^S \mathbf{u}^\alpha + \bar{\boldsymbol{\tau}} \, \nabla_{\bar{x}}^S \mathbf{u}^\alpha \right] \tag{5.8}
$$

With this we write the incremental form of the boundary value problem as

$$
\mathbf{L}_{\bar{x}}(\mathbf{u}^\alpha) = \Delta\lambda \, \mathbf{f} \tag{5.9}
$$

where $\mathbf{f}$ is the applied load. Let $\mathbf{u}$ denote the exact solution of (5.2) and let $\mathbf{u}_h$ denote the discrete FEM-solution. With

$$
\mathbf{e} = \mathbf{u} - \mathbf{u}_h \tag{5.10}
$$

we define the error in the displacement field. Here the error due to the penalty regularization will not be discussed, for details see e.g. Kikuchi and Oden (1988).

5.2 Error estimation for nonlinear problems

In order to treat problems involving nonlinear load paths, error estimation procedures, which initially have been developed for linear problems in Babuska and Rheinboldt (1978) and Zienkiewicz and Zhu (1987) have to be modified in a sufficient way. Following an approach pursued by Rheinboldt (1985) and Stein et al. (1994) we develop a procedure for nonlinear contact problems by applying error estimators to the linearized problem at an equilibrium point, related to the state $\bar{\varphi}$. The formal argumentation to this idea is summarized below.

Let $\mathbf{G}$ be a nonlinear operator projecting $\varphi \in \mathbf{V}$ from the space of deformations $\mathbf{V}$ into the space of forces $\mathbf{V}^*$

$$\mathbf{G}(\varphi) = \lambda \mathbf{P} . \tag{5.11}$$

where λ is a scalar load parameter and $\mathbf{P} \in \mathbf{V}^*$ a given load. In general we obtain the vector valued function $\mathbf{G}$ from (5.2) by carrying out the variations. Excluding limit and bifurcation points we assume that the inverse of the FRECHET derivative exists and is bounded for each φ with $\|\varphi - \bar{\varphi}\| \leq \delta_2$. Thus we have

$$\|D\,\mathbf{G}^{-1}(\varphi)\| \leq C_1 \qquad \text{and} \qquad \|D^2\,\mathbf{G}(\varphi)\| \leq C_2 , \tag{5.12}$$

where $(\bar{\varphi}\,,\bar{\lambda})$ is the exact equilibrium point which denotes the state where the linearization will be performed. With these assumptions, Rheinboldt (1985) obtained the result

$$\|\bar{\varphi}_h - \bar{\varphi}\|(1+c) = \|\bar{\varphi}_h - \mathbf{u}\| , \tag{5.13}$$

where $|c| \leq \frac{1}{2}\,C_1\,C_2\,\delta_2$. Thus with relation (5.13) we can approximate the discretization error $\|\bar{\varphi}_h - \bar{\varphi}\|$ of the nonlinear problem by the difference between $\bar{\varphi}_h$ and the solution of the linearized problem $\mathbf{u}$, provided c is small. Hence, with the assumptions made in (5.12), we can apply the error estimators of the linear theory described in the following sections also to nonlinear problems.

The residuals or stresses required for the error estimation at a computed equilibrium point $(\bar{\varphi}_h, \bar{\lambda})$ are obtained from the solution of the linear problem (5.5)

$$D\,G(\bar{\varphi}_h, \bar{\lambda}) \cdot \mathbf{u}_h^{\epsilon} = -R(\bar{\varphi}_h, \bar{\lambda} + \delta\lambda) , \tag{5.14}$$

$D\,G(\bar{\varphi}_h, \bar{\lambda})$ denotes the linearization as defined in (5.6) where the contact interface Γ_c^1 is left unchanged. δ is a small parameter and therefore $\bar{\lambda} + \delta\lambda$ with ($\delta\lambda << \bar{\lambda}$) describes a perturbed loading state in the vicinity of $(\bar{\varphi}_h, \bar{\lambda})$. The residuals or stresses computed from the increase of displacements $\mathbf{u}_h^{\delta}$ are then used for error estimation.

5.3 Residual based error estimator for contact

Error estimation can be found for frictionless contact problems of linear elastic bodies in e.g. Kikuchi and Oden (1988) or Hlavacek et al. (1988). To derive error estimates for contact we follow in this section the approach which was developed in Johnson (1991) for the unilateral membrane problem and extend it to the case of frictionless contact between elastic bodies. The error estimator which is formulated is an extension of the error estimator in Babuska and Miller (1987).

Contact problem and discretization Assume that two bodies come into contact. In that case the non–penetration condition is given by (2.3) as $g_N \geq 0$. Let us recall the variational inequality (3.3) for linear elasticity

$$a(\mathbf{u}, \mathbf{v} - \mathbf{u}) \geq f(\mathbf{v} - \mathbf{u}),$$

which describes the frictionless contact problem in the case of small deformations. The problem $(\mathbf{P})$ is now to find $\mathbf{u} \in \mathbf{K}$ such that (3.3) is fulfilled for all $\mathbf{v} \in \mathbf{K}$ with

$$\mathbf{K} = \{\mathbf{v} \in \mathbf{V} \mid (\mathbf{v}^2 - \bar{\mathbf{v}}^1) \cdot \bar{\mathbf{n}}^1 + G_N \geq 0 \text{ on } \Gamma_c\}$$

where $\mathbf{V}$ is already defined in (3.7). Due to the inequality constraint on the displacement field this problem is nonlinear.
The penalty method is now applied to solve $(\mathbf{P})$. This technique replaces $(\mathbf{P})$ by an unconstrained problem $(\mathbf{P}_\epsilon)$ with regard to the contact constraint (2.3) as follows: Find $\mathbf{u}_\epsilon \in \mathbf{V}$ such that

$$a(\mathbf{u}_\epsilon, \mathbf{v}) + c^-(\mathbf{u}_\epsilon, \mathbf{v}) = f(\mathbf{v}) \quad \forall \mathbf{v} \in \mathbf{V}, \tag{5.15}$$

where $\mathbf{V}$, $a(\mathbf{u}_\epsilon, \mathbf{v})$ and $f(\mathbf{v})$ are defined as above and

$$c^-(\mathbf{u}_\epsilon, \mathbf{v}) = \int_{\Gamma_c} \varepsilon\, u_{\epsilon N}^- \, v_N \, d\Gamma. \tag{5.16}$$

The gap function $u_{\epsilon N}^-$ has already been defined in (2.3) and $v_N = \mathbf{v} \cdot \bar{\mathbf{n}}^1$. The penalty parameter ϵ_N is a positive constant. It can be shown, see e.g. Kikuchi and Oden (1988), that the solution of $(\mathbf{P}_\epsilon)$ will converge to the solution of $(\mathbf{P})$ as ε tends to infinity.
To discretize $(\mathbf{P}_\epsilon)$, we divide Ω into non overlapping finite elements T of diameter h_T and introduce a standard finite element space

$$\mathbf{V}_h = \{\mathbf{v} \in \mathbf{V} \mid \mathbf{v} \in C(\Omega),\ \mathbf{v}|_T \in [P(T)]^{dim},\ \forall T\}, \tag{5.17}$$

where $P(T)$ is a space of polynomials of degree p_T on T and p_T is a positive integer. The discrete finite element problem $(\mathbf{P}_h)$ for $(\mathbf{P}_\epsilon)$ is now: Find $\mathbf{u}_h \in \mathbf{V}_h$ such that

$$a(\mathbf{u}_h, \mathbf{v}) + c^-(\mathbf{u}_h, \mathbf{v}) = f(\mathbf{v}) \quad \forall \mathbf{v} \in \mathbf{V}_h. \tag{5.18}$$

Error estimator for frictionless contact Let $\mathbf{u}_\epsilon$ and $\mathbf{u}_h$ denote the exact penalty solution of $(\mathbf{P}_\epsilon)$ and the discrete finite element solution of $(\mathbf{P}_h)$, respectively. With

$$\mathbf{e} = \mathbf{u}_\epsilon - \mathbf{u}_h \tag{5.19}$$

we define an error measure as follows:

$$r = \alpha \|\nabla \mathbf{e}\|^2_{L2(\Omega)} + c^-(\mathbf{e}, \mathbf{e}), \qquad \alpha > 0. \tag{5.20}$$

Due to the ellipticity of $a(\mathbf{e}, \mathbf{e})$, see e.g. Johnson and Hansbo (1992), we can estimate

$$r \leq a(\mathbf{e}, \mathbf{e}) + c^-(\mathbf{e}, \mathbf{e}). \tag{5.21}$$

By subtracting (5.18) from (5.15) we derive the condition

$$a(\mathbf{e},\mathbf{v}) + c^-(\mathbf{e},\mathbf{v}) = 0 \quad \forall\, \mathbf{v} \in \mathbf{V}_h\,. \tag{5.22}$$

Adding (5.22) with $\mathbf{v} = \mathbf{u}_h - \mathbf{w}$ to the right side of (5.21) we obtain

$$r \leq a(\mathbf{e}, \mathbf{u}_\epsilon - \mathbf{w}) + c^-(\mathbf{e}, \mathbf{u}_\epsilon - \mathbf{w})\,. \tag{5.23}$$

Inserting (5.19) and rearranging terms provides

$$r \leq a(\mathbf{u}_\epsilon, \mathbf{u}_\epsilon - \mathbf{w}) + c^-(\mathbf{u}_\epsilon, \mathbf{u}_\epsilon - \mathbf{w}) - a(\mathbf{u}_h, \mathbf{u}_\epsilon - \mathbf{w}) - c^-(\mathbf{u}_h, \mathbf{u}_\epsilon - \mathbf{w})\,. \tag{5.24}$$

The first two terms on the right side of inequality (5.24) are according to (5.15) equal to $f(\mathbf{u}_\epsilon - \mathbf{w})$.

Next we choose $\mathbf{w} = I_h \mathbf{u}_\epsilon$, where I_h is a projection onto $\mathbf{V}_h$ (e.g. the nodal interpolation if $\mathbf{u}_\epsilon$ is smooth), and find that $\mathbf{u}_\epsilon - \mathbf{w} = \mathbf{e} - I_h\mathbf{e}$. Elementwise integration by parts of the third term under consideration of the constitutive equation then leads to

$$\begin{aligned} r \leq & \sum_T \int_T \hat{\boldsymbol{b}} \cdot (\mathbf{e} - I_h\mathbf{e})\, d\Omega + \sum_{\partial T \subseteq \Gamma_\sigma} \int_{\partial T} \hat{\boldsymbol{t}} \cdot (\mathbf{e} - I_h\mathbf{e})\, d\Gamma + \\ & + \sum_T \int_T \operatorname{div} \boldsymbol{\sigma}_h \cdot (\mathbf{e} - I_h\mathbf{e})\, d\Omega + \sum_{\partial T} \int_{\partial T} (-\boldsymbol{\sigma}_h \mathbf{n}) \cdot (\mathbf{e} - I_h\mathbf{e})\, d\Gamma - \\ & - \sum_{\partial T \subseteq \Gamma_c} \int_{\partial T} \epsilon\, \mathrm{u}_{N\,h}^-(\mathrm{e} - I_h \mathrm{e})_N\, d\Gamma\,. \end{aligned} \tag{5.25}$$

The last part denotes the contribution due to the penalization of the contact constraint equation. Here we have defined the normal projection $(\mathrm{e} - I_h\mathrm{e})_N = (\mathbf{e} - I_h\mathbf{e}) \cdot \bar{\mathbf{n}}^1$.

We now introduce Γ_i as the set of faces of elements which are not contained in Γ. Thus we can write

$$\begin{aligned} r \leq & \sum_T \int_T (\operatorname{div} \boldsymbol{\sigma}_h + \hat{\boldsymbol{b}}) \cdot (\mathbf{e} - I_h\mathbf{e})\, d\Omega + \\ & + \sum_{\partial T \subseteq \Gamma_i} \int_{\partial T} \tfrac{1}{2}[-\boldsymbol{\sigma}_h \mathbf{n}] \cdot (\mathbf{e} - I_h\mathbf{e})\, d\Gamma + \sum_{\partial T \subseteq \Gamma_\sigma} \int_{\partial T} (\hat{\boldsymbol{t}} - \boldsymbol{\sigma}_h \mathbf{n}) \cdot (\mathbf{e} - I_h\mathbf{e})\, d\Gamma + \\ & + \sum_{\partial T \subseteq \Gamma_c} \int_{\partial T} (-\boldsymbol{\sigma}_h \mathbf{n}) \cdot (\mathbf{e} - I_h\mathbf{e})\, d\Gamma - \sum_{\partial T \subseteq \Gamma_c} \int_{\partial T} \epsilon\, \mathrm{u}_{hn}^-(\mathrm{e} - I_h\mathrm{e})_N\, d\Gamma\,. \end{aligned} \tag{5.26}$$

The square brackets denote the jump of a function over the element interfaces. Using the symmetry of the inner product $(\mathrm{e}-I_h\mathrm{e})_N$, (5.26) can be rewritten as

$$r \leq \sum_T \int_T (\operatorname{div}\boldsymbol{\sigma}_h + \hat{\boldsymbol{b}}) \cdot (\mathbf{e} - I_h\mathbf{e})\, d\Omega +$$

$$+ \sum_{\partial T \subseteq \Gamma_i} \int_{\partial T} \tfrac{1}{2}[-\boldsymbol{\sigma}_h \mathbf{n}] \cdot (\mathbf{e} - I_h\mathbf{e})\, d\Gamma + \sum_{\partial T \subseteq \Gamma_\sigma} \int_{\partial T} (\hat{\boldsymbol{t}} - \boldsymbol{\sigma}_h \mathbf{n}) \cdot (\mathbf{e} - I_h\mathbf{e})\, d\Gamma +$$

$$+ \sum_{\partial T \subseteq \Gamma_c} \int_{\partial T} (-\epsilon\, \mathrm{u}_{hn}^{-} \bar{\mathbf{n}}^1 - \boldsymbol{\sigma}_h \mathbf{n}) \cdot (\mathbf{e} - I_h\mathbf{e})\, d\Gamma\,. \tag{5.27}$$

The right side of (5.27) can be estimated with the absolute value and therefore

$$r \leq \sum_T \Big| \int_T (\operatorname{div}\boldsymbol{\sigma}_h + \hat{\boldsymbol{b}}) \cdot (\mathbf{e} - I_h\mathbf{e})\, d\Omega \Big| +$$

$$+ \sum_{\partial T \subseteq \Gamma_i} \Big| \int_{\partial T} \tfrac{1}{2}[\boldsymbol{\sigma}_h \mathbf{n}] \cdot (\mathbf{e} - I_h\mathbf{e})\, d\Gamma \Big| + \sum_{\partial T \subseteq \Gamma_\sigma} \Big| \int_{\partial T} (\hat{\boldsymbol{t}} - \boldsymbol{\sigma}_h \mathbf{n}) \cdot (\mathbf{e} - I_h\mathbf{e})\, d\Gamma \Big| +$$

$$+ \sum_{\partial T \subseteq \Gamma_c} \Big| \int_{\partial T} (\epsilon\, \mathrm{u}_{hn}^{-} \bar{\mathbf{n}}^1 + \boldsymbol{\sigma}_h \mathbf{n}) \cdot (\mathbf{e} - I_h\mathbf{e})\, d\Gamma \Big|\,. \tag{5.28}$$

Using the CAUCHY–SCHWARZ inequality and an estimate for the interpolation error $(\mathbf{e} - I_h\mathbf{e})$, see Carstensen et al. (1999), we obtain

$$r \leq C \Big\{ \sum_T h_T^2\, \|\operatorname{div}\boldsymbol{\sigma}_h + \hat{\boldsymbol{b}}\|^2_{L2(T)} +$$

$$+ \sum_{\partial T \subseteq \Gamma_i} h_T\, \|\tfrac{1}{2}[\boldsymbol{\sigma}_h \mathbf{n}]\|^2_{L2(\partial T)} + \sum_{\partial T \subseteq \Gamma_\sigma} h_T\, \|\hat{\boldsymbol{t}} - \boldsymbol{\sigma}_h \mathbf{n}\|^2_{L2(\partial T)} +$$

$$+ \sum_{\partial T \subseteq \Gamma_c} h_T\, \|\epsilon\, \mathrm{u}_{hn}^{-} \bar{\mathbf{n}}^1 + \boldsymbol{\sigma}_h \mathbf{n}\|^2_{L2(\partial T)} \Big\}^{\frac{1}{2}} \|\nabla \mathbf{e}\|_{L2(\Omega)}\,, \tag{5.29}$$

where C is a positive constant.
Division of (5.29) by $\|\nabla\mathbf{e}\|_{L2(\Omega)}$ provides the error estimator

$$\alpha \|\nabla \mathbf{e}\|_{L2(\Omega)} + \frac{c^-(\mathbf{e}, \mathbf{e})}{\|\nabla \mathbf{e}\|_{L2(\Omega)}} \leq C \Big\{ \sum_T h_T^2\, \|\operatorname{div}\boldsymbol{\sigma}_h + \hat{\boldsymbol{b}}\|^2_{L2(T)} +$$

$$+ \sum_{\partial T \subset \Gamma_i} h_T\, \|\tfrac{1}{2}[\boldsymbol{\sigma}_h \mathbf{n}]\|^2_{L2(\partial T)} + \sum_{\partial T \subseteq \Gamma_\sigma} h_T\, \|\hat{\boldsymbol{t}} - \boldsymbol{\sigma}_h \mathbf{n}\|^2_{L2(\partial T)} +$$

$$+ \sum_{\partial T \subseteq \Gamma_c} h_T\, \|\epsilon\, \mathrm{u}_{hn}^{-} \bar{\mathbf{n}}^1 + \boldsymbol{\sigma}_h \mathbf{n}\|^2_{L2(\partial T)} \Big\}^{\frac{1}{2}}\,. \tag{5.30}$$

Inequality (5.30) yields an upper bound for the error, which consists of two parts: an error in the strain components and an error resulting from the not exactly fulfilled contact constraint (2.3).

The error is bounded by the deviation of the discrete solution from equilibrium and the element size. The first and the third term of the right hand side contribute to the error bound if local equilibrium is violated. Further the term $-\epsilon\, u_{hN}^{-}\bar{\mathbf{n}}^1$ can be interpreted as the contact pressure on Γ_c. Therefore the second and the fourth term also correspond to a deviation from the equilibrium condition.

REMARK: There are different constants which are related to each of the contributions in (5.29). Thus using only one constant, C, is not optimal for error control. However a good estimation of these different constants needs additional effort. Two different techniques have to be mentioned in this context. The first one is related to the computation of eigenvalue problems to estimate C, see e.g. Johnson and Hansbo (1992). The second approach uses the sequence of meshes which are generated within the adaptive refinement process. In this case, the last generated meshes can be used to estimate the constant C.

Now the error E_T can be computed for each element in the finite element mesh as follows

$$E_T^2 \;=\; h_T^2 \int_T |\,\mathrm{div}\boldsymbol{\sigma}_h + \hat{\mathbf{b}}\,|^2 d\Omega \;+\; h_T \int_{\partial T \cap \Omega} 1/2\,|\,[\mathbf{t}_h]\,|^2 d\Gamma \;+ \tag{5.31}$$

$$h_T \int_{\partial T \cap \Gamma_\sigma} |\,\hat{\mathbf{t}} - \mathbf{t}_h\,|^2\, d\Gamma \;+\; h_T \int_{\partial T \cap \Gamma_c} |\,\epsilon_N\, g_N^{-}\,\bar{\mathbf{n}}^1 - \mathbf{t}_h|^2\, d\Gamma \tag{5.32}$$

Inequality (5.30) yields an upper bound for the error which is bounded by the deviation of the discrete solution from equilibrium and the element size.
The error estimator described above yields a measure between the exact penalty solution of (5.15) and its finite element approximation (5.18). What really is needed is the error between the exact solution of (3.3) and the approximate finite element solution (5.18). So far there do not exist computable error bounds for contact problems in elasticity. But we can make use of a result derived by Kikuchi and Oden (1988) to change the penalty parameter in such a way that an optimal convergence rate of the method is achieved.

5.4 Error indicator for contact based on projection

Another possibility to derive an error estimator for elastic contact problems starts directly from the complementary elastic energy norm. A simple, but in many cases efficient, error estimator is now provided by the superconvergent–stress–recovery technique which is due to Zienkiewicz and Zhu (1987). The equivalence of such error measures with the residual based error estimators of the last section has been shown in Verfürth (1996). The idea to derive these error estimators is based on the fact that many finite element meshes have superconvergence properties which means that there exists points in which the stresses are approximated with higher accuracy. By using projection procedure the stresses $\boldsymbol{\sigma}^*$ can be computed from the superconvergent points. It should be noted in passing that the stress–recovery error estimators work also well in if the sampling points are not superconvergence points, see Babuska et al. (1994). An especially efficient projection technique is provided by the lumped L^2–projection which is described in detail in Zienkiewicz and Taylor (1989). In general the projection procedures assume that the projected stresses do not have jumps which needs some special considerations in the contact interface, see below. Denoting by $I\!P$ a projection operator we obtain $\boldsymbol{\sigma}^*$ from

$$\int_\Omega I\!P\,[\,\boldsymbol{\sigma}^* - \boldsymbol{\sigma}_h\,]\,d\Omega = \mathbf{0} \tag{5.33}$$

and can then compute an approximation of the error using the complementary elastic energy norm

$$\|\,\boldsymbol{\sigma} - \boldsymbol{\sigma}_h\,\|^2_{E^{-1}} \le \int_\Omega (\,\boldsymbol{\sigma}^* - \boldsymbol{\sigma}_h\,)\cdot \mathbf{C}_0^{-1}\,[\boldsymbol{\sigma}^* - \boldsymbol{\sigma}_h]\,d\Omega \tag{5.34}$$

This error estimator can be evaluated in an efficient way and has been shown to be robust, see Babuska et al. (1994). Equation (5.34) does not include special boundary terms for the contact contributions. The estimate for the contact area is in this case included implicitly since the evaluation of the integral in (5.34) has to be done with respect to $\Omega = \Omega^1 \cup \Omega^2$ and thus includes also the contact interface. However a special projection has to be performed since it makes sense to treat normal and tangential stress components in the contact interface differently. Hence in the frictionless contact the normal component, given by $p_N = \bar{\mathbf{n}}^1 \cdot \boldsymbol{\sigma}\,\bar{\mathbf{n}}^1$, has to be projected using all elements connected to a point in the contact interface, e. g. from both bodies. Since we deal here with frictionless contact, we can neglect the tangential stresses for the moment. Thus first this special projection scheme has to be used and then the normal components of the stresses are transformed back to $\boldsymbol{\sigma}^*$. This technique works as follows. Since the

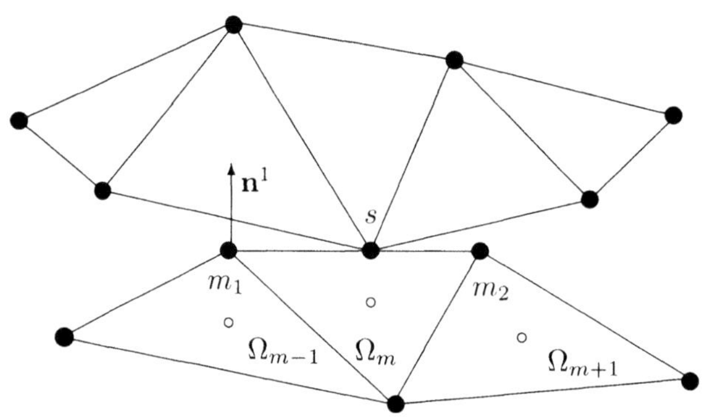

Figure 9. Element patches for stress projection on Γ_c

normal stress is continuous over Γ_c in the frictionless case it is given by $p_N = \bar{\mathbf{n}}^1 \cdot \boldsymbol{\sigma}\,\bar{\mathbf{n}}^1$ we can refer to p_N as the principal stress on the interface since all tangential components of the stress tensor are zero in the frictionless case. A simple procedure which yields a projected and normal field is as follows. We define a patch ω_{m_α} associated with node m_α $(\alpha = 1, 2)$ on the master surface, see Fig. 9, by all elements belonging to the node (for m_1 these are here Ω_{m-1} and Ω_m). Now a projection procedure like Z^2 or another, which is more accurate for boundary nodes, see e. g. Zienkiewicz and Taylor (1989), can be applied. It yields the stress tensor at the node m_α

$$I\!P(\boldsymbol{\sigma}_{\omega_{m_\alpha}}) \longrightarrow \boldsymbol{\sigma}^*_{m_\alpha} \, . \tag{5.35}$$

The normal and tangential components of the stress vector can be computed from the stresses $\boldsymbol{\sigma}^*_{m_\alpha}$ via Cauchys theorem by using the normal and tangential vectors defined following (4.10) for the two-dimensional case, see Fig. 9,

$$p^*_{N\,\alpha} = (\boldsymbol{\sigma}^*_{m_\alpha}\, \mathbf{n}^1) \cdot \mathbf{n}^1 \, , \tag{5.36}$$

$$t^*_{T\,\alpha} = (\boldsymbol{\sigma}^*_{m_\alpha}\, \mathbf{n}^1) \cdot \mathbf{a}^1_1 \, . \tag{5.37}$$

In case of frictionless contact we are only interested in the normal stress and can define with $p^*_{N\,\alpha}$ now a linear interpolation of the normal contact stress

$$p^*_N(\xi) = (1-\xi)\, p^*_{N\,1} + \xi\, p^*_{N\,2} \, . \tag{5.38}$$

As in (5.34) this result can now be applied to compute the error associated with the contact stresses. This leads to the additional error in $\partial\Omega^\gamma \cap \Gamma_c = \sum_T \partial T \cap \Gamma_c$

$$\| \mathbf{e} \|^{C\,\gamma_N} = \sum_T \| \mathbf{e} \|^{C\,\gamma_N}_{\partial T} \tag{5.39}$$

with the error related to the segment m_1—m_2

$$\| \mathbf{e} \|^{C\,\gamma_N}_{\partial T} = \left\{ \int_0^1 \frac{1}{\epsilon_N} \, [\, p^*_N(\xi) - p_{N\,h} \,]^2 \, l \, d\xi \right\}^{\frac{1}{2}} \tag{5.40}$$

where $p_{N\,h}$ is obtained according to (4.17) for the slave s node which is in contact with the master surface m_1—m_2, see Fig. 9.
The error within the whole domain is then computed by the sum over all elements T with Ω being the union of all elements. Thus we have

$$\| \boldsymbol{\sigma} - \boldsymbol{\sigma}_h \|_{E^{-1}} \approx \sum_T \| \mathbf{e} \|_T \tag{5.41}$$

with

$$\| \mathbf{e} \|_T = \int_T (\boldsymbol{\sigma}^* - \boldsymbol{\sigma}_h) \cdot \mathbf{C}_0^{-1} \, [\, \boldsymbol{\sigma}^* - \boldsymbol{\sigma}_h \,] \, dT \, . \tag{5.42}$$

Equation (5.40) provides the additional term due to contact which has to be included in case that the element includes a contact interface.

$$\| \mathbf{e} \|^C_T = \| \mathbf{e} \|_T + \| \mathbf{e} \|^{C\,\gamma_N}_{\partial T} \tag{5.43}$$

REMARK:

1. In the stick case a penalty stiffness is used to prevent relative tangential movements in the contact area. Thus we can derive an error measure for stick using the interpolation (5.38) together with (5.37) $\mathbf{t}_T^*(\xi) = (1-\xi)\,\mathbf{t}_{T\,1}^* + \xi\,\mathbf{t}_{T\,2}^*$ which yields

$$\|\,\mathbf{e}\,\|_{\partial T}^{C\,\gamma T} = \left\{ \int_0^1 \frac{1}{\epsilon_T}\,[\,\mathbf{t}_T^*(\xi) - \mathbf{t}_{T\,h}\,]^2\, l\, d\xi \right\}^{\frac{1}{2}} \tag{5.44}$$

   which has to be added to (5.43), $\mathbf{t}_{T\,h}$ follows for the two–dimensional case from (4.17).
2. If the penalty parameter ϵ_N has a very large value compared to the elasticity constants of the bodies in contact then terms (5.41) and (5.44) are small compared to (5.34). In this case it makes sense to scale the penalty parameter such that it has the same magnitude as the elasticity constants in (5.34) of the bodies being contact.
3. So far the error estimators have been developed for small elastic deformations. In case of large elastic strains there is no mathematically sound basis. However it should be noted that existence results exist for polyconvex materials, see e.g. the overview in Ciarlet (1988). These results have been extended to contact problems in Ciarlet (1988), Kikuchi and Oden (1988) and Curnier et al. (1992). Thus there is at least an existence result available for contact problems. The question of uniqueness, also needed for the derivation of error estimators, can of course not be solved since problems undergoing large elastic deformations may exhibit as well material as geometrical instabilities (e.g. limit points or bifurcations) see also section 5.2. If we now formulate all equations associated with the variational inequality in the tangent space of a given deformation map and exclude within this configuration instabilities, then the information from these incremental equations can be used for an error estimate within the incremental step. This means that we can exchange the stresses in (5.34) by the appropriate nonlinear stress measures, appearing in (5.4 and (5.7), when using the residual based error estimator. If we employ the error estimator based on the superconvergent recovery technique then the incremental constitutive tensors $\mathbb{C}_{\bar{x}}$ have to be used in (5.34) and (5.40) to compute the error (5.43).

A detailed discussion for the projection scheme can be found in Wriggers and Scherf (1997) and Wriggers et al. (2000).

5.5 Error estimators based on dual principles

The adaptive mesh refinement strategies described in the previous sections are based on a posteriori error estimates in the global energy or L^2 norm involving local residuals or postprocessed stresses of the computed solution. However for practical purposes the energy or L^2 norm may be of minor importance. In general the variables of interest are displacements or stresses at some particular points. Lately procedures for estimation of various kinds of local error functionals have been introduced in a general framework by in Becker and Rannacher (1996) or Rannacher and Suttmeier (1997). This approach is developed here for contact problems involving large elastic strains. The error estimator is evaluated solving the set of equations for an additional right hand side and applying the classical error estimators (based on residuals or superconvergence properties). The combination of the discretization errors for the initial and the dual problem gives the desired error quantity. The formal approach is given in the following.

Displacement error control As for residual norm error estimators the starting point is again the differential equation (5.8) for the discretization error $\mathbf{e} = \mathbf{u} - \mathbf{u}_h$

$$\mathbf{L}_{\bar{x}}(\mathbf{u} - \mathbf{u}_h) = \mathbf{L}_{\bar{x}}\mathbf{e} = \mathbf{b} - \mathbf{L}_{\hat{x}}\mathbf{u}_h := \mathbf{R}_1 \tag{5.45}$$

with the formal differential operator $\mathbf{L}$ and the element internal residuals $\mathbf{R}_1$.
With $\boldsymbol{\eta}$ denoting the test functions and $a(\, , \,)$ as defined above we obtain the weak form

$$a(\mathbf{e}, \boldsymbol{\eta}) = \sum_T \left\{ \int_{\Omega_T} (\operatorname{div} \bar{\boldsymbol{\tau}}_h + \hat{\mathbf{f}})\boldsymbol{\eta}\, d\Omega + \int_{\partial T \cap \Gamma_\sigma} (\hat{\mathbf{t}} - \bar{\boldsymbol{\tau}}_h \mathbf{n}) \cdot \boldsymbol{\eta}\, d\Gamma + \int_{\partial T \cap \Omega} \frac{1}{2} [\mathbf{t}_h] \boldsymbol{\eta}\, d\Gamma \right\} \tag{5.46}$$

The first integral represents the virtual work done by element internal residuals whereas the second term represents the virtual work done by jumps in the tractions at the force boundaries.
For edges apart from the structure boundaries the jumps in the tractions of both neighboring elements are the edge residuals represented in the third term. The jumps are split by the factor 1/2 into the two elements sharing the considered edge.
For simplicity we use $\mathbf{R}_1$ and $\mathbf{R}_2$ containing internal and jump residuals, respectively.

$$a(\mathbf{e}, \boldsymbol{\eta}) = \sum_T \{ (\mathbf{R}_1, \boldsymbol{\eta})_{\Omega_T} + (\mathbf{R}_2, \boldsymbol{\eta})_{\Gamma_T} \} \tag{5.47}$$

In order to estimate the error of a specific displacement in the component i at point $\mathbf{x} = \bar{\mathbf{x}}$ we additionally consider the following dual problem

$$\operatorname{div} \boldsymbol{\tau}(\mathbf{G}) + \boldsymbol{\delta}_i(\bar{\mathbf{x}}) = \mathbf{0}\,, \tag{5.48}$$

or in weak form

$$a(\mathbf{G}, \boldsymbol{\eta}) = (\boldsymbol{\delta}_i, \boldsymbol{\eta})\,, \tag{5.49}$$

where $\boldsymbol{\delta}_i$ is the Dirac delta (unit point load vector) in the direction i and $\mathbf{G}$ denotes Green's function. Applying the principle of Betti-Maxwell to the error problem (5.47) and the dual problem (5.49) yields the following relation

$$(\mathbf{e}, \boldsymbol{\delta}_i) = \sum_T \{ (\mathbf{R}_1, \mathbf{G})_{\Omega_T} + (\mathbf{R}_2, \mathbf{G})_{\Gamma_T} \}\,. \tag{5.50}$$

The term on the left hand side is the work of a unit load with the error function $\mathbf{e}$ and is equal to the error $e_i(\bar{\mathbf{x}})$ of the i-th component of the displacement at point $\bar{\mathbf{x}}$.
Now inserting $\mathbf{G}$ in equation (5.47) instead of the test function $\boldsymbol{\eta}$ the local error can be expressed by the bilinear form

$$e_i(\bar{\mathbf{x}}) = a(\mathbf{e}, G)\,. \tag{5.51}$$

Of course the solution of the dual problem is not known, but it can be computed also numerically based on the same discretization. It is simply another load case. Using the Galerkin orthogonality with $\mathbf{G}_h$ as finite element approximation of $\mathbf{G}$

$$e_i(\bar{\mathbf{x}}) = a(\mathbf{e}, \mathbf{G} - \mathbf{G}_h) \tag{5.52}$$

and applying the Cauchy-Schwarz inequality, the local error is estimated by the error energy $a(\mathbf{e}, \mathbf{e})$ of the primal problem (5.47) weighted by the error energy $a(\mathbf{G} - \mathbf{G}_h, \mathbf{G} - \mathbf{G}_h)$ of the dual problem (5.49):

$$e_i^2(\bar{\mathbf{x}}) \leq a(\mathbf{e}, \mathbf{e})\, a(\mathbf{G} - \mathbf{G}_h, \mathbf{G} - \mathbf{G}_h) \tag{5.53}$$

The second term serves as weighting function and filters out the influence of the overall residuals over the displacement error of interest. Inequality (5.53) can be computed elementwise, applying the estimators (residual based or postprocessed) described in the previous sections on primal and dual problem, respectively

$$e_i^2(\bar{\mathbf{x}}) \leq \sum_T a(\mathbf{e}, \mathbf{e})_T\, a(\mathbf{G} - \mathbf{G}_h, \mathbf{G} - \mathbf{G}_h)_T \tag{5.54}$$

Applying the stress-recovery error estimator $a(\mathbf{e}, \mathbf{e})_T$ and $a(\mathbf{G} - \mathbf{G}_h, \mathbf{G} - \mathbf{G}_h)_T$ are computed by

$$a(\mathbf{e}, \mathbf{e})_T = \int_{\Omega_T} (\bar{\boldsymbol{\tau}}^*(\mathbf{u}_h) - \bar{\boldsymbol{\tau}}(\mathbf{u}_h)) \cdot \mathbf{C}_{\bar{x}}^{-1} [\bar{\boldsymbol{\tau}}^*(\mathbf{u}_h) - \bar{\boldsymbol{\tau}}(\mathbf{u}_h)]\, d\Omega \tag{5.55}$$

and

$$a(\mathbf{G} - \mathbf{G}_h, \mathbf{G} - \mathbf{G}_h)_T = \int_{\Omega_T} (\bar{\boldsymbol{\tau}}^*(\mathbf{G}_h) - \bar{\boldsymbol{\tau}}(\mathbf{G}_h)) \cdot \mathbf{C}_{\bar{x}}^{-1} [\bar{\boldsymbol{\tau}}^*(\mathbf{G}_h) - \bar{\boldsymbol{\tau}}(\mathbf{G}_h)]\, d\Omega\,. \tag{5.56}$$

With the same argumentation as in the previous section, the stresses required to evaluate (5.55) and (5.56) are obtained from the solution of the corresponding linear problems. Therefore, we compute $\bar{\boldsymbol{\tau}}(\boldsymbol{\varphi}_h)$ and $\bar{\boldsymbol{\tau}}^*(\boldsymbol{\varphi}_h)$ from the increase of displacements of a perturbed loading state $\bar{\lambda} + \delta\lambda$ in the vicinity of the current equilibrium point $(\bar{\boldsymbol{\varphi}}_h, \bar{\lambda})$.
In the same way $\bar{\boldsymbol{\tau}}(\mathbf{G}_h)$ and $\bar{\boldsymbol{\tau}}^*(\mathbf{G}_h)$ are obtained from the increase of displacements by applying $\boldsymbol{\delta}_i$ as an additional small load on the current loading state. Since for a small load we will not observe any chance of the size of the contact area, this procedure is also consistent with the assumption that the contact area Γ_c^h does not change when $\boldsymbol{\delta}_i$ is applied.

REMARK: Although the internal energy of an elastic structure loaded by point loads is infinite, the error for the displacement of a specific finite element node can be computed simply by applying a nodal force in the dual problem. The smearing effect of elements with finite length leads to the most simple way of regularization bounding the energy.

Stress error control Errors for stress variables at a specific point $\bar{\mathbf{x}}$ can be estimated in a straight forward manner using the same concept as in the previous section. In contrast to the local displacement error control a discontinuity on the related displacement variable must be applied in the dual problem:

$$\operatorname{div} \boldsymbol{\tau}(\mathbf{z}) + \frac{\partial}{\partial x_j} \boldsymbol{\delta}_i(\bar{\mathbf{x}}) = \mathbf{0} . \tag{5.57}$$

Applying again the reciprocal theorem and subsequently the Galerkin orthogonality, the error in the displacement gradient or the associated stress value can be computed by

$$(\mathbf{e}, \frac{\partial}{\partial x_j} \boldsymbol{\delta}_i) = \frac{\partial e_i(\bar{\mathbf{x}})}{\partial x_j} = \sum_T \{(\mathbf{R}_1, \mathbf{z})_{\Omega_T} + (\mathbf{R}_2, \mathbf{z})_{\Gamma_T}\} = a(\mathbf{e}, \mathbf{z} - \mathbf{z}_h) . \tag{5.58}$$

In order to apply a discontinuity to an individual displacement in two or three dimensions, some sort of regularization must be applied. The simplest way in the finite element method is to substitute the displacement jump by equilibrated forces acting at two neighboring nodes to point $\bar{\mathbf{x}}$, see also last remark.
The above described concept for local quantities can be easily extended to arbitrary integral variables like reaction forces or pressure quantities on a contact interface. In order to estimate the error in boundary reactions, the displacement discontinuity can be imitated by a line load in the vicinity of the boundary.

5.6 Adaptive Mesh Refinement Strategy

Based on knowledge of the error distribution a refined mesh can be constructed which yields a better approximate solution. In general there exist several possibilities to obtain refined meshes. One approach is called the hierarchical method where elements of an existing mesh are subdivided at parts where the error measure indicates it. Another method is to define an error density function an to use this for the generation of a completely new refined mesh. To obtain an optimal mesh which means we want the error to be of the same order in all mesh parts, it is desirable to design the mesh such that the error contributions of the elements are equidistributed over the mesh.
The object of an adaptive algorithm is usually stated as a nonlinear optimization problem: construct a mesh such that the associated FEM–solution satisfies

$$\| \boldsymbol{\sigma} - \boldsymbol{\sigma}_h \|_{E^{-1}} \leq C \sum_T [E_T(h_T, \mathbf{u}_h, \hat{\mathbf{b}}_T)]^2 \leq TOL , \tag{5.59}$$

with TOL being a given tolerance. Furthermore the expense to compute $\mathbf{u}_h$ or $\boldsymbol{\sigma}_h$ should be nearly minimal. The measure E_T can be for the residual or the projection method either

$$E_{T\,1}^2 = E_T^2 \text{ from equation (5.32) or} \tag{5.60}$$

$$E_{T\,2}^2 = \|\mathbf{e}\|_T^C \text{ from equation (5.43).} \tag{5.61}$$

Or it is defined in case of the dual method by (5.53) leading to $E_{T\,3}^2$. As a measure of computational work the total number of degrees of freedom is chosen. Since the exact

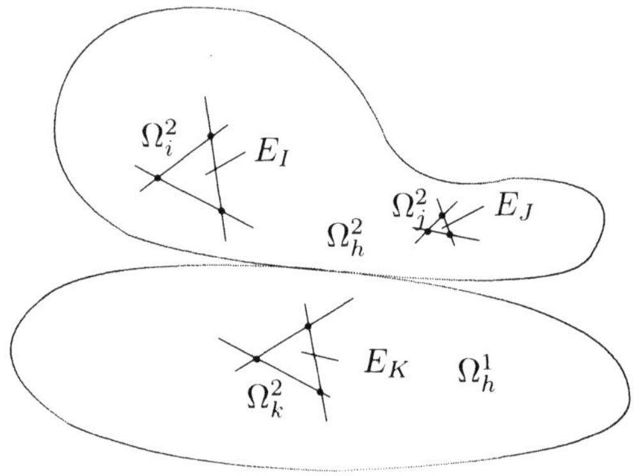

Figure 10. : Equidistributed error in finite element mesh.

solution $\mathbf{u}$ is not known we demand that the error contributions of all elements yields

$$\sum_T E_{T_k}^2 \leq TOL\,. \tag{5.62}$$

(5.62) serves as a stopping criterion in the adaptive process. To minimize the number of degrees of freedom during refinement, we require that the mesh is an optimal mesh, i.e. that the error E_T^2 is equally distributed between elements, see Fig. 10, ideally we then have for the elements I, J K

$$E_I = E_J = E_K\,. \tag{5.63}$$

With this requirement we can write for the error contribution of all elements

$$\sum_T E_{T_k}^2 = N\,E_I^2\,. \tag{5.64}$$

N denotes the number of elements in the mesh and E_I the error which is now associated with all elements according to (5.63). Finally (5.62) yields together with (5.59) the refinement criterion

$$E_{T_k}^2 \leq \frac{TOL}{N}\,. \tag{5.65}$$

Now we state the overall algorithm of the h adaptive method for contact problems. The algorithm includes the following steps:

1. Set initial values: $l = 0$, $\lambda_0 = 0$, $\Delta\lambda$, $i = 0$
2. Generation of start mesh: $\mathcal{M}_i$
3. Loop over load increments : $\lambda_{l+1} = \lambda_l + \Delta\lambda$
   a) IF $\lambda_{l+1} > \lambda_{max} \Longrightarrow$ STOP
   b) Iteration loop to solve contact problem

c) Mesh optimization
- Compute $E^2_{T_k}$
- IF $\sum E^2_{T_k} < TOL \Longrightarrow$ GOTO 3.
- IF $E^2_{T_k} > TOL\,/\,N \Longrightarrow$ refine element T
- Set $i = i + 1$
- Generate new mesh $\mathcal{M}_i$
  - Delaunay triangularization
  - Smoothing, if necessary
- Interpolate displacement and history variables the new mesh
- GOTO 3 (b)

The new mesh is assumed to be generated by a Delaunay triangularization, but also different generation techniques like the advancing front method or others can be applied.

5.7 Numerical examples

Different numerical examples are considered to discuss and compare the above derived error measures.

The introduced adaptive algorithm has been implemented in an advanced version of the finite element program (FEAP), described in Zienkiewicz and Taylor (1989). The mesh is defined via a parametric surface description of the boundaries. All loads, boundary conditions and contact constraints are defined with respect to these surfaces. Then a Delaunay triangularization is used to create the successive meshes during the adaptive process; for the associated algorithm, see Sloan (1987). Throughout the computations we use linear triangular elements.

Hertzian problem In this example we apply the error estimation procedures to solve the well known 2D Hertzian problem of an elastic cylinder (Young's modulus $E = 7000$, Poisson's ratio $\nu = 0.3$) contacting a plane rigid surface. The cylinder with a radius $r = 1$ and infinite length is loaded by an overall load of $F = 100$. Due to the symmetry we discretize one half of the problem using linear triangular elements. For this problem an exact solution for the contact pressure can be obtained analytically. Hence we compare the results computed by the adaptive methods directly to the exact solution. To omit problems related to a point load in elasticity the load is distributed over a small surface on top of the cylinder. In order to simulate the rigid surface we set Young's modulus to $E = 100000$ and Poisson's ratio to $\nu = 0.45$ in the finite element model. The initial mesh is depicted in fig. 11. Within the duality based method we assume that the maximum contact pressure at the bottom of the cylinder is the quantity to be computed with best accuracy.

Fig. 13 and 14 show the meshes optimized with respect to the local stresses and the global energy norm, respectively. While the local error control focuses on the point of interest, the global energy adaptive strategy leads also to refinements in the vicinity of load initiation. The maximal contact pressure obtained for the two different finite element models is compared with the analytical solution ($p_{max} = 495$) in Fig. 12. We observe a significant higher effectivity of the mesh optimized with respect to the maximum contact pressure, which reaches the exact value with an amount of half of the elements required

Figure 11. Initial mesh

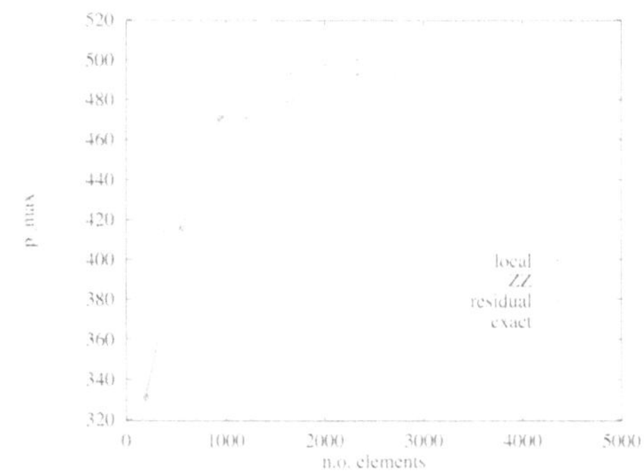

Figure 12. Convergence behaviour

Figure 13. Global SPR

Figure 14. Duality based SPR

within the global method.

Crossing tubes In this example we consider two rubber tubes which are perpendicular to each other ($E = 3000$, $\nu = 0.3$). Each tube has a length of $L = 100$mm and radii of $r_o = 50$mm , $r_i = 30$mm. Due to symmetry only one fourth of the model is used. The mesh adaption is performed with respect to the global error energy norm and with a prescribed relative error tolerance of $\eta_{err} = 15\,\%$.

Fig. 18 and Fig. 19 depict the mean stress field for the initial mesh and the adapted mesh. One can observe that the accuracy of the stress field is substantially improved after the adaption process. Mesh generation has been performed by using the tetrahedron mesh generator from Schöberl (1997).

Figure 15. Initial geometry

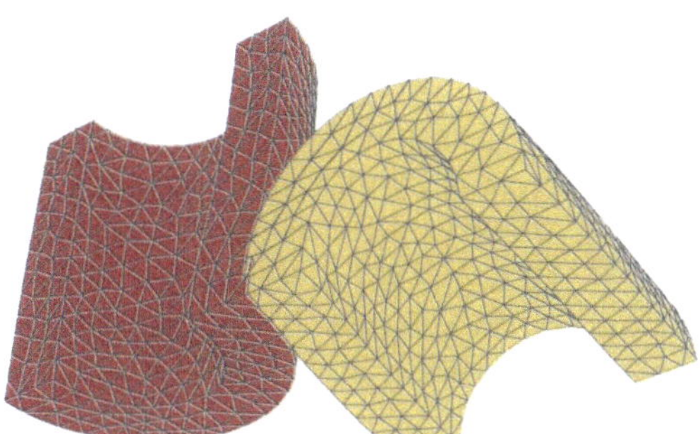

Figure 16. One fourth model,initial mesh

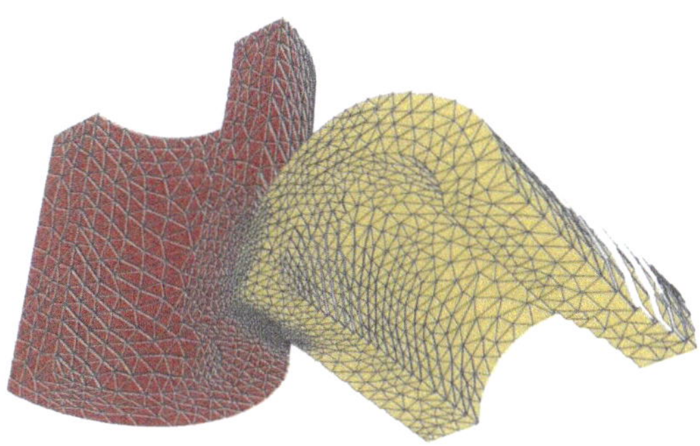

Figure 17. adapted mesh

5.8 Error indicator for frictional problems

In the previous section additional terms for the error associated with normal contact have been introduced for frictionless contact. Now we want to establish the same sort of error indicators also for contact with friction. Here the mathematical theory is not so much developed and error estimators are only provided for frictional contact with known normal pressure field, see Hlavacek et al. (1988). Thus we will use in the case of frictional contact where we do not assume a known pressure field within the iteration process measures which indicate the error in the contact interface. Again we can base

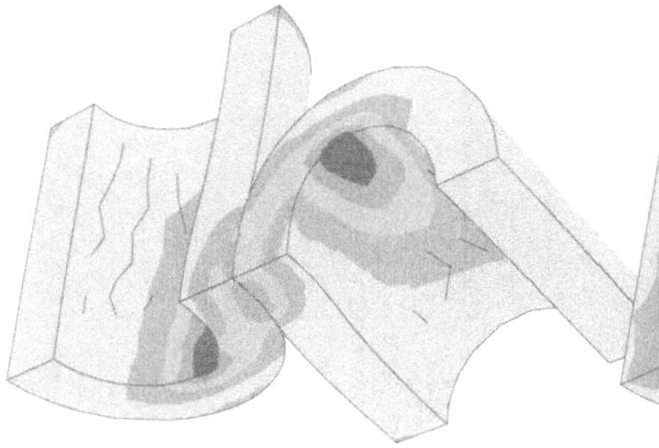

Figure 18. $\boldsymbol{\sigma}_1$. initial mesh

Figure 19. $\boldsymbol{\sigma}_1$, adapted mesh

the error computations on either residual error measures or error indicators based on superconvergence properties. both will be developed in the following.
Let us recall the residual error estimator for frictionless contact for geometrical linear problems (5.30)

$$|\, \epsilon_N \, g_{N\,h}^{-}\, \bar{\mathbf{n}}^1 - \mathbf{t}_h | \quad \text{on } \partial T \cap \Gamma_c$$

which corresponds to the local equilibrium in the contact interface. The term $\epsilon_N \, g_{N\,h}^{-}\, \bar{\mathbf{n}}^1$ can be interpreted as the contact pressure on Γ_c. The stress vector is given by $\mathbf{t}_h = \boldsymbol{\sigma}_h \, \bar{\mathbf{n}}^1$. In analogy to the derivation leading to (5.30) we can now introduce an error measure for the frictional part of the contact in the same way

$$R_4(\boldsymbol{\sigma}_h, \mathbf{u}_h) = |\, \mathbf{t}_{T\,n+1\,h} - \mathbf{t}_h | \quad \text{on } \partial T \cap \Gamma_c \tag{5.66}$$

In this relation, the frictional stress $\mathbf{T}_{T\,n+1\,h}$ is computed via the update formula (3.24) and (3.26).
Adding the frictionless contact contribution in (5.30) and (5.66) to equation (5.32) yields an error indicator for the frictional contact problem. Now E_T can be computed for each element in the finite element mesh as follows, see Wriggers et al. (1994)

$$\begin{aligned} E_T^2 \;=\; & h_T^2 \int\limits_T |\, \text{div}\boldsymbol{\tau}_h + \hat{\mathbf{b}}\,|^2 d\Omega \;+\; h_T \int\limits_{\partial T \cap \Omega} \frac{1}{2}\, |\, [\mathbf{t}_h]\,|^2 d\Gamma \;+ \\ & h_T \int\limits_{\partial T \cap \Gamma_\sigma} |\, \hat{\mathbf{t}} - \mathbf{t}_h \,|^2 \, d\Gamma \;+\; h_T \int\limits_{\partial T \cap \Gamma_c} |\, \epsilon_N \, g_{N\,h}^{-}\, \bar{\mathbf{n}}^1 + \mathbf{t}_{T\,n+1\,h} - \mathbf{t}_h |^2 \, d\Gamma \end{aligned} \tag{5.67}$$

Since frictional processes are path, and with this, time dependent, an estimation for the magnitude of the time step used to integrate the frictional constitutive equations (3.15) to (3.20) is still missing. In this study we will neglect this fact and prescribe the time steps from the outside of the adaptive algorithm.
Now we will also state the error indicator which can be derived using the superconvergence properties of the projected stress values at the boundary. As before we can define the

projected stresses by any procedure at the nodes, see Fig. 9. With this we obtain the stress tensor at the node m_α, see (5.35)

$$I\!P(\boldsymbol{\sigma}_{\omega_{m_\alpha}}) \longrightarrow \boldsymbol{\sigma}^*_{m_\alpha} .$$

The normal and tangential components of the stress vector follow then from (5.37). As in case of frictionless contact we can define with $p^*_{N\,\alpha}$ and $t^*_{T\,T}$ now a linear interpolation of the contact stresses for the two–dimensional case

$$\begin{aligned} p^*_N(\xi) &= (1-\xi)\,p^*_{N\,1} + \xi\,p^*_{N\,2}\,, \\ t^*_T(\xi) &= (1-\xi)\,t^*_{T\,1} + \xi\,t^*_{T\,2}\,. \end{aligned} \tag{5.68}$$

As in (5.41) this result can now be applied to compute the error associated with the contact stresses and we arrive at an expression for the additional contact error in $\partial\Omega^\gamma \cap \Gamma_c = \sum_T \partial T \cap \Gamma_c$

$$\|\, \mathbf{e}\,\|^{C\,\gamma} = \sum_T \|\, \mathbf{e}\,\|^{C\,\gamma_N}_{\partial T} + \sum_T \|\, \mathbf{e}\,\|^{C\,\gamma_T}_{\partial T} \tag{5.69}$$

with the error related to the segment m_1—m_2

$$\begin{aligned} \|\, \mathbf{e}\,\|^{C\,\gamma_N}_{\partial T} &= \left\{ \int_0^1 \frac{1}{\epsilon_N}\,[\,p^*_N(\xi) - p_{N\,h}\,]^2\, l\, d\xi \right\}^{\frac{1}{2}} , \\ \|\, \mathbf{e}\,\|^{C\,\gamma_T}_{\partial T} &= \left\{ \int_0^1 \frac{1}{c_T}\,[\,t^*_T(\xi) - t_{T\,n+1\,h}\,]^2\, l\, d\xi \right\}^{\frac{1}{2}} \end{aligned} \tag{5.70}$$

where $p_{N\,h}$ is obtained according to (4.17) for the slave s node which is in contact with the master surface m_1—m_2, see Fig. 9. The tangential stress $t_{T\,n+1\,h}$ follows from the algorithmic update formula (3.24).
The error within the whole domain including contact is computed from the sum over all elements T with Ω being the union of all elements. Thus we have with (5.42)

$$\|\, \boldsymbol{\sigma} - \boldsymbol{\sigma}_h \,\|_{E^{-1}} = \sum_T \|\, \mathbf{e}\,\|_T \tag{5.71}$$

Equation (5.69) provides the additional term due to normal and tangential contact stress which has to be included (5.71) in case that the element includes a contact interface.

$$\|\, \mathbf{e}\,\|^{C}_{T} = \|\, \mathbf{e}\,\|_T + \|\, \mathbf{e}\,\|^{C\,\gamma}_{\partial T} \tag{5.72}$$

As discussed before, we will state the adaptive method as a nonlinear optimization problem: construct a mesh such that the associated FEM–solution satisfies

$$\|\, \boldsymbol{\tau} - \boldsymbol{\tau}_h \,\|_{E^{-1}} = \|\mathbf{u} - \mathbf{u}_h\|_E \le C \sum_T [E_T(h_T, \mathbf{u}_h, \hat{\mathbf{b}}_T)]^2 \le TOL\,, \tag{5.73}$$

with TOL being a given tolerance. For cases with large deformations or involving inelastic materials the displacement and history variables have to be transferred to the new mesh, see e.g. Ortiz and Quigley (1991).

Numerical example We consider an example of a sealing in which an elastic block is pressed between to almost rigid parts. The problem and its initial mesh are depicted in Fig. 20. The refined mesh and the convergence behaviour of the method are depicted in Fig. 21 to 23.

Fig. 24 shows the contribution of the tangential stress. This has been confirmed with a uniformly refined mesh of over 16.000 degrees of freedom. We see that the adaptive method is more efficient since it needs less unknowns to arrive at the same result. This is also confirmed by Fig. 25 which shows the evolution of a characteristic displacement in the contact area.

Figure 20. Problem with initial mesh

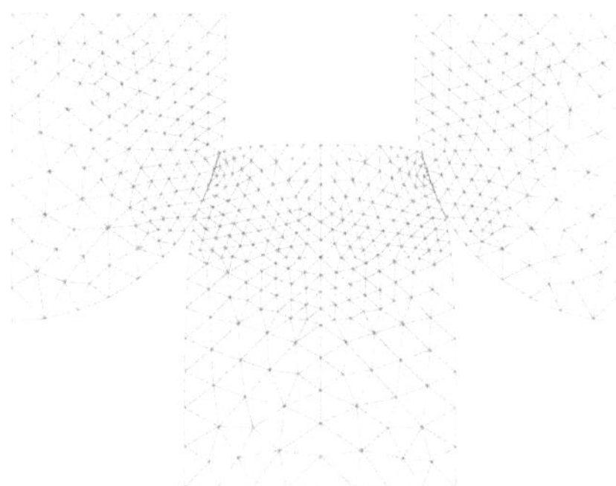

Figure 21. First refined mesh

Bibliography

Alart, P. and Curnier, A. (1991). A mixed formulation for frictional contact problems prone to Newton like solution methods. *Computer Methods in Applied Mechanics and Engineering*, 92:353–375.

Babuska, I. and Miller, A. (1987). A feedback finite element method with a posteriori error estimation: Part i. the finite element method and some basic properties of the a posteriori error estimation. *Computer Methods in Applied Mechanics and Engineering*, 61:1–40.

Babuska, I. and Rheinboldt, W. (1978). Error estimates for adaptive finite element computations. *SIAM Journal on Numerical Analysis*, 15:736–754.

Babuska, I., Strouboulis, T., Upadhyay, C. S., Gangaraj, S. K., and Copps, K. (1994). Validation of a posteriori error estimators by numerical approach. *International Journal for Numerical Methods in Engineering*, 37:1073–1123.

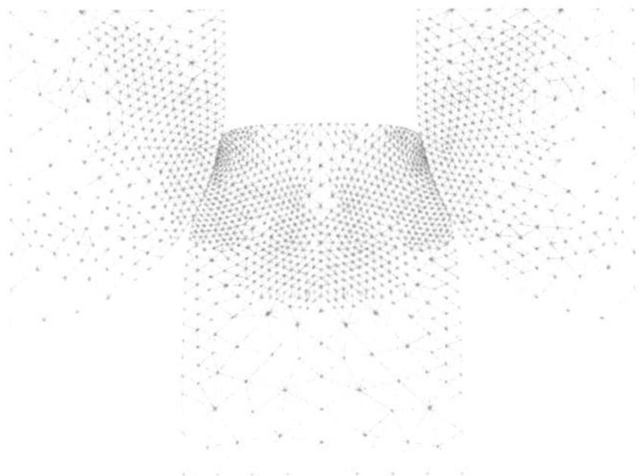

Figure 22. Second refined mesh

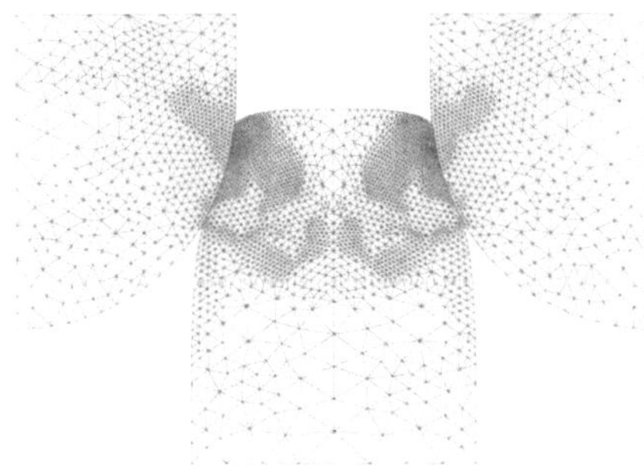

Figure 23. Third refined mesh

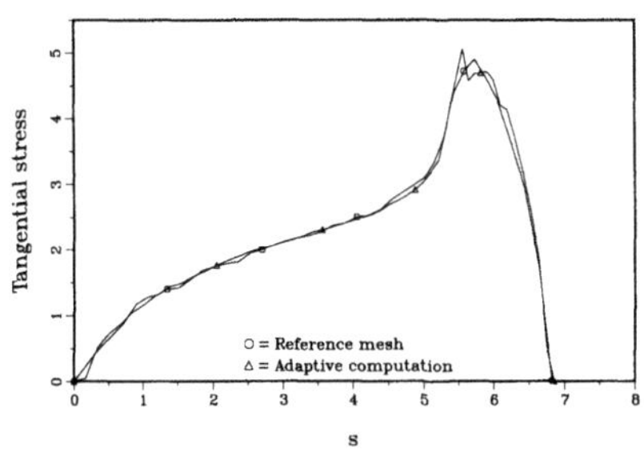

Figure 24. Tangential stress in contact interface

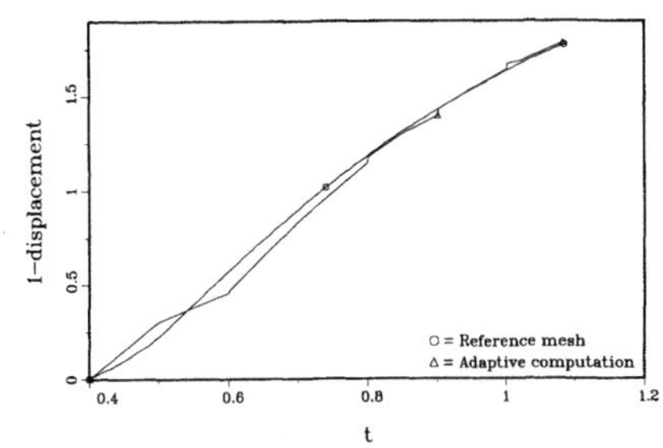

Figure 25. Evolution of the vertical displacement

Barthold, F. J. and Bischoff, D. (1988). Generalization of Newton type methods to contact problems with friction. *Journal Mec. Theor. Appl.*, pages 97–110.

Bathe, K. J. (1986). *Finite-Elemente-Methoden, Matrizen und lineare Algebra. Die Methode der finiten Elemente. L"osung von Gleichgewichtsbedingungen und Bewegungsgleichungen; Deutsche "Ubersetzung von P. Zimmermann.* Springer-Verlag, Berlin-Heidelberg-New York.

Bathe, K. J. and Chaudhary, A. B. (1985). A solution method for planar and axisymmetric contact problems. *International Journal for Numerical Methods in Engineering,*

21:65–88.

Becker, R. and Rannacher, R. (1996). A feed-back approach to error control in finite element methods: Basic analysis and examples. *EAST-WEST J. Numerical Mathematics*, 4:237–264.

Bertsekas, D. P. (1984). *Constrained Optimization and Lagrange Multiplier Methods*. Academic Press, New York.

Björkman, G., Klarbring, A., Sjödin, B., Larsson, T., and Rönnqvist, M. (1995). Sequential quadratic programming for non-linear elastic contact problems. *International Journal for Numerical Methods in Engineering*, 38:137–165.

Carstensen, C., Scherf, O., and Wriggers, P. (1999). Adaptive finite elements for elastic bodies in contact. *SIAM Journal of Scientific Computing*, 20:1605–1626.

Ciarlet, P. G. (1988). *Mathematical Elasticity I: Three-dimensional Elasticity*. North-Holland, Amsterdam.

Crisfield, M. A. (1991). *Non-linear Finite Element Analysis of Solids and Structures*, volume 1. Wiley. Chichester.

Curnier, A. and Alart, P. (1988). A generalized newton method for contact problems with friction. *J. Mec. Theor. Appl.*, 7:67–82.

Curnier, A., He, Q. C., and Telega, J. J. (1992). Formulation of unilateral contact between two elastic bodies undergoing finite deformation. *C. R. Acad. Sci. Paris*, 314:1–6.

Duvaut, G. and Lions, J. L. (1976). *Inequalities in Mechanics and Physics*. Springer Verlag, Berlin.

Francavilla, A. and Zienkiewicz, O. C. (1975). A note on numerical computation of elastic contact problems. *International Journal for Numerical Methods in Engineering*, 9:913–924.

Giannokopoulos, A. E. (1989). The return mapping method for the integration of friction constitutive relations. *Computers and Structures*, 32:157–168.

Glowinski, R. and Le Tallec, P. (1984). Finite element analysis in nonlinear incompressible elasticity. In *Finite Element, Vol. V: Special Problems in Solid Mechanics*. Prentice-Hall, Englewood Cliffs, New Jersey.

Hallquist, J. O., Goudreau, G. L., and Benson, D. J. (1985). Sliding interfaces with contact-impact in large-scale lagrange computations. *Computer Methods in Applied Mechanics and Engineering*, 51:107–137.

Hallquist, J. O., Schweizerhof, K., and Stillman, D. (1992). Efficiency refinements of contact strategies and algorithms in explicit fe programming. In Owen, D. R. J., Hinton, E., and Onate, E., editors, *Proceedings of COMPLAS III*, pages 359–384. Pineridge Press.

Hansson, E. and Klarbring, A. (1990). Rigid contact modelled by CAD surface. *Engineering Computations*, 7:344–348.

Heegaard, J.-H. and Curnier, A. (1993). An augmented Lagrange method for discrete large-slip contact problems. *International Journal for Numerical Methods in Engineering*, 36:569–593.

Hlavacek, I., Haslinger, J., Necas, J., and Lovisek, J. (1988). *Solution of Variational Inequalities in Mechanics*. Springer-Verlag, Berlin, New York.

Hughes, T. R. J., Taylor, R. L., and Kanoknukulchai, W. (1977). A finite element method for large displacement contact and impact problems. In Bathe, K. J., editor, *Formulations and Computational Algorithms in FE Analysis*, pages 468–495, Boston. MIT Press.

Johnson, C. (1987). *Numerical Solution of Partial Differential Equations by the Finite Element Method.* Cambridge University Press.

Johnson, C. (1991). Adaptive finite element methods for the obstacle problem. Technical report, Chalmers University of Technology, Göteborg.

Johnson, C. and Hansbo, P. (1992). Adaptive finite element methods in computational mechanics. *Computer Methods in Applied Mechanics and Engineering*, 101:143–181.

Ju, W. and Taylor, R. L. (1988). A perturbed lagrange formulation for the finite element solution of nonlinear frictional contact problems. *Journal of Theoretical and Applied Mechanics*, 7:1–14.

Kikuchi, N. and Oden, J. T. (1988). *Contact Problems in Elasticity: A Study of Variational Inequalities and Finite Element Methods.* SIAM, Philadelphia.

Klarbring, A. (1986). A mathematical programming approach to three-dimensional contact problems with friction. *Computer Methods in Applied Mechanics and Engineering*, 58:175–200.

Laursen, T. A. and Simo, J. C. (1993). A continuum-based finite element formulation for the implicit solution of multibody, large deformation frictional contact problems. *International Journal for Numerical Methods in Engineering*, 36:3451–3485.

Lee, C. Y., Oden, J. T., and Ainsworth, M. (1991). Local a posteriori error estimates and numerical results for contact problems and problems of flow through porous media. In Wriggers, P. and Wagner, W., editors, *Computational Methods in Nonlinear Mechanics*, pages 671–689, Berlin. Springer.

Luenberger, D. G. (1984). *Linear and Nonlinear Programming.* Addison-Wesley, Reading, Mass., second edition.

Oden, J. T. (1981). Exterior penalty methods for contact problems in elasticity. In Wunderlich, W., Stein, E., and Bathe, K. J., editors, *Nonlinear Finite Element Analysis in Structural Mechanics*, Berlin. Springer.

Ortiz, M. and Quigley, J. J. (1991). Adaptive mesh refinement in strain localization problems. *Computer Methods in Applied Mechanics and Engineering*, 90:781–804.

Papadopoulos, P. and Taylor, R. L. (1992). A mixed formulation for the finite element solution of contact problems. *Computer Methods in Applied Mechanics and Engineering*, 94:373–389.

Parisch, H. (1989). A consistent tangent stiffness matrix for three-dimensional nonlinear contact analysis. *International Journal for Numerical Methods in Engineering*, 28:1803–1812.

Ramm, E. and Cirak, F. (1997). Adaptivity for nonlinear thin-walled structures. In Owen, D. R. J., Hinton, E., and Onate, E., editors, *Proceedings of COMPLAS 5*, pages 145–163, Barcelona. CIMNE.

Rannacher, R. and Suttmeier, F. T. (1997). A posteriori error control in finite element methods via duality techniques: Application to perfect plasticity. Technical Report 97-16, Institut for Applied Mathematics, SFB 359, University of Heidelberg.

Rheinboldt, W. (1985). Error estimates for nonlinear finite element computations. *Computers and Structures*, 20:91–98.

Schöberl, J. . (1997). Netgen – an advancing front 2d/3d mesh generator based on abstract rules. *Computer Visualization Science*, 1:41–52.

Simo. J. C. and Laursen. T. A. (1992). An augmented Lagrangian treatment of contact problems involving friction. *Computers and Structures*, 42:97–116.

Simo. J. C. and Taylor. R. L. (1985). Consistent tangent operators for rate-independent elastoplasticity. *Computer Methods in Applied Mechanics and Engineering*, 48:101–118.

Simo, J. C., Wriggers, P., and Taylor, R. L. (1985). A perturbed Lagrangian formulation for the finite element solution of contact problems. *Computer Methods in Applied Mechanics and Engineering*. 50:163–180.

Sloan. S. W. (1987). A fast algorithm for constructing delaunay triangularization in the plane. *Adv. Eng. Software*. 9:34–55.

Stadter, J. T. and Weiss, R. O. (1979). Analysis of contact through finite element gaps. *Computers and Structures*. 10:867–873.

Stein. E.. Seifert. B.. Ohnimus. S.. and Carstensen, C. (1994). Adaptive finite element analysis of geometrically nonlinear plates and shells. especially buckling. *International Journal for Numerical Methods in Engineering*, 37:2631–2655.

Strang. G. and Fix, G. J. (1973). *An Analysis of the Finite Element Method*. Prentice-Hall, Inc., Englewood Cliffs.

Verfürth, R. (1996). *A Review of A Posteriori Error Estimation and Adaptive Mesh-Refinement Techniques*. Wiley, Teubner. Chichester, New York, Stuttgart, Leipzig.

Wriggers. P. (1987). On consistent tangent matrices for frictional contact problems. In Pande. G. and Middleton. J., editors. *Proceedings of NUMETA 87*, Dordrecht. M. Nijhoff Publishers.

Wriggers. P. (2001). *Nichtlineare Finite Elemente*. Springer. Berlin.

Wriggers. P. and Imhof. M. (1993). On the treatment of nonlinear unilateral contact problems. *Ing. Archiv*, 63:116–129.

Wriggers, P. and Miehe, C. (1992). On the treatment of contact contraints within coupled thermomechanical analysis. In Besdo, D. and Stein, E., editors, *Proc. of EUROMECH, Finite Inelastic Deformations*, Berlin. Springer.

Wriggers, P. and Miehe, C. (1994). Contact constraints within coupled thermomechanical analysis - a finite element model. *Computer Methods in Applied Mechanics and Engineering*. 113:301–319.

Wriggers. P.. Rieger. A.. and Scherf. O. (2000). Comparison of different error measures for adaptive finite element techniques applied to contact problems involving large elastic strains. *Computer Methods in Applied Mechanics and Engineering*.

Wriggers. P. and Scherf. O. (1997). Adaptive methods for contact problems. In Gastecki, A.. editor. *Proceedings of PCCMM XIII*, Poznan.

Wriggers, P. and Scherf, O. (1998). Different a posteriori error estimators and indicators for contact problems. *Mathematics and Computer Modelling*, 28:437–447.

Wriggers, P., Scherf, O., and Carstensen, C. (1994). Adaptive techniques for the contact of elastic bodies. In Hughes, J., Onate, E., and Zienkiewicz, O., editors, *Recent Developments in Finite Element Analysis*, pages 78–86, Barcelona. CIMNE.

Wriggers, P. and Simo, J. (1985). A note on tangent stiffnesses for fully nonlinear contact problems. *Communications in Applied Numerical Methods*, 1:199–203.

Wriggers, P., Simo, J., and Taylor, R. (1985). Penalty and augmented lagrangian formulations for contact problems. In Middleton, J. and Pande, G., editors, *Proceedings of NUMETA Conference*, Rotterdam. Balkema.

Wriggers, P., Van, T. V., and Stein, E. (1990). Finite-element-formulation of large deformation impact- contact -problems with friction. *Computers and Structures*, 37:319–333.

Wriggers, P. and Zavarise, G. (1993a). On the application of augmented lagrangian techniques for nonlinear constitutive laws in contact interfaces. *Communications in Applied Numerical Methods*, 9:815–824.

Wriggers, P. and Zavarise, G. (1993b). Thermomechanical contact – a rigorous but simple numerical approach. *Computers and Structures*, 46:47–53.

Zienkiewicz, O. C. and Taylor, R. L. (1989). *The Finite Element Method, 4th Ed.*, volume 1. McGraw Hill, London.

Zienkiewicz, O. C. and Taylor, R. L. (1991). *The Finite Element Method, 4th Ed.*, volume 2. McGraw Hill, London.

Zienkiewicz, O. C. and Zhu, J. Z. (1987). A simple error estimator and adaptive procedure for practical engineering analysis. *International Journal for Numerical Methods in Engineering*, 24:337–357.

GPSR Compliance

The European Union's (EU) General Product Safety Regulation (GPSR) is a set of rules that requires consumer products to be safe and our obligations to ensure this.

If you have any concerns about our products, you can contact us on ProductSafety@springernature.com

In case Publisher is established outside the EU, the EU authorized representative is:

Springer Nature Customer Service Center GmbH
Europaplatz 3
69115 Heidelberg, Germany

Batch number: 10584556

Printed by Printforce, the Netherlands